Skin Models

Models to Study Function and Disease of Skin

Edited by
Ronald Marks and Gerd Plewig

With 217 Illustrations

Springer-Verlag
Berlin Heidelberg New York Tokyo

Ronald Marks
Department of Medicine, University of Wales College of Medicine, Heath Park
Cardiff CF4 4XN, United Kingdom

Gerd Plewig
Universitäts-Hautklinik Düsseldorf
Moorenstr. 5, D-4000 Düsseldorf 1, FRG

ISBN-13:978-3-642-70389-8 e-ISBN-13:978-3-642-70387-4
DOI: 10.1007/978-3-642-70387-4

Library of Congress Cataloging in Publication Data.
Skin models. Includes index. 1. Dermatology, Experimental. 2. Skin-Models. 3. Skin-Diseases-Animal models. 4. Biological models. I. Marks, Ronald. II. Plewig, Gerd, [DNLM: 1. Models, Biological. 2. Skin-physiology. 3. Skin Diseases. WR 102 S628]
RL79.S55 1986 616.5′00724 85-26128
ISBN-13:978-3-642-70389-8 (U.S.)

Typesetting: Elsner & Behrens GmbH, 6836 Oftersheim

2127/3145-543210

Introduction

In the last fifty years dramatic progress has been made in the understanding of skin and skin diseases. Although we are still someway off understanding the ultimate causes of such disorders as psoriasis, atopic dermatitis and the congenital disorder of keratinization, we now have considerable information on the physiological disturbances in various diseases. This has permitted and encouraged a rational approach to treatment. The successful use of antimitotic agents, immunomodulators and retinoids may be cited as examples. A major reason for this improvement may be the fact that researchers accept models for the investigation of skin diseases. Increasing numbers of them have become available in the past years. So many have been described that it is doubtful whether any one researcher is aware of all the other models described – even in his own field of interest. This book is a challenge for those involved in the study of skin and its disorders to use the sundry models of skin that have proven helpful. It would be impossible for this work to be all-embracing but it is hoped that the choice of models offered in this publication will be stimulating and helpful in the solution of knotty skin questions.

April, 1986

Ronald Marks, Cardiff

Gerd Plewig, Düsseldorf

Table of Contents

In Vivo Models

In Vitro Models

List of Contributors

Akhter, S. A.
Postgraduate School of Studies in Pharmacy, University of Bradford, Bradford, West Yorkshire, United Kingdom

Ashton, R. E.
Department of Dermatology, UCLA School of Medicine, University of California, Los Angeles, CA 90024, USA

Barbenel, J. C.
Bioengineering Unit, University of Strathclyde, Wolfson Centre, 106 Rottenrow, Glasgow G4 0NW, Scotland, United Kingdom

Barry, B. W.
Postgraduate School of Studies in Pharmacy, University of Bradford, Bradford, West Yorkshire BD7 1DP, United Kingdom

Basketter, D. A.
Environmental Safety Laboratory, Unilever Research, Colworth House, Sharnbrook, Bedfordshire MK44 1LQ, United Kingdom

Bennett, S. L.
Postgraduate School of Studies in Pharmacy, University of Bradford, Bradford, West Yorkshire, United Kingdom

Bladon, P. T.
Department of Biochemistry, University of Leeds, Leeds LS2 9JT, United Kingdom

Botham, P. A.
Imperial Chemical Industries, Macclesfield, United Kingdom

Breeding, J.
Department of Dermatology, UCLA School of Medicine, University of California, Los Angeles, CA 90024, USA

Brummitt, L. C.
Leeds Dermatological Research Foundation, Department of Dermatology, The General Infirmary at Leeds, Great George Street, Leeds, United Kingdom

Budtz, P. E.
Zoophysiological Laboratory A, 13 Universitetsparken, DK-2100 Copenhagen Ø, Denmark

Carr, N. M.
Department of Dermatology, University of Edinburgh, Edinburgh, United Kingdom

Chung, J.-C.
Department of Dermatology, University of Innsbruck, Anichstr. 35, A-6020 Innsbruck, Austria

Cooper, N. F.
Department of Biochemistry, University of Leeds, Leeds LS2 9JT, United Kingdom

Corbett, M. F.
University of Wales College of Medicine, Heath Park, Cardiff, CF4 4XN, United Kingdom

Cunliffe, W. J.
Department of Dermatology, The General Infirmary at Leeds, Great George Street, Leeds LS1 3EX, United Kingdom

Dalziel, K.
Department of Medicine, University of Wales College of Medicine, Heath Park, Cardiff CF4 4XN, United Kingdom

Deodhar, S. D.
Department of Immunopathology, The Cleveland Clinic Foundation, Cleveland, OH 44106, USA

Diffey, B. L.
Regional Medical Physics Department, Dryburn Hospital, Durham DH1 5TW, United Kingdom

Dykes, P. J.
Department of Medicine, University of Wales College of Medicine, Cardiff CF4 4XN, United Kingdom

Edwards, C.
Department of Instrumentation and Analytical Science UMIST, P.O. Box 88, Manchester M60 1QD, United Kingdom

Elias, P. M.
Dermatology Service (109), Veterans' Administration Medical Center, 4150 Clement St., San Francisco, CA 94121, USA

Emeis, J. J.
Department of Dermatology, Electron Microscopy, University Medical Centre Leiden, Gaubius Institute TNO Rijnsburgerweg 10, NL-2333 Leiden, The Netherlands

Fitzmaurice, M.
Department of Immunopathology, The Cleveland Clinic Foundation, 9500 Euclid Avenue, Cleveland, OH 44106, USA

Foreman, M. I.
Data Analysis and Research (DAR) Ltd. The Bell Tower, New Lanark, Lanarkshire ML 119DH, Scotland, United Kingdom

Foglia, A.
Department of Dermatology, University of Pennsylvania School of Medicine, Philadelphia, PA 19104, USA

Fritsch, P. O.
Department of Dermatology, University of Innsbruck, Anichstr. 35, A-6020 Innsbruck, Austria

Galosi, A.
Hauptplatz 1, D-8068 Pfaffenhofen a. d. Ilm, Federal Republic of Germany

Gawkrodger, D. J.
Department of Dermatology, University of Edinburgh, Edinburgh, United Kingdom

Gaylarde, P. M.
Department of Dermatology, The Royal Free Hospital, Hampstead, London, NW3, United Kingdom

van Genderen, J.
Medical Biological Laboratory TNO, P.O. Box 45, NL-2280 AA Rijswijk, The Netherlands

Gerber, H. A.
Department of Pathology, University of Berne,
CH-3000 Berne, Switzerland

Gowland, G.
Leeds Dermatological Research Foundation, Department of Immunology, The General Infirmary at Leeds, Great George Street, Leeds, United Kingdom

Hacking, C. J.
Department of Instrumentation and Analytical Science UMIST, P.O. Box 88, Manchester M60 1QD, United Kingdom

Havekes, L.
Department of Dermatology, University Hospital and Gaubius Institute TNO, NL-2333 Leiden, The Netherlands

Hess, M.
Department of Pathology, University of Berne,
Ch-3000 Berne, Switzerland

Hetherington, A. M.
Department of Dermatology, University of Dundee, Dundee,
United Kingdom

Hiernickel, H.
Department of Dermatology, University of Cologne, Josef-Stelzmann-Str. 9, D-5000 Köln 41, Federal Republic of Germany

Hölzle, E.
Department of Dermatology, University of Düsseldorf, Moorenstr. 5, D-4000 Düsseldorf 1, Federal Republic of Germany

Hunter, J. A. A.
Department of Dermatology, University of Edinburgh,
Edinburgh EH3 9YW, United Kingdom

Hunziker, T.
Department of Dermatology, University of Berne,
CH-3000 Berne, Switzerland

Ilchyshyn, A.
Department of Dermatology, Central Out-Patients Department,
Hartshill Road, Stoke-on-Trent, Staffordshire ST4 7PA,
United Kingdom

Ilderton, E.
Skin Department, North Staffordshire Hospital Centre, Stoke-on-Trent, Staffordshire ST4 7PA, United Kingdom

Jenner, L. A.
Department of Medicine, University of Wales College of Medicine, Cardiff CF4 4XN, United Kingdom

Johnson, B. E.
Department of Dermatology, University of Dundee, Dundee DD1 95Y, United Kingdom

Kempenaar, J.
Department of Dermatology, University Hospital and Gaubius Institute TNO, NL-2333 Leiden, The Netherlands

Kieny, M.
Laboratoire de Zoologie & Biologie Animale, Université Scientifique et Médicale de Grenoble, F-38000 Grenoble, France

Kingsbury, J.
Department of Biological Sciences, University of Keele, Staffordshire, United Kingdom

Kleihues, P.
Division of Neuropathology, Institute of Pathology, University of Zürich, CH-8091 Zürich, Switzerland

Kligman, A. M.
Department of Dermatology, University of Pennsylvania School of Medicine, Philadelphia, PA 19104, USA

Kuehnl-Petzoldt, Ch.
Division of Neuropathology, Institute of Pathology, University of Freiburg, D-7800 Freiburg i. Br., Federal Republic of Germany

Lapière, Ch. M.
Department of Dermatology, University of Liège, B-4020 Liège, Belgium

Ledolter, A.
Department of Dermatology, University of Düsseldorf, Moorenstr. 5, D-4000 Düsseldorf 1, Federal Republic of Germany

Luepke, N. P.
Institute of Pharmacology and Toxicology, University of Münster, Domagkstr. 12, D-4400 Münster, Federal Republic of Germany

Lowe, N. J.
Department of Dermatology, UCLA School of Medicine, University of California, Los Angeles, CA 90024, USA

Maloney, M. E.
Dermatology Service, Veterans' Administration Medical Center, Department of Dermatology, University of California School of Medicine, San Francisco, CA 94121, USA

Marks, R.
Department of Medicine (Dermatology), University of Wales College of Medicine, Heath Park, Cardiff CF4 4XN, United Kingdom

Martineau, G. P.
Faculty of Veterinary Medicine, University of Montréal, Montréal, Canada

Matias, J. R.
The Animal Sciences Laboratory, Orentreich Foundation for the Advancement of Science, 910 Fifth Avenue, New York, NY 10021, USA

McGinley, K. J.
Department of Dermatology, University of Pennsylvania School of Medicine, Philadelphia, PA 19104, USA

Mc Vittie, E.
Department of Dermatology, University of Edinburgh, Edinburgh, United Kingdom

Mitrani, E.
Embryology Section, The Hebrew University of Jerusalem, Jerusalem 91904, Israel

Mommaas-Kienhius, A. M.
Department of Dermatology, Electron Microscopy, University Medical Center Leiden, Gaubius Institute TNO Rijnsburgerweg 10, NL-2333 Leiden, The Netherlands

van Neste, D.
Unit for Occupational and Environmental Dermatology. Louvain University, Clos Chapelle-aux-Champs 30, B-1200 Brussels, Belgium

Norris, J. F. B.
Skin Department, North Staffordshire Hospital Centre, Stoke-on-Trent,
Staffordshire ST4 7PA, United Kingdom

Nydegger, U. E.
Blood Transfusion Service SRC, Central Laboratory,
CH-3000 Berne, Switzerland

Orentreich, N.
Department of Dermatology, New York University Medical Center,
550 Fifth Avenue, New York, NY 10016, USA

Ortonne, J. P.
Department of Dermatology, C.H.U., F-06000 Nice, France

Payne, P. A.
Department of Instrumentation and Analytical Science UMIST,
PO Box 88, Manchester M60 1QD, United Kingdom

Pearse, A. D.
Department of Medicine, University of Wales College of Medicine,
Heath Park, Cardiff CF4 4XN, United Kingdom

Plewig, G.
Department of Dermatology, University of Düsseldorf, Moorenstr. 5,
D-4000 Düsseldorf 1, Federal Republic of Germany

Ponec, M.
Department of Dermatology, University Hospital and Gaubius
Institute TNO, Rijnsburgerweg 10, NL-2333 Leiden, The Netherlands

Priestley, G. C.
University Department of Dermatology, The Royal Infirmary,
Edinburgh EH3 9YW, United Kingdom

Ross, J. A.
Department of Dermatology, University of Edinburgh, Edinburgh,
United Kingdom

Rüger, R.
Department of Dermatology, University of Munich,
Frauenlobstr. 9–11, D-8000 München 2, Federal Republic of Germany

Schaefer, H.
Centre International de Recherches Dermatologiques (CIRD),
Sophia-Antipolis, F-06560 Valbonne, France

Schalla, W.
Centre International de Recherches Dermatologiques (CIRD),
Sophia-Antipolis, F-06560 Valbonne, France

Schultz-Ehrenburg, U.
Division of Dermatology, Histopathology and Angiology,
Ruhr University, St. Josef-Hospital, Gudrunstr. 56, D-4630 Bochum,
Federal Republic of Germany

Sengel, P.
Laboratoire de Zoologie & Biologie Animale, Université Scientifique
et Médicale de Grenoble, F-38000 Grenoble, France

Späth, P. J.
Blood Transfusion Service SRC, Central Laboratory,
CH-3000 Berne, Switzerland

Staquet, M. J.
Department of Dermatology, Hôpital Edouart Herriot, F-69001 Lyon,
France

Stewart, I. C.
Department of Respiratory Medicine, University of Edinburgh,
Edinburgh, United Kingdom

Stüttgen, G.
Department of Dermatology, Free University, Rudolf-Virchow-
Krankenhaus, Augustenburger Platz 1, D-1000 Berlin 65,
Federal Republic of Germany

Summerly, R.
Skin Department, North Staffordshire Hospital Centre, Stoke-onTrent,
Staffordshire ST4 7PA, United Kingdom

Thomas, S.
Department of Medicine (Dermatology), University of Wales College
of Medicine, Heath Park, Cardiff CF4 4XN, United Kingdom

Turnbull, F. W.
Solid State Sensors Group, Honeywell Control Systems Ltd.,
Newhouse, Lanarkshire, United Kingdom

Uozumi, A.
Division of Neuropathology, Institute of Pathology, University
of Freiburg, D-7800 Freiburg i. Br., Federal Republic of Germany

Vermeer, B. J.
Department of Dermatology, University Hospital and Gaubius Institute TNO, Rijnsburgerweg 10, NL-2333 Leiden, The Netherlands

Vogel, H. G.
Hoechst AG, Pharma-Forschung, D-6230 Frankfurt/Main 80, Federal Republic of Germany

Volk, B.
Division of Neuropathology, Institute of Pathology, University of Freiburg, D-7800 Freiburg i. Br., Federal Republic of Germany

Walker, E. M.
Department of Dermatology, University of Dundee, Dundee United Kingdom

Wiesmann, U.
Department of Pediatrics, University of Berne, CH-3000 Berne, Switzerland

Wijsman, M. C.
Department of Dermatology, Electron Microscopy, University Medical Centre Leiden, Gaubius Institute, Rijnsburgerweg 10, NL-2333 Leiden, The Netherlands

Williams, D. L.
Department of Medicine (Dermatology), University of Wales College of Medicine, Heath Park, Cardiff CF4 4XN, United Kingdom

Williams, M. L.
Dermatology Service, Veterans' Administration Medical Center, Department of Dermatology, University of California School of Medicine, 4150 Clement St., San Francisco, CA 94121, USA

Wolthuis, O. L.
Medical Biological Laboratory TNO, PO Box 45, NL-2280 AA Rijswijk, The Netherlands

Wood, E. J.
Department of Biochemistry, University of Leeds, Leeds, LS2 9JT, United Kingdom

Woodford, R.
School of Pharmacy, Portsmouth Polytechnic, Portsmouth, Hants., United Kingdom

Yardley, H. J.
Department of Biological Sciences, University of Keele, Keele,
Staffordshire, United Kingdom

In Vivo Models

Human Model for Acne

L. C. Brummitt, W. J. Cunliffe and G. Gowland

The main stimulus to acne is the over production of sebum [8, 30] but for the development of individual lesions, other events are necessary. These include hyperkeratosis of the sebaceous follicular duct [21, 36], some change in microbial ecology [16, 24, 26] and the production of inflammation [10, 14, 41].

Although acne is not alone in having virtually no animal counterpart, this deficiency does make the relevance of certain animal studies somewhat difficult to interpret. The sebaceous follicle in man is unique. The duct systems in most animals are short in contrast to the relatively long follicular duct of man and microscopically in many lower mammals, lipid globules can often be seen appearing on the surface of their skin. The sebaceous secretion is also unique to man, in that it contains squalene and certain fatty acids probably not found elsewhere in the animal kingdom. The microbial population of the sebaceous rich areas of man are also quite unlike other species [28]. Inflammatory responses also differ between species. Despite these differences considerable use of other mammals has been made to understand the physiology of the sebaceous follicle. However, these fundamental differences between man and other animals do indicate the need for a human model for acne. The function of such a human model would be to improve our understanding of acne and of the effect of therapy.

Animal Models

Sebaceous Gland

The reader is referred to the many detailed references [7, 21, 35]. Many experiments performed in animals cannot readily be carried out on human subjects. In particular this applies to the classical endocrine studies of ablation and hormonal replacement. For these purposes many authors have used the sebaceous gland of the rat [35], but Plewig and Luderschmidt have found the ventral surface of the Syrian hamster ear most useful because of its apparent greater similarity to the human species [29]. The hamster flank organ [38] has also been used. Some authors have also investigated sebaceous gland-like structures, such as the preputial gland, which has some similarity to the sebaceous gland in its response to hormones [37].

Ductal Cornification

The rabbit ear readily produces open comedones and this phenomenon has been utilised to investigate comedogenesis. Possible comedogenic agents are applied to the ear and the development of comedones in the treated ear as compared with the untreated side allows a comedogenic index to be determined [17, 22, 23]. This technique has shown for example that squalene and certain free fatty acids will induce hyperkeratosis, as will certain substances such as stearic acid and butyl alcohol, which may be responsible for cosmetic acne. More recently the original technique has been improved and criticised [12] and the technique has also been extended to man. In human subjects the agent is applied under occlusion for up to six weeks [27]. The human model has also shown that the rabbit ear model is over predictive. The rhino mouse has also been used to investigate comedogenesis but it should be emphasised that in neither the rabbit nor the mouse are whiteheads found. In humans whiteheads occur much more frequently than blackheads. These animal models also differ from acne in patients in that they do not proceed to inflamed lesions.

Bacteria and Inflammation

There are virtually no animal models for investigating bacterial involvement in acne, or for the production of inflammation. This is presumably due to the fact that the bacterial flora is quite different in animals. *Propionibacterium acnes* (*P. acnes*) is rarely found in species other than man. Furthermore, the host response to different microbiological stimuli may differ between species.

Human Models Already in Use

There is therefore, a need to develop human models for acne. Some models do already exist but improvements are required, especially for a better understanding of the role of bacteria in acne and the production of inflammation.

Sebum excretion rate (SER) in man is easily measured and the effect of different endocrine diseases on sebaceous gland function can readily be assessed [3, 7, 13]. Another way of assessing sebaceous gland function is to measure sebaceous lipogenesis [2, 4]. This technique is also applicable to the investigation of drug effects and is of particular help in distinguishing central from local effects.

Surprisingly, the use of SER measurements, although used extensively once a drug has been developed [5, 11, 15], have been little used in the early development of anti-seborrhoeic drugs (Fig. 1). The development of new drugs is a costly procedure and we wish to stress to the pharmaceutical industry the real benefit of using human SER measurements to provide useful facts on the clinical potential of new drugs. For many topical therapies there usually exists the possibility of using several bases and so SER measurement in man can often help to decide the choice of the base. Any potentially useful sebum suppressive drug need only be applied to a small area of the forehead

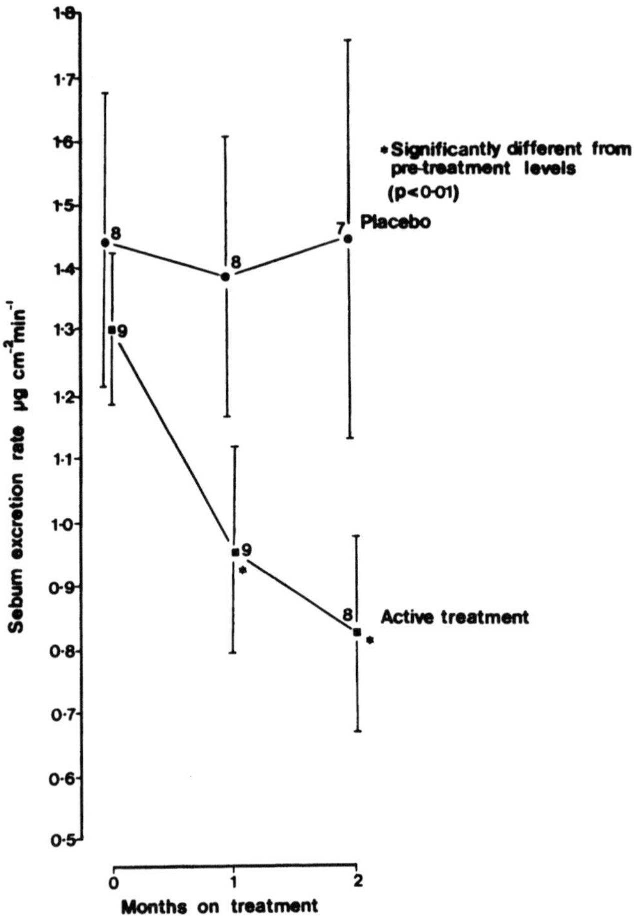

Fig. 1. Results obtained from a lipid inhibitor during its early development

skin. For effective sebaceous gland inhibition, only two months therapy is required and significant systemic absorption of a drug following its application to this small area is probably negligible, Thus, limited toxicity studies only would be required at this stage of a drugs development. The use of this approach is not limited to anti-seborrhoeic drugs; it can be usefully employed to investigate drugs which may inhibit *P. acnes* – the main organism involved in the aetiology of acne. With some clinically effective drugs (such as tetracycline 0.5 g/day) there is no reduction in viable bacterial counts but there is a significant reduction in the skin surface free fatty acid (FFA) concentration. Although we do not consider that FFA are relevant to the development of acne lesions [25, 31, 32] measurement of skin surface FFA is a useful way of assessing the function of *P. acnes. Propionibacterium acnes* produces lipases which convert sebaceous triglycerides to FFA; a reduction in bacterial function will significantly reduce the lipase production and/or function and thereby reduce skin surface FFA [6, 9] (Fig. 2). Topical application of a drug suspected of having an antimicrobial effect will, if it is to be of clinical benefit, reduce the surface FFA. These approaches

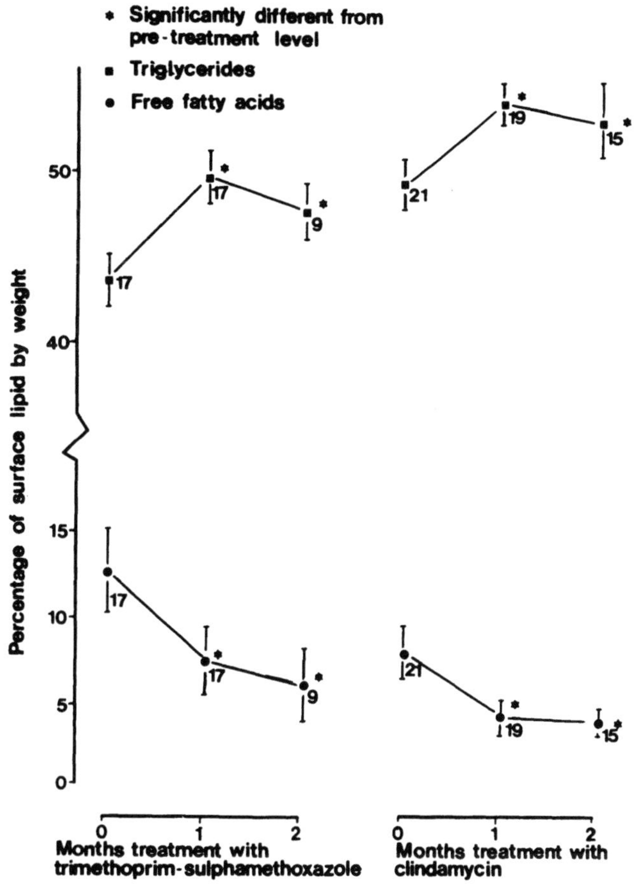

Fig. 2. The effect of two antibiotics on triglyceride and free fatty acid levels in skin surface lipid

could save the pharmaceutical industry considerable money as limited toxicity studies only would be required before using this human model.

New Human Models

The remainder of this chapter will discuss two techniques currently in use and being developed at our institution.

Human Skin Grafts Transplanted to Hairless Mice

This technique can be considered as a human model and is likely to provide useful information on sebaceous gland control, ductal cornification and bacterial function. It is unlikely to benefit studies on inflammation because of the different inflammatory response in the mouse as compared to man. However, it is likely to be of considerable benefit in studying the penetration of drugs into the pilosebaceous duct. Certain radioactive markers, chemical mutagens, pathogenic organisms and even certain antibiotics cannot be even considered for use in humans *in vivo* but can be considered in this model.

The methods of free skin grafting and immunological suppression have been described in detail previously [1, 18, 19]. Briefly, human cadaver skin from the mid-thoracic region is cut to 1.5–2.0 mm thickness with a sterile Castroviejo electrokeratotome and placed on filter paper soaked in sterile 0.9% (w/v) NaCl solution. The skin is obtained within 36 h of death, during which time the cadaver is maintained at 4 °C. The sheet of skin is cut into graft-sized pieces (approximately 1 cm^2) which are transferred to a graft bed cut into the dorso-lateral thoracic surface of a hairless mouse, covered with a sterile vaseline-impregnated tulle gras, and the whole thoracic region encased by a plastercast. The percentage take at 7 days varies between 60–95%.

The bacterial profile of pregrafted cadaveric skin does not differ significantly from the normal human microflora whereas murine skin exhibits quantitative and qualitative differences from human flora, in particular by the complete absence of *P. acnes.* The normal microbial profile of the human grafts is maintained throughout the experimental period despite the novel environmental milieu. There is little contamination of the grafts from the normal murine flora. It is concluded that the grafted human skin would provide a realistic model for studying the ecology of human cutaneous microorganisms. For example, grafts cut to a deapth of 0.6 mm to exclude the sebaceous glands show a significant reduction in surface *P. acnes* but not in the staphylococci [20].

In this model bacterial colonisation could be used to investigate the penetration of topically applied antimicrobial drugs. Alternatively, a radio-labelled drug could be applied to the surface of the grafted skin and its penetration noted by assessing the radioactivity in sections cut parallel to the skin surface. Comparison of the penetration of an aromatic retinoid (in a gel base with 3% azone added) 4 h after application to cadaveric skin left on the bench and transplanted skin (Gowland G, data to be published) demonstrates the value of this model (Fig. 3).

Inflammatory Model of Acne

The main problem of acne, as far as the patient is concerned, is usually the incessant development of inflamed lesions which represent the end result of a series of biological events – the sequence of which is unknown. Thus, a better understanding of their development could give an insight into possible new therapeutic ideas and a greater knowledge of the mechanism of action of existing treatments.

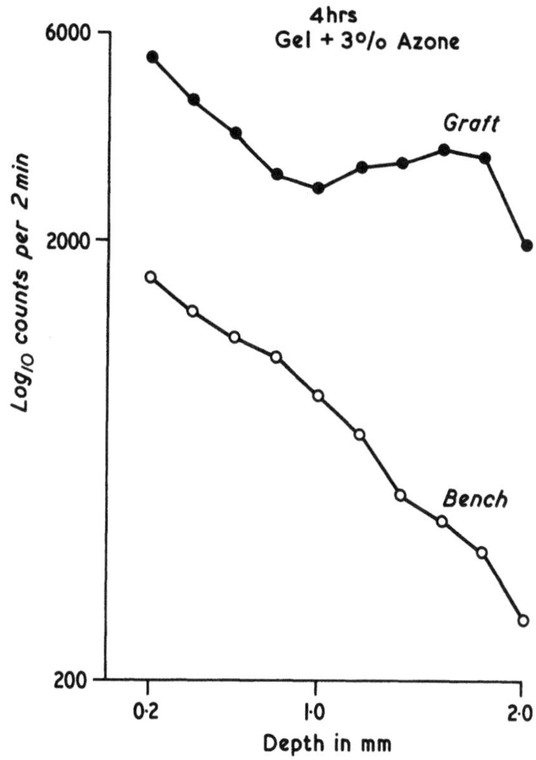

Fig. 3. The penetration of an aromatic retinoid through cadaveric skin either on the bench or transplanted to the hairless mouse

It is known from *in vitro* studies that *P. acnes* produces several biological substances which may mediate inflammation either through complement activation [39, 40] or by chemoattractant mechanisms [33, 41]. *In vivo* early inflamed lesions show evidence of the stimulation of both the alternate and classical complement pathways [10, 34]. Some investigators have failed to recognise the spectrum of inflamed lesions which may range from a small (less than 1 mm diameter) pink, hardly palpable papule, to an extensive nodule 2–3 cm in diameter. Table 1 demonstrates the many types of lesions that are readily identifiable. The inflamed lesions represent a compromise between the production of biologically active mediators and the host response to these substances. This may involve a whole range of reactions, recognised and unrecognised, which in turn may determine whether a lesion progresses or resolves.

Table 1. Types of Lesion Identified

Open comedone	Active papule
Closed comedone	Nodule
Erythema – non-palpable	Less active pustule
Erythema – follicle visible	Active pustule
Erythema – slightly palpable	Less active macule
Less active papule	Active macule

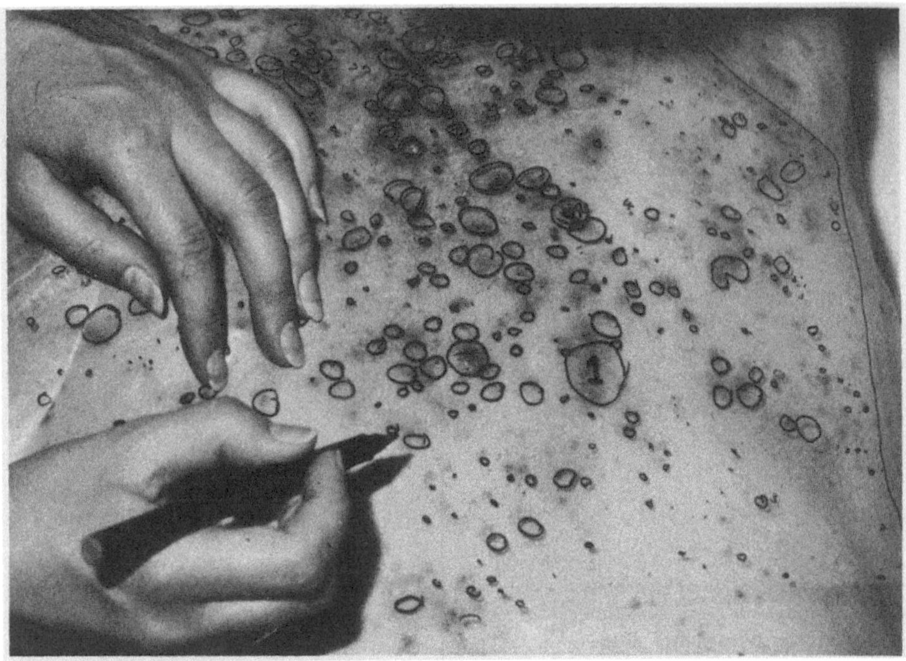

Fig. 4. Marking the position of acne lesions onto a cellulose acetate sheet

Assessment of the mechanisms involved in inflammation depends upon an understanding of their morphogenesis. By mapping lesions onto cellulose acetate sheets (Fig. 4) at the patients' initial and subsequent visits, the observer can determine the following factors:

From which lesion, if any, an inflamed lesion arises
Rate of generation of new spots
The varied appearance of spots during their lifespan
The time course of individual lesions
Variables which may influence factors 1–4, and such variables include:

- age and sex of patient
- duration of acne
- the effect of the menstrual cycle
- effect of known anti-inflammatory therapies and the effect of possible new treatments

Three important uses for this acne model for inflammation are evident.

The question as to why acne resolves has never really been adequately investigated. It is not due to a reduction in sebum excretion [8]. A change in the host response to mediators of inflammation seems a likely possibility.

A potential anti-inflammatory drug should, if successful, produce, compared to its placebo, a reduction in the rate of development of new lesions and an increased rate at which they resolve. A well controlled study on a small number of subjects may help

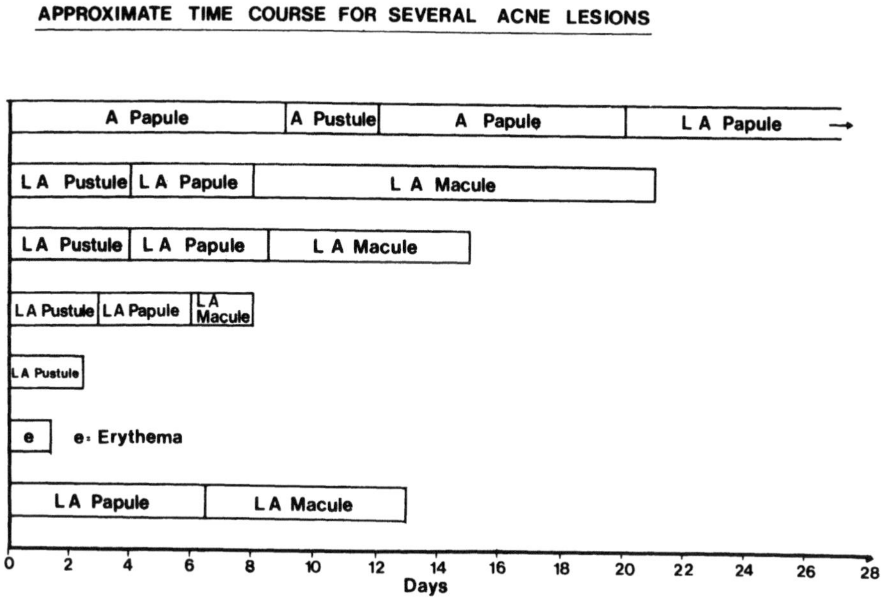

Fig. 5. Examples of the time taken for several lesions to complete their life-cycle. LA = less active

to determine the most active anti-inflammatory drug currently under investigation. The most appropriate base could also be chosen using this model.

It is pointless attempting to induce acne lesions *in vivo* by injecting, for example, extracts of the stratum corneum, if the observer does not know the usual morphogenesis of an acne lesion. A logical sequel of this suggestion is the need to know the histological sequence of events in a lesion at each step of its clinical appearance. With the advent of monoclonal antibodies to various lymphocyte subsets a dynamic understanding of acne morphogenesis can be developed.

The technique of assessing the morphogenesis of acne lesions is, in essence, quite simple, but technically requires considerable patience on behalf of the observer, good, standardised lighting and maximum patient compliance. Other obvious drawbacks include the fact that, once biopsied, the eventual life history of a particular lesion will never be known.

Figs. 5 and 6 indicate some basic data obtained using this technique. The rate of genesis of early low grade inflammatory lesions in subjects with mild and moderate acne is the same, but generally lesions in the former group are less inflamed than those in the latter. These less inflamed lesions, not surprisingly, generally last for a shorter length of time than the highly inflamed lesions common in severe acne sufferers. There is considerable variation in the lifespan of similar looking lesions. For example, a less active papule may last from 2.5 to 25 days and may represent three stages of development. It can represent the major type of inflammation for that lesion, or it may represent a more active lesion either on its way to resolution or en route through to a more active phase, although very inflamed lesions usually arise quickly, reaching

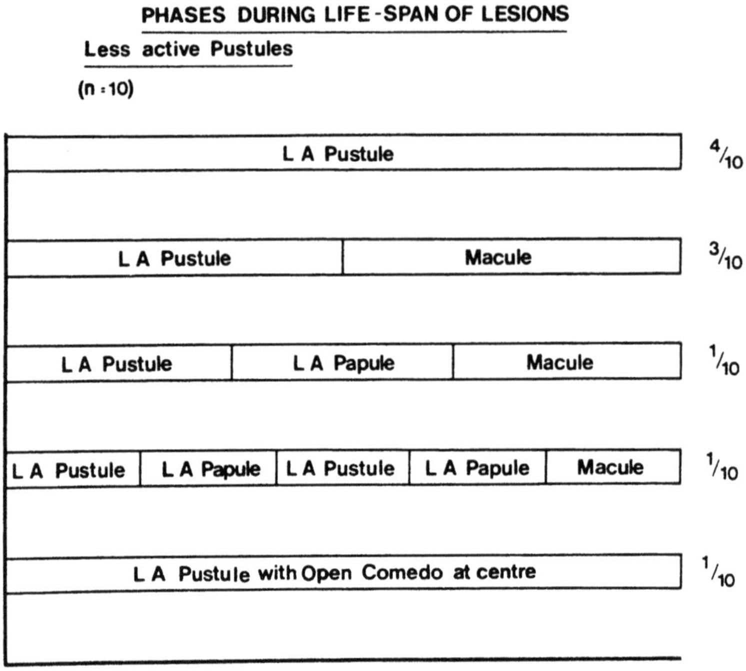

Fig. 6. The stages of development and resolution of ten acne lesions which were first viewed as pustules. LA = less active

their peak of inflammation in 24–48 h. Apparent also from our preliminary data is the very varied nature of macules. Macules are the end stage of inflammation and may last for 2–40 days, and so markedly contribute to the overall degree of inflammation.

Conclusions

The four methods described offer considerable potential for investigating the aetiology and treatment of acne in humans. Measurement of SER, or measurement of bacterial numbers and function, early on in the development of a drug could well save considerable money for the pharmaceutical industry. Cadaveric skin transplanted to mouse should give valuable information on sebaceous gland function and bacterial ecology as well as penetration of the drug into the pilosebaceous duct. The absence of an animal model for looking at the inflammation of acne could well be filled by the human model for assessing acne inflammation, but much more work on this particular model is required before its full potential is realised.

References

1. Billingham RE, Mewawar PB (1951) The technique of free skin grafting in mammals. J Exp Biol 28:385–402
2. Bowden PE, Meddis D, Cooper MF, Thody AJ, Shuster S (1976) Effects of 5α-reduced androgens on preputial-gland size and lipogenic activity. Biochem Soc Trans 4:795–797
3. Burton JL, Libman LJ, Cunliffe WJ, Wilkinson R, Hall R, Shuster S (1972) Sebum excretion in acromegaly. Br Med J 1:406–408
4. Cooper MF, Hay JB, McGibbon D, Shuster S (1976) Sebaceous lipogenesis and androgen metabolism in acne. Biochem Soc Trans 4:793–796
5. Cotterill JA, Cunliffe WJ, Williamson B (1971) Sebum excretion rate and biochemistry in patients with acne vulgaris treated by oral fenfluramine. Br J Dermatol 85:127–129
6. Cotterill JA, Cunliffe WJ, Williamson B, Bulusu L (1972) Further observations on the pathogenesis of acne. Br Med J 3:444–446
7. Cunliffe WJ, Cotterill JA (eds) (1975) The Acnes. WB Saunders, London
8. Cunliffe WJ, Shuster S (1969) The pathogenesis of acne. Lancet I:685–687
9. Cunliffe WJ, Cotterill JA, Williamson B (1972) The effect of clindamycin in acne – a clinical and laboratory investigation. Br J Dermatol 87:37–41
10. Dahl MGC, McGibbon DH (1979) Complement C_3 and immunoglobulin in inflammatory acne vulgaris. Br J Dermatol 101:633–640
11. Farrell LN, Strauss JS, Stranieri AM (1980) The treatment of severe cystic acne with 13-cis retinoic acid: evaluation of sebum production and the clinical response in a multi-dose trial. J Am Acad Dermatol 3:602–611
12. Frank SB (1982) Is the rabbit ear model in its present state prophetic of acnegenicity? J Am Acad Dermatol 6:373–377
13. Goolamali SK, Burton JL, Shuster S (1973) Sebum excretion in hypopituitarism. Br J Dermatol 89:21–24
14. Gowland G, Ward RM, Holland KT, Cunliffe WJ (1978) Cellular immunity to *P. acnes* in the normal population and patients with acne vulgaris. Br J Dermatol 99:43–47
15. Greenwood R, Jones DH, Brummitt L (1983) Comparison of isotretinoin and cyproterone acetate – a clinical and laboratory study. In: Cunliffe WJ, Miller AJ (eds) Retinoid Therapy. MTP Press Limited, Lancaster, pp 287–292
16. Holland KT, Cunliffe WJ, Roberts CD (1978) The role of bacteria in acne vulgaris: a new approach. Clin Exp Dermatol 3:253–257
17. Kanaar P (1971) Follicular keratogenic properties of fatty acids in the external ear canal of the rabbit. Dermatology 142:14–16
18. Kearney JN, Gowland G, Holland KT, Cunliffe WJ (1982a) Transplantation model for study of microbiology of human skin. Lancet I:334–335
19. Kearney JN, Gowland G, Holland KT, Cunliffe WJ (1982b) Maintenance of the normal flora of human skin grafts transplanted to mice. J Gen Microbiol 128:2431–2437
20. Kearney JN, Harnby D, Gowland G, Holland KT (1984) Follicular distribution and abundance of resident bacteria on human skin. J Gen Microbiol 130:797–801
21. Kligman AM (1974) An overview of acne. J Invest Dermatol 62:268–287
22. Kligman AM, Katz AG (1968) Pathogenesis of acne vulgaris. Comedogenic properties of human sebum in the external ear canal of the rabbit. Arch Dermatol 98:53–58
23. Kligman AM, Kwong T (1979) An improved rabbit ear model for assaying comedogenic substances. Br J Dermatol 100:699–702
24. Leyden JJ, McGinley KJ, Mills OH, Kligman AM (1975) *Propionibacterium* levels in patients with and without acne vulgaris. J Invest Dermatol 65:382
25. Marples RR, Downing DT, Kligman AM (1971) Control of free fatty acids on human surface lipid by *Corynebacterium acnes.* J Invest Dermatology 56:127–131
26. Marples RR, McGinley KJ, Mills OH (1973) Microbiology of comedones in acne vulgaris. J Invest Dermatol 60:80–83
27. Mills OH, Kligman AM (1982) A human model for assaying comedolytic substances. Br J Dermatol 107:543–548

28. Noble WC (ed) (1981) Microbiology of human skin. Lloyd-Luke, London
29. Plewig G, Luderschmidt CH (1977) Hamster ear model for sebaceous glands. J Invest Dermatol 68:171–176
30. Pochi PE, Strauss JS (1974) Endocrinologic control of the development and activity of the human sebaceous gland. J Invest Dermatol 62:191–201
31. Puhvel SM, Sakamoto M (1977) A re-evaluation of fatty acids as inflammatory agents in acne. J Invest Dermatol 68:93–97
32. Puhvel SM, Sakamoto M (1977) An *in vivo* evaluation of the inflammatory effect of purified comedonal components in human skin. J Invest Dermatol 69:401–406
33. Puhvel SM, Sakamoto SM (1978) The chemoattractant properties of comedonal contents. J Invest Dermatol 71:324–329
34. Scott DG, Cunliffe WJ, Gowland G (1979) Activation of complement – a mechanism for the inflammation in acne. Br J Dermatol 101:315–320
35. Shuster S, Thody A (1974) The control and measurement of sebum secretion. J Invest Dermatol 62:172–190
36. Strauss JS, Kligman AM (1960) The pathologic dynamics of acne vulgaris. Archs Dermatol 82:779–790
37. Thody AJ, Shuster S (1971) Pituitary control of sebum secretion in the rat. J Endocrinol 51:6–7
38. Vermorken AJM, Goos CMAA, Wirtz P (1982) Evaluation of the hamster flank organ test of the screening of antiandrogens. Br J Dermatol 106:99–101
39. Webster GF, McArthur WP (1982) Activation of components of the alternative pathway of complement by *Propionibacterium acnes* cell well carbohydrate. J Invest Dermatol 79:137–140
40. Webster GF, Leyden JJ, Nilsson UR (1979) Complement activation in acne vulgaris: consumption of complement by comedones. Infect Immun 26:183–186
41. Webster GF, Leyden JJ, Tsai CC, Baehni DDS, McArthur WP (1980) Polymorphonuclear leukocyte lysosomal release in response to *Propionibacterium acnes in vitro* and its enhancement by sera from patients with inflammatory acne. J Invest Dermatol 74:398–401

Models to Study Follicular Diseases

G. Plewig

There is a great variety of follicular diseases in man related to the pilosebaceous apparatus. They include excessive production of sebum, accumulation of keratinized material (comedones) in the follicular infundibula, and inflammatory lesions (papules, pustules, nodules). Clinical expressions of these diseases are seborrhea, acne vulgaris, contact acne due to industrial compounds or toiletries to name but a few [21]. Models to study follicular diseases can be divided into two, the keratinizing and the sebaceous gland ones. In the keratinizing models etiopathogenesis of follicular diseases can be studied as well as harzardous compounds leading to follicular impactions as in the case with some industrial compounds, cosmetic products or even acne remedies. In the sebaceous gland models etiopathogenesis of diseases associated with seborrhea can be investigated as well as compounds directed against the production of sebum (sebum-suppressive agents). It is important to realize the scope of these models and also to be aware of their shortcomings. This is elucidated by the differentiation of, e.g., comedogenicity versus irritancy, or sebum suppression versus irritancy. Asking the right question in the proper models helps to avoid controversies.

The Rabbit Ear Assay: Study of Keratinization and Comedolysis

The by now classical model to study keratinization is the rabbit ear. It was introduced by the Kligman group in Philadelphia, was improved by the same author [8] and modified by others who used a stereomicroscopic technique of viewing peeled off but otherwise unsectioned epidermal sheets with follicular comedones attached.

The rabbit ear model is very well suited for assessing comedogenic compounds as are encountered in the etiopathogenesis of acne vulgaris, acne cosmetica, chloracne, cutting oil acne, pitch acne, pomade acne, tar acne etc. [2] (Fig. 1).

In a classical study it was shown that human sebum is comedogenic [9]. Industrial compounds and modern toiletries can be studied in this model to identify and ban substances which may lead to follicular hyperkeratinization. Two controversies can be quoted in this context, the comedogenesity of sulfur [16, 26] and the more recent debate about cosmetics, lanolins, surfactants and detergents, pigments, moisturizers, therapeutic agents and miscellaneous ingredients [3]. It is said that the rabbit ear model is specific and sensitive [8]. Known acneigens elicit comedones in proportion to their established potency. Unfortunately true comedogenicity has been mismatched with irritancy leading to follicular hyperkeratosis. A final word how to use this rabbit

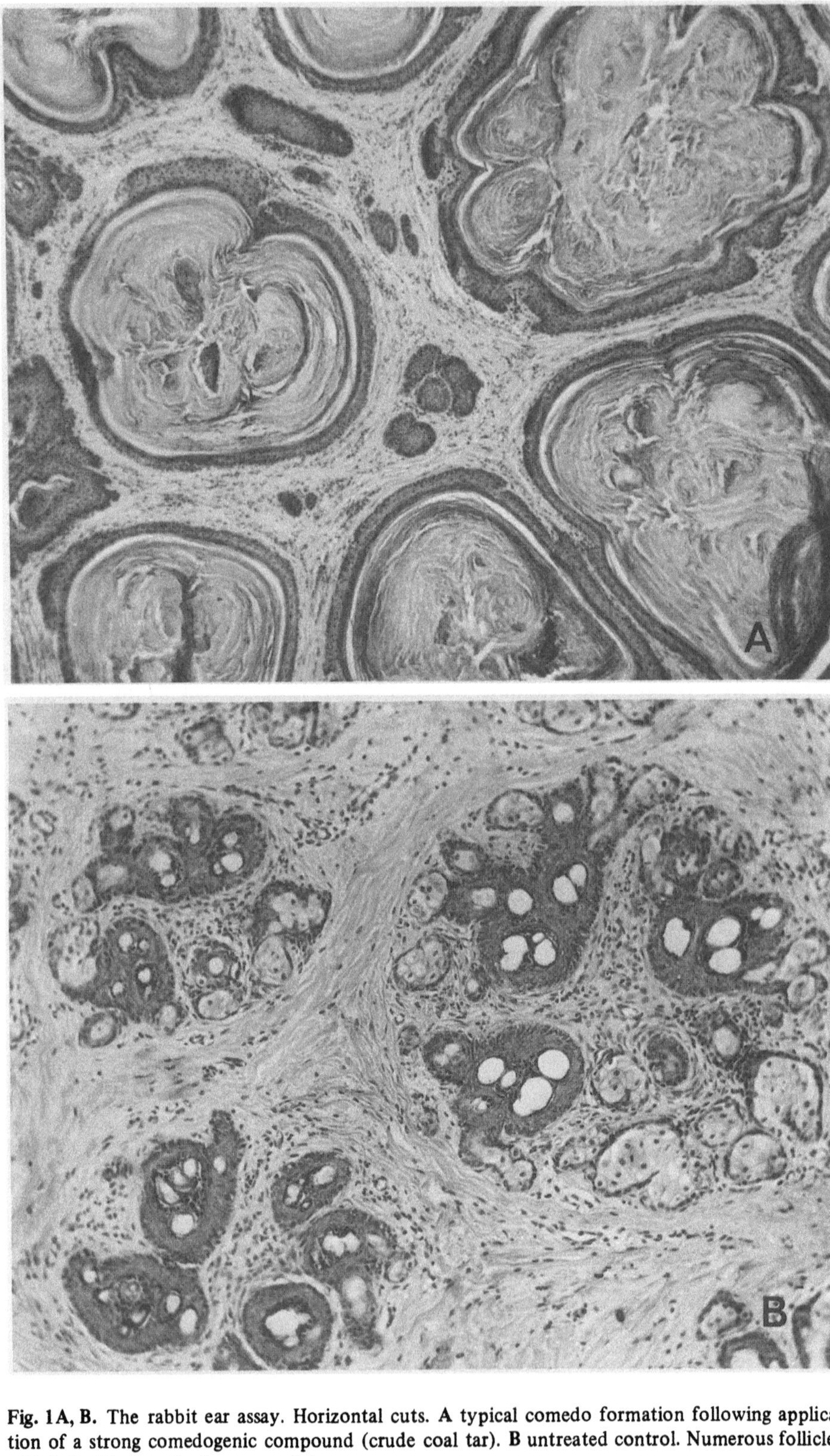

Fig. 1A, B. The rabbit ear assay. Horizontal cuts. **A** typical comedo formation following application of a strong comedogenic compound (crude coal tar). **B** untreated control. Numerous follicles with infundibula and sebaceous acini

ear model, how to apply, read and interpret it is expected soon by experts of the field. Likewise the rabbit ear model is very well suited for studying comedolytic agents. The aim of this model is to judge the prevention or uprooting of experimentally induced comedones. Keratolytic or comedolytic agents are all-trans retinoic acid, benzoyl peroxide, and salicylic acid, almost inactive are resorcinol or sulfur suspensions.

Enhancement of comedogenicity is possible by in-vivo ultraviolet radiation to the rabbit ear [18] or by in-vitro ultraviolet radiation of test compounds prior to their application to the rabbit pinna [19].

The Rhino Mouse Model

A new model for the study of agents which influence keratinization and exfoliation is the rhino mouse. The skin of the rhino mouse which is an allelic of the hairless mouse contains deep dermal cysts and a huge number of utriculi filled with keratinous debris which resemble comedones. In a preliminary study Van Scott investigated the effect of various anti-acne compounds on the rhino mouse. Kligman and Kligman used this model extensively, and applied various retinoids to this shortlived deficient animal [7]. They conclude that the rhino mouse is a suitable model for assessing chemicals which effect epithelial differentiation (retinoids) or which promote loss of cohesion between horny cells (descaling agents).

Sebaceous Gland Assay in Animals

Various animals have been used to study the morphology and cellular kinetics of sebaceous acini or the effects of drugs on sebaceous glands. Many authors prefer the rat, either the glands attached to the fur hair or to the preputial gland. We – like many others – prefer the Syrian hamster [11–13, 20].

The Hamster Flank Organ. In the hamster there are two areas suitable for these studies. One is the costovertebral flank organ. The relevance of the flank organ assay, in particular for antiandrogens was described by Gomez and Frost [5]. An improved histolog-

The Hamster Ear Model. Our own group described the architecture and sebocyte kinetics of the numerous sebaceous follicles to be found on the ventral side of the earlobe of the Syrian hamster [22] (Fig. 2). Matias and Orentreich examined the same model by planimetry of peeled off skin instead of routine vertical histological sections [15]. Antiandrogens as cyproterone acetate are well studied in this model [13]. Like almost all other models the sebaceous gland assay has its pitfalls. There are claims that benzoyl peroxide reduces sebum production in men and in hamsters. No sebum suppression was reported by others [4]. True sebum suppression as, e.g., from the systemic administration of isotretinoin [23] may not be confused with unspecific

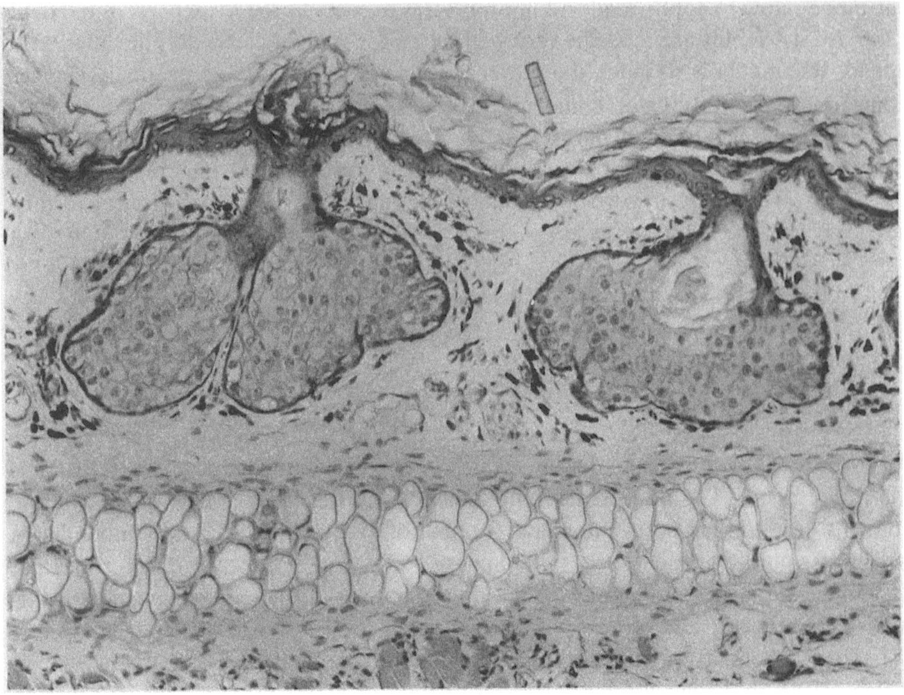

Fig. 2. Sebaceous gland assay in the Syrian hamster. Numerous large sebaceous acini on the ventral side (above cartilage) of a testosterone-stimulated female hamster. Uptake of radioactive material in basal cells of epidermis and sebaceous acini. 3H-thymidine, 45 min. Hematoxilin

effects on sebaceous glands due to irritancy and dedifferentiation. Irritancy causes rapid proliferation of sebocytes which consequently turn into less differentiated cells which look like keratinocytes. Sebaceous acini shrink into epithelial-like buds, a rather unspecific effect of irritancy. The same phenomenon is seen after topical application of other irritants (e.g. all-trans-retinoic acid) or physical stimuli, e.g., one minimal erythema dose UV-B daily [1]. The same caution applies for this model as has been mentioned above for comedogenicity or irritancy in the rabbit ear model.

The Human Model

The human skin contains acne-prone areas with numerous huge sebaceous follicles – the ideal site for acne vulgaris and acne-vulgaris-like follicular diseases (Fig. 3, 4), for acne cosmetica or cutting oil acne. The literature on this subject is comprehensive.

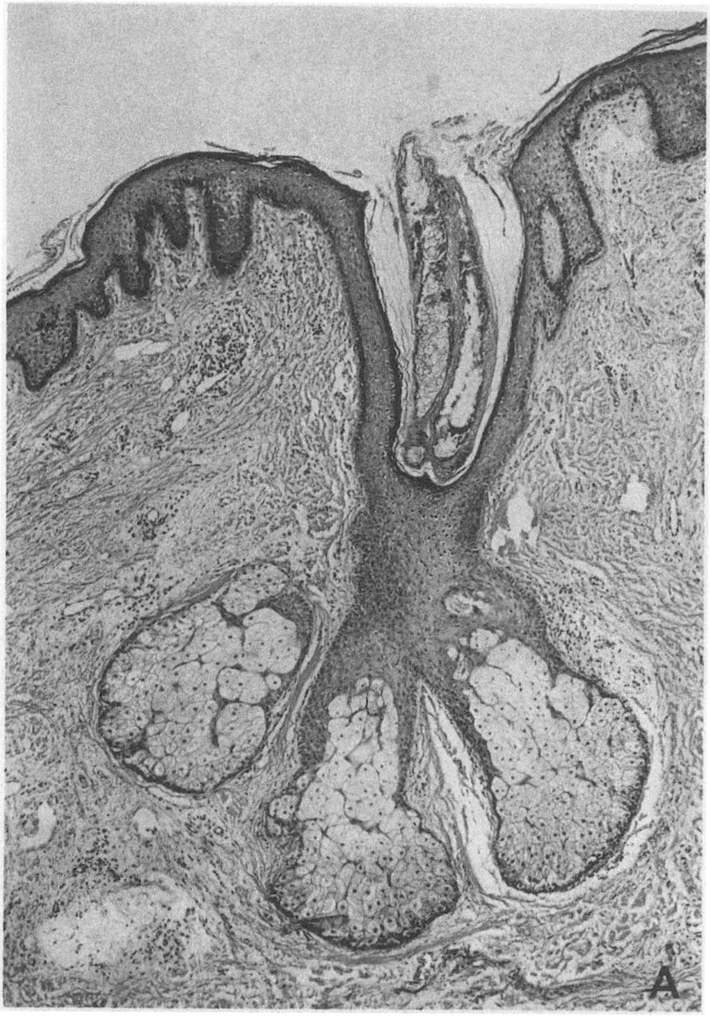

Fig. 3A–D. The sebaceous follicle assay in the human (microcomedones and comedones)
A = Large sebaceous follicle of the upper back in a young adult with prominent follicular skin pores and seborrheic skin. The amount of impacted corneocytes is still normal; the widened space between the follicular filament and the infundibulum is a histotechnical artefact. H & E

The Follicular Model: Non-invasive Cyanoacrylate Technique

Marks and Dawber expanded the ancient technique to remove horny cell layers (tape strips by Wolf, and the sticky slide technique by Goldschmidt and Kligman) by representing the very potent glue cyanoacrylate. They termed it skin surface biopsy [14]. Their major interest was skin surface, the stratum corneum. This technique was extended for the excavation of follicular contents (Fig. 4C), analogous to the more painful and less controlled squeezing of sebaceous filaments [21, pp. 56–57]. It has

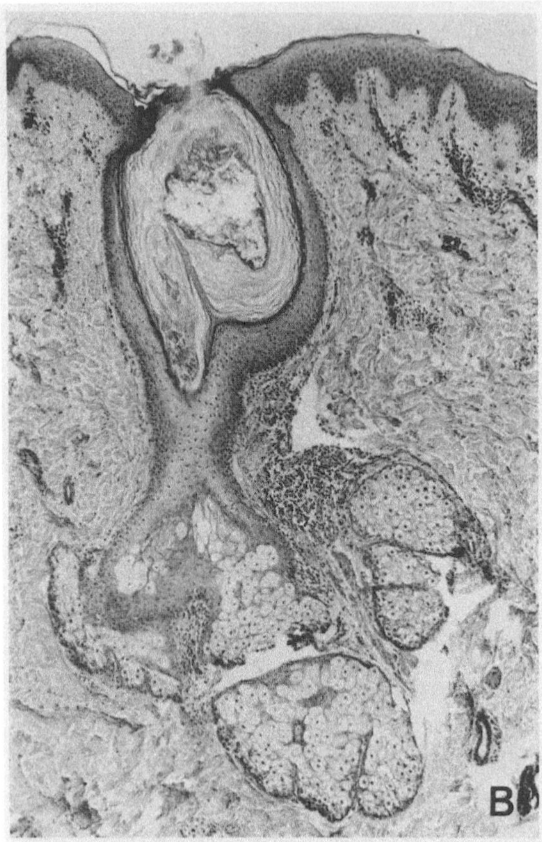

Fig. 3. B = Early comedo formation in a sebaceous follicle of the upper back induced in a young adult by continuous application of a surgical tape (Clear Tape^{R}) for eight weeks. H & E

been used for studying follicular filaments [24], microcomedones, trichostasis spinulosa [21, pp. 102–103] and demodex folliculorum mites with the light microscope. Electron microscopy of horny cells (Fig. 4B) as well as bacterial and yeast colonization is also possible [21]. Early events of comedo formation in relation to the bacterial contents were found out by Lavker et al. [10]. The cyanoacrylate technique is appreciated by those who study the quantitative bacteriology and the effects of drugs, e.g., benzoyl peroxide [25]. Others use the calcium chloride technique which requires a biopsy [6].

Mills and Kligman inaugurated a model in which they can assay comedolytic substances [17]. Microcomedones are induced on the back of adult males by a two week occlusive exposure to 10% crude coal tar. The test agents are then applied for two weeks and the reduction in density of microcomedones determined by the non-invasive cyanoacrylate technique.

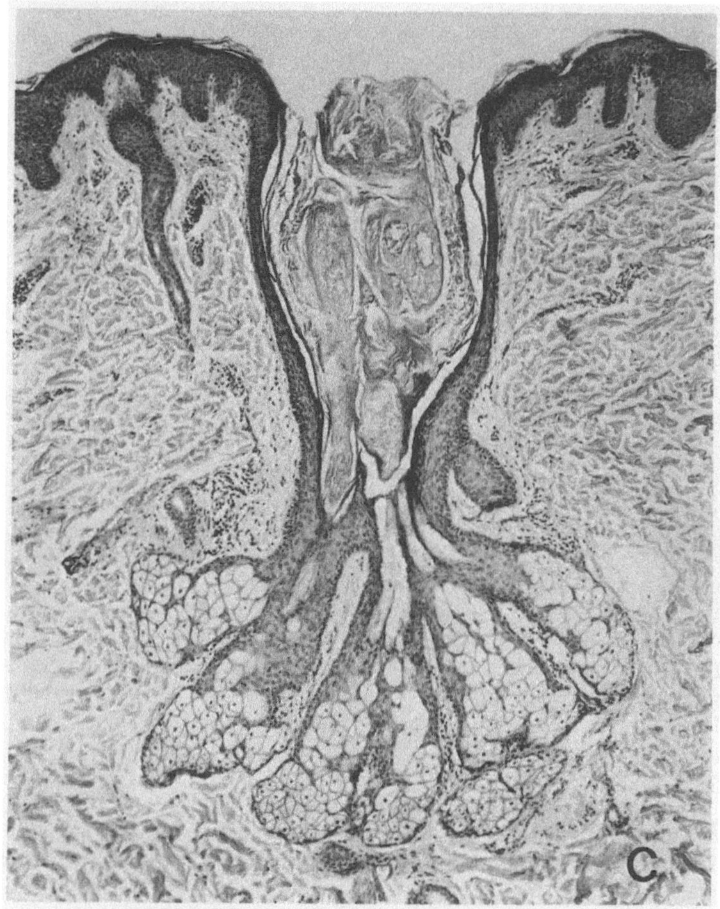

Fig. 3. C = Early comedo formation in the acro- and infrainfundibulum of a large sebaceous follicle on the back of a young adult. 20 days of 0.5% fluocinolone acetonide solution under polyethylene film. H & E

Skin Biopsy: The Invasive Technique

Often a full thickness biopsy containing several sebaceous follicles is wanted. This technique was used routinely in the past (Fig. 3, 4). The recent observation of sebum suppression after orally given isotretinoin was documented by serial biopsies from patients receiving this drug [23]. Sebaceous follicles shrank within weeks by 50 to 90% of their size compared to pre-treatment values. Bipsies are inevitable if sebaceous glands (sebocytes) are to be investigated.

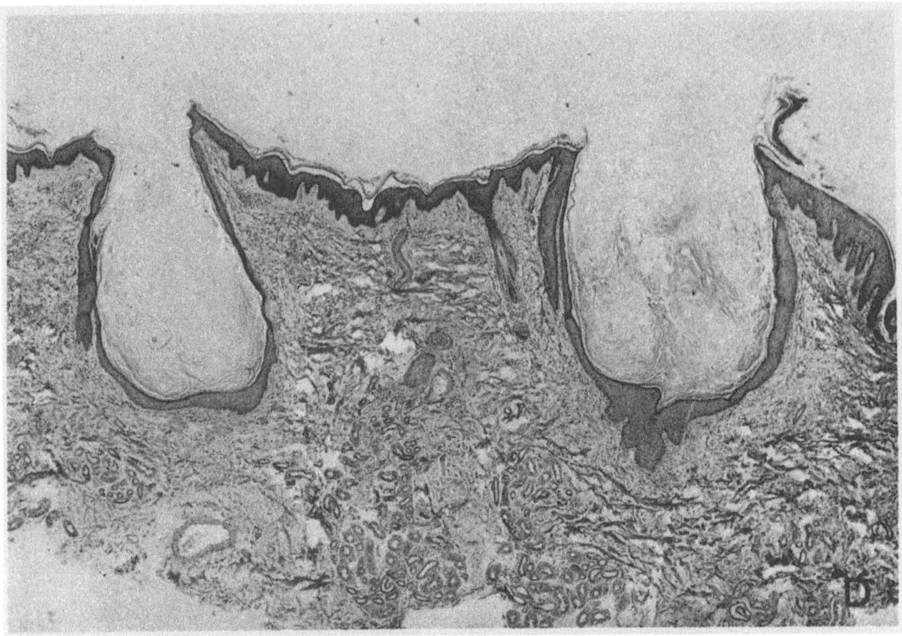

Fig. 3. D = Compact comedones in entire sebaceous follicles with complete dedifferentiation of sebaceous acini on the back of a young adult. Crude coal tar under polyethylene film for 20 days. H & E

Acknowledgements
The technical assistance of Miss A. Rufus MTA, Mrs. A. Ledolter MTA, and Miss E. Januscke MTA, is acknowledged.

This study was supported by a grant by the Deutsche Forschungsgemeinschaft Pl 58/5–7.

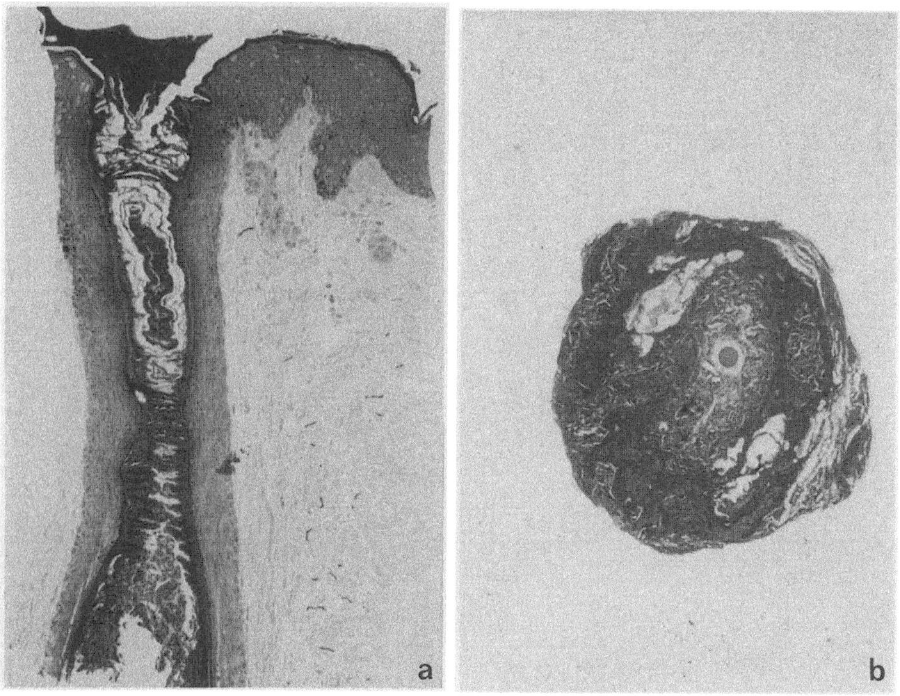

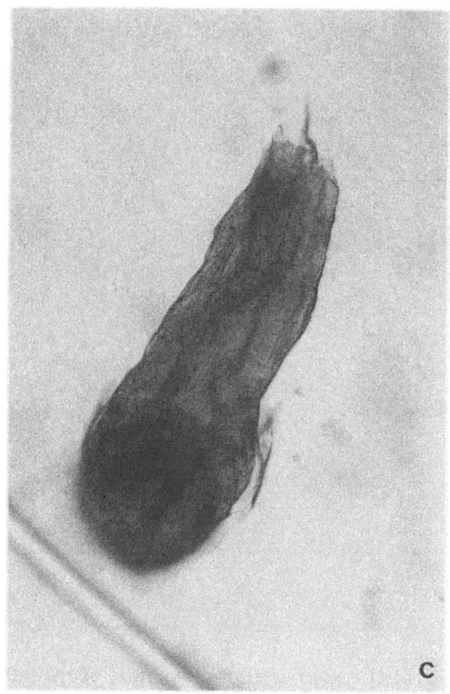

Fig. 4a–d. The follicular canal and the sebaceous filament (follicular cast). **a** = Semithin section of a biopsy showing epidermis above, acro- and infrainfundibulum, corneocytes, cellular debris and bacteria. Methylene blue. **b** = Semithin section of a squeezed out sebaceous filament with vellus hair in center and concentric corneocyte lamellae. Methylene blue. **c** = Whole mount of a sebaceous filament removed with cyanoacrylate. The vellus hair in the center of the keratinous debris. Unstained, immersed in oil

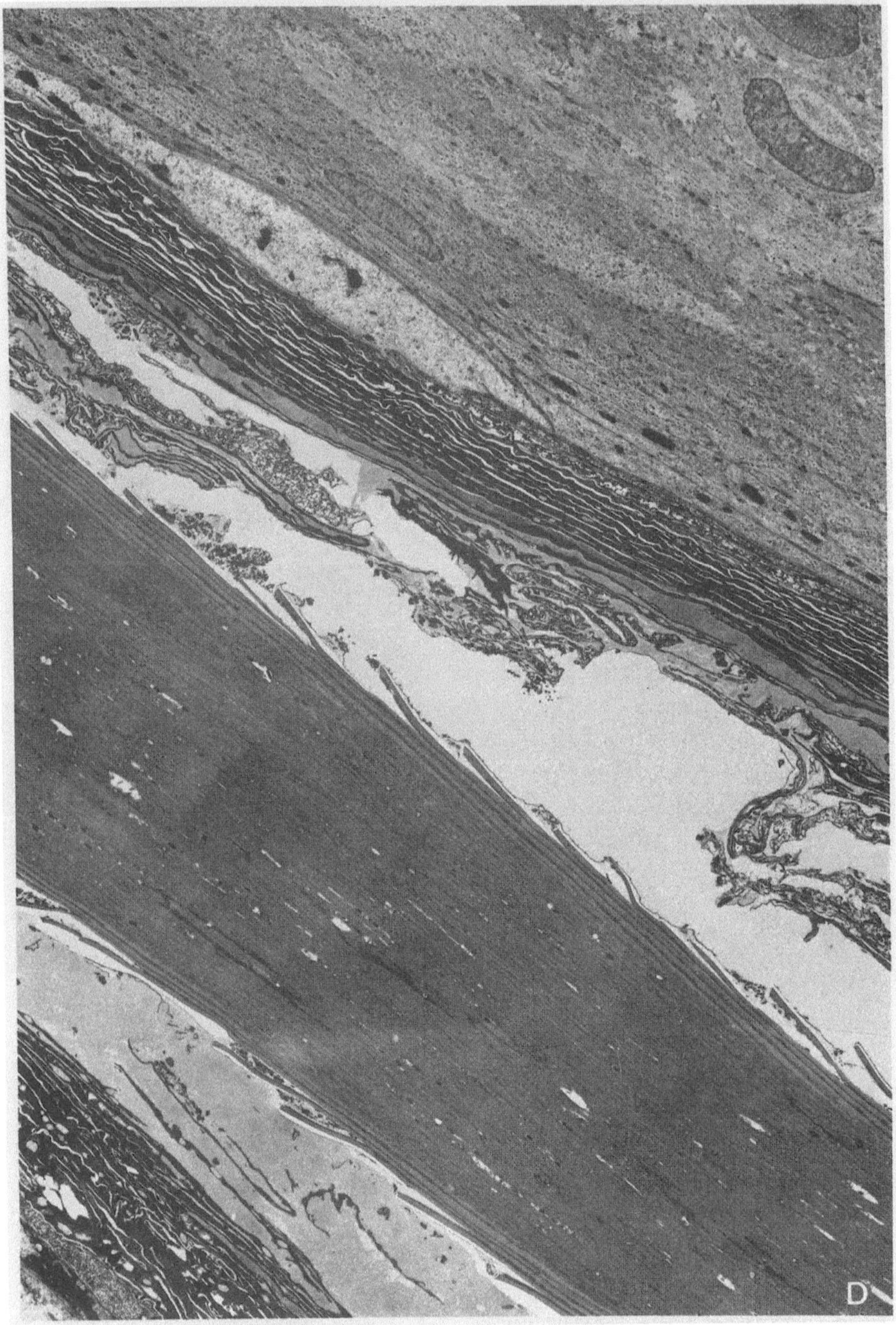

Fig. 4. D = Overview of mid-portion of the infundibulum of a sebaceous follicle from the upper back of a young adult (biopsy). The follicular epithelium is both to the upper right and lower left corner. Approximately 15 layers of corneocytes line the vellus hair (center), with sebum and cellular debris in the follicular canal. Electron microscopy, ×4,000

References

1. Dachs U (1984) Effekte physikalischer, chemischer und mechanischer Behandlung (Ultraviolettlicht, Vitamin-A-Säure, Epilation) auf Haut und Adnexe am Tiermodell des Syrischen Hamsters. Inaugural Dissertation, Department of Dermatology, Munich
2. Frank SB (1982) Is the rabbit ear model, in its present state, prophetic of acnegenicity? J Am Acad Dermatol 6:373–377
3. Fulton JE, Pay SR, Fulton III JE (1984) Comedogenicity of current therapeutic products, cosmetics, and ingredients in the rabbit ear. J Am Acad Dermatol 10:96–105
4. Goldstein JA, Pochi PE (1981) Failure of benzoyl peroxide to decrease sebaceous gland secretion in acne. Dermatologica 162:287–291
5. Gomez EC, Frost P (1975) Hamster flank organ: relevance studies with topically applied antiandrogens. In: Animal models in dermatology. Churchill Livingstone, Edinburgh London New York, pp 190–202
6. Holland KT, Roberts CD, Cunliffe WJ, Williams M (1974) A technique for sampling microorganisms from pilo-sebaceous duct. J Appl Bacteriol 37:289–296
7. Kligman LH, Kligman AM (1979) The effect on rhino mouse skin of agents which influence keratinization and exfoliation. J Invest Dermatol 73:354–358
8. Kligman AM, Kwong T (1979) An improved rabbit ear model for assessing comedogenic substances. Br J Dermatol 100:699–702
9. Kligman AM, Wheatley VR, Mills OH (1970) Comedogenicity of human sebum. Arch Dermatol 102:267–275
10. Lavker RM, Leyden JJ, McGinley KJ (1981) The relationship between bacteria and the abnormal follicular keratinization in acne vulgaris. J Invest Dermatol 77:325–330
11. Luderschmidt C, Bidlingmaier F, Plewig G (1982) Inhibition of sebaceous gland activity by spironolactone in Syrian hamster. J Invest Dermatol 78:253–255
12. Luderschmidt C, Eiermann W, Jawny J (1983) Steroid hormone receptors and their relevance for sebum production in the sebaceous gland ear model of the Syrian hamster. Arch Dermatol Res 275:175–180
13. Luderschmidt C, Plewig G (1977) Effects of cyproterone acetate and carboxylic acid derivates on the sebaceous glands of the Syrian hamster. Arch Dermatol Res 258:185–191
14. Marks R, Dawber RPR (1971) Skin surface biopsy: An improved technique for the examination of the horny layer. Br J Dermatol 84:117–123
15. Matias JR, Orentreich N (1983) The hamster ear sebaceous glands. I. Examination of the regional variation by stripped skin planimetry. J Invest Dermatol 81:43–46
16. Mills OH, Kligman AM (1972) Is sulphur helpful or harmful in acne vulgaris? Br J Dermatol 86:620–627
17. Mills OH, Kligman AM (1982) A human model for assaying comedolytic substances. Br J Dermatol 107:543–548
18. Mills OH, Porte M, Kligman AM (1978) Enhancement of comedogenic substances by ultraviolet radiation. Br J Dermatol 98:145–150
19. Motoyoshi K (1983) Enhanced comedo formation in rabbit ear skin by squalene and oleic acid perioxides. Br J Dermatol 109:191–198
20. Plewig G (1980) Der Einfluß des aromatischen Retinoids Ro 10-9359 und der 13-cis-Retinsäure Ro 4-3780 auf die Talgdrüsen des Syrischen Hamsters. Arch Dermatol Res 268:239–246
21. Plewig G, Kligman AM (1975) Acne. Morphogenesis and treatment. Springer, Berlin
22. Plewig G, Luderschmidt C (1977) Hamster ear model for sebaceous glands. J Invest Dermatol 68:171–176
23. Plewig G, Nikolowski J, Wolff HH (1982) Action of isotretinoin in acne, rosacea and gram-negative folliculitis. J Am Acad Dermatol 6:766–785
24. Plewig G, Wolff HH (1976) Follikel-Filamente. Arch Dermatol Res 255:9–21
25. Puschmann M (1982) Klinisch-experimentelle Untersuchungen zum Wirkungsnachweis von Benzoylperoxid. Hautarzt 33:257–265
26. Strauss JS, Goldman PH, Nacht G, Gans EH (1978) A reexamination of the potential comedogenecity of sulphur. Arch Dermatol 114:1340–1342

Models for Wound Healing

R. Marks, D. Williams and A. D. Pearse

Care of wounds and ulcers has always been an important aspect of the practice of medicine. Yet despite our profession's early assumption of this responsibility, over the years there have been very few changes in the way that these lesions are treated. It is true that the availability of efficient antibacterial agents has enabled the complications of infection to be removed but we are still unable to accelerate the healing of large denuded areas of skin. Even the comparatively recent realization of the roles of venous hypertension and the pericapillary deposition of fibrin in gravitational ulcers [1–3] and the significance of pressure induced ischaemia in decubitus ulcers [19], has not much alleviated these clinical problems. A major reason for our comparatively ineffectual management of these lesions is our lack of understanding of the biochemistry and pharmacology of wound healing which of necessity has led to an empirical approach to treatment.

The Need for Models

The development and use of good models of wound healing should enable this Cinderella of medicine to cast off the ugly rags of traditional type dressings and go to the modern pharmacological ball. Models are needed both to investigate the biology of wound healing and for the assessment of new forms of treatment. Models are especially necessary for the study of chronic ulcers of the skin because of the difficulty of research employing the lesions themselves in man. The main problem in human studies is the difficulty in obtaining identifiably similar groups of patients receiving the comparison treatment. It is infrequent that there is just one major cause for the persistence of an ulcer – complicating factors and compounding pathologies are common (e.g. atherosclerotic disease accompanying gravitational ulceration is far from uncommon). Secondly, it is almost impossible to avoid the use of some adjunctive treatment (such as a dressing or cleansing agent) which may complicate the interpretation of the results of an experiment. Then there is the issue of ulcer size. It seems quite likely that the tissue process involved and the metabolic consequences of an ulcer of 200 cm^2 are quite different from those involved in an ulcer of 20 cm^2 (Fig. 1).

The length of time an ulcer has been present may also alter its capacity to heal within a particular period of time; it is difficult to transfer information gained from the study of stinking, deep, ragged lesion of 50 cm^2 present for the past seven years in an anaemic, diabetic patient of 70 with a gravitational ulcer, who also has atherosclero-

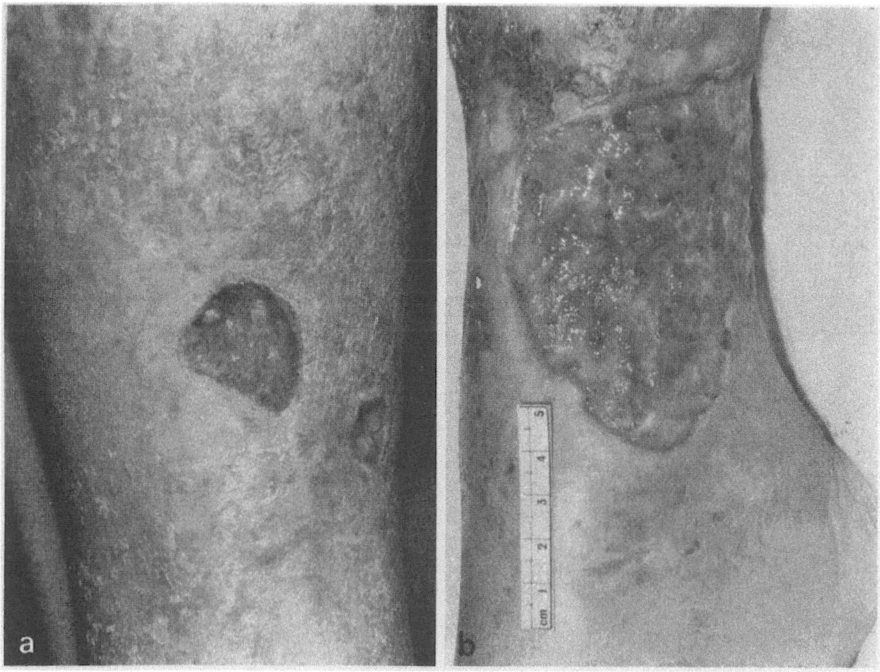

Fig. 1a, b. Two gravitational ulcers: a shows a gravitational ulcer of approximately 2 cm^2 and b shows a gravitational ulcer of approximately 80 cm^2. Although the rate of re-epithelialisation may be very similar, the biological problems involved are quite different

sis, with that obtained from a clean gravitational ulcer of 9 cm^2 in an otherwise fit man of 49.

Two other difficulties are worth considering. The first concerns the lack of good objective measures of healing of chronic ulcers *in vivo* in man. The second is the group of patients afflicted with the lesions under discussion. They are mostly elderly, frail, and socially deprived. They often have considerable difficulty in complying with even what one may consider quite straightforward instructions, and, because of immobility, it may be difficult for them to attend regularly for the necessary assessments. Because of these problems it seems almost impossible to conduct a useful clinical trial in patients with chronic ulcers.

This is not to suggest that important data cannot be obtained from clinical investigation. Information concerning the pathodynamics and intimate pathogenesis is probably best obtained by patient studies. As an example one may quote the contribution made to the understanding of the haemodynamics and role of the deposition of fibrin around the small blood vessels in the gravitational syndrome [1–3].

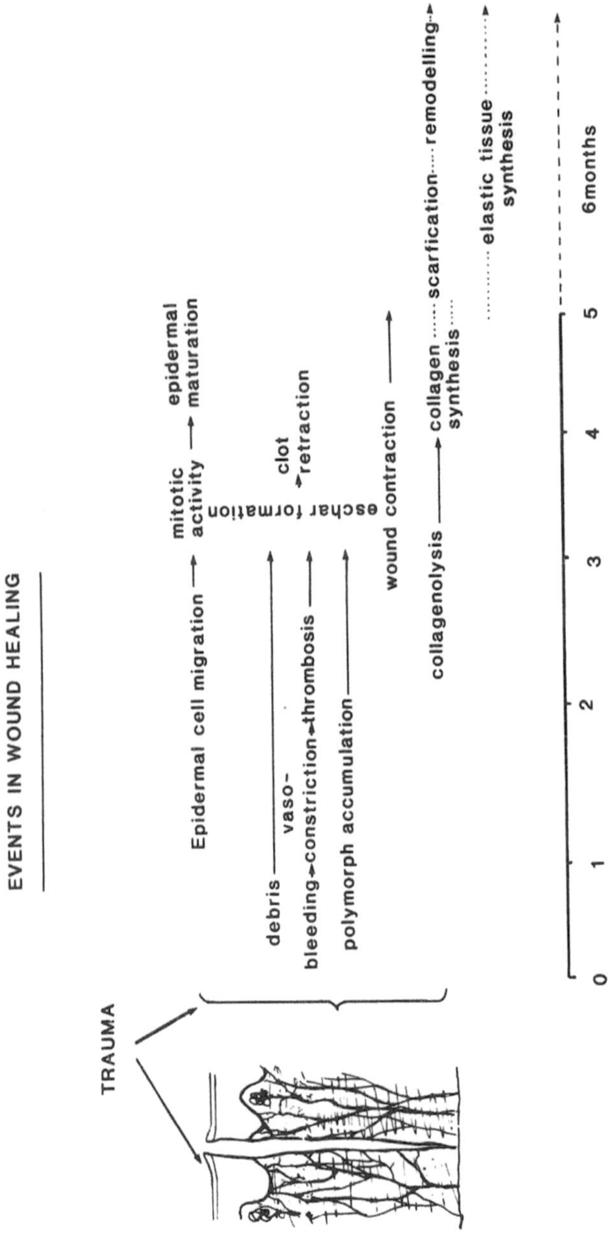

Fig. 2. Diagram to illustrate the events in wound healing

Factors influencing choice of model

Granted that models are required for wound healing studies, what features of wound healing may we want to emulate in a model? The major events that may need modelling include re-epithelialization, angiogenesis, wound contraction, wound strength and

Table 1. Numbers of migrating cells from human skin explants in short term organ culture after addition of different fractions of serum or whole normal human serum or no addition made to the medium. The fractions of serum were obtained by sephadex G-200 column chromatography of serum after 50% ammonium sulphate precipitation. The most effective fraction (2) contains the albumin and proteins travelling with albumin

Addition to medium	Mean number of migrating cells per site of migration
Fraction 1	0
Fraction 2	200
Fraction 3	25
Fraction 4	0
Normal human serum	200
No addition	0–10

the restoration of skin function of the wounded area (Fig. 2). In addition, the degree of inflammation, its type and its timing in relation to the wounding stimulus and the treatment used, may also need to be modelled. Furthermore it may be important to attempt to simulate the vascular changes which result in tissue hypoxemia (e.g. fibrin deposition around small blood vessels. It must be emphasized that it is most unlikely that any one model will adequately serve to model all the features seen in a non-healing wound or ulcer of the skin in man. Models should be chosen which are most relevant to the particular therapeutic aim or process under investigation. A large number of models of wound healing have been describe, and a complete review would be outside the scope of this paper.

In Vitro Models

In vitro models may give excellent information concerning some aspect of the early phases of wound healing – particularly the process of re-epithelialisation. Various groups, including ourselves [8, 13–15], and more recently Stenn [16], have used simple short term organ culture of skin to investigate factors controlling epidermal cell migration (Table 1). Others have used fibroblast cultures to examine the pharmacological effects on connective tissue to learn about the process of restoration of dermal integrity and how it may be influenced [9, 22]. Some have found it useful to examine endothelial cell cultures to look at chemical agents that modulate angiogenesis and many influence new vessel formation during wound healing [20], but there are few reported studies of this type. Corneal vascularization [6], the vascularization of chick chorioallantoic membrane [4] and hamster cheek pouch [24] have also been used to investigate this aspect of the pharmacology of wound healing, and promise to yield important information on the vascularization of wounds.

Epidermal cell migration alone does not constitute re-epithelialization: neither does collagen formation entirely represent restoration of a functional dermis, nor does endothelial culture necessarily give much information concerning blood flow in healing tissue. Clearly cell and even organ culture techniques can only inform on one small aspect of the response to wounding, and even this information can only be transferred to healing wounds *in vivo* if it is known that *in vivo* influences don't swamp the cellular change observed *in vitro.*

Human in Vivo Models

Human models have the obvious advantage for observation of the wound healing response in the target organ of interest. They have several disadvantages, not the least of them being that it is difficult to produce anything other than a clean, rapidly healing wound in human volunteer subjects. There are also practical difficulties which include logistical problems due to the large number of wounds sometimes needed in studies, ethical considerations, and cost. We have used a biopsy/rebiopsy technique in several studies with novel dressing materials [25]. This is easily accomplished by using a 3 or 4 mm diameter trephine to produce the initial wound and then removal of the entire wound with a 6 mm diameter trephine at different time intervals subsequent to treatment in different wounds. Another similar approach that we have employed involves producing a shallow keratotome wound on the thigh and the subsequent trephine biopsy of portions of the healing area at differing times. The wounds produced in these ways are mostly healed by five days, and clearly this fact imposes a major limitation. It is quite difficult to make quantitative assessments in these experiments and the only proper measurement we have been able to make is of the extent of re-ephithelialization in the trephine wound experiments (Table 2). We also routinely use scoring systems on sections viewed blind, to grade the degree of inflammation present and the amount of slough in the wound cavity as well as the degree of new vessel formation in the wounded area (Table 3).

The main use of these human models is to determine the comfort, practicability and tolerance of wounded areas to the topical agent or dressing being tested, and the type of reaction generated at the wound/dressing interface. It is also theoretically possible to learn whether any of the agents used have an accelerating or retarding effect on the healing processes.

Table 2. Measurement of re-epithelialisation made on experimental trephine wounds in human volunteer subjects after treatment with Silastic foam. The measurements were derived from sections cut in a standardised manner using a digitizing tablet to obtain data

Assessment	Day 1	Day 2	Day 4
Wound width (mm)	2.14 ± 1.45	2.18 ± 0.25	2.48 ± 0.33
Epithelial tongue length (mm)	0.225 ± 0.155	0.50 ± 0.09	0.85 ± 0.10
Epithelial tongue area (mm^2)	0.012 ± 0.0054	0.038 ± 0.019	0.098 ± 0.037

Adapted from: Williams, Dykes and Marks [26]

Table 3. Data from visual analogue scale evaluation on experimentally produced trephine wound in human volunteer subjects at various times after treatment of the wounds with Silastic foam

Assessment	Day 1	Day 2	Day 4
Inflammation	1.40 ± 0.84	0.78 ± 0.34	1.40 ± 1.20
Slough	1.85 ± 2.32	2.60 ± 1.14	0.28 ± 0.21
Vascularisation	0.60 ± 0.32	0.93 ± 0.33	2.75 ± 2.40

Adapted from Williams, Dykes and Marks [26]

Animal in Vivo Models

For many purposes animal models are more appropriate than other types of model, as a much wider range of questions can be answered using them. For example, it has been possible to make observations on wound strength and healing in jaundiced rats [17, 18] using experimental manoeuvres that are obviously not possible in man. Mechanical stress was found to improve the strength of healing wounds in pigs [10]. This study involved the insertion of an inflatable bladder beneath the wound and subsequent testing of the mechanical properties of the healing area, and obviously this type of study is only possible in experimental animals. Similarly, determination of the roles of inflammatory cells and complement factors in wound healing proved possible in a series of experiments in small mammals which could not have been performed by Ross and his co-workers in human subjects [21]. It is also impossible to use a range of materials whose toxicity has been uncharacterized in animal models of wound healing [23].

Determination of the mechanical load necessary to cause rupture of healing tissue *in vivo* or *in vitro* can also be accomplished using experimental animals, as in a recent experiment in which we were interested in the effects of factor XIII on wound strength.

A larger number of observations is possible when animals are employed than when human subjects are studied. Eaglstein and colleagues have used large numbers of shallow wounds in domestic pigs to assess the effect on healing of antibiotics, corticosteroids and a wide range of other topical medicaments [5, 7]. They have measured the proportions of wounds that healed at various time points and clearly this approach is only possible if a large number of wounds are observed. Furthermore, it is possible to assess the effects of a wide variety of new agents in wounds in experimental animals when it could be unjustifiable to use them in man.

The Use of Models to Assess Treatments for Chronic Wounds and Ulcers

In vivo models are often used to investigate the effects of one or another treatment on healing, with the objective of finding something that will accelerate the closure of non-healing wounds and chronic ulcers. As the great majority of small, clean, simple wounds heal within four or five days, it seems intrinsically unlikely that this type of wound will be able to demonstrate any enhancement in the rate of healing after an experimental treatment. Indeed it may well be that wound closure is proceeding maximally in such clean lesions. It is for this reason that we wanted to produce and use a lesion whose healing was compromised and which had some of the features of the sloughy ulcers that we see clinically. In our search we investigated thermal burns, cold injury and the damaging effect of various chemicals on the skin of small mammals. In most instances the wounds were irregularly shaped and variable in size and depth. The only manoeuvre that proved satisfactory was the intracutaneous injection of sodium tetradecyl sulphate (STD) [24]. This substance is a sclerosant and is used by surgeons to thrombose haemorrhoids and oesophageal varices.

The Sodium Tetradecyl Sulphate (STD) Model

When STD is injected intracutaneously into the skin of either rat or guinea pig, a reproducibly sized area of necrosis of approximately 0.5 cm diameter results. This wound takes approximately ten days to heal compared to four or five days for an equivalent sized excisional wound. The STD destroys hair follicle tissue in the centre of the lesions produced and seems to have a particular irritative action on the small dermal blood vessels.

Clearly it cannot be claimed that the STD lesion is completely similar to a persistent ulcer due to venous hypertension or ischaemia. However, it certainly does heal slowly, and although the vascular basis of the STD model is not venous hypertension or long-standing ischaemia, there is undoubted vascular inflammation in the STD-induced lesion and presumably diminished flow rates as a result of this. The central necrosis produced by the STD results in a sloughy mass which simulates some human wounds quite closely. In order to quantify the changes, histometric measurements of step sections taken through these ulcers are made, using either the Quantimet 720 image analysis system or a microcomputer linked digitizing tablet. The degree of re-ephithelialisation is expressed as millimetres of new epidermis, and the amount of slough present is also measured. We have had to resort to less sophisticated methods to estimate the density and types of inflammation present, and have found that scoring on a 10 cm visual analogue scale is quite reproducible when the assessments are performed blind and by one experienced observer.

Using this system we have evaluated a wide range of substances for the topical treatment of ulcerated lesions, including dextran polymer beads, enzymes employed for debridement, polymer films, povidone iodine, pectincarboxymethyl cellulose viscous masses, antibacterial lotions and insulin solutions. We have also evaluated factor XIII

Table 4. Re-epithelialisation of sodium tetradecyl sulphate induced wounds in guinea pig skin demonstrating that streptokinase/streptodornase preparation appears to accelerate the re-epithelialisation taking place

	Length of re-epithelialising epidermis (mm). Mean ± s.d.	
	Day 4 ulcers	Day 6 ulcers
Streptokinase/streptodornase	0.596 ± 0.36	0.857 ± 0.19
Control (saline)	0.407 ± 0.17	0.488 ± 0.32

Adapted from Manna, Bem and Marks [11]

Table 5. Effect of topically applied insulin on the rate of re-epithelialisation in the guinea pig STD model

Treatment	Mean length of re-epithelialisation (mm)
Diluent	0.81 ± 0.24
Insulin (80 iu/ml)	0.95 ± 0.32*
Insulin (40 iu/ml)	0.84 ± 0.29

* Significantly different from diluent control ($p < 0.05$)

Adapted from Marks (1985) [12]

and hydroxyethyl rutosides. The only materials that have seemed to accelerate the process of re-epithelialization have been streptokinase/streptodornase solution and indulin (Tables 4, 5).

Conclusions

In conclusion it may be said that there are no ideal models for evaluating the treatment and all the processes involved in the healing of chronic wounds and ulcers. It should be borne in mind that particular questions may well require particular models and several models may have to be used to provide useful data with regard to a proposed new treatment. It is certain that whatever models are used they should employ objective parameters of the healing process.

References

1. Browse NL, Jarrett PEM, Morland M, Bernand K (1977) Treatment of liposclerosis of the leg by fibrinolytic enhancement: A preliminary report. Br Med J August 1977:434–435
2. Burnand KG, Clemenson G, Whimster I, Gaunt J, Browse NL (1976) Extravascular fibrin deposition in response to venous hypertension – the cause of venous ulcers. Br J Surg 63:660–661
3. Burnand KG, Clemenson G, Morland M, Jarret PEM, Browse NL (1980) Venous lipodermatosclerosis: treatment by fibrinolytic enhancement and elastic compression. Br Med J 280 (6206):7–11
4. Cherry G (1984) Wound occlusion and angiogenesis in the experimental model. International Symposium, An Environment for Healing, The Role of Occlusion. J Royal Soc Med (in press)
5. Eaglstein WH, Mertz PM (1978) New method for assessing epidermal wound healing: the effects of triamcinolone acetonide and polyethylene film occlusion. J Invest Dermatol 71:382–384
6. Fromer CH, Klintworth GK (1976) An evaluation of the role of leukocytes in the pathogenesis of experimentally induced corneal vascularization. III. Studies related to the vasoproliferative capability of polymorphonuclear leukocytes and lymphocytes. Am J Pathol 82:157–167
7. Geronemus RG, Mertz PM, Eaglstein WH (1979) Wound healing: The effects of topical antimicrobial agents. Arch Dermatol 115:1311–1314
8. Hashimoto T, Marks R (1984) Factor XIII inhibits epidermal cell migration in vitro. J Invest Dermatol 83:441–444
9. Johnson RL, Ziff M (1976) Lymphokine stimulation of collagen accumulation. J Clin Invest 58:240–252
10. Langrana NA, Alexander H, Strauchler I, Mehta A, Ricci J (1983) Effect of mechanical load in wound healing. Anals Plastic Surgery 10(3):200–208
11. Manna V, Bem J, Marks R (1982) An animal model for chronic ulceration. Brit J Dermatol 106:169–181
12. Marks R (1985) The use of models for the study of wound healing. Proc: An Environment for Healing. The Role of Occlusion. In: L'Etang HJCJ (ed) International Congress and Symposium Series 88. Oxford University Press, Oxford, pp 21–29
13. Marks R, Nishikawa T (1973) Active epidermal cell movement *in vitro.* Br J Dermatol 88: 245–248
14. Mitrani E, Marks R (1978) Towards charcterisation of epidermal cell migration promotion activity in serum. Brit J Dermatol 99:513–518
15. Nishikawa T, Marks R (1973) The effect of cytochalasin B on epidermal cell migration in vitro. Br J Dermatol 88:466–474
16. Stenn KS (1981) Epibolin: a protein of human plasma that supports epithelial cell movement. Proc Natl Acad Sci USA 78:6907–6911
17. Than-Than (1976) Skin prolyl hydroxylase and tensiometry in jaundice. Thesis, University of Glasgow
18. Than-Than, Evans, JH, Ryan CJ, Smith DA, Harper AM, Blumgart LH (1979) Rupture strength of skin wounds in jaundiced rats. Br J Exp Pathol 60:107–112
19. Thiyagarajan C, Silver JR (1984) Aetiology of pressure sores in patients with spinal cord injury. Br Med J 289:1487–1490
20. Thorgeirsson G, Robertson AL (1978) Platelet factors are not required for endothelial cell proliferation and migration. Part 2. Such factors are not required for endothelial cell proliferation and migration. Atherosclerosis 31:231–238
21. Wahl SM, Arend WP, Ross R (1974) The effect of complement depletion on wound healing. Am J Pathol 75:73–89
22. Werb Z, Aggeler J (1978) Proteases induce secretion of collagenase and plasminogen activator by fibroblast. Proc Natl Acad Sci USA 75:1839–1843
23. Weringer EJ, Kelso JM, Tamai IY, Aquilla ER (1982) Effects of insulin on wound healing in diabetic mice. Acta Endocrinologic 99:101–108
24. Wolf JE, Harrison RG (1973) Demonstration and characterization of an epidermal angiogenic factor. J Invest Dermal 61:130–141

25. Williams DL, Dykes PJ, Marks R (1984) The effect of hydrogel dressings on objective and subjective parameters of healing. Proc: Oxford Gelliperm Symposium, September 1983 (in press)
26. Williams DL, Dykes PJ, Marks R (1985) Effects of a new hydrocolloid dressing on healing of full thickness wounds in normal volunteers. Proc: An Environment for Healing: The Role of Occlusion. J Royal Soc Med (in press)

Sunscreen Evaluation by Mouse Spectrophotometric and Human Assays

N. J. Lowe and J. Breeding

The ideal sunscreen should protect against all biological effects of ultraviolet (UV) radiation (UVR) not just simply inhibit UV-induced skin erythema. Therefore an important property for such a sunscreen would be that in reduces UVR damage to the epidermis.

UVB (290–320 nm) effects on epidermal DNA are well documented [1] and while a definite cause and effect relationship between such damage and skin carcinogenesis is not certain, UVB-induced DNA damage is thought to be an important component. We have therefore, investigated the ability of sunscreens to protect against suppression of UVB-induced epidermal DNA synthesis suppression as measured by epidermal cell incorporation of tritiated thymidine [4, 13].

Induction of the polyamine biosynthesis enzyme ornithine decarboxylase (ODC) has been shown to be an important biochemical event involved in tumour promotion [9]. This enzyme is also induced by UVB radiation [8] and it has been shown that topical drugs which inhibit UVB-induced ODC also inhibit UVB-induced skin carcinogenesis [6]. We have therefore used the ability to inhibit UVB-induction of epidermal ODC as an assay for sunscreen protectiveness.

Following UVB irradiation, hairless mice develop skin oedema. This oedema possibly is caused by release of vasoactive substances released following UVB irradiation and results in increased skin thickness. This can be measured using a rachet micrometer to evaluate the thickness of UVB irradiated dorsal skin in sunscreen treated and control mice. We have utilized measurements of dorsal skin oedema to assess the protectiveness of different sunscreening products.

The UV absorbance of sunscreening chemicals in solution can be used as a predictor of sunscreen efficacy [10] and if sunscreens of known human sun protection factor (SPF) are compared with experimental sunscreens in this spectrophotometric absorption assay, a protective factor can be predicted. This spectrophotometric assay, while not measuring in vivo epidermal damage has also been studied since the time and expense required to perform the tests are so minimal, it is potentially an excellent first estimate for new compounds or formulations that predicts reasonably human SPF.

Finally, the evaluation of the human erythema following UVR is the accepted method of evaluating sunscreen products [2, 10–12, 14]. These studies may utilize sunlight or solar simulating light sources. They are performed by determining the ratio of the minimal erythema dosages (MED) in protected skin to the MED in unprotected skin. The human SPF assay is a required assay for formal human SPF rating of sunscreens [12]. These experimental animal and in vitro assays were compared with human SPF assay to determine the relative usefulness of prehuman predictive assays of sunscreen protectiveness.

Materials and Methods

UVB Irradiation of Mice

Five to eight week old skh/HR-1 female mice were sedated 15 min prior to UVR using chloralhydrate (600 mg/kg body weight). The mice were placed 15 cm beneath a bank of FS40 sunlamps fitted with cellulose triacetate sheets to filter out wavelengths below 290 nm. Irradiation was monitored by an international Light (IL) sensor fitted with a 313 nm narrow band filter connected to an IL700 research radiometer and an IL720 photodosimeter. A feedback circuit connected to the lamps automatically terminate the irradiation at the predetermined dose. One MED (minimal erythema dose) was taken to be 0.05 J/cm^2 with this dosimetry equipment as determined by previous human skin testing [7].

Sunscreens

Those sunscreens used were commercially available and assayed under coded double-blind conditions. In addition, sunscreens with determined human SPF were included in all assays as controls.

Application of Sunscreens to Mice

Sunscreen solutions were applied by micropipette at 2 $\mu l/cm^2$ one hour before irradiation.

Epidermal DNA Synthesis Assay

The DNA synthesis suppression assay has been used previously to predict sunscreen efficacy [4, 13]. A modification of an epidermal disc method was used to measure epidermal incorporation of tritiated thymidine (Amersham, Arlington Heights, Illinois 2 Ci/mmol) [5]. A 17 mm punch was used to score two circles through the epidermis

Table 1. Example of one series of sunscreens tested

Coded sunscreen	Active ingredients
A	Octyl dimethyl PABA, oxybenzone
B	PABA, oxybenzone
C	PABA
D	Homomenthylsalicylate (standard SPF 4)

of each dorsal mouse skin four hours post irradiation (or from unirradiated controls). The whole back skin was then placed in a constant temperature water bath at 58 °C for 30 s, removed to ice, and the epidermis removed from within the inscribed circles by scraping. The epidermal samples were then fixed in 10% formalin, dehydrated using baths of increasing alcohol concentration and cleared with xylene in an Autotechnicon to remove unincorporated ^{3}H-thymidine, water and alcohol. The samples were then placed in vials and counted in OCS scintillation cocktail (Amersham, Arlington Heights, Illinois). The percentage suppression of thymidine incorporation was calculated using the mean of duplicate samples from groups of four mice. Four or more dose levels were used for each sunscreen. The amount of UVB required to achieve a 50% suppression of thymidine incorporation in treated and untreated mice was compared as a ratio to determine the protective factor (PF).

Assay of Epidermal Ornithine Decarboxylase Activity

Enzyme activity was determined by measuring the release of $^{14}CO_2$ from L-[1-^{14}C] ornithine 57 mCi/mmol, Amersham) as previously described [6, 9, 8] with modifications.

At 24 h following a single UVB exposure, mice were sacrificed by cervical dislocation and the back skin removed. The epidermis was separated from the dermis by a brief heat treatment (55 °C for 30 s), removed to ice water, placed flat on a piece of glass and the epidermis removed by scraping with a scalpel. The epidermal scrapings were then placed in ice-cold buffered saline (1 ml) and homogenized using a Brinkman polytron homogenizer at level 6 for 15 s. The homogenates were then centrifuged at 30,000 x g for 30 min at 4 °C and the supernatants recovered and frozen at −70 °C until assayed for ODC activity. Protein content was determined by the Bio-Rad Protein Assay. The ODC was carried out in 17 mm x 105 mm Pyrex (9820) tubes tilted with 5 mm below the lip of the tube. A # 11-hole stopper was slipped over the sidearm to accomodate the mouth of a standard scintillation vial. The main tube opening was filled with a #14 sleeve stopper to close the system. The incubation mixture was prepared as previously described and maintained on ice. 100 μl NCS (Amersham) was placed in each of the scintillation vials before attachment to the reaction tube. When all reagents had been added and the tubes capped, the whole rack of tubes was placed in a pre-heated shaking water bath at 37 °C and incubated for one hour at approximately 100 cycles/min.

The rack of tubes was then removed from the water bath, cooled in ice water for 5 min and 0.5 cm^3 2 M citric acid was injected through the top past the sidearm into reaction mixture to halt the reaction. The tubes were then returned to the shaking water bath and incubation allowed to continue for one more hour. At this time, the scintillation vials were removed from the sidearms and 10 ml OCS Amersham was added to each vial. This vials were counted in a scintillation spectrometer. The activity of ODC was expressed in nmol CO_2 h/mg protein.

Epidermal ODC induction showed a linear increase with increasing amounts of UVB radiation.

After establishing linearity, 4 mice each were treated with sunscreen and irradiated with 3 MED more than the estimated SPF. The 24 h time point was used to establish an ODC activity vs. MED curves for treated and untreated mice. The relative slopes of these lines taken as a ratio determined the PF for each screen [3].

$$PF = \frac{\text{slope untreated}}{\text{slope treated}}$$

UVB-Induced Dorsal Skin Oedema Assay

Oedema was estimated by the increase in double skin fold thickness measured by a caliper at 24 h post irradiation. The relative sunscreen effectiveness was estimated using a standard curve of skin thickness measurements from UVB-irradiated mice 24 h after 0, 1, 2, 3, 4 and 5 MED. The differences between the amount of UVB delivered to sunscreen protected skin and the amount of UVB needed to produce the same amount of oedema in unprotected animals was expressed as the protective index.

PF = MED delivered – equivalent MED in unprotected mice (From standard curve obtained from the known SPF sunscreens).

Spectrophotometric Absorption Assay

The sunscreens were evaluated in diluted solutions following a modification of the method of Sayre et al. [10].

Sunscreens were dissolved in reagent alcohol (Scientific Products), 0.2 μl/ml and read in a scanning spectrophotometer against an alcohol blank. The optical transmission (T) between 290 and 320 nmol was weighted to the erythemal efficiency spectrum x solar intensity spectrum as per Sayre [10] and the inverse of the natural log (ln) of the weighted sum of Sayre's transformation was multiplied by a correction factor (CF) determined by a standard curve derived by testing 2 known SPF sunscreens (4 and 15.4) which had been determined by prior human testing. Therefore, the following formula was used to predict the PF:

$$PF = \ln E_{290}^{320} \frac{\text{Correction Factor}}{[\%T \times \text{Erythemal Efficiency Spectrum} \times \text{Solar Intensity Spectrum}]}$$

Human Sun Protection Factor

The subjects were healthy male and female caucasians (skin types I, II and III) aged 18 to 50 years. No subjects were taking medications. Areas of previously non-sun exposed mid or lower back skin were chosen as test sites.

UV Source
For the human studies, the UV source was a 5,000 watt xenon arc solar simulator equipped with UVR reflecting dichroic mirrors, absorbing water filter and sunlens diffuser giving spectral distribution at skin level between 290 to 410 nm. Intensity was measured before each test using a IL sensor fitted with a 313 nm calibrated filter connected to an IL700 research radiometer. Spectral characteristics were determined by Optronics spectral radiometer to be solar simulating.

MED Estimations
The MED was determined in each individual by administering a series of exposures in 25% dose increments. The MED was the smallest exposure necessary to produce discernible erythema over the irradiated area 24 h later.

Determination of the Human SPF Factor

Test agents were applied uniformly at a dose of 2 $\mu l/cm^2$ to rectangular areas of the skin. Fifteen minutes later, small areas within the sunscreen protected skin were exposed to 25% increments of radiation, beginning with a sub-threshold dose based on an estimated SPF value. Twenty-four hours later, the MED in protected skin the degree of erythema was recorded and the SPF calculated as the ratio of the MED in protected skin to the MED in unprotected skin.

Results

Estimation of Protection Against Changes of Mouses Epidermal DNA Synthesis. These results are summarized in Table 2. This assay consistently underestimated the SPF's of PABA ester containing sunscreens. This is consistent with the reduced absorption by PABA esters at shorter wavelengths of UVB, the wavelengths that DNA is most susceptible to damage.

Estimation of Protection Against UV-Induced Epidermal Ornithine Decarboxylase. These results are summarized in Table 2.

This assay shows a close correlation with human SPF estimation for the sunscreens tested (r = 0.937). The action spectra for UV induction of epidermal ODC is not known at present, but the protection index predicted using UVB in this assay correlates well with human erythema SPF, mouse skin oedema and in vitro spectrophoto metric absorption.

Table 2. Summary of protection factors (PF) of sunscreens in five different prediction assays

Coded sunscreen	Human sun SPF	Spectro-photo-metric predicted	Mouse: Skin oedema	Mouse: ODC induction	Mouse: Modulation of DNA synthesis suppression
A Contains PABA ester	17.6	13.70	15.20	15.00	10.00
B Contains PABA	15.4	15.20	14.30	17.30	20.00
C Contains PABA	9.2	7.30	10.10	9.20	8.70
D Homosalate	4.7	4.6	4.10	4.20	3.20
Correlation coefficient relative to human SP		0.96	0.93	0.94	0.60

Estimation of Protection Against Mouse Skin Oedema. The results are shown in the Table 2, showing a high correlation (r = 0.948) with human SPF results.

Spectrophotometric Absorption Assay. The results of analyzing the sunscreens by this method (Fig. 1) relative to their SPF values are shown in Table 2 show a correlation coefficient of 0.96. We have therefore included this fast, simple and inexpensive assay even though it does not take into consideration solution-skin interactions and cannot be used to ascertain specific biological effects nor to ascertain wash-off characteristics.

Human Sun Protection Factor Assay. All the sunscreens tested in this assay correlated well with the other assays except for the case of epidermal DNA synthesis suppression assay where SPF'S of PABA ester-containing sunscreens were consistently under-estimated.

Discussion

The results reported confirm the possibility of using different mouse assays to determine a protective index for sunscreens. These results of the different assays correlate closely with the results achieved by human SPF testing with the possible exception of the DNA synthesis suppression assay.

These non-human assays may be useful initial assay procedures for novel sun protective agents. Also, there is less need for toxicological testing prior to use in animal assay than for human testing. A wide range of different concentrations of chemicals and different vehicle delivery systems can be studied in the animal screening assays readily and without the need for human safety considerations. It is possible to study the water-resistence, wash-off resistance persistence of sunscreens using in vivo animal assay systems and to study the effects of different vehicles and chemicals on improving water-resistance.

The assays of epidermal DNA synthesis and epidermal ornithine decarboxylase activity are measurements of biochemical responses to UVB-induced epidermal damage. These measurements are potentially more relevant to the ability of sunscreens to prevent UV-B induced skin carcinogenesis than is the inhibition of UV-B induced erythema utilized in the human SPF assay.

While Sayre et al. [10], comparing in vitro and in vivo methods of predicting human SPF values, concluded that absorption curves of tested compounds did not predict human values well over a wide range, log transformation of their data by us, however, yielded a spectrophotometric/human SPF correlation coefficient of 0.96 ($p < 0.0005$), as high as the epidermal spectrophotometric method Sayre et al. recommended.

Assays of sunscreens using the DNA synthesis suppression assay produced interesting results. The UV-absorption of PABA is greater over shorter wavelengths than is that of the PABA esters. The protection index afforded by the PABA, derived from its ability to protect against the effects on DNA synthesis, were significantly greater than the PABA esters compared to their human SPF values. This is presumably a measure of the ability of PABA to absorb the ultraviolet wavelengths most damaging to DNA.

While we did not use solar-simulating light sources for the hairless mouse studies, plans are now to use the 5,000 Watt xenon arc solar stimulator which has a sufficient area of high intensity irradiance to expose sufficient mice at a time. The advantages of the Westinghouse lamps are those of inexpensiveness and ease of irradiance. Therefore, while these lamps are not solar simulating, they do contain the carcinogenic wave-lengths of UVB, in addition to UVA.

Oedema was evaluated 24 h after UV irradiation at which time the greatest increase in dorsal skin thickness occurs. The increase in skin fold thickness at this time point probably is mainly a dermal response because epidermal hyperplasia has not yet occurred. The dermal oedema may be due to the release of mediators of inflammation into the dermis following UV irradiation.

In summary, we present evidence that there are satisfactory alternative in vivo and in vitro assays that measure the ability of different sunscreens to protect against the effects of ultraviolet irradiation on the epidermis and dermis. The ability of sunscreens to protect against UV-induced epidermal damage may be observed with the assays of ornithine decarboxylase induction and modulation of epidermal DNA synthesis following UVR. In addition, the ability of sunscreens to protect against the effects of UVR on skin oedema may reflect reduced vasoactive mediator release in sun-protected compared to non-sun-protected animals.

The spectrophotometric absorption assay, while not evaluating sunscreen skin contact gives a simple and rapid in vitro evaluation of the UV absorbance of different chemicals.

The ability of PABA-containing sunscreens to provide higher PF's in the DNA synthesis assay compared with PABA ester-containing sunscreens, suggests the need for protective sunscreen chemicals with a wider range of UVB absorbance than some PABA esters.

The assays described may therefore be used to investigate mechanisms and develop alternative compounds and formulations.

Acknowledgement

This chapter has been published with permission from Models in Dermatology, Vol 2. Maibach HI and Lowe NJ. Karger, Basel, Switzerland 1985

References

1. Epstein JH, Fukuyama K, Fye K (1970) Effects of ultraviolet radiation on the mitotic cycle and DNA, RNA and protein synthesis in mammalian epidermis in vivo. Photochem Photobiol 12:57–64
2. Fitzpatrick TB, Pathak MA, Parrish JA (1974) Protection of human skin against the effects of sunburn ultraviolet (190–320 nm). In: Fitzpatrick TB, Pathak MA, Harber L, et al (eds) Sunlight and Man. University of Tokyo Press, Tokyo, pp 751–765
3. Gange RW, Mendelson R (1982) Sunscreens block the induction of epidermal ornithine decarboxylase by ultraviolet-B radiation. Br J Dermatol 107:215–220
4. Lowe NJ, Breeding J (1980) Evaluation of sunscreen protection by measurement of epidermal DNA synthesis. J Invest Dermatol 74:181–182
5. Lowe NJ, Breeding J, Wortzman M (1982) New coal tar extract and coal tat shampoos. Evaluation by epidermal cell DNA synthesis suppression assay. Arch Dermatol 118:481–489
6. Lowe NJ, Connor MJ, Breeding J, Chalet M (1982) Inhibition of ultraviolet induced epidermal ornithine decarboxylase and carcinogenesis by topical antiinflammatory drugs. Cancer Res 42:3941–3943
7. Lowe NJ, Koo L, Breeding J (1980) Ultraviolet light induces human epidermal ornithine decarboxylase. Br J Dermatol 103:18
8. Lowe NJ, Verma AK, Boutwell RK (1978) Ultraviolet light induced epidermal ornithine decarboxylase activity. J Invest Dermatol 71:417–419
9. O'Brien TG (1976) The induction of ornithine decarboxylase as an early possibly obligatory event in mouse skin carcinogenesis. Cancer Res 36:2644–2653
10. Sayre RM, Agin PP, LeVee G, Marlowe E (1979) A comparison of in vivo and in vitro testing of sunscreening formulas. Photochem Photobiol 29:560–566
11. Sayre RM, Marlowe E, Agin PP, LeVee GJ, Rosenberg EW (1979) Performance of six sunscreen formulations on human skin. A comparison. Arch Dermatol 115:46-49
12. Sunscreen products for over-the-counter use. (1978) Fed Reg 43:28206–28269
13. Walter JF (1981) Evaluation of seven sunscreens on hairless mouse skin. Arch Derm 117: 547–550
14. Wilson, PD, Kaidbey KH, Kligman AM (1981) Ultraviolet light sensitivity and prolonged UVR-erythema. J Invest Dermatol 77:434–436

The Rhino Mouse: Retinoid Effects on a Model of Keratin Accumulation

R. E. Ashton and N. J. Lowe

The skin of the rhino mouse is characterized by numerous large cysts of two types: superficial cysts termed utricles, and deeper cysts. The utricles are derived from the hair canals and open to the surface. The squamous epithelium of these is continuous with the surface epidermis, and produces keratin which is found in the cyst cavity. A sebaceous gland lies at their base. The underlying cysts situated in the subcutaneous fat are derived from the hair root sheath epithelium and are lined with keratinizing epithelium [5].

Kligman and Kligman [6] found that topical retinoic acid produced a striking and marked reduction in the size of the utricles, but had little effect on the deeper cysts. These are unchanged by retinoid therapy unless fed vitatmin A from birth. Kligman and Kligman suggested that the change in the size of the utricles could be used to assess the relative potency of retinoids. Bonne et al. [3] have quantified changes in utricle size by evaluation of a "comedo profile", the ratio of the diameter at the centre of the utricle to the diameter at the mouth. Measuring diameters from vertical sections is limited in that the plane of the section may not intersect the maximum diameter of the utricle. To overcome difficulty, Mezick et al. [7] have developed a method of examining sheets of epidermis from the skin of the rhino mouse horizontally, and directly measuring the diameter of the utricles.

This method has been used to observe the effect of various retinoids on utricle size and to correlate this with changes seen in conventional histology.

Methods

The retinoids were applied topically in acetone in all experiments, and the solutions were made up immediately before use. 0.1 ml of the retinoid solution was applied to the dorsal skin using micropipettes over an area of 20 cm^2. Groups of 4–5 mice were treated daily for ten days (Monday to Friday on two consecutive weeks). The following retinoids and doses were studied:

1) Vehicle (acetone);
2) 0.0001, 0.001, 0.01 and 0.1% retinoic acid;
3) 0.0001, 0.001, 0.01 and 0.1% 13-cis-retinoic acid;
4) 0.001, 0.01 and 0.1% aromatic retinoid;
5) 0.0001 and 0.001% arotinoid.

On the third day after the last application of retinoid, the mice were killed by cervical dislocation, and the dorsal skins removed. Each skin was soaked in 0.5% acetic acid at 4 °C over-night, and the epidermis peeled off [7]. This epidermal sheet was dehydrated by immersion in increasing concentrations of reagent alcohol (70, 80, 95 and 100%), followed by xylene, and mounted on glass slides using Pro-Texx. The mean utricle diameter for each sample was obtained by measuring the diameters of 25 utricles in five optical fields using a mechanical eyepiece micrometer.

Biopsies of dorsal skin were fixed in 10% formalin, sectioned at 6 μm vertically, and stained with haematoxylin and eosin. The cell layers and thickness of the epidermis and stratum granulosum were measured at five interfollicular sites. The number of cell layers in the walls of the utricles was measured at the base of the utricle and at opposite sides midway up the utricle. Those utricles selected for measurement were those with an opening to the surface since these were more likely to have been sectioned through the center.

Results

All the four retinoids tested influenced the size and appearance of the utricles in a dose dependent manner. The change of diameter of utricle with increasing concentration of the retinoids is shown in Fig. 1. Arotinoid was the most potent of retinoids tested followed by aromatic retinoid, retinoic acid and 13-cis-retinoic acid.

Acetone treated controls had 2–3 cell layers on histological examination, with a thin stratum granulosum of one cell layer. Changes after retinoid treatment in the number of cell layers in the epidermis and stratum granulosum are shown in Fig. 2. An increasing concentration of retinoid resulted in an increased number of cell layers

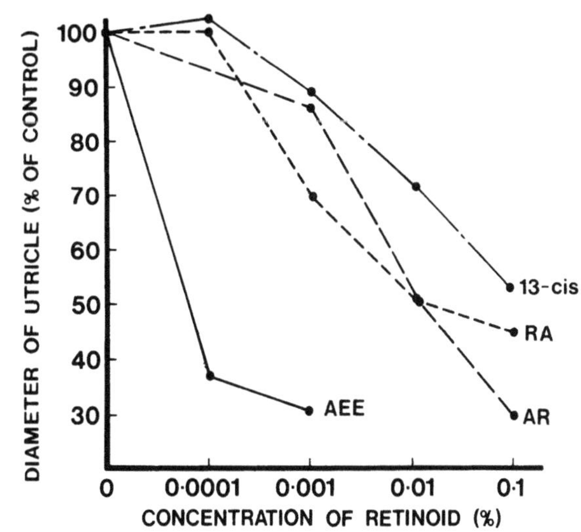

Fig. 1. Changes in utricle diameter following ten topical applications of retinoid

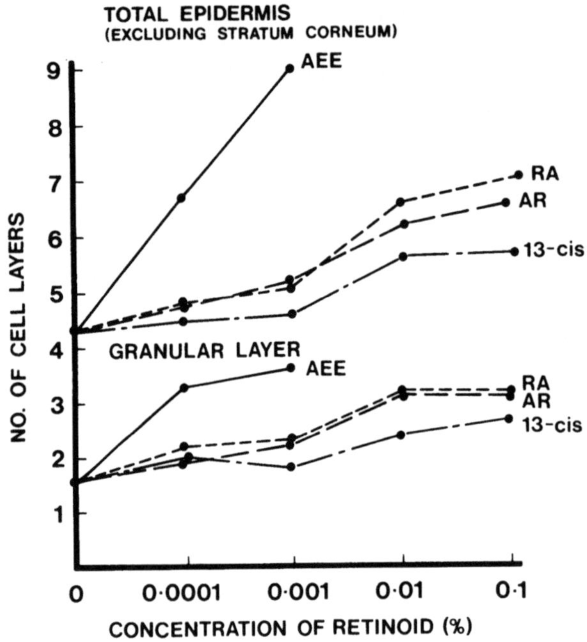

Fig. 2. Changes in number of interutricular epidermal and granular cell layers following ten topical applications of retinoid

in the epidermis and stratum granulosum. Following retinoid treatment, the stratum granulosum became more prominent.

The increase in inter-follicular epidermal cell layers is paralleled by a corresponding increase in the number of cell layers in the walls of the utricles (Fig. 3).

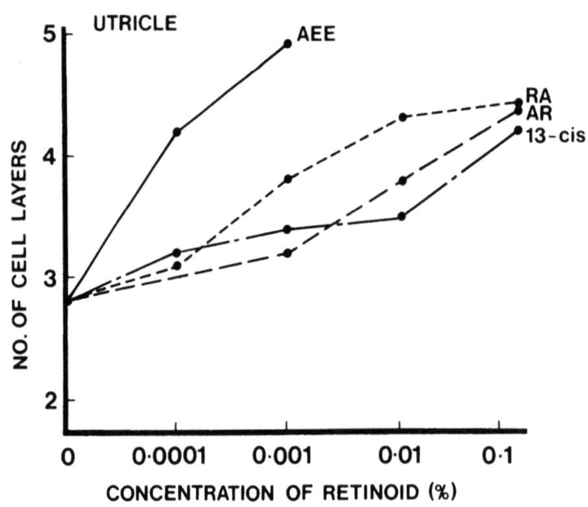

Fig. 3. Changes in number of cell layers in utricle wall after ten topical applications of retinoid

Discussion

These results demonstrate that the size and structure of the utricles found in the skin of the rhino mouse are reduced in a dose dependent manner. Arotinoid was 1,000 times as potent as retinoic acid in producing a similar reduction in utricle diameter.

Retinoids are able to produce epidermal hyperproliferation [4], and we have demonstrated a similar effect in rhino mouse skin. The reduction in utricle diameter occurs only when both epidermal and utricle wall hyperplasia was induced, suggesting that the changes in utricle diameter are closely related to the hyperplastic effect of the retinoids.

Retinoid induced epidermal hyperplasia occurred at doses well below those required to induce manifest toxicity, assessed by weight loss during treatment and severity of scaling [2]. The reduction of utricle diameter is accompanied by the transformation of the utricles into more normal looking pilar units, and an increase in the stratum granulosum layers, both examples of the differentiating effect of retinoids [1]. This effect was also seen only after doses of retinoid capable of inducing epidermal hyperplasia, and suggests that the induction of hyperplasia may be important in the mechanism of action of retinoids on the utricles of the rhino mouse.

References

1. Ashton RE, Connor MJ, Lowe NJ (1984) Histologic changes in the skin of the rhino mouse (hr^{rh} hr^{rh}) induced by retinoids. J Invest Dermatol 82:632–635
2. Bollag W (1974) Therapeutic effect of an aromatic retinoic acid analog on chemically induced skin papillomas and carcinomas of mice. Europ J Cancer 10:732–737
3. Bonne C, Zeziola F, Secchi J, Saurat J (1981) A new model for the assay of comedolytic activity. Int J Cosmetic Sci 3:23–28
4. Connor MJ, Lowe NJ (1983) Induction of ornithine decarboxylase activity and DNA synthesis in hairless mouse epidermis by retinoids. Cancer Res 43:5174–5177
5. Howard A (1940) "Rhino", an allele of hairless in the house mouse. J Hered 31:467–470
6. Kligman LH, Kligman AM (1979) The effect on rhino mouse skin of agents which affect keratinization and exfoliation. J Invest Dermatol 73:354–358
7. Mezick JA, Chabria MC, Thorne EG, Capetola RJ (1982) Topical and oral (P.O.) effects of 13-cis-retinoid acid and all-trans-retinoic acid on horn-filled utriculi (pseudocomedones) size in the Rhino mouse. J Invest Dermatol 78:350A

Further Studies with Models for the Transepidermal Elimination Process

K. Dalziel and R. Marks

Transepidermal elimination is an epidermal response to the presence of foreign material or damaged dermal components in the subepidermal zone. These materials are engulfed in an apparently purposeful way by the epidermis and eventually expelled to the outside during a particularly interesting form of dermo-epidermal interaction. It occurs spontaneously in a number of skin disorders including elastosis perforans serpiginosa [7], perforating folliculitis [6] and necrobiosis lipoidica [8].

Transepidermal elimination also occurs in reactions to foreign materials, for example in tattoos [5], cutaneous shistosomiasis and blastomycosis [11], and to abnormal deposits such as occur in calcinosis cutis [1], amyloid [9] and immune deposits [4].

Transepidermal elimination has been previously investigated in a guinea pig model [2, 10]. It appears that certain factors are important if a material is to be transepidermally eliminated. The process occurs maximally when substances are introduced immediately subepidermally, above the level of the hair follicle. The particle size and the degree of inflammatory response invoked by a material are important in determining which will be eliminated in this way [10]. It was found that particles of activated charcoal (particle diameter 1–100 μm) injected intracutaneously into guinea pig flank regularly and reproducibly invoked transepidermal elimination. The involvement of follicular epithelium in the transepidermal elimination of charcoal was a prominent feature. However, the process could also occur from hairless sites such as the footpad.

Other substances such as glass beads, dextran polymer beads and colloidal carbon produced varying degrees of inflammation and tissue necrosis but were not eliminated. Marking ink, in the form of a particulate suspension, did produce transepidermal elimination in the guinea pig model but less reliably than charcoal.

Attempts to influence transepidermal elimination in the guinea pig model with antihistamines and indomethacin have been described [3]. Indomethacin reduced the degree of epidermal hyperplasia overlying the site of transepidermal elimination but none of the drugs used inhibited the process itself.

The present experiments were designed firstly to try to develop a human model of the transepidermal elimination process and secondly to determine whether simple physical manipulations of the animal model might provide further information on it aetiopathogenesis. Sterile black ink was the injected substance for two reasons. Firstly it is readily available pre-sterilised, and secondly because of our observations in a patient with a disorder characterised by transepidermal elimination.

A 38-year-old patient with cystinosis was started on treatment with penicillamine 1.8 g daily 19 years previous to investigation. In 1982 he was commenced on tiopronin (N-[2 mercapto-propionyl]-glycine, a synthetic thiol compound. This has a similar

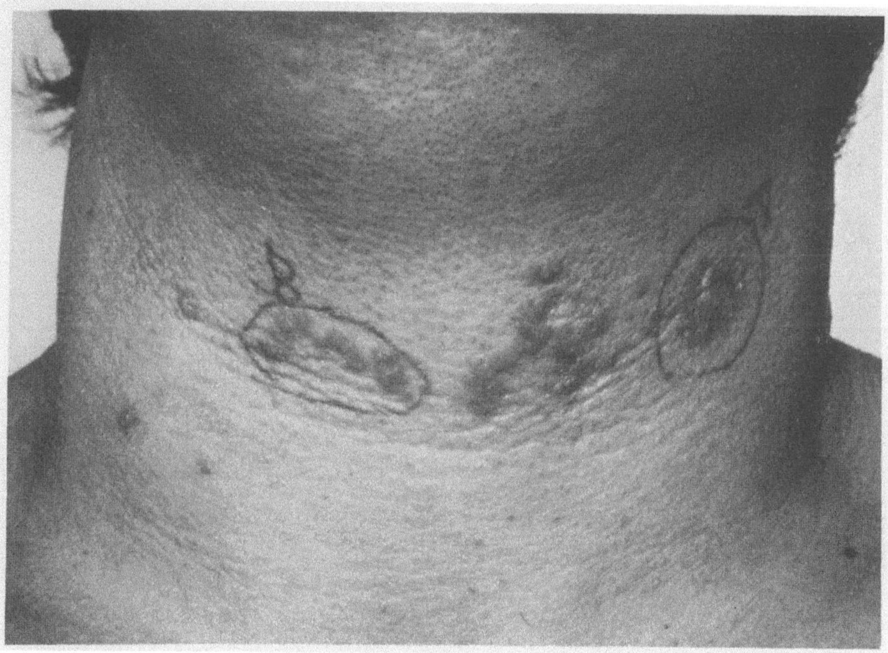

Fig. 1. Lesions of penicillamine induced elastosis perforans serpiginosa

action to penicillamine in the treatment of cystinuria. Good control of symptoms was obtained at a dose of 400 mg tds.

In 1983 he presented with a five year history of papules appearing around the back of the neck (Fig. 1). These had gradually extended around to the front of the neck. Examination showed both linear and individual papular lesions characteristic of elastosis perforans serpiginosa. This was felt to be secondary to penicillamine treatment. Biopsy of the lesions confirmed the diagnosis, showing epidermal hyperplasia with an increased amount of elastic tissue in the upper dermis and dermal papillae. Transepidermal elimination of this material via the hair follicle was prominent (Fig. 2).

Method

Human Model

0.02 ml sterile black ink was injected subcutaneously into non lesional forearm skin of the patient. The site was biopsied one week later and the biopsy fixed in formalin and prepared histologically. Serial sections were cut and stained with haematoxylin and eosin. 0.02 ml sterile black ink was injected subcutaneously into each of two sites on the forearm of 12 Caucasian adult volunteers (age 18–50 years) who had given their informed consent. The sites were biopsied at time intervals between 24 h and 14 days.

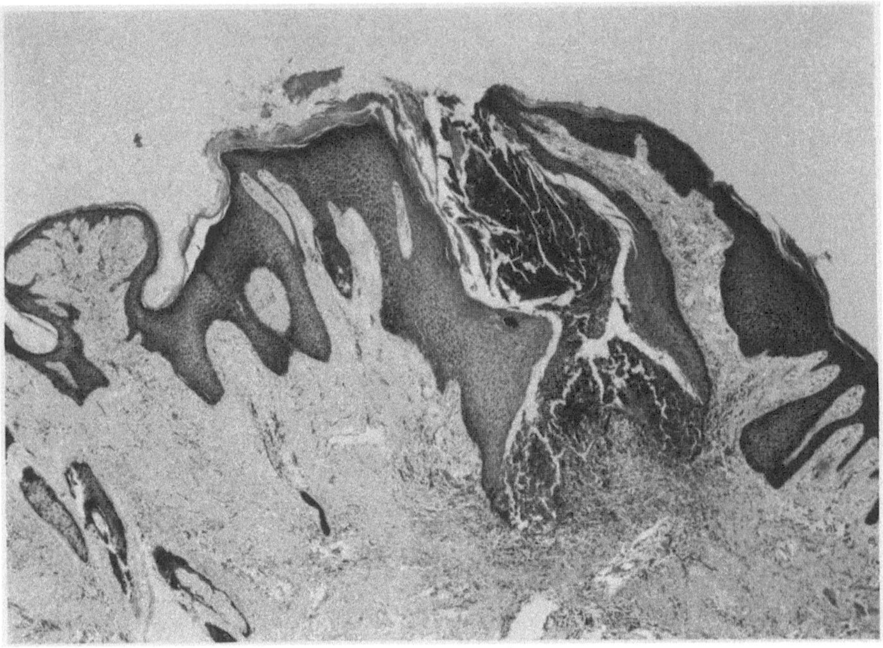

Fig. 2. Transepidermal elimination of effete connective tissue in elastosis perforans serpiginosa

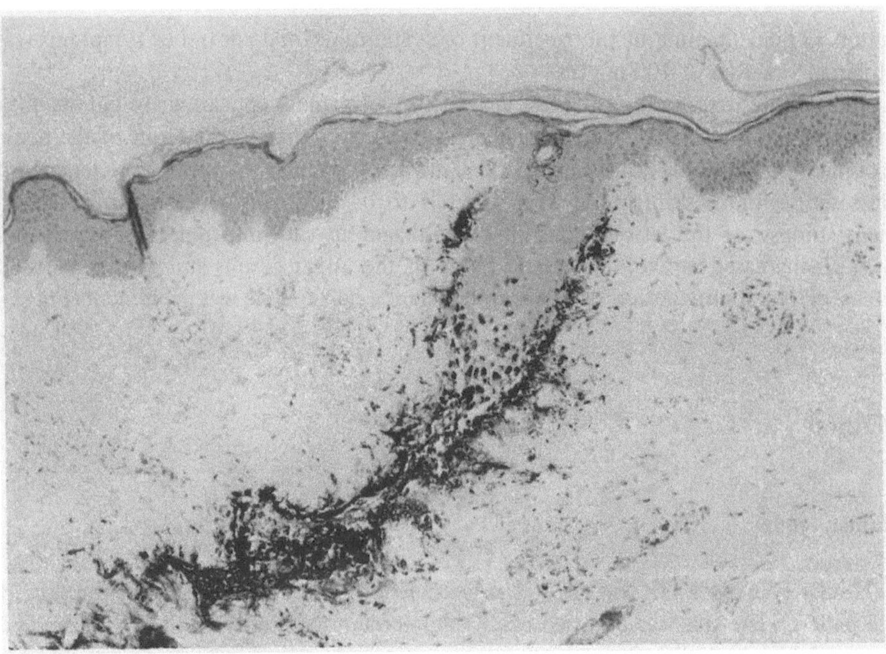

Fig. 3. Transepidermal elimination of sterile ink in patient with elastosis perforans serpiginosa

To determine the role of intratissue tension in the transepidermal elimination process, using the guinea pig model, 0.02 ml of the sterile ink was injected intracutaneously into each of four sites on the epilated flanks of anaesthetized adult albino guinea pigs. Immediately after injection, a circular cut was made around half the injection sites using a 6 mm disposable trephine. The pigs were sacrificed after seven days and the injections sites excised and fixed in formalin. They were prepared histologically and serial sections stained with H & E.

Results

Human Model

In biopsies taken from the patient with elastosis perforans serpiginosa, transepidermal elimination of sterile black ink was a prominent feature. As in the guinea pig model, involvement of the hair follicle epithelium was a prominent feature (Fig. 3).

In biopsies from normal volunteers, transepidermal elimination was virtually absent at all time points. Particles of ink could be seen engulfed in dermal macrophages but even when ink particles were in close proximity to hair follicle epithelium, transepidermal elimination was not invoked. Occasional solitary particles of ink could be seen within epidermal cells. In only one out of the 12 volunteers did transepidermal elimination via the hair follicles appear to be occurring (Fig. 4).

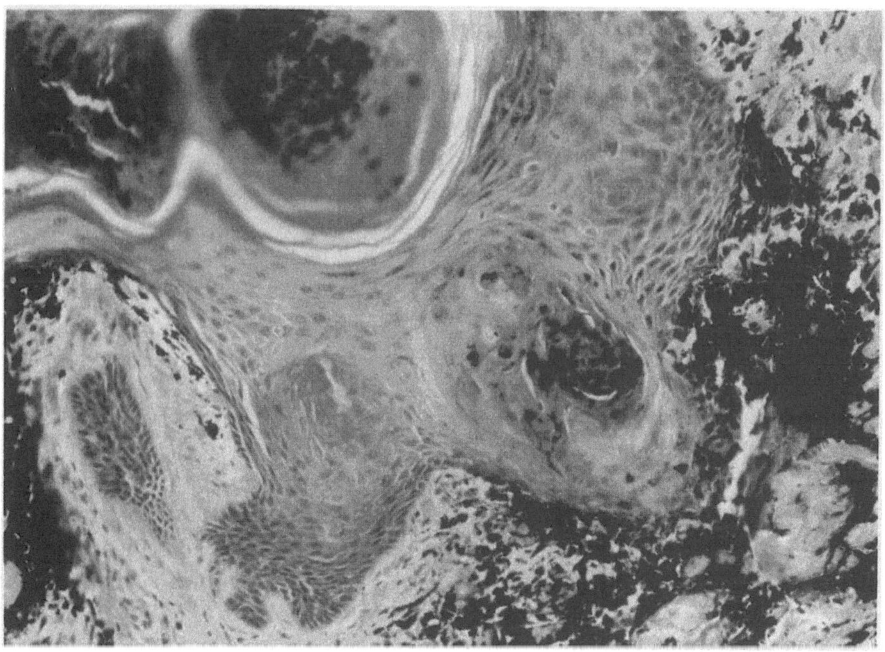

Fig. 4. Transepidermal elimination via hair follicle epithelium seen in only one normal volunteer

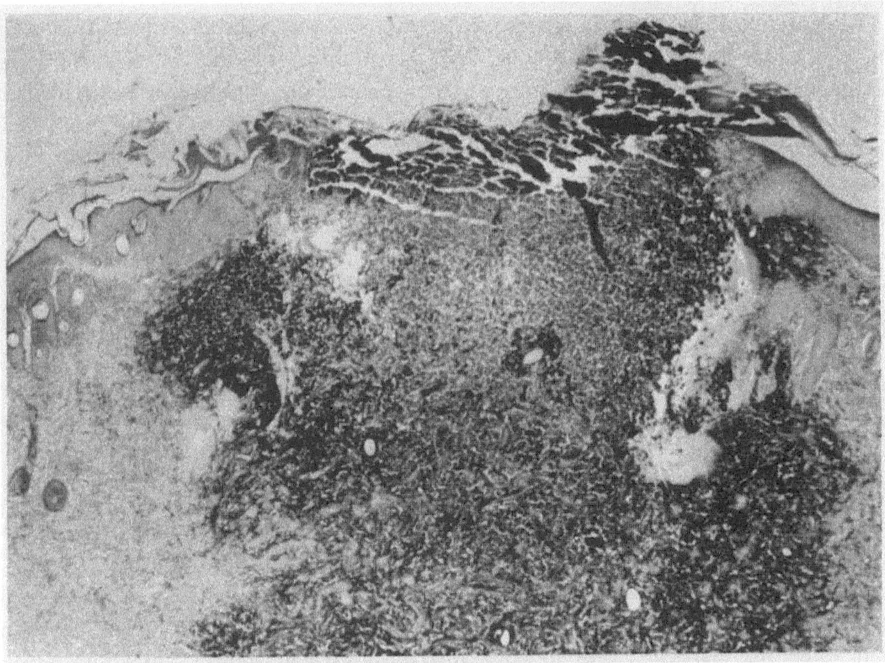

Fig. 5. Transepidermal elimination in animal model from site without encircling incision

Animal Model

Transepidermal elimination occurred more readily in the guinea pig model although still to a much lesser extent than it had with charcoal. The rise in intratissue tension produced by the intradermal injection of ink caused necrosis and ulcertain of the overlying epidermis in two out of four sites where no encircling incision had been made. No ulceration was seen in sites where an encircling incision had been made. Ink particles were visible both within and between epidermal cells in the periphery of the ulcerated sites (Fig. 5) and also in the epithelial cells of some hair follicles deeper in the dermis. In biopsies from sites with an encircling incision, transepidermal elimination of ink particles via hair follicle epithelium was seen to a small degree.

Discussion

Since sterile ink provoked marked transepidermal elimination in the nonlesional skin of a patient with elastosis perforans serpiginosa, it was thought that it would be a suitable substance to study in a human model. However, in normal volunteers transepidermal could not be regularly evoked with this substance.

It is an interesting observation that a patient with a disease characterised by transepidermal elimination in lesional skin should demonstrate transepidermal elimination of an apparently inert substance from clinically normal skin. Two possible mechanisms suggest themselves. It may be possible that the epidermis can become primed by certain stimuli and will then eliminate autologous or foreign material much more rapidly. Alternatively there may be a population of eliminators, possibly genetically determined, who will demonstrate the phenomenon if they, for example, develop any of the conditions in which transepidermal elimination is known to occur.

The experiment to determine the role of the rise of intratissue tension in the development of transepidermal elimination is also difficult to interpret. Transepidermal elimination occurred to a greater extent in biopsies from injection sites not surrounded by an incision. This would suggest that increased tissue pressure may be important in promoting the reaction. In the sites where there was necrosis and ulceration of the overlying epidermis, an inflammatory infiltrate was present. It was previously noted in the guinea pig model for transepidermal elimination that the process occurred more readily in the presence of some inflammatory infiltrate than when no inflammation occurred. It may be, therefore, that the increased transepidermal elimination seen in the sections which had not had encircling incisions was induced more by the presence of inflammation than by the rise in tissue tension.

References

1. Bardach H (1976) Dermatosen mit transepithelialer Perforation. J Cutan Pathol 4:99–104
2. Bayoumi A-HM, Gaskell S, Marks R (1978) Development of a model for transepidermal elimination. Br J Dermatol 99:611–620
3. Bayoumi A-HM, Marks R (1980) Transepidermal elimination: studies with an animal model. Br J Exp Path 61:560–566
4. Freudenthal W (1930) Amyloid in der Haut. Arch Dermatol Syphil 162:40–94
5. Marks R (1981) Transepidermal elimination. Japan Dermatol J 91(13):1692–1697
6. Mehregan AH, Coskey RJ (1968) Perforating folliculitis. Arch Dermatol 97:394–399
7. Miescher G (1955) Elastoma interpapillare perforans verruciforme. Dermatologica 11):254–266
8. Parra CA (1977) Transepithelial elimination in necrobiosis lipoidica. Br J Dermatol 95:83–86
9. Schoenfeld RJ, Grelain JN, Mehregan A (1965) Calcium deposition in the skin. Neurology 15: 477–480
10. Vecki H, Kubo M, Masuda T, Nohara N (1976) Transepidermal elimination of immune deposits in the skin. An immunohistochemical study on immune deposits using horse radish peroxidase as an antigen. J Dermatol (Tokyo) 3:209–213
11. Wood MG, Spolovitz H, Schetman D (1976) Schistosomiasis. Paraplegia and ectopic skin lesion as admission symptoms. Arch Dermatol 92:559–560

Induction of Cutaneous Melanocytic Tumours by N-Ethyl-N-Nitrosourea in the Mongolian Gerbil (Meriones Unguiculatus)

Ch. Kuehnl-Petzold, A. Uozumi, B. Volk and P. Kleihues

There is an increasing incidence of malignant melanomas in man but the factors involved in the development of this tumour are not well characterized. Two questions are of special interest; the first of which concerns the relationship between melanocytic nevi and melanomas. Most individuals have melanocytic nevi, but only few people develop malignant melanoma. Many patients report the sudden growth of a longstanding mole, but it is not assumed that benign nevi become malignant. More likely, melanomas may exist for years with minimal increase in size, until they show themselves by rapid growth [9]. The other major point of interest focuses on environmental factors possibly involved in the etiology of melanoma in man. Sunlight is considered as the most important single cause [7], although a large proportion of tumours are located at sites not usually exposed to ultraviolet irradiation (e.g. eye, vulva, gastrointestinal mucose, brain). Therefore, a systemic effect of UV-irradiation has been postulated [6], but conclusive evidence for malignant transformation by this mechanism is still lacking. Melanomas have been observed following exposure to chemical carcinogens both in man [1] and experimental animals [3] and the possibility exists that malignant transformation of melanocytes may be due to systemic rather than topical effects of environmental factors.

Material and Methods

Experiments were carried out on 22 Mongolian gerbils (*Meriones unguiculatus*) of both sexes. *N*-ethyl-*N*-nitrosourea (ENU) was dissolved in 3 mM sodium citrate buffer (pH 6.0) and administred as a single i.p. injection of 100 mg/kg on the 7th postnatal day. Animals were examined every week and tumours with a diameter greater than 1.0 mm were noted.

For semithin sections and electron microscopy, tumors were fixed with 2.5% glutaraldehyde (pH 7.4) and embedded in araldite. Semithin sections were stained with toluidine blue. Thin sections were contrasted with uranyl acetate and lead citrate. Before sacrifice, some animals received 17 i.p. injections (0.1 mCi each) of (^{3}H-methyl)-thymidine at intervals of 3 h. Autoradiographs were prepared from 5 μm paraffin sections, using the stripping film technique (Kodak AR10).

Results

Incidence and latency period. Within one year, 96% of animals developed multiple melanocytic tumours. The first neoplasms were observed after three months. The 50% incidence was reached after five months. The number of tumours per tumour bearing animal increased gradually to an average of 4.2 at 12 months, i.e., the total number of tumours observed within one year after ENU administration was 89. After 18–24 months, three animals died of widespread metastases.

Location and morphology. Tumours developed preferentially at sites with little hair, i.e., snout (13%), eyelids (15%), forefeet (8%), hind feet (36%), and ears (27%). In these areas, gerbils have a dense network of dermal melanocytes. All tumours were black and sharply demarcated. Small neoplasms were only slightly elevated and had a smooth surface. Tumours with a diameter of more than 5 mm developed superficial scaling followed by ulcerations (Fig. 1).

Histopathology and electron microscopy. Tumours originated from subepidermal melanocytes and grew slowly. In small neoplasms the epidermis and the underlying

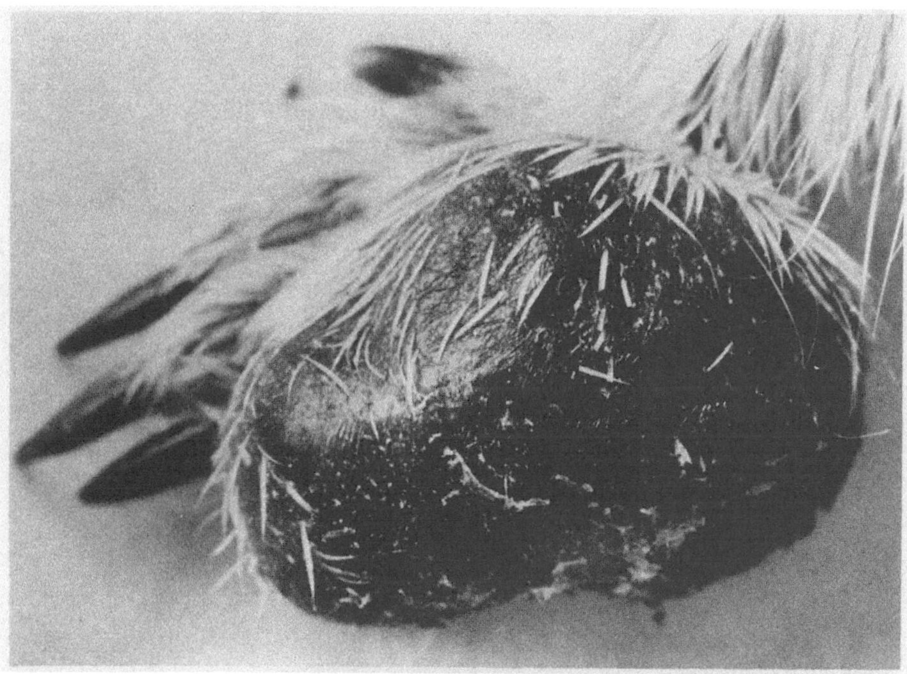

Fig. 1. Large, superficially ulcerating melanoma at the right hind foot of a gerbil 17 months after injection of ENU

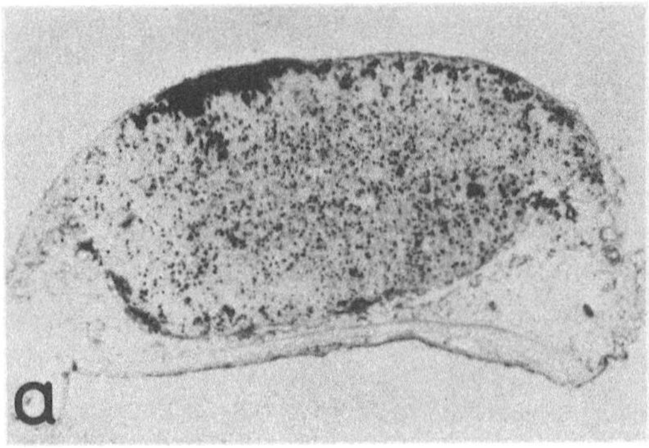

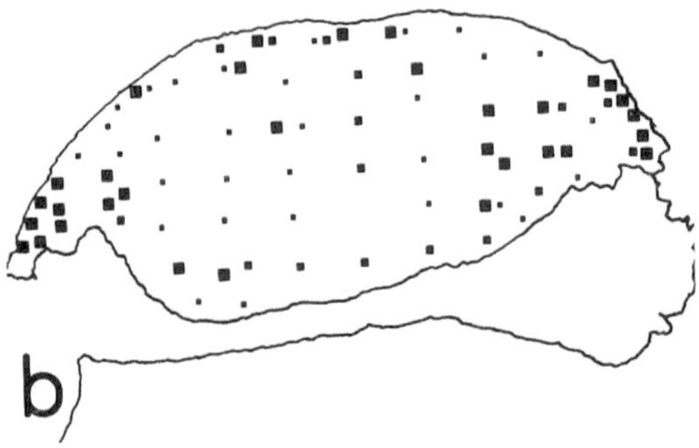

Fig. 2a, b. Light microscopy and growth of an ENU-induced melanocytic tumour. Before sacrifice, animals received 17 i.p. injections of ^{3}H-thymidine at 3 h intervals. Square size indicates 1–4, 5–7 and 8–10 labelled nuclei per visual field (x40). Note that the accumulation of labelled cells in the superficial lateral portions of the tumour

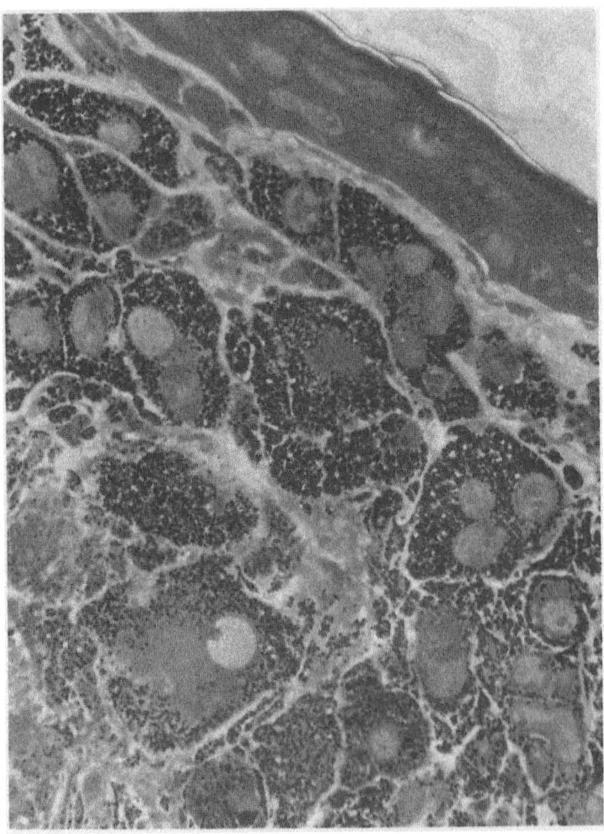

Fig. 3. Detail of cutaneous melanocytic tumour in the gerbil. Tumour cells which arise from dermal melanocytes are often multinucleated and do not invade the epidermis. Semithin section, phase contrast, toluidine blue (x1,040)

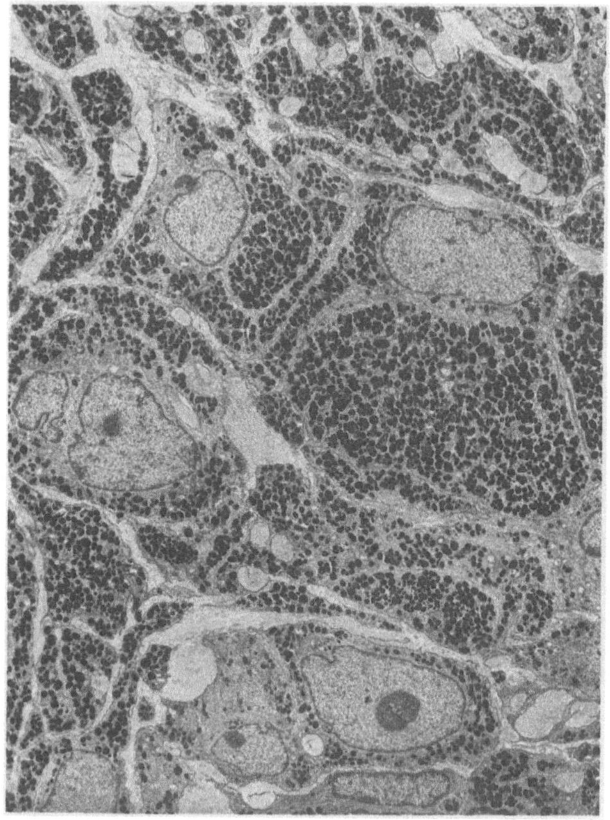

Fig. 4. Superficially located tumour cells are usually diffusely pigmented throughout the cytoplasm. Electron microscopy (x4,500)

structures (bones and cartilage) remained unaffected (Fig. 2a). Autoradiographic studies (Fig. 2b) prepared after multiple injections of ^{3}H-thymidine indicate that tumour growth is primarily directed laterally rather than towards the subcutis and adjacent structures. Intraepidermal neoplastic melanocytes were only found in large, ulcerated tumours. The subepidermal tumour cells showed strong pigmentation, evenly distributed throughout the cytoplasm (Figs. 3, 4). In deeper portions, the cells revealed degenerative changes with large confluent melanin droplets (Fig. 5).

Discussion

Cutaneous melanocytic tumours have been induced by chemical carcinogens in a variety of laboratory rodents [3]. The Syrian golden hamster and the Mongolian gerbil seem to be most susceptible. In hamsters, a tumour incidence of up to 100% was observed following treatment with polycyclic aromatic hydrocarbons. The tumours induced did not metastasize and were classified by the authors as blue-nevus-like-tumours [10].

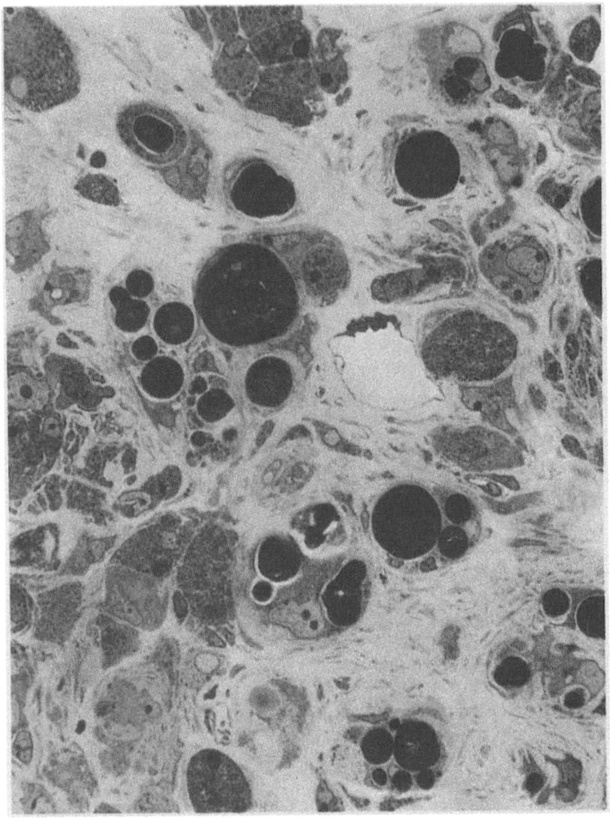

Fig. 5. In deeper portions of the tumour, numerous giant melanosomes are present. Semithin section (x670)

In some experiments, transplantation of tumour material led to metastatic spread [2]. Metastases from primary tumours were only observed in guinea pigs, but this species develops melanocytic tumours at a rather low incidence [8]. We have previously shown that postnatal administration of 50 mg ENU/kg to gerbils produces cutaneous melanocytic tumours, although at a low incidence and after a mean survival time of more than 800 days [4]. In the present study melanocytic tumours were induced by a single dose of 100 mg/kg of *N*-ethyl-*N*-nitrosourea on the 7th postnatal day.

This model is characterized by a high tumour incidence (96%), short latency period (50% incidence after five months), malignant transformation selectively of dermal melanocytes (no tumours in other tissues), and delayed metastatic spread.

Most of these tumours appear to be benign nevi. They originate from dermal melanocytes and are located in the dermis, similar to the human blue nevus. Metastasizing malignant melanomas developed only in 14% of animals. The tumours which metastasized were larger. At the time of metastatic spread they had a diameter of greater than 5.0 mm and showed ulcerations. Metastatic spread occurred 12–18 months after clinical manifestation of the primary tumour. Delayed metastatic spread was also observed when initiation by ENU was followed by topical administration of the tumour promoter, 12-0-tetradecanoyl-phorbol-13-acetate (TPA) but the latter treatment did not accelerate tumour induction [5].

References

1. Bahn AK, Rosenwaike I, Herrmann N, Grover P, Stellmann J, O-Oleary C (1976) Melanoma after exposure to PCBs. N Engl J Med 295:450
2. Goerttler K, Loehrke H, Schweizer J, Hesse B (1980) Two stage tumorigenesis of dermal melanocytes in the back skin of the Syrian golden hamster using systemic initiation with 7,12-dimethylbenz(a)anthracene and topical promotion with 12-0-tetradecanoylphorbol-13-acetate. Cancer Res 40:155–161
3. Kuehnl-Petzoldt C (1983) Tiermodelle für die chemische Induktion melanozytärer Tumoren. In: Luger A, Gschnait F (Hrsg) Dermatologische Onkologie. Urban & Schwarzenberg, Wien München Baltimore, pp 17–22
4. Kleihues P, Bücheler J, Riede UN (1978) Selective induction of melanomas in gerbils (*Meriones Unguiculatus*) following postnatal administration of *N*-ethyl-*N*-nitrosourea. J Natl Cancer Inst 61:859–863
5. Kleihues P, Volk B, Kovacs G, Kuehnl-Petzoldt C (1985) Development and biology of ethylnitrosourea-induced melanomas in the Mongolian gerbil (*Meriones Unguiculatus*). In: Bagnara J, Klaus SN, Paul E, Schartl M (eds) Pigment Cell 1985. Biological, molecular and clinical aspects of pigmentation. University of Tokyo Press, Tokyo, pp 465–470
6. Lee JAH, Merill JM (1970) Sunlight and the aetiology of malignant melanoma. A synthesis. Med J Austr 2:846–851
7. McGovern VJ (1977) Epidemiological aspects of melanoma: A review. Pathology 9:233–241
8. Pawlowski A, Haberman HF, Menon IA (1980) Skin melanoma induced by 7,12-dimethylbenzanthracene in albino guinea pigs and its similarities to skin melanoma of humans. Cancer Res 40:3652–3660
9. Paul E (1980) Growth dynamics of malignant melanoma. Arch Dermatol 116:182–185
10. Rappaport H, Pietra G, Shubik P (1961) The induction of melanotic tumors resembling cellular blue nevi in the syrian white hamster by cutaneous application of 7,12-dimethylbenz(a)-anthracene. Cancer Res 21:661–666

Amphibian Skin as a Model in Studies on Epidermal Homeostasis

P. E. Budtz

Models to study function and disease of skin may for instance imply the use of a particular animal species or preparation as especially useful in studies of a particular function, studies on a particular (biochemical, physiological or morphological) parameter as indicator of a particular function, or computer simulating models. The present paper will deal with amphibian and particularly toad skin as a suitable model to study epidermal homeostasis defined as maintenance of tissue (population) size. This intriguing question as to how the appropriate proportions of proliferating and differentiating cells are maintained to ensure a constant epidermal cell pool is of general biological importance (Fig. 1). According to the general concept of tissue homeostasis, in a renewing tissue like the epidermis, the efflux in terms of keratinization and exfoliation must – over a period of time – be counterbalanced by a corresponding number of cell divisions, when the tissue is in equilibrium. Later we shall see, that the three parameters of paramount importance in studies on epidermal homeostatis: influx, cell pool size, and efflux, can quite easily be assessed in toad skin.

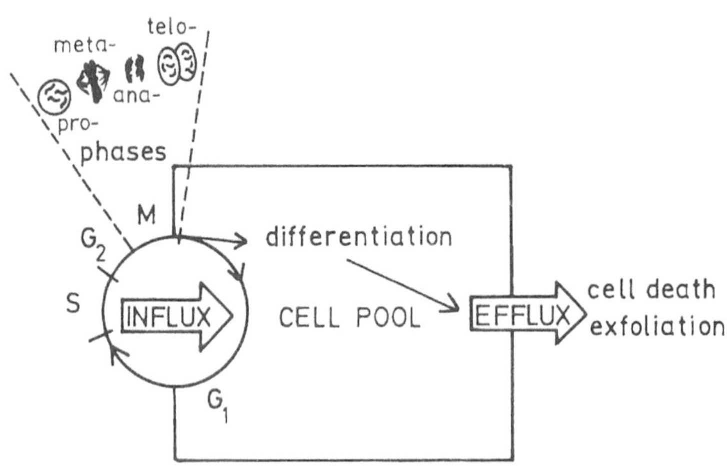

Fig. 1. Diagram indicating the important parameters in studies on epithelial tissue homeostasis

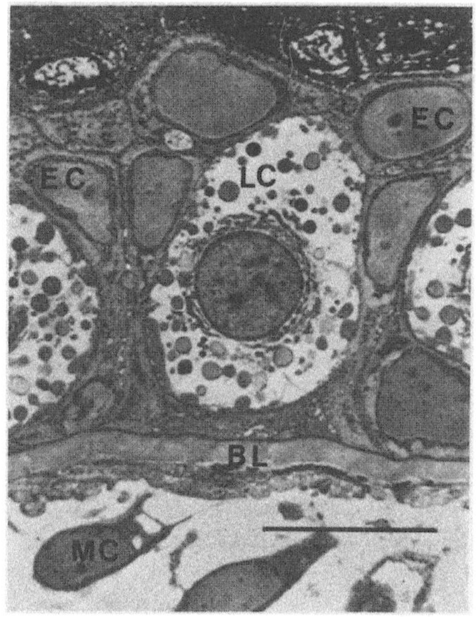

Fig. 2. Larval skin of the axolotl (*Ambystoma mexicanum*). EC, epidermal cell; LC, Leydig cell; BL, basal lamella; MC, mesenchymal cell. Scale: 25 μm

Structure and Function of Amphibian Skin

Larval Skin

A characteristic feature of most amphibians is their development from larvae through metamorphosis to juveniles and adults. At metamorphosis the entire body undergoes profound changes, which are manifested also in structural and functional changes of the skin [20].

The epidermis of the newly hatched tadpole is only one cell thick, but in most species it soon becomes stratified, throughout larval life consisting of two or three cell layers (Fig. 2). Initially the epidermis is ciliated, but ciliation gradually regresses during larval life, and only basal bodies and ciliary rodlets may be recognized in late prometamorphosis [19]. The larval skin is unkeratinized. The epidermal cells contain abundant tonofilaments and various numbers of mucous granules and superficially they are laterally joined by zonulae occludentes; deeper in the epidermis desmosomes are frequent. A characteristic cell type of the larval skin, especially in that of the urodele, is the large Leydig cell, containing some mucoid material produced in the perinuclear endoplasmic reticulum. Release of this material is questioned [38], and the physiological role of these cells is unclear; it has been suggested that they may provide a fluid reserve for the subsurface intercellular pool [25].

This cell type decreases in number during late larval life and disappears at metamorphosis [20]. Epidermal melanophores may be present, and various other cell types may be found in the larval epidermis, including Merkel cells and presumptive immigrant granular cells, probably polymorphonuclear neutrophils [19]. Young anuran larvae possess epidermal unicellular glands [36] but no multicellular glands are present until close to metamorphosis, when the development of these take place. According to some authors they are formed by aggregations of cells present within the epidermis, but in vivo and in vitro studies involving pharmacologic blocking of mitotic activity indicate that the glands are formed by divisions of single precursor cells and their progonies [34].

The larval epidermis is resting upon a basement lamella (Fig. 2), a few μm thick, consisting of orthogonally arranged collagen fibrils. Preceding metamorphosis profound alterations in biochemical properties occur in the skin. At the tadpole stage hyaluronate is virtually the sole acid mucopolysaccharide, but during metamorphosis this glycosaminoglycan rapidly declines and becomes replaced by chondroitin sulphate and dermatan sulphate, the former becoming predominant in some species, the latter in others [33]. During metamorphosis mesenchymal cells migrate into the basal lamella to form the stratum spongiosum of the postmetamorphic dermis, a process which involves simultaneous secretion of hyaluronidase from the mesenchymal cells and collagenase from the epidermal cells [15]. Probably the dermis of the adult is a result of integrated phagocytosis of the larval lamellar collagen by mesenchymal macrophages and synthesis of new collagen [19].

Adult Skin

Following metamorphosis, the epidermis of all amphibians becomes keratinized (in some species the keratinization process is initiated already during late metamorphosis) and multicellular glands are formed. In adult amphibians there is no principal difference between the structure of the skin of urodeles and anurans, but differences may exist between even closely related species, as well as between different parts of the body. Generally the skin consists of a dermis and, separated from this by a basal lamina and an adepidermal space, the epidermis. The dermis is separated into a stratum compactum and a stratum spongiosum. The stratum compactum consists of a few sheets of dense connective tissue and is closely attached to the underlying skeletal muscles, except along those areas, where the capacious lymph sinuses occur; here a tela subcutanea forms the outer lining of the subcutaneous lymph sacs. The bulk of the dermal connective tissue is loose, however, and the main part of the fibres in this stratum spongiosum consists of collagen fibres. In many anurans a 5–30 μm thick layer rich in calcium and acid mucopolysaccharides, particularly chondroitin sulphate, is found interposed between the stratum compactum and the stratum spongiosum. The calcium is present as calcium phosphate and concentrations between 0.5 and 1.0 M have been demonstrated in some species [57]. The physiological role of this layer is by no means clear. Amphibians easily loose water by evaporation since the epidermis is very permeable to water, and it has been suggested that the layer may play a role in defence against desiccation by acting like a sponge, absorping large quantities of water easily, giving it up only slowly during dehydration [17]. Others, however, believe it to be a calcium reservoir for the whole animal or being of importance for maintenance of normal permeability and/or structural properties of the skin [57]. Blood vessels, nerve fibres and chromatophores are situated in the upper part of the stratum spongiosum and so are the two types of integumentary glands found in postmetamorphic amphibians: the granular (serous, poison) glands and the mucous glands [37]. The granular glands produce in different amphibian species a variety of pharmacologically potent substances, including cardio- and neurotoxins, hemolytic, vasocon-

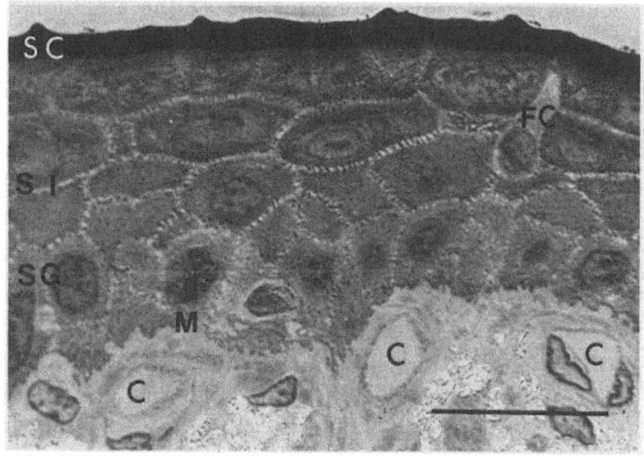

Fig. 3. Adult skin of the toad, *Bufo bufo*. SG, stratum germinativum; SI, stratum intermedium; SC, stratum corneum; FC, flask cell; M, mitotic figure; C, cappillaries in the upper part of stratum spongiosum. Scale: 25 μm

stricting, and hallucinogenic substances, which chemically include alkaloids, biogenic amines, indolalkylamines, more or less complicated peptides, and the steroid-derived bufogenins and bufotoxins [21]. The mucous gland secretion is shown histochemically to be acid mucins containing carboxyl and sulphate groups as well as vicinal glycol groups [10, 14], and probably also the electrolytes demonstrated in frog skin secretions [11] may originate in mucous glands. Recently alveolar lipid-secreting glands have been demonstrated in certain frogs from South America, the secretary product containing, in addition to a water soluble fraction, a lipid component, which is primarily composed of triglycerides with smaller amounts of sterol or wax esters [2]. These glands play an important role in protection against desiccation; thus the rate of evaporative water loss in frogs possessing these glands has been shown to be 0.6 to 0.9 mg g^{-1} h^{-1}, whereas closely related species, which lack these glands, under the same evaporative conditions showed water losses of 10–12 mg g^{-1} h^{-1} [43].

The epidermis is separated from the dermis by a basal lamina and an adepidermal space of about 0.2 μm. The epidermis (Fig. 3) is a stratified epithelium, which in most amphibians is 4 to 10 cells thick. It consists basally of the stratum germinativum and outermost of the stratum corneum; cell layers in between these have been designated the stratum intermedium [8]. Structurally the cells of the stratum germinativum are rather uniform in amphibians, and usually there are only few and very low dermal papillae. Basally the cells are attached to the dermis by hemidesmosomes [42] and pinocytotic vesicles are frequently observed at membranes facing the dermis. Laterally the basal cells are jouned to one another by moderate numbers of desmosomes. The basal cells contain predominantly fine bundles of filaments and a moderate number of mitochondria, wheres endoplasmic reticulum and Golgi apparatus are sparse. The mitotic activity takes place predominantly in this layer although in some species mitoses have been reported occasionally to occur also in the stratum intermedium [23]. The main structural unit of the stratum intermedium is the Malphigian cell or the keratinocyte. In the lower stratum intermedium the keratinocytes contain fine filament bundles, but as the cells migrate upwards the filament bundles become more coarse and may possible also increase in number although the main synthesis appears to take place in in the lower epidermis [9]. The amphibian keratinocytes do not contain the typical mammalian keratohyaline granules, but two types of (probably mucous) granules are an outstanding feature of the amphibian keratinocyte although little is known of their chemistry; they may probably also contain bound phospholipids [44]. The small granules are possibly released into the intercellular space, and some of the large ones are released into the subcorneal space whereas others appear to be involved in the keratinization process [4]. The outermost layer of the stratum intermedium is often referred to as the transitional or replacement layer because after shedding it becomes the new stratum corneum. Structurally it differs from the lower stratum intermedium by the presence of tight junctions or zonulae occludentes at the latero-distal parts of the cells [18] and physiologically by its polarity with regard to membrane properties [30]. A characteristic, although less frequent, cell type is the flask cell, which is found in the epidermis of all metamorphosed amphibians investigated, constituting 5–10% of the epidermal cell population [47]. Flask cells (Fig. 3) are pear shaped or slender cells, rich in mitochondria and with relative little endoplasmic reticulum and few filaments; their function is still a matter of discussion. Merkel cells, which are also found in the larvae [48] have in the frog Rana pipiens been calculated to constitute about 0.3% of the epidermal cell population [35]; their possible function (s) in amphibian skin have been discussed by Budtz [4].

The stratum corneum of amphibians is usually only one cell thick but may in some species be multi-layered, e.g., in the giant salamander Cryptobranchus, where it may be up to six cells thick [45]. The cells are laterally held together by strong modified desmosomes and distally by zonulae occludentes [18]. Details of the cell interior is usually obscured by a dense matrix precipitated between numerous filaments. Histochemically cystine can be shown to be concentrated peripherally as a thin band, whereas bound sulphydryl groups, bound phospholipids, protein-bound calcium, and free fats are more uniformly distributed within the stratum corneum cells [44]. The stratum corneum cells and their zonulae occludentes, and the outer membranes of the replacement layer and their zonulae occludentes, are the main diffusion barriers of the amphibian skin [32].

In all postmetamorphic amphibians investigated the stratum corneum is periodically shed and the shed corneal layer replaced by cells of the upper layer of the stratum intermedium. Little is known, however, about the physiology and biochemistry of the amphibian keratinization process. X-ray patterns of shed stratum corneum indicate that amphibian keratin belongs to the α-type [39]. Knowledge of structural manifestations of keratinization is mainly derived from two types of approach: comparison of the different epidermal layers at a given time, considering the inner and the outer layers to be in successive stages of differentiation [28] or studies on particular layers at different times during the moulting cycle [8, 9]. When it comes to details, differences exist between species, but as a general pattern keratinization is accomplished by an initial process, which involves lytic digestion of cell organelles not retained in the stratum corneum and formation of a dense peripheral band, and a final keratinization process during which a dense interfibrillar matrix is laid down. The formation of the peripheral shell is a rapid process, which in toads occurs immediately after shedding and with a high degree of synchronization [9], but in other amphibian species it appears to be less synchronous [45]; the interfibrillar lay-down of matrix takes longer time and is less synchronous [9].

Moulting Cycles

Keratinization of the outermost epidermis takes place from the time of metamorphosis, and with the formation of the first stratum corneum also moulting is initiated. In all amphibians investigated the stratum corneum is shed at regular intervals, and the time between two moults (the intermoult period) is dependent upon the temperature and may also be influenced by light and amount of food; in the toad Bufo bufo it is shorter in newly metamorphosed animals than in full-grown toads [27].

The moulting cycle includes separation of the old stratum corneum from the underlying epidermis, the actual shedding of the stratum corneum (the slough), and keratinization of the replacement layer into a new stratum corneum. A prerequisite of shedding is separation of the stratum corneum from the underlying epidermis. This process is gradual and apparently accomplished by a chemical change within the desmosomes between these layers which leads to breakage within the desmosomes, and/or by rupture of the distal cell membrane of the replacement layer [3]. It is not known what determines the site of breakage and causes the desmosomes to break. Shedding is in some species, and notably in the toad Bufo bufo, accompanied by a copious slime secretion and a moulting behaviour, by which the slough is removed from the body and usually eaten.

Control of Moulting

Probably a moulting rhythm, which is dependent upon hormones, is inherent in the skin [3]. Hypophysectomized amphibians, or those deprived of the pars distalis of the pituitary gland, stop moulting, and unshed layers of stratum corneum accumulate in many species, even with an increased rate, e.g., in the Bufonidae [4] whereas the piling up is less pronounced in frogs (Ranidae) [27]. Moulting, abolished by hypophysectomy, can in urodeles (newts and salamanders) be elicited by TSH or thyroxin and in (some) toads by ACTH or adrenocorticosteroids. However, hormones do not elicit a moult when injected into intact animals, and probably their action is permissive rather than inductive [27]. Thus the pituitary-thyroid glands and the pituitary-adrenocortical system exert the principal endocrine influence on moulting of urodels and (some) toads, respectively. However, other hormones, e.g., prolactin in urodeles, may be involved, too. This hormone appears to be an important regulator of epidermal proliferation by stimulating mitotic activity in the epidermis of the newt Notophthalmus viridescens [23], and it can in the hypophysectomized newt Triturus cristatus, although in rather high dosis, elicit a moult [46]. The relationship between the control of moulting and proliferation is not well investigated. Proliferation, differentiation (keratinization), separation, and shedding are distinct processes, which are not necessarily coupled, and are probably differently controlled.

In larval epidermis, which neither keratinizes nor moults, there is convincing evidence that the proliferation is hormonally controlled, thyroxine and prolactin appearing to be the main factors influencing the epidermal proliferative pattern [12, 16, 49–52, 56].

Methodology

When our knowledge of epidermal tissue homeostatic mechanisms is relatively sparse, this is probably de due to fundamental deficiencies in experimental methods. The epidermal cell pool, and the influx and efflux to and from this pool must be measureable directly and independently of each other and this is difficult in mammalian skin, partly due to the architecture of the epidermis (dermal papillae) which makes it difficult to assess the epidermal cell pool size, but particularly due to the continuous desquamation, which makes it almost impossible to assess the cell loss quantitatively with any degree of accuracy. Amphibian skin, and particularly toad skin is a good model to study epidermal homeostasis as the epidermis is quite simple (few tiers, few and low dermal papillae (Fig. 3)) and the stratum corneum is single-layered and the entire corneal sheet is shed at regular intervals, at moult.

Assessment of Epidermal Cell Pool

The number of epidermal cells may conveniently be counted from photomicrographs on which the length of the stratum corneum is measured and the number of cells below it counted. The number of cells in the stratum germinativum and in the stratum intermedium may be counted separately. The stratum corneum cells are usually too densely stained to allow identification of individual cells, but in these sections the number of corneal cells may be estimated from the number of cells in the replacement layer as it has been shown that the number of cells in these two layers is the same [4].

It is important to recognize that the epidermis is a three-dimensional structure, and since it has been pointed out that two-dimensional measurements may be misleading [6, 13, 54] all parameters should be expressed per square unit of surface. The number of cells per mm^2 (N) in the different epidermal compartments may be calculated according to equation 1

$$N = 0.75 \frac{n^2}{L} \tag{1}$$

where n = the number of cells in the compartment per mm stratum corneum, L = the number of layers in the compartment, and 0.75 a correction factor due to the geometry of the cells. Considering a single cell layer, some cells in a histological section may be cut only partly, others in their maximal width. Assuming that the number of cells per mm^2 in that layer (N_1) would be equal to n_1^2, where n_1 = the number of counted cell profiles per mm, the number of cells per unit area would be overestimated, the degree of fault depending upon the geometry of the cells in question, and must therefore be corrected for by an experimentally determined correction factor. Thus Budtz [4] for the stratum corneum of toad epidermis calculated a correction factor of 0.75 as the number of cells per square unit was overestimated by approximately 25% by the above assumption.

The number of cells in stratum germinativum plus stratum intermedium, expressed per mm^2, is designated the stratum corneum recruitment cell pool (SCRP) [4] since all epidermal cells below the stratum corneum are potential corneal cells.

Assessment of Cell Loss by Moulting (Efflux)

Toads moult at intervals, which may vary from toad to toad but which is fairly regular for the individual toad, at 20 °C usually every 6–10 days. The moulting process is accompanied by a characteristic behaviour with a well defined sequence of events. Prior to shedding, gaping and eye movements are seen, accompanied by a characteristic moulting posture. By movements of the body and legs, coincident with a copious slime secretion, the detached stratum corneum (the slough) is then pulled into the mouth and eventually swallowed [27]. As this entire sequence of events is usually accomplished within a couple of hours or less and often takes place during the night, the occurrence of a moult may be recorded by marking the toads with lipstick, which does not penetrate the stratum corneum and consequently disappears after a moult [1]. By daily observations and repeated markings the intermoult period can be determined. Thus knowing the number of corneal cells per mm^2 and hence the number of cells lost at each moult, the efflux from the SCRP in terms of the number of cells lost per hour can be calculated by equation 2

$$\mathrm{CLM} = \frac{N_{\mathrm{s.corn.}}}{\mathrm{i.p.} \times 24} \tag{2}$$

where CLM = cells lost per mm^2 per h by moulting, $N_{\mathrm{s.\,corn.}}$ = number of corneal cells per mm^2, and i.p. = the intermoult period in days. Although the CLM is expressed per h for comparison with the cell birth rate, it should be remembered that moult is a periodic event.

In hypophysectomized toads, or in those deprived of the pars distalis of the pituitary gland, where the stratum corneum is not shed the efflux may be calculated on basis of the number of cells in one corneal layer and the increase in number of corneal layers between two samplings. It is therefore important that none of the often loose corneal layers are lost during histological processing. A small device consisting of two rings, the one supplied with a groove, the other with a ring attached to a steel clamp, has been constructed (Fig. 4). Clamped between the two rings, the skin sample may be processed from the fixative right through all solutions to the embedding medium, without any loss of eventually detached corneal layers.

Assessment of Epidermal Proliferative Capacity (Influx)

The epidermal proliferative capacity is often assessed by the labelling index (LI) or the mitotic index (MI). The LI is the number of labelled cells per 100 basal cells as seen after pulse-labelling with isotope-labelled thymidine [40, 41] and it indicates the number of cells which, at a given time, is synthesizing DNA. The MI is the number of mitotic figures per 100 basal cells and indicates the number of cells, which at a

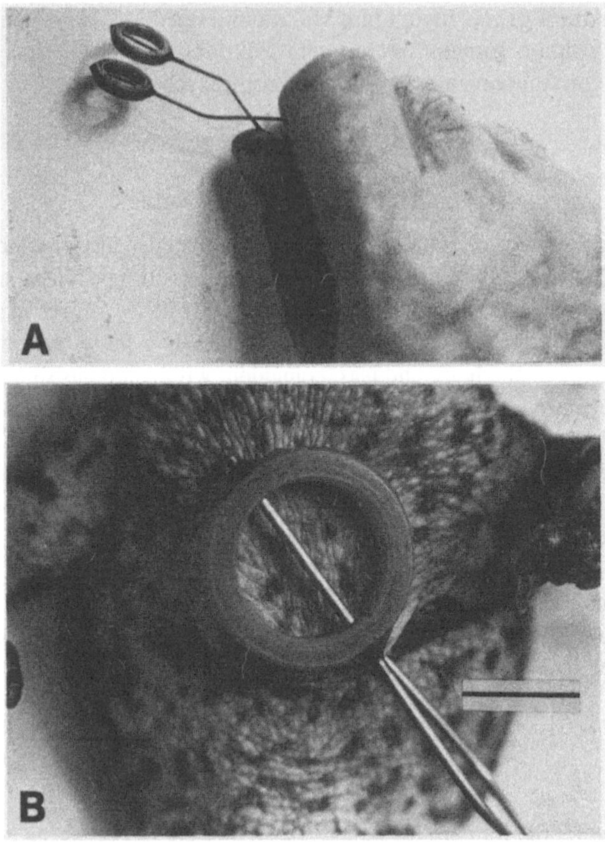

Fig. 4. A Device to hold skin samples during histological processing. **B** An incision has been made and the rings on the clamp brought in position, after which the skin has to be cut along the outer edges of the rings and processed, clamped between the rings. Scale: 1 cm

given time is in cell division. Thus both of these indices are static in nature, and as they are furthermore subject to changes with alterations in the duration of the S-phase or mitosis, respectively, they are by themselves poor parameters of the proliferative capacity. Furthermore, the use of LI in studies on proliferation presupposes that all cells which go through the S-phase also divide, but this may not necessarily be the case [6, 24].

The rate of proliferation, or the cell birth rate, may be directly assessed by means of the metaphase-arrest technique. By this method, cells to go through mitosis are arrested within the metaphase by means for instance of colchicine, vinblastine or vincristine. When the proper precautions [53] are taken, the rate of entry into metaphase is a fairly good measure of the rate of proliferation. For toad skin vinblastine sulphate in a dosage of 50 μg/g toad has proved optimal as have metaphase index readings at one and four h after the injection [6]. The metaphase index is the number of metaphases per 1,000 basal cells. Knowing the number of basal cells per mm surface, the metaphase index may be recalculated to the number of metaphases per mm of surface. Then the number of metaphases per mm^2 (M) may be calculated according to equation 3 [6].

$$M = 0.75 \times m \times n^2_{s.\,germ.} \times 10^{-3} \qquad (3)$$

where 0.75 is the correction factor mentioned above, m = number of metaphases per 1,000 basal cells and $n_{s.\,germ.}$ = the number of cells in stratum germinativum per mm of surface.

Let M_1 and M_4 be the mean number of metaphases per mm^2 one and 4 h after vinblastine injection and SEM_1 and SEM_4 their respective standard errors, the cell birth rate expressed per mm^2 per h (CBR) ± SEM may be calculated according to equation 4 [6].

$$CBR = \frac{M_4 - M_1}{3} \pm \frac{\sqrt{SEM_1^2 + SEM_4^2}}{3} \qquad (4)$$

Equation 3 is based upon the assumption that the factor 0.75, which is experimentally determined for (the geometry of) stratum corneum is also applicable for (the geometry of) stratum germinativum cells. This may not necessarily be so since the cells are smaller and the correction factor (cf) may therefore be larger ($0.75 \leqslant cf \leqslant 1.00$). It would be very difficult, however, to evaluate this possibility since it is not feasible to study the stratum germinativum in sheets like the shed stratum corneum. Obviously a larger cf would increase the calculated cell birth rate, and since it has not been proven that the factor of 0.75 is not applicable to the stratum germinativum, it may be considered reasonable to use it for stratum germinativum, too, in order to avoid a possible overestimate of the CBR.

Tissue Homeostatic Parameters in Epidermis of Intact Toads

In the context of tissue homeostasis, the living cells within the stratum germinativum plus stratum intermedium (SCRP) may be regarded as the cell pool (Fig. 1) and is calculated according to equation 1. The influx to this pool is by means of cell divisions in stratum germinativum and is calculated according to equation 4. Formation of dead corneal layers and their shedding during moults as in intact toads, or their piling up as in hypophysectomized or pars distalis deprived toads, are both considered as the efflux from the SCRP.

The size of the SCRP of adult toads has repeatedly been shown to be fairly constant (25,000–30,000 cells mm^{-2}) [4–7] indicating a thoroughly controlled balance between influx and cell loss by moulting. In recent experiments designed to assess the cell birth rate at various hours of the day it was shown, however, that the cell birth rate in 8 groups of 10 toads tested during a period of 24 h in all instances were much higher than the calculated cell loss by moulting [6]. By summation of the cell productions during the 8 periods of 3 h, a daily production of 620 cells mm^{-2} was calculated, corresponding to 6,400 cells mm^{-2} during an intermoult period of 10.3 days on average, while only about 2,400 cells mm^{-2} were lost at moult. It was further shown that the cells produced and not used for formation of corneal cells could not be permanently accomodated within the living epidermis because this in consequence would lead to a doubling of the SCRP within 7 1/2 moulting cycle or 77 days, which seems most unlikely since the toads were adult, and since a SCRP exceeding about

40,000 cells mm^{-2} was neither seen in these nor in other toads studied [6]. Thus the surplus production of about 400 cells mm^{-2} day^{-1}, corresponding to about 60% of the cells produced, must consequently be deleted in one way of another. It could be by apoptosis as suggested by Kerr et al. [26] or by emigration in the "wrong direction" to be eliminated within the dermis or via the vascular system.

Tissue Homeostatic Parameters in Epidermis of Toads Deprived of the Pars Distalis of the Pituitary Gland

Pars distalis ablation in toads (Bufo bufo) leads to abolishment of moulting but also to an increased rate of formation of corneal layers; in spite of this, no significant depletion of the SCRP could be osberved [4]. In accordance with the general concept of tissue homeostasis this finding was interpreted as indicative of an increased rate of proliferation following pars distalis ablation. In a subsequent cell kinetic study [5] an analysis of the cell kinetic parameters studied (SCRP, MI, LI, and mean grain count) showed the results incompatible with an increased rate of proliferation. Since the actual cell division rates were not measured in that investigation, this has recently been done [7]. The purpose was to compare an eventual change in the SCRP in 14 days following pars distalis ablation as directly assessed (difference between the size of the SCRP at start and after 14 days of pars distalis deprivation) with that assessed by flux measurements (difference between total influx and efflux in 14 days). The daily and total influxes and effluxes are shown in Table 1. As can be seen from the Table, throughout the experiment there was a decrease in the influx and an increase in the efflux which must lead to a depletion of the SCRP. This can be calculated to about 4,300 cells mm^{-2} with lower and upper limits of about 3,000 and 5,700 cells mm^{-2}, respectively, presupposing that all cells produced were used for formation of corneal cells. Assuming a cell deletion rate – beyond that of corneal cell formation – as in intact toads, an

Table 1. Daily influx to and efflux from the epidermal cell pool (SCRP) as a function of time after pars distalis ablation (Data from [7])

	Time after pars distalis ablation (days)				Total in 14 days
	0–1	1–3	3–7	7–14	
Influx* (cells mm^{-2} day^{-1})	879 (863–895)	556 (216–895)	192 (167–216)	209 (153–251)	4,219 (3,032–5,306)
Efflux* (cells mm^{-2} day^{-1})	396 (384–410)	419 (382–438)	524 (501–547)	747 (744–750)	8,559 (8,361–8,721)
Expected depletion (cells mm^{-2})					4,340 (3,055–5,689)

* Average values; in brackets lower and upper limits

expected depletion of the SCRP in 14 days of about 11,700 cells mm^{-2} can be calculated (lower and upper limits 10,400 and 13,100 cells mm^{-2}, respectively). Direct assessment showed an actual depletion of the SCRP in 14 days of about 5,000 cells mm^{-2}, which evidently is much lower than these latter values but consistent with the data obtained by means of flux measurements. It is therefore reasonable to conclude that the cell deletion seen in intact toads, after removal of the pars distalis is abolished or at least strongly reduced.

Mechanisms of Epidermal Tissue Homeostasis

In toad skin it is possible to assess independently of each other the size of the epidermal cell pool (SCRP) and the influx to and efflux (rate of formation of corneal layers, whether shed or not) from this cell pool. In intact toads the influx is 2.7 times that needed to replace cells lost through moult. In addition it was shown that, in an experimental situation where at the same time the efflux is increased and the influx is decreased, namely pars distalis ablation, a surplus cell production is most likely abolished, all cells produced being used for formation of corneal layers. These observations lend support to a revised concept of the mechanism of epidermal tissue homeostasis (Fig. 5), originally proposed by Kerr et al. [26] and elaborated upon by Marks [31], according to which controlled cell deletion is a tissue homeostatic mechanism complementary to controlled cell divisions. Teleologically this mechanism appears useful as it tends to prevent large fluctuations in epidermal cell pool size in response to large fluctuations in stimulatory or inhibitory factors influencing epidermal mitotic activity and/or efflux in terms of rate of formation of corneal cells. Obviously a crucial point in this concept is the fate of the deleted cells. They could be removed by apoptosis [26] but apoptotic cells are in toad skin seen only rarely. Assuming cell deletion in toad skin to occur by apoptosis, a cell deletion rate of about 400 cells mm^{-2} day^{-1} would correspond to an incidence of 0.17 cells mm^{-1} h^{-1} in histological sections. Obviously the frequency by which apoptotic cells would be seen depends upon the duration of the apoptotic process, but it may not be surprising that they are seen only rarely since "apoptosis is a process by which large numbers of cells may be deleted in a short time, yet leaving intact the overall stromal-parenchymal organization of the tissue" [55]. The possibility should not be excluded that basal cells may squeeze out of the epidermis in the wrong direction to become eliminated within the dermis or via the vascular system.

The exit arrow at G_2 in Fig. 5 needs a short explanation. Levi and Nielsen [29] in a study of toad epidermal cell kinetics by means of ^{3}H-thymidine incorporation, also found a cell production rate exceeding the rate of cell loss through moulting, but by a factor of about 6. The discrepancy between 6 and the factor of 2.7 shown in the present study may be explained in the light of a recent finding by Keiding et al. [24] who showed by means of double-labelling experiments a diurnal mean S-phase influx and efflux of 0.5% of the hamster cheek pouch epithelial cell pool per h, whereas in the same epithelium the influx to mitosis as obtained by colcemid was found to be

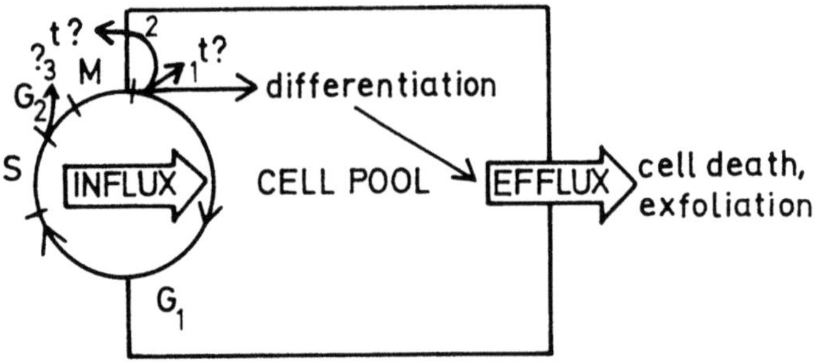

Fig. 5. Proposed revised concept of epidermal tissue homeostatic mechanisms. Arrow 1 indicates the possible cell deletion by apoptosis, arrow 2 the possible emigration to the dermis and arrow 3 a possible arrest or elimination of cells in the G_2 phase (for details, see text)

only 0.3% of the epithelial cell pool per h. Thus it seems as if about 50% of the hamster cheek pouch cells passing the S-phase are arrested or disappear in the G_2-phase, and if this applies to toad epidermis as well it may explain the difference between Levi and Nielsen's factor of 6 and that of 2.7, the former based in principle on entry into S-phase, the latter on entry into mitosis.

References

1. Bendsen J (1956) Shedding of the skin of the common toad, *Bufo bufo*. Vidensk Medd dansk nat hist. Foren 11:211–225
2. Blaylock LA, Ruibal R, Platt-Aloia K (1976) Skin structure and wiping behaviour of phyllomedusine frogs. Copeia:283–295
3. Budtz PE (1977) Aspects of moulting in Anurans and its control. Symp Zool Soc Lond 39:317–334
4. Budtz PE (1979) Epidermal structure and dynamics of the toad, *Bufo bufo*, deprived of the pars distalis of the pituitary gland. J Zool Lond 189:57–92
5. Budtz PE (1982) Time-dependent effects of removal of the pars distalis of the pituitary gland on toad epidermal cell and tissue kinetic parameters. Cell Tissue Kinet 15:507–519
6. Budtz PE (1985) Epidermal tissue homeostasis. I. Cell pool size, cell birth rate, and cell loss by moulting in the intact toad, *Bufo bufo*. Cell Tissue Kinet 18:521–532
7. Budtz PE (1985) Epidermal tissue homeostasis. II. Cell pool size, cell birth rate, and cell loss in toads deprived of the pars distalis of the pituitary gland. Cell Tissue Kinet 18:533–542
8. Budtz PE, Larsen LO (1973) Structure of the toad epidermis during the moulting cycle. I. Light microscopic observations in *Bufo bufo*. Z Zellforsch 144:353–365
9. Budtz PE, Larsen LO (1975) Structure of the toad epidermis during the moulting cycle. II. Electron microscopic observations on *Bufo bufo*. Cell Tiss Res 159:459–483
10. Breckenridge WR, Murugapillai R (1974) Mucous glands in the skin of *Ichthyophis glutinosus* (Amphibia: Gymnophiona). Ceylon J Sci (Bio Sci) 11:43–52
11. Campbell JP, Aiyawar RM, Berry ER, Huf EG (1967) Electrolytes in frog skin secretions. Comp Biochem Physiol 23:213–223
12. Chiakulas JJ, Scheving LE (1967) The effects of the presence or absence of the pituitary gland on the daily rhythmicity of mitotic rates in urodele larval tissues. In: Mayersbach H von (ed) Cellular aspects of biorhythms. Springer, Berlin, p 155

13. Clarke RM (1973) Progress in measuring the epithelial turnover in the villus of the small intestine. Digestion 8:161
14. Dapson RW (1970) Histochemistry of mucus in the skin of the frog, *Rana pipiens*. Anat Rec 166:615–626
15. Delsol M, Flatin J (1971) Tissus conjonctifs et métamorphose chez les Batrachiens. Lyon Med 226:615–626
16. Dournon C, Chibon P (1974) Influence de la temperature, de l'age et des conditions hormonales (thyroxine) sur la proliferation cellulaire chez la jeune larve et pendant la métamorphose due crapaud *Bufo bufo* L. (Amphibian Anoure). Wilh Roux Archiv 175:27–47
17. Elkan E (1976) Ground substance: an anuran defence against desiccation. In: Lofts B (ed) Physiology of the Amphibia, vol 3. Academic Press, New York San Francisco London, p 101
18. Farquhar MG, Palade GE (1965) Cell junctions in amphibian skin. J Cell Biol 26:263–291
19. Fox H (1977) The anuran tadpole skin: changes occurring in it during metamorphosis and some comparisons with that of the adult. Symp zool Soc Lond 39:269–289
20. Fox H (1981) Cytological and morphological changes during amphibian metamorphosis. In: Gilbert LI, Frieden E (ed) Metamorphosis: a problem in developmental biology, 2nd edn. Plenum Publ Corp, p 327
21. Habermehl GG (1974) Venoms of amphibia. Chem Zool 9:161–183
22. Hoffman CW, Dent JN (1977) Hormonal effects on mitotic rhythm in the epidermis of the red-spotted newt. Gen Comp Endocr 32:512–521
23. Hoffman CW, Dent JN (1977) Hormonal regulation of cellular proliferation in the epidermis of the red-spotted newt. Gen Comp Endocr 32:522–530
24. Keiding N, Hartmann NR, Møller U (1984) Diurnal variation in influx and transition intensities in the S phase of hamster cheek pouch epithelium cells. In: Edmunds LN (ed) Cell cycle clocks. Dekker, New York, p 135
25. Kelly DE (1966) Fine structure of desmosomes, hemidesmosomes and an adepidermal globular layer in developing newt epidermis. J Cell Biol 28:51–73
26. Kerr JFR, Wyllie AH, Currie AR (1972) Apoptosis: a biological phenomenon with wide ranging implications in tissue kinetics. Br J Cancer 26:239–257
27. Larsen LO (1976) Physiology of moulting. In: Lofts B (ed) Physiology of amphibians, vol 3. Academic Press, New York San Francisco London, p 53
28. Lavker RM (1974) Horny cell formation in the epidermis of *Rana pipiens*. J Morphol 142: 365–378
29. Levi H, Nielsen A (1982) An autoradiographic study of cell kinetics in epidermis of the toad *Bufo bufo bufo* (L.). J Invest Dermatol 79:292–296
30. Lindemann B, Voute C (1976) Structure and function of the epidermis. In: Llinas R, Precht W (ed) Frog neurobiology. Springer, Berlin Heidelberg New York, p 169
31. Marks R (1980) Is epidermal homeostasis a necessity? – comments on epidermal growth control. Br J Dermatol 103:697–702
32. Martinez-Palomo A, Erlij D, Bracho H (1971) Localization of permeability barriers in the frog skin epithelium. J Cell Biol 50:277–287
33. Mathews MB (1975) Connective tissue. Macromolecular structure and evolution. Springer, Berlin Heidelberg New York (Molecular biology biochemistry and biophysics, vol 19)
34. McGarry MP, Vanable JW (1969) The role of cell division in *Xenopus laevis* skin gland development. Develop Biol 20:291–303
35. Nafstad PHJ, Baker RE (1973) Comparative ultrastructural study of normal and grafted skin in the frog, *Rana pipiens*, with special reference to neuroepithelial connections. Z Zellforsch 139:451–462
36. Picard JJ (1976) Ultrastructure of the cement gland of *Xenopus laevis*. J Morphol 148:193–208
37. Quay WB (1972) Integument and the environment: glandular composition, function, and evolution. Am Zoologist 12:95–108
38. Rosenberg M, Lewinson D, Warburg MR (1982) Ultrastructural studies of the epidermal Leydig cell in larvae of *Salamandra salamandra* (Caudata, Salamandrida). J Morphol 174: 275–281

39. Rudall KM (1947) X-ray studies of the distribution of protein chain types in the vertebrate epidermis. Biochem Biophys Acta 1:549–562
40. Schultze B (1969) DNA synthesis. In: Schultze B (ed) Autoradiography at the cellular level, 2nd edn. Academic Press, New York London, p 76 (Pollister AW (ed) Physical techniques in biological research, vol 3)
41. Schultze B (1981) Double labeling autoradiography. Cell kinetic studies with ^{3}H- and ^{14}C-thymidine. J Histochem Cytochem 29:109–116
42. Shienvold FL, Kelly DE (1976) The hemidesmosome: new fine structural features revealed by freeze-fracture technique. Cell Tissue Res. 172:289–307
43. Shoemaker VH, McClanahan LL (1975) Evaporative water loss, nitrogen excretion and osmoregulation in phyllomedusine frogs. J Comp Physiol 100:331–345
44. Spearman RIC (1968) Epidermal keratinization in the salamander and a comparison with other amphibia. J Morphol 125:129–144
45. Spearman RIC (1973) The integument. Cambridge University Press
46. Vellano C, Lodi G, Gani G, Sacerdote M, Mazzi V (1970) Analysis of the integumentary effect of prolactin in the hypophysectomized crested newt. Monit Zool Ital 4(NS):115–146
47. Whitear M (1975) Flask cells and epidermal dynamics in frog skin. J Zool Lond 175:107–149
48. Whitear M (1977) A functional comparison between the epidermis of fish and amphibians. Symp zool Soc Lond 39:291–313
49. Wright ML (1973) DNA synthesis during differentiation of tadpole shank epidermis. J Exp Zool 186:237–256
50. Wright ML (1977) Regulation of cell proliferation in tadpole limb epidermis by thyroxine. J Exp Zool 202:223–234
51. Wright ML, Majerowski MA, Lukas SM, Pike PA (1979) Effect of prolactin on growth, development, and epidermal cell proliferation in the hindlimb of the *Rana pipiens* tadpole. Gen Comp Endocr 39:53–62
52. Wright ML, Sicbaldi EM, Loveridge KM, Pike PA, Majerowski MA (1981) Cell population kinetics in tadpole limb epidermis during thyroxine-induced, spontaneous, and prolactin-inhibited metamorphosis. Gen Comp Endocr 43:451–461
53. Wright NA, Appleton DR (1980) The metaphase arrest technique. A critical review. Cell Tissue Kinet 13:643–663
54. Wright NA, Irwin M (1982) The kinetics of villus cell populations in the mouse small intestine. I. Normal villi: the steady state requirement. Cell Tissue Kinet 15:595–609
55. Wyllie AH (1981) Cell death: a new classification separating apoptosis from necrosis. In: Bowen ID, Lockshin RA (ed) Cell death in biology and pathology. Chapman and Hall, London New York, p 9
56. Yoshizato K, Yasumasu I (1972) Effect of prolactin on the tadpole fin. V. Stimulatory effect of prolactin on the incorporation of ^{3}H-thymidine into DNA of the tadpole tail fin. Developm Growth Diff 14:129–132
57. Zadunaisky J, Lande MA (1972) Calcium content and exchange in amphibian skin and its isolated epithelium. Am J Physiol 222:1309–1315

Transplantation Studies on Solar Keratoses in Nude Mice

S. Thomas, A. D. Pearse and R. Marks

Solar keratoses are premalignant lesions of the epidermis and occur on the chronically sun exposed areas of individuals with a fair complexion. Solar keratoses can develop into squamous cell carcinomas. Montgomery and Dörffel [11] estimated that 20% of patients with solar keratoses develop squamous cell carcinomas. However, according to Lund [9] and Muller et al. [12] these squamous cell carcinomas do not metastasize. Solar keratoses can also completely resolve, and the so-called lichenoid keratoses [16] may represent lesions in which immunological rejection has begun. An important question for carcinogenesis in general and for solar keratoses in particular concerns the determinants that influence the fate of these premalignant lesions. Are such lesions more or less likely to transform to squamous cell carcinoma when isolated from their usual immunological milieu? Transplantation to an immunologically privileged site would be one way of answering these questions.

Athymic mice have been used for transplantation of normal [10] and diseased [3–5] human skin. Cell lines have also been transplanted to nude mice [1, 7, 14].

As a first step in our studies we have attempted to transplant solar keratoses on to athymic nude mice (Nu Nu) to establish whether they could be maintained and studied in this situation.

Methods

Grafting of Solar Keratoses

Tissue was collected at time of biopsy, and divided into two. Half was placed in formalin for routine histological studies, and the remainder placed in DMEM supplemented with FCS, and transported to the animal house. The grafting technique used was a modification of that of Briggaman and Wheeler [2]. Briefly, a graft bed approximating to the size of the human lesion was prepared on the dorsal flank region of the nude mouse after removal of full thickness skin from the site. The segment of the lesion was placed on the bed, covered with vaseline, gauze and plaster, and left in place for 14 days.

Immunofluorescent Staining of Grafts for Involucrin

Using the method of Banks-Schlegel and Green [1], tissue was fixed in formalin and embedded in paraffin; sections were dewaxed and slides incubated with involucrin antiserum raised in rabbits (a gift from Dr. Fiona Watts, Kennedy Institute) for 30 min at 37 °C. Sections were washed three times in phosphate buffered saline (pH 7.2) FITC labelled sheep anti-rabbit IgG was added and the slides incubated for 30 min at 37 °C, washed three times in phosphate buffered saline, mounted and viewed using a Nikon Apophot fluorescence microscope fitted with a high pressure mercury lamp.

Labelling with [^{3}H]thymidine

[^{3}H] TdR, specific activity 25 Ci/mmol (Amersham International) was injected intraperitoneally into the nude mice at a concentration of 1 μCi/g one hour prior to sacrifice. Tissue was removed, placed in formalin and processed by routine histological methods. 5 μm paraffin sections were prepared and autoradiography was performed using a modified version of the dipping film technique of Joftes and Warren [6]. The sections were stained through the emulsion with haematoxylin and eosin, and the number of labelled basal and suprabasal cells was counted and expressed as a percentage of the total number of basal and suprabasal cells for each specimen.

Results

The acceptance rates of the grafted lesions (Table 1) were 75% for solar keratoses and 33% for basal cell carcinomas over periods of from 4 to 16 weeks. All transplanted lesions maintained a histological appearance resembling that of the human lesions in vivo.

The lesions shown in Fig. 1a is a solar keratosis from the back of the hand of a 73-year-old female patient. It can be seen that there is hyperkeratosis and patchy parakeratosis. The epidermis is thickened and shows irregular downward proliferations. The epidermal cells are irregular in size and show nuclear heterochromia, atypia, and loss of polarity with a generally disordered arrangement. In the dermis there is a dense inflammatory cell infiltrate consisting of lymphocytes and histiocytes. There is also much solar elastotic degenerative change present in the upper and papillary dermis.

Table 1. Take rates of xenografts

	No. of implants	Number surviving
Solar keratoses	12	9
Basal cell carcinomas	3	1

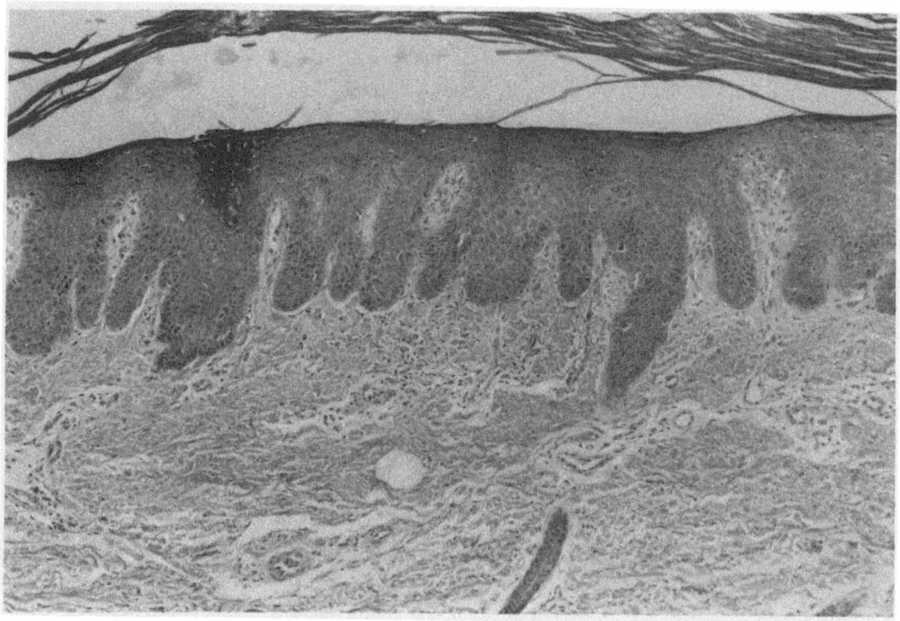

Fig. 1a. Pre-graft solar keratosis exhibiting hyperkeratosis and patchy parakeratosis, a thickened epidermis and loss of polarity of cells, with solar elastotic degenerative changes present in the upper papillary dermis

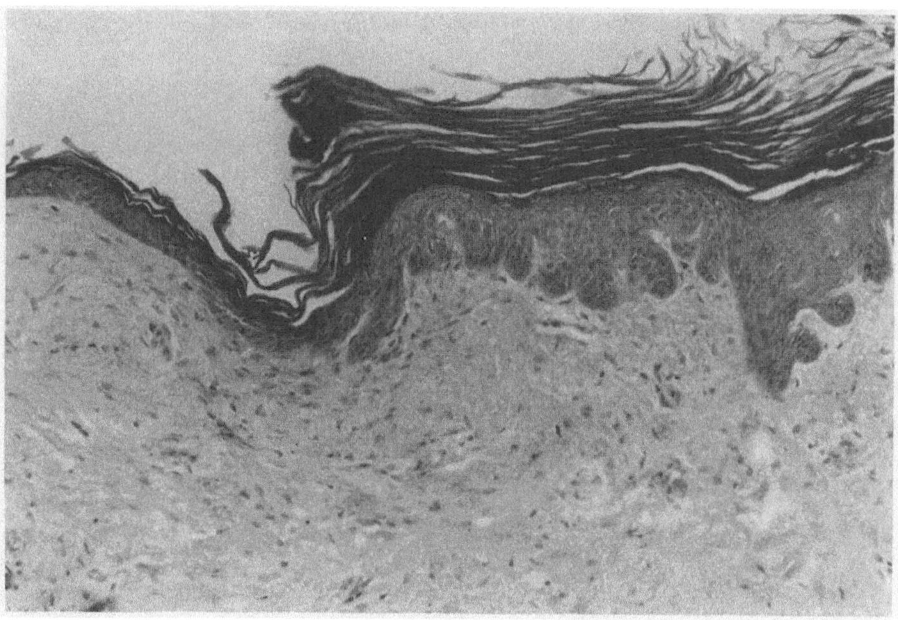

Fig. 1b. Post graft solar keratosis on the nude mouse for 16 weeks, exhibiting many of the features of the original lesion in Fig. 1a

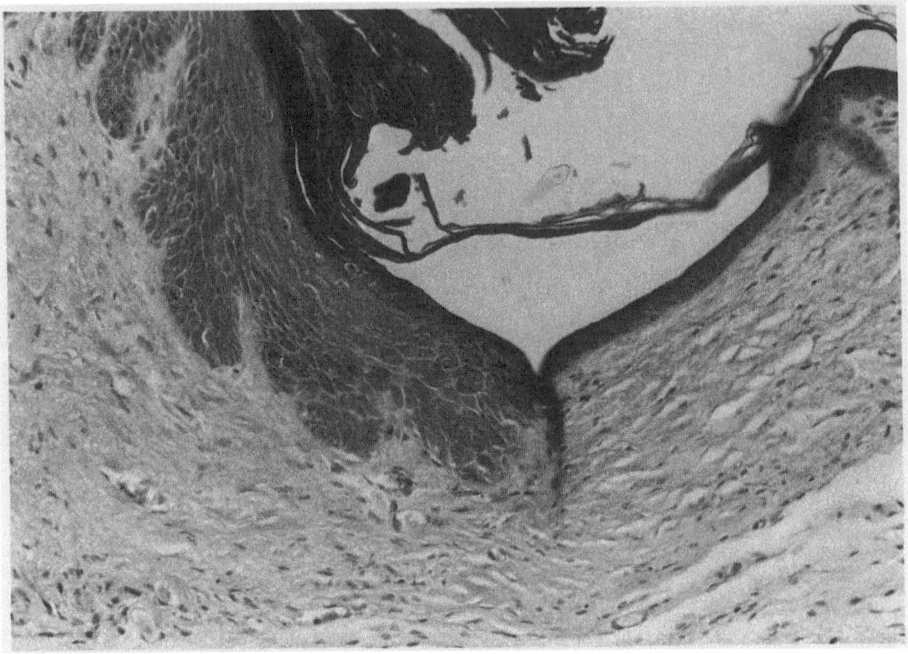

Fig. 1c. Junction zone between mouse skin and human lesion

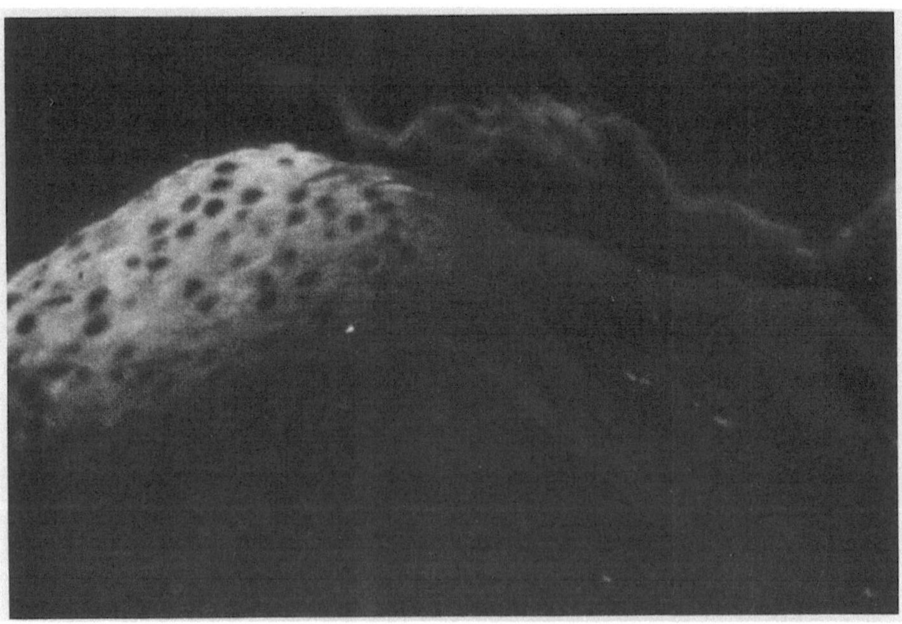

Fig. 2. Section of grafted lesion stained with antiserum to human envelope precursor protein. Note that the immunofluorescence is confined to the outer half of the human epidermis

Table 2. Labelling index. [^{3}H] TdR was injected i.p. into nude mice one hour prior to sacrifice (1 μCi/g)

Lesion	Mean LI ± SD (%)
Solar keratoses	11.2 ± 1.9
Basal cell carcinomas	10.9
Adjacent mouse skin	3.6 ± 1.0

Table 3. Measurements of mean epidermal thickness (MET) and mean keratinocyte height (MKH)

	Lesion	Graft
MET (μm)	96.0 ± 39.0	55.0 ± 28.0
MKH	12.3 ± 1.7	9.7 ± 0.85

The other half of the lesion from this patient (Fig. 1b) was transplanted to a nude mouse and retained in situ for 16 weeks. It maintained a similar histological appearance to the original lesion, possessing most of the features mentioned above. The human lesion is quite distinct from the mouse skin, the junction between the two being clearly discernible (Fig. 1c).

Although the grafted lesion is usually easily distinguishable from the mouse host skin, we wished to confirm the human origin of the tissue at the site of the graft using an antibody specific for human involucrin [15]. This envelope precursor is species specific and does not cross-react.

When examined by immunofluorescence technique with involucrin antiserum, the human epidermis stained well but the stratum corneum did not (Fig. 2). The sharp junction between the non-staining mouse epidermis and the human lesion indicated that the epidermis was of human origin.

The labelling indices (Table 2) of the grafted lesions (11.2 ± 1.9%) were higher than those usually observed in normal human skin (4 to 8% depending on area and technique) and significantly different when compared to normal nude mouse epidermis (3.6 ± 1.9%).

Measurements of the mean epidermal thickness (Table 3) indicated that the grafted lesion has a thinner epidermis than the original human lesion. The mean keratinocyte height, which is a measure of cell size indicates that the cells are smaller in the grafted lesion compared to the original human lesion.

Discussion

These studies have shown that premalignant human skin lesions can be grafted on to nude mice and maintain an overall morphology similar to the original human lesion.

Our acceptance rate of 75% compares well with those of other workers. Kondo and Aso [7] were only able to grow one tumour after numerous attempts at implanting squamous cell carcinoma cells into nude mice. Similarly, only 8% of basal cell carcinomas could be transplanted by Krueger and Briggaman [8]. However, Briggaman and Wheeler [3] reported a 75% success rate for transplantation of psoriatic epithelium.

Our results also suggest that the immunological identity of the grafted solar keratoses is preserved, as the lesions remained positive for the envelope precursor protein involucrin which is specific to human skin. Many further studies are needed to be certain that they retain an identical antigenic profile to the in vivo state.

The grafted lesions showed elevated tritiated thymidine autoradiographic labelling indices compared to areas of normal skin. They also showed similar labelling indices to those found in solar keratoses in vitro by Gregg and Mansbridge [4] and our own previous studies [13].

No progression of the grafted solar keratoses to frank malignancy was observed. However, the longest period of observation in the nude mouse thus far has been 16 weeks and this may have been too short a time for the transformation to occur, as the lesion must adjust initially to its new environment and establish vascular continuity with the host.

We believe these preliminary studies indicate the suitability of this model for the study of the biology of premalignant epithelial lesions.

References

1. Banks-Schlegel S, Green H (1980) Formation of epidermis by serially cultivated human epidermal cells transplanted as an epithelium to athymic mice. Transplantation 29:308–313
2. Briggaman RA, Wheeler CE (1976) Lamellar ichthyosis: Long term graft studies on congenitally athymic nude mice. J Invest Dermatol 67:567–572
3. Briggaman RA, Wheeler CE (1978) Nude mouse – human skin graft model III. J Invest Dermatol 74:262A
4. Gregg K, Mansbridge J (1982) Epidermal characteristics related to skin cancer susceptibility. J Invest Dermatol 79:178–182
5. Haftek M, Ortonne JP, Staquel MJ, Viac J, Thivolet J (1981) Normal and psoriatic human skin grafts on nude mice. Morphological and immunochemical studies. J Invest Dermatol 76: 48–52
6. Joftes DL, Warren S (1955) Simplified liquid emulsion autoradiograph. J Biol Photo Ass 23:145–150
7. Kondo S, Aso K (1981) Establishment of a cell line of human skin squamous cell carcinoma in vitro. Br J Dermatol 105:125–132
8. Krueger GC, Briggaman RA (1982) The nude mourse in the pathology of the skin. In: Fogh J, Gioucinella BS (eds) The nude mouse in experimental and clinical research, vol 2. Academic Press, New York
9. Lund HZ (1965) How often does squamous cell carcinoma of the skin metastasize? Arch Dermatol 92:635–637
10. Manning DD, Reed ND, Schaffer CF (1973) Maintenance of skin xenografts of widely divergent phylogenetic origin on congenitally athymic nude mice. J Expt Med 138:488–494
11. Montgomery H, Dörffel J (1932) Verruca senilis und keratoma senile. Arch Dermatol Syphil 166:286–296

12. Muller SA, Wilhelmj CM, Harrison EG, Winkelmann RK (1964) Adenoid squamous cell carcinoma. Arch Dermatol 89:589–594
13. Pearse AD, Marks R (1977) Actinic keratoses and the epidermis in which they arise. Br J Dermatol 96:45–50
14. Rheinwald JG, Beckett MA (1981) Tumorigenic keratinocyte lines requiring anchorage and fibroblast support cultured from squamous cell carcinomas. Cancer Res 41:1657–1663
15. Rice RH, Green H (1979) Presence in human epidermal cells of a soluble protein precursor of the cross-linked envelope. Activation of the cross-linking by calcium ions. Cell 18:681–694
16. Tan CY, Marks R (1982) Lichenoid solar keratoses – Prevalence and immunological findings. J Invest Dermatol 79:365–367

Modelling for the Inflammation of Acne

K. Dalziel, P. J. Dykes and R. Marks

There has been no model, animal or human, of the inflammation which accompanies acne. However, it is the inflammatory component of this disease which is responsible for most of the clinical manifestations of the disease. The papules, pustules, cysts and scars are all caused by the inflammation which occurs when there is leakage from, or rupture of a plugged, dilated follicle. Initially, numerous polymorphonuclear leucocytes accumulate around the damaged follicle. Later mononuclear cells, macrophages and giant cells surround the inflammatory mass producing the granulomatous picture characteristic of acne inflammation.

The pathogenesis of this inflammation is uncertain. Various mechanisms have been proposed including the leakage of sebum hydrolysis products from ruptured follicles [8], the activation of complement [7], and the release of keratinous debris into the dermis. We found this latter concept appealing for several reasons.

Our observations of the histology of inflammatory acne papules has shown that early in the reaction, polymorphonuclear leucocytes are most numerous around and within the horny plug inside the follicle or horny debris in the dermis. It might be expected that sebum and fatty acids would be found in similar locations and could be responsible for the inflammation seen. However, during and after therapy with isotretinoin, a drug known to dramatically reduce sebum secretion rates [9] and cause a marked decrease in sebaceous gland size [4], the morphology of the remaining inflamed papules was unaltered [2]. This suggests that sebum and its breakdown products are not solely responsible for the inflammation in acne.

In other situations where stratum corneum is released into the dermis, for example after rupture of epidermoid cysts, a very similar granulomatous reaction is seen to that in acne.

Puhvel and Sakamoto [6] injected physiological amounts of various comedonal components into normal human skin. They found the insoluble keratinous fraction to be the most uniformly inflammatory material.

These observations seem to indicate that the exposure of stratum corneum to the dermis following rupture of a comedone may be largely responsible for the inflammation in acne.

Initial experiments were designed to test this premise in a guinea pig model and, more recently, a human model has been devised.

Details of Experiments

Animal Model

Methods. The guinea pig model has been described previously [1]. Human stratum corneum, derived from foot callus, was implanted or injected into guinea pig dermis. Biopsies of these sites were taken at time intervals between 4 days and 12 weeks. These were examined histologically and by immunofluorescence.

Results. In biopsies taken early (2–7 days) the stratum corneum had evoked a brisk inflammatory response characterised by the presence of large numbers of polymorphonuclear leucocytes around callus fragments. After 14 days, the picture altered with less neutrophils and increasing numbers of mononuclear cells, macrophages and multinucleate giant cells. Transepidermal elimination of corneocytes and inflammatory debris was prominent. Overall, the morphology of the inflammatory reaction closely resembled that seen in acne.

Direct immunofluorescent examination revealed deposits of all immunoglobulins in individual inflammatory cells and as small aggregates between connective tissue bundles. C3 was detected only in individual cells. There was no overall pattern to the immunofluorescence.

Human model

Clearly, if any data obtained from the animal model is to be useful in studying human disease, it must be shown that the inflammation is similarly mediated. In particular, it is important to establish whether the use of guinea pig rather man has influenced the nature of the inflammatory response or whether the presence of heterologous stratum corneum within the dermis has modulated the inflammation via immune mechanisms. To study this, we have injected autologous stratum corneum into human dermis and examined the inflammatory reaction.

Methods. Stratum corneum was collected as heel callus from consenting healthy human volunteers using an aseptic technique. After paring away and discarding the superficial layers of callus, the skin was carefully washed with chlorhexidine gluconate and then with 70% isopropyl alcohol. A Scholl foot scraper was used to collect callus, producing a fine powder. Samples were sent for microbiological testing. After collection the callus powder from each individual was homogenised in phosphate buffered saline in a sterile cabinet. The number of particles/ml was calculated using a hemocytometer. Microscopically, the homogenate contained a mixture of corneocytes and corneocyte fragments. The suspension was centrifuged to leave a pellet of homogenized stratum corneum. This was stored for a maximum of 6 h at 4 °C. Before injection, the pellet was resuspended in 1 ml of sterile normal saline and 0.5 mls drawn into each of two syringes. These were inverted, allowing the stratum corneum to settle on the plunger. The saline could then be ejected leaving a small volume of homogenate. Two intracutaneous injections of 0.05 mls of homogenate were made into the forearm.

Each injection contained a mean of 1.2×10^9 corneocytes. In some volunters, a control injection of 0.05 mls of sterile saline was also made. The injection sites were biopsied at time intervals of between 24 h and 12 days, using sterile disposable 4 mm trephines. The biopsies were prepared for histological examination and 5 μ sections cut every 100 μ of tissue.

Results. In none of the samples of stratum corneum sent for microbiological testing was a pathogenic organism detected, and most were bacteriologically sterile. In most cases an erythematous papule appeared within eight hours at the site of stratum corneum injection. No such papules appeared at the control saline sites. This reaction persisted for up to 72 h, and then remained as a palpable, slightly erythematous area for up to 10 days. Histological examination of biopsies taken at 24–72 h after injection showed many polymorphonuclear leucocytes in a discrete inflammatory focus within the dermis. In some sections particles of stratum corneum were obvious in the inflamed area.

By 7 days, many neutrophils were still present but there were more mononuclear cells and macrophages at the periphery of the inflammatory focus. Occasional giant cells were seen. After 12 days, very few acute inflammatory cells remained in the centre of the involved area. Around these was a mass of macrophage and multinucleate giant cells. These contained amorphous material and occasional vacuoles.

Modifying the Inflammation with Anti-inflammatory Agents

Having ascertained that the human and guinea pig models show a very similar pattern of inflammatory response to implanted stratum corneum and that this resembles that seen in human acne, we have attempted to manipulate the animal model with anti-inflammatory drugs.

Method. Benoxaprofen was chosen because of its known immunomodulatory effects [5] and because there were anecdotal resports of its usefulness in nodulocystic acne [3]. Eight groups of four guinea pigs were injected intracutaneously with 0.05 mls of corneocyte suspension (80 mg/ml). Groups then received either 1, 3 or 20 mg/kg/day benoxaprofen or vehicle alone by intraperitoneal injection for 5 or 14 days. The corneocyte injection site was excised at sacrifice of the guinea pig one day after the course of treatment.

Biopsies were prepared for histological examination, serially sectioned and assessments made "blind" of the volume of inflammation using a microscope projection technique, the density of the inflammatory cell infiltrate on a 10 cm analogue scale and the proportion of individual cell types on a simple 0–5 scoring system.

Prednisolone: Further groups of 6 guinea pigs were injected intracutaneously with 0.05 mls corneocyte suspension at four sites. Each group then received either 1 or 5 mg/kg/day prednisolone or vehicle alone by intraperitoneal injection for 4 days. The pigs were sacrified at 7 days and corneocyte injection sites biopsies. Histological sections were prepared and step sections examined. Measurements were made of the maximum number of high power microscope fields involved in the inflammatory

Table 1. Effect of benoxaprofen on inflammation caused by stratum corneum

Time		Volume of inflammation (mm^3)			
		Vehicle control	Benoxaprofen		
			1 mg/kg	3 mg/kg	20 mg/kg
5 days	Mean (4 animals, 8 biopsies)	5.09	2.39	3.85	2.31
	SD	2.32	0.77	1.45	1.00
15 days	Mean (4 animals, 8 biopsies)	3.99	1.67	2.94	3.04
	SD	3.02	1.29	2.57	0.39

Table 2. Effect of prednisolone on inflammation caused by stratum corneum

		Vehicle control	Prednisolone	
			1 mg/kg	5 mg/kg
Number of microscopic fields		1.18 ± 0.42	0.99 ± 0.54	0.95 ± 0.15
Density of infiltrate		1.99 ± 0.82	1.57 ± 0.77	1.54 ± 0.25
Proportions of	Histiocytes	1.17 ± 0.55	1.29 ± 0.79	1.48 ± 0.26
	Giant cells	1.14 ± 0.76	1.73 ± 1.08	1.15 ± 0.45
	Neutrophils	1.94 ± 1.40	0.91 ± 0.77	1.05 ± 0.46

process and the density of the cellular infiltrate and proportion of individual inflammatory cells on 10 cm analogue scales.

Results. Benoxaprofen. The quantitative assessment of the volume of inflammation showed a reduction after 5 days significant at the 1% level (Table 1). There was a non-statistically significant decrease at day 14. The semiquantitative assessments of the intensity of inflammation at both day 5 and 14 showed a non statistically significant decrease.

Prednisolone produced a non significant decrease in the number of high power fields involved and in the density of the infiltrate (Table 2).

Overall, it can be said that benoxaprofen and prednisolone appeared to reduce the amount of inflammation present. However, the decrease was not dose dependent, was not great in degree and did not produce a change in the morphology of the inflammation.

Extraction Procedures

Provisional experiments were devised to try to remove lipid or water soluble components from the stratum corneum in order to determine whether this method altered the inflammatory response.

Methods. Minced callus fragments were extracted in 50 mls of either chloroform methanol or normal saline. These were agitated at room temperature for 24 h and the supernatant changed 8 hourly. After filtration, the callus was dried and used in the implantation procedure described, in 12 guinea pigs. Implantation sites were biopsied at 4, 7 and 12 days and 4 weeks and examined histologically.

Results. No difference was found in the morphology or intensity of the inflammation after extraction as compared with control.

Comment

These experiments have demonstrated that the implantation of stratum corneum into the dermis produces a pattern of inflammatory response similar to that seen in acne. This is irrespective of whether the implanted stratum corneum is autologous or homologous.

Clearly, stratum corneum alone is unlikely to be responsible for all the inflammation which follows comedonal rupture, sebum breakdown products and bacteria may contribute to the reaction. However, these factors alone have never been shown to produce the granulomatous type of inflammation seen in acne. The inflammation demonstrated in both the animal and human models described sufficiently resembles that of acne to make further studies on the mediators of the response worthwhile.

References

1. Dalziel K, Dykes PJ, Marks R (1984) Inflammation due to intracutaneous implantation of stratum corneum. Br J Exp Path 65:107–117
2. Dalziel K, Kingston T, Marks R (1985) The effect of isotretinoin on the pathology of early acne papules. Clin Exp Dermatol 10:365–370
3. Hindson C, Lawlor F, Wacks H (1982) Benoxaprofen for nodular acne. Lancet Jun 19:1(8286): 1415
4. Landthaler M, Kummermehr J, Wagner A, Plewig G (1980) Inhibitory effects of 13-cis retinoic acid on human sebaceous glands. Arch Derm Res 269:297–309
5. Meacock SCR, Kitchen EA (1979) Effects of the non-steroidal anti-inflammatory drug benoxaprofen on leucocyte migration. J Pharm Pharmacol 31:366–370
6. Puhvel SM, Sakamoto M (1978) An in vivo evaluation of the inflammatory effect of purified comedonal components in human skin. J Invest Dermatol 69:401–406
7. Scott DG, Cunliffe WJ, Gowland G (1979) Activation of complement: a mechanism for the inflammation in acne. Br J Dermatol 101:315–320
8. Strauss JS, Pochi PE (1965) Intracutaneous injection of sebum and comedones: histological observations. Arch Dermatol 92:443–456
9. Strauss JS, Stranieri AM (1982) Changes in long term sebum production from isotretinoin therapy. J Am Acad Dermatol 6:751–754

New Models for Testing Skin Toxicity

J. van Genderen and O. L. Wolthuis

It is generally accepted that the stratum corneum acts as an effective barrier to the penetration of a wide variety of substances. When sufficient quantities of a toxic agent pass this barrier, epidermal or dermal damage may result or the agent may pass into the blood stream resulting in possible damage elsewhere in the body.

To develop a methodology for measuring *in vivo* penetration in animals and to compare the percutaneous absorption in animals and man, various animal species have been used [2, 10, 19]. From these studies it has been concluded that the permeability of the skin of weanling swines and rhesus monkeys comes closest to that of human skin. However, no animal species exists that will simulate the penetration in humans for all compounds [19].

For ethical as well as practical reasons it is not possible to use humans to obtain information about percutaneous absorption of toxic compounds and skin damage caused by these agents. Congenitally athymic (nude) mice have been shown to accept skin grafts from several species, including normal and pathological human skin, because of their impaired immunological response [3, 5, 6, 8, 12, 13, 15–17, 20]. The human skin retains its features for months without alterations. On the basis of these observations a project was started to investigate whether the nude mouse with a human skin graft is a suitable model to examine human dermatotoxicity and whether the grafted human skin simulates the penetration of compounds in man.

Materials and Methods

Nude Mice

Male mice, ten to twelve weeks old, of three genetic backgrounds (Balb/c, B10LP and C57B1) were used. The animals were purchased from the REP-institutes TNO, Rijswijk, The Netherlands and the Central Institute for Breeding of Laboratory Animals TNO, Zeist, The Netherlands, and were kept under specific pathogen-free conditions. The mice were housed in Makrolon^{R}, type II cages (ITL, The Netherlands) with filter tops and the provision of sterile bedding, autoclaved food pellets and acidified-chlorinated water.

Skin Grafting

Mammary or abdominal skin was obtained from patients undergoing cosmetic operations in the St. Elizabeth Hospital, Leiderdorp, The Netherlands. The mammary skin was prepared by gently scraping away all subcutaneous fascia. To be successful with the abdominal skin split thickness grafts were used. Varying amounts of the dermis were trimmed so that the thickness of the human grafts approximated 0.5 mm. The prepared skin may be stored for as long as 5 days in a gauze soaked in Earle's Minimal Essential Medium (Flow, USA) at 4 °C until use.

The mice were anaesthetized with sodium hexabarbital (Evipan^R; Bayer AG, Leverkusen, FRG). Graft beds were prepared on the dorsal thorax. After thorough disinfection of the skin surface with Hibitane^R (chlorohexidine gluconate 20% g/v, ICI-Farma, The Netherlands) a circular piece of skin of 15 to 20 mm in diameter was removed to the level of the panniculus carnosus by using very sharp, fine curved scissors and fine, toothed forceps. Care was taken not to injure this layer, because it is essential for vascularisation of the graft. After the human skin was fitted into the prepared graft site it was secured by stitching (Perma-Hand^R Seide, 5–0, Ethicon, FRG). Thereafter, the graft was dressed with Op-Site (T. J. Smith and Nephew Limited, England) and Microfoam (3M, USA), which was firmly applied around the entire thorax. Mice with grafts were housed individually until the bandage was removed 7 days later. Only one graft was done on each animal. Generally, these animals were used two months later for penetration and skin damage studies.

Immunofluorescent Staining

With the exception of some minor modifications the staining procedure was similar to the technique of Banks-Schlegel and Green [1].

Grafts to be examined histologically were removed along with the surrounding mouse skin and the underlying musculature. The tissue was frozen in liquid nitrogen and six micrometer-thick cryostat sections were prepared. One section was fixed in methanol/ether (1 : 1) and stained with haematoxylin, azophloxine and saffron alcoholic. Another non-fixed section was stained by immunofluorescence using rabbit antiserum against human epidermal and human myeloma cells. The frozen sections were thawed, dried, and washed three times in phosphate buffered saline (PBS, pH 7.2). Thereafter, 100 μl of a 1/5 dilution of the absorbed rabbit antiserum in PBS (pH 7.2) was added to the sections and these were incubated for 30 min at room temperature in a moistened Petri-dish. The sections were then washed three times for 15 min with PBS. Subsequently, the sections were treated with 100 μl of a 1/50 dilution of pig anti-rabbit FITC or with 100 μl of a 1/50 dilution of goat anti-rabbit TRITC (Nordic Immunological Laboratories, The Netherlands) and incubated for 30 min at toom temperature. Finally, three successive 15-min washes were carried out and the sections were covered with a glass cover slip using glycerol/PBS (pH 7.5). Photographs were taken with a Zeiss fluorescence microscope using a Kodak Ektochrome^R film (400 ASA).

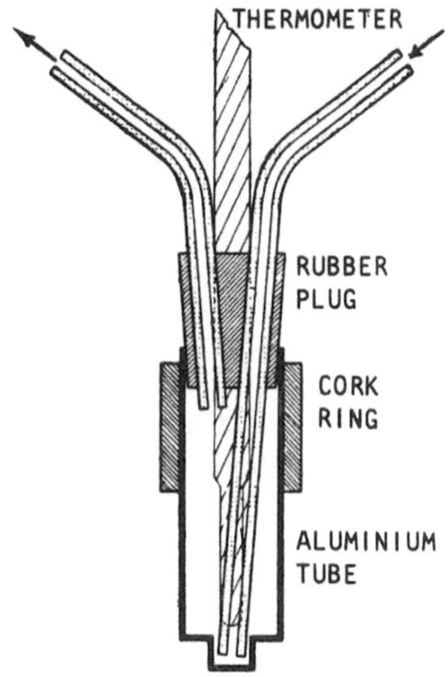

Fig. 1. Apparatus to elicit heat blisters. The aluminium tube is heated by a water flow of the desired temperature (according to Burbach [4])

Blister Formation

Sulphur Mustard

Minute drops of undiluted sulphur mustard (dichlorodiethyl sulphide) were weighed on aluminium disks (1 cm^2) with a Cahn Electrobalance. The disks were applied upside down to the grafted human skin and fixed with adhesive tape. In this way the disk prevented evaporation of the agent. The dose range was from 3 μg to 40 μg. After one hour the disk was removed and the skin was decontaminated with Fuller's earth (Leyland and Birmingham Rubber Company Ltd, England). The grafts were inspected repeatedly and 24 h after decontamination, lesions were photographed in a standardized way and the animals sacrificed. The grafts were removed to be examined histologically; the tissue was fixed in 4% formaldehyde, dehydrated, and embedded in paraffin. Sections were stained with hematoxylin and eosin. Care was taken to embed the grafts in such a way that sections could be cut perpendicular to the skin surface.

Heating

Human skin grafts and the skin of the recipients were heated with the apparatus described in Fig. 1. This apparatus was held firmly on the skin for 15, 20, 25 or 30 s, the temperature of the water flow through the aluminium tube was 60 °C. Macroscopic and histologic examination was the same as described above.

Percutaneous Absorption of Soman

As the sulphur mustard, soman (0-pinacolyl-methylphosphonylfluoridate) was synthesized by Dr. H. P. Benschop and C. de Borst of the Chemical Research Department of the Prins Maurits Laboratory TNO. The chemical purity was 97.5–98.5%. Immediately after arrival a stock solution of the nerve gas in a concentration of 0.1 M in anhydrous and peroxide free isopropanol was prepared, which was stored at –20 °C. Under these conditions the soman can be kept for several months with only a few per cents decomposition.

Soman was applied to human skin grafts, mouse skin grafts (autografts) and to the skin of non-treated mice in a dose of 100 $\mu g \cdot kg^{-1} \cdot cm^{-2}$. The desired concentration was prepared shortly before applying the solution to the skin. Immediately after application, the soman solution was covered by an aluminium disk of 1 cm^2 to prevent evaporation. The aluminium disk was kept in place by adhesive tape. At several times after contamination blood samples of 1 µl were taken from the tail vessels of the mice. The irreversible inhibition of the enzyme acetylcholinesterase was chosen as a measure of the rate of the percutaneous absorption of soman.

Radiometric Assay for Cholinesterase

In principle, the assay was based on the technique of Johnson and Russell [11], but included a few modifications.

Labelled Acetylcholine Substrate

[^{3}H]acetylcholine chloride (batch TRA.277, 591 mCi/mmol) was supplied by Amersham International, England. One mCi of the labelled substrate was dissolved in 5 ml H_2O and stored frozen at –20 °C. Samples contained 2–5% [^{3}H]acetate. This background was reduced to 0.2–0.3% by the following method: 10 µl glacial acetic acid was added to 0.2 ml of the stock solution of [^{3}H]acetylcholine which had been diluted with 0.8 ml H_2O (pH 4). [^{3}H]acetate was repeatedly extracted with 5 ml toluene/isoamyl alcohol (10:1). To remove isoamyl alcohol from the aqueous phase repeated extraction with 5 ml ether was carried out and finally residual ether was removed by evaporation in a nitrogen stream. The [^{3}H]acetylcholine chloride solution was diluted with a freshly prepared solution of unlabelled 0-acetylcholine perchlorate (10^{-2} M, BDH Chemicals Ltd, England). Substrate solutions may be stored for as long as 3 days at 4 °C.

Standard Assay Procedure

One microliter of whole blood was added to 0.5 ml of a solution, containing 50 mM sodium phosphate buffer (pH 7), 50 mM sodium chloride and 2 mM magnesium sulphate. One per cent Triton X-100 was added to accomplish complete haemolysis. Thereafter, 5 µl of the substrate solution was added and incubation was carried out at 25 °C for 10 min. The reaction vessels were miniature polyethylene vials (Packard-Becker B.V., The Netherlands) with plastic screw caps, suitable for insertion into standard widemouth scintillation vials. The reaction was stopped by adding 0.1 ml 10 M acetic acid, immediately followed by 4 ml of the scintillation mixture (10% isoamyl alcohol added to a standard toluene-based scintillation fluid of 0.4% PPO, 0.01% POPOP in toluene). The vials were capped, shaken (extraction of labelled acetate into the organic phase), and inserted into widemouth scintillation vials for counting in a scintillation counter.

Results and Discussion

Graft Survival

At this moment a total of 250 mice have received skin grafts. Initial failures were due to a combination of technical shortcomings, i.e., the use of an excessively thick graft with subsequent inadequate perfusion, the use of a cyanoacrylate cement (Histoacryl^{R} blue, B. Braun Melsungen AG, FRG) to secure the grafts, the wrong dressing and keeping of the mice without precautions to decrease the incidence of infections to which athymic (nude) mice are very susceptible. Currently the success rate is approximately 90%. In accordance with the study of Rygaard [16, 17] the acceptance of the human skin seemed to be independent of the genetic background of the nude mice. From a practical point of view C57B1 mice were chosen to carry out dermatotoxicity studies. Contraction of the graft bed was almost completely prevented by using Op-site. A result of these grafting experiments is shown in Fig. 2.

Histology

On gross inspection, the grafted skin had the appearance of normal human skin. A number of grafts (n = 10) were left *in situ* for 6 months and were still in perfect shape. This was satisfactory in view of the plans to use the mice with skin grafts 6–8 weeks after transplantation. Microscopically, the pre- and post-grafted human skin showed no differences. The grafted human skin maintained its characteristic histological pattern. A clear demarcation between mouse and human tissue could always be

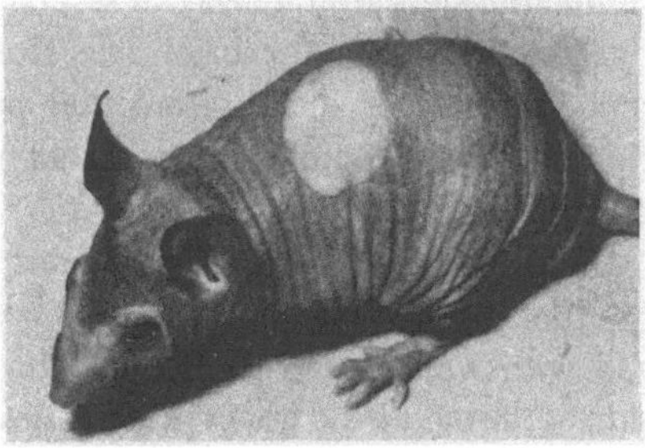

Fig. 2. Nude mouse of C57B1 background with accepted human skin graft 60 days after transplantation

Fig. 3. Histologic section of human mammary skin on a nude mouse 70 days after transplantation. Arrow points to distinct junction between human skin graft (right) and surrounding mouse skin (left) (x25)

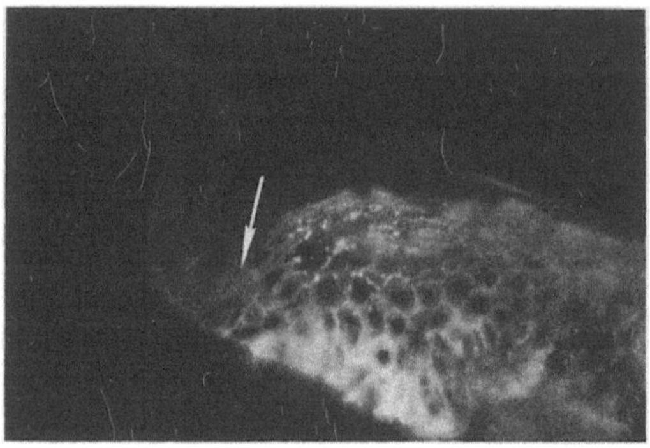

Fig. 4. Section of human mammary skin and mouse skin stained with antiserum to human epidermal cells and human myeloma cells. The arrow points to the junction between human skin and mouse skin (x16)

seen at the graft junction (Fig. 3). This distinction could also be shown by immunofluorescent staining using rabbit antiserum against human epidermal and human myeloma cells. The antiserum stained the Malpighian layer and the granular layer, but not the stratum corneum of the human epidermis (Fig. 4). The antiserum did not stain any part of the mouse epidermis.

Blister Formation

Although suction blisters can be produced on the clipped rat skin [14], it is not possible to elicit blisters on animal skin by sulphur mustard [7]. In human skin sulphur mustard produces lesions varying from mild erythema to vesicles, or large scale necrosis [18]. Blister formation usually occurs when sulphur mustard in a dose as low as 0.5–0.7 μg is applied to the ventral side of the human forearm. These small amounts did not cause a visible blister in a human skin graft. Histological examination of the grafted skin revealed that an increase of the dose to 40 μg resulted in the formation of microblisters after a delay of several hours (Fig. 5). These microblisters were formed by a separation of the basal cells from the basal lamina. Disappearance of the basophilic staining of the keratinocytes pointed at alterations in the cell nucleus. These findings resemble previously reported observations in man (Papirmeister 1984, personal communication).

It was also possible to elicit blisters on the human skin grafts by heating. This could be done by holding the apparatus described in Fig. 1 firmly onto the skin for 20 s and with water of 60 °C flowing through the tube. Parallel heating of the adjacent skin of the mouse did not induce a vesicle; the only result was a severe necrosis of the skin.

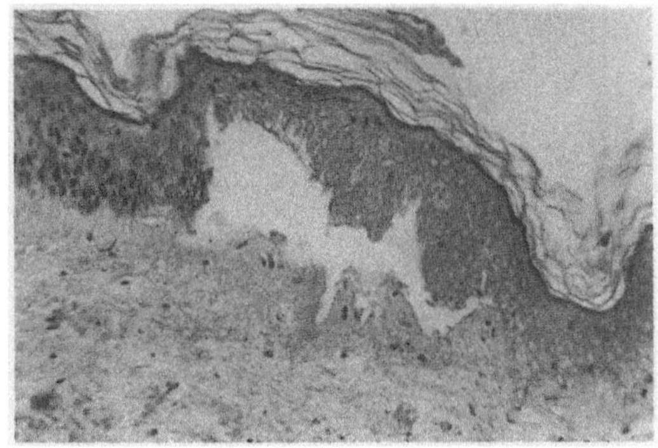

Fig. 5. Histological section of grafted human skin 25 h after contamination with 40 μg sulphur mustard

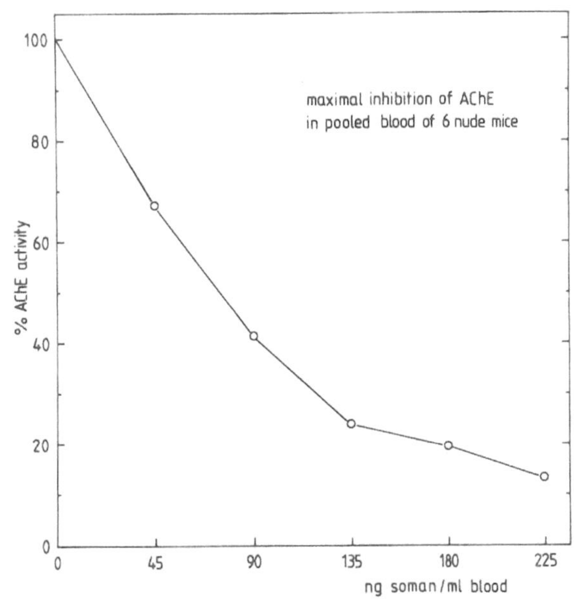

Fig. 6. Inactivation of acetylcholinesterase in whole blood *in vitro*. Soman was added to pooled blood of 6 mice. Data are maximal inhibitions at several dose levels of soman

A great advantage of using nude mice with skin grafts is the lack of innervation of the grafts, so that invasive manipulations can be carried out painlessly.

Penetration of Soman

As shown above certain human characteristics are maintained after grafting human skin. These findings were encouraging enough to start an investigation on the penetration characteristics of the organophosphorous cholinesterase inhibitor soman. In this study the irreversible inhibition of the enzyme acetylcholinesterase in blood was chosen as a parameter for the rate of percutaneous absorption. This approach was used successfully in a study on the penetration of the nerve gas sarin through the skin of the rabbit [9]. These authors used a colorimetric method. The radiometric

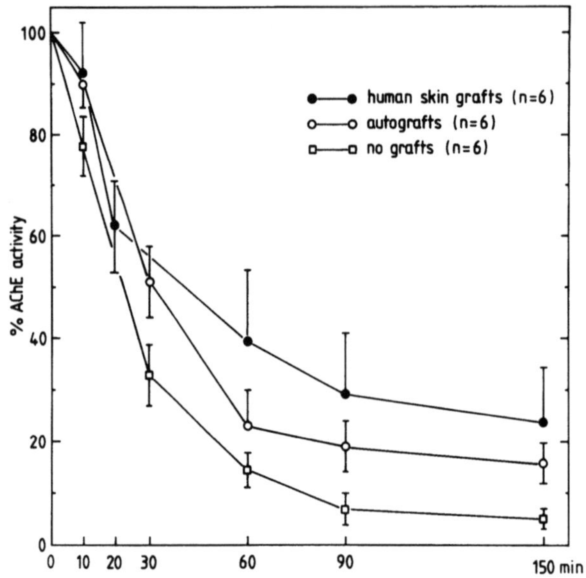

Fig. 7. Inactivation of acetylcholinesterase in whole blood by percutaneous absorption of 100 $\mu g \cdot kg^{-1} \cdot cm^{-2}$ versus time

assay for cholinesterase developed by Johnson and Russell [11] makes it possible to follow acetylcholinesterase-inhibition within one mouse in the course of time after application of a given quantity of soman to the skin. This method can be used because *in vitro* the percentage inactivation of acetylcholinesterase in whole blood appears to be quantitatively related to the amount of soman added (Fig. 6).

Figure 7 shows the drop in blood acetylcholinesterase activity at various times following the application of soman to the skin of animals without grafts, to grafted human mammary skin and to grafted mouse skin (autografts). The autografts served to examine the influence of grafting on the percutaneous absorption. These preliminary results suggest small differences in the penetration of soman through the grafted human skin and the mouse skin. The results indicate also that post-grafting alterations in the vascularisation probably do not influence the percutaneous absorption of soman.

Acknowledgements

The authors are grateful to Dr. A. N. Posma and Dr. R. J. Zeeman and their co-operators of the St. Elizabeth Hospital at Leiderdorp for their co-operative spirit. Many thanks are due to Dr. O. Brocades Zaalberg of the Medical Biological Laboratory TNO for preparing the antiserum. Conny Dekker, Sophie Gravenstein and Dory van de Meent are gratefully acknowledged for their diligent technical assistance.

References

1. Banks-Schlegel S, Green H (1980) Formation of epidermis by serially cultivated human epidermal cells transplanted as an epithelium to athymic mice. Transplantation 29:308–313
2. Bartek MJ, LaBudde JA, Maibach HI (1972) Skin permeability in vivo: comparison in rat, rabbit, pig and man. J Invest Dermatol 58:114–123
3. Briggaman RA, Wheeler CE (1976) Lamellar ichthyosis: long-term graft studies on congenitally athymic nude mice. J Invest Dermatol 67:567–572
4. Burbach JPE (1957) Bijdrage tot de kennis van blaarvorming in de huid. Thesis, University of Amsterdam, Amsterdam
5. Duprez A (1979) Etude macroscopique et microscopique de la reprise des greffes de peaux de cadavres humains chez la souris athymique nu/nu. C R Hebd Seances Acad Sci Ser D Sci Nat 288:1505–1508
6. Duprez A (1980) Greffes chez la souris athymique nu/nu de peaux humaines porteuses de naevi ou de naevocarcinomes superficiels: étude macroscopique et microscopique. C R Hebd Seances Acad Sci Ser D Sci Nat 290:461–464
7. Elskamp DMW, Verzantvoort CAM, Cohen EM (1973) Skin penetration and decontamination in man and animals. MBL-report 1973-9
8. Fraki JE, Briggaman RA, Lazarus GS (1983) Transplantation of psoriatic skin onto nude mice. J Invest Dermatol 80:31s–35s
9. Griesemer LD, Blank IH, Gould E (1958) The penetration of an anticholinesterase agent (sarin) into skin. III. A method for studying the rate of penetration into the skin of the living rabbit. J Invest Dermatol 31:255–258
10. Hunziker N, Feldmann RJ, Maibach HI (1978) Animal models of percutaneous penetration: Comparison in Mexican hairless dogs and man. Dermatological 156:79–88
11. Johnson CD, Russel RL (1975) A rapid, simple radiometric assay for cholinesterase, suitable for multiple determinations. Anal Biochem 64:229–238
12. Krueger GG, Manning DD, Malouf J, Ogden B (1975) Long-term maintenance of psoriatic human skin on congenitally athymic (nude) mice. J Invest Dermatol 64:307–312
13. Manning DD, Reed ND, Schaffer CF (1973) Maintenance of skin xenografts of widely divergent phylogenetic origin on congenitally athymic (nude) mice. J Exp Med 138:488–494
14. Middleton MC (1980) Evaluation of cellular injury in skin utilizing enzyme activities in suction blister fluid. J Invest Dermatol 74:219–223
15. Reed ND, Manning DD (1973) Long-term maintenance of normal human skin on congenitally athymic (nude) mice. Proc Soc Exp Biol Med 143:350–353
16. Rygaard J (1974) Skin grafts in nude mice. 3. Fate of grafts from man and donors of other taxonomic classes. Acta Path Microbiol Scand Section A 82:105–112
17. Rygaard J (1974) Skin grafts in nuce mice. 1. Allografts in nude mice of three genetic backgrounds (Balb/c, C3H, C57/Bl). Acta Path Microbiol Scand Section A 82:80–92
18. Weger N (1980) Alkylantien Vergiftungen. In: Rebentisch E (ed) Wehrmedizin. Urban und Schwarzenberg, München, pp 261–266
19. Wester RC, Maibach HI (1983) In vivo percutaneous absorption. In: Marzulli FN, Maibach HI (eds) Dermatotoxicology. Hemisphere Publishing Corporation, Washington New York London, pp 131–146
20. Yuspa SH (1979) Maintenance of human skin on nude mice for studies of chemical carcinogenesis. Cancer Lett 6:301–310

Human Cutaneous Pharmacokinetics In Vivo

H. Schaefer and W. Schalla

Usually, pharmacokinetics of topical application are performed on normal skin. We will orient skin pharmacokinetics to *diseased human* skin, its layers and appendages. Permeation through the skin is – in contrast to pharmacokinetics – a key phenomenon in *systemic* toxicity of substances; permeation studies are mostly concerned with the horny layer of skin having an intact stratum corneum.

Methods

For the purposes of skin pharmacology, only those methods are relevant to pharmacokinetics which allow the measurements of concentration relative to time in the skin itself after topical application as concentration gradients are formed. Such profiles are important for the understanding of pharmacological effects. They cannot be measured by indirect methods like urinary excretion or blood levels, these indirect parameters being only of interest for systemic toxicity as they estimate composite kinetics of skin and systemic distribution.

In vitro methods are relevant to kinetics within the horny layer. If the penetration time under consideration is relatively short (up to 5 h) this also applies to the epidermis.

Due to the absence of a drainage system of skin in vitro, dermal concentration levels found in vitro are always too high relative to in vivo conditions: nevertheless, in vitro the dermis behaves usually like a sink.

Since the morphology and composition of human skin is distinctly different from other species (pig skin structure being near to that of human skin) animal experiments are a priori of little relevance. The necessity to establish models of diseased (eczematous) skin laboratory animals, however, represents an important exception. Apart from this investigations in human skin in vitro are more useful than those in animal skin in vivo. Of great interest in this respect is the study of penetration on human skin, which has been transplanted to athymic mice [4].

In general, the amount of a substance penetrating into the skin is so small that radiolabelled drugs have to be used. The resulting concentrations are nevertheless quite high because the skin is a very thin tissue.

Recent investigations revealed a close correlation between the barrier function of the horny layer, its reservoir function and the resulting penetration into the skin [2, 5]. Penetration into human skin in vivo can thus be predicted from drug quantification in horny layer strippings. This allows sensitive – non radioactive – methods like high performance liquid chromatography to be applied. For metabolism studies radiolabelling is often indispensable.

Many investigators have tried to use quantitative assays of pharmacologic or microbiologic reactions of drugs such as vasoconstriction or the growth inhibition of microorganisms as indirect measurements of human skin pharmacokinetics in vivo. It has to be kept in mind, however, that pharmacodynamic processes imply at least two pharmacokinetic factors, i.e., drug transport and drug metabolism, plus the mode of action and its corresponding dose-response-effects. Unless the kinetics of metabolism and the action profile have been quantified, conclusions on the pharmacokinetics are invalid. It has to be clearly stated that this kind of investigation, although often very useful for therapeutic range finding, do not allow conclusions as to drug concentrations in the skin.

Most investigators only bear in mind topical drug application when discussing skin pharmacokinetics. Even more challenging is the problem of transport to the skin after systemic administration of drugs. Investigations in this respect are rare, though they are urgently needed for a better understanding of skin pharmacology.

Nevertheless it has to be remembered that topical medication has obvious advantages over systemic treatment in that smaller amounts of drug are more effective since they are applied directly to the target organ and far higher target concentrations are established.

In addition, the diseased skin is in most of the common dermatoses much more accessible than the surrounding uninvolved skin because of disturbed horny layer function.

In Vitro/in Vivo Investigations

Penetration studies on hydrocortisone [11] (Fig. 1) reveal similar concentrations in vitro and in vivo in the horny layer and in the upper layers of the epidermis. In the dermis, however, the steroid accumulates in vitro, obviously because of the lack of vessel function. Therefore, in vitro studies can be useful the target is situated in the horny layer or in the epidermis.

Animal Models

Approximately 99% of the barrier against penetration is in the horny layer (Table 1). Barrier function depends – especially in its selectivity for the commonly used drugs – on the composition of the horny layer lipids. There are distinct differences between human stratum corneum lipids and those of rodents, i.e., the most frequently studied species. Although there is some parallelism between results on human and on animal skin for many substances [1], there are sufficient differences to make prediction hazardous.

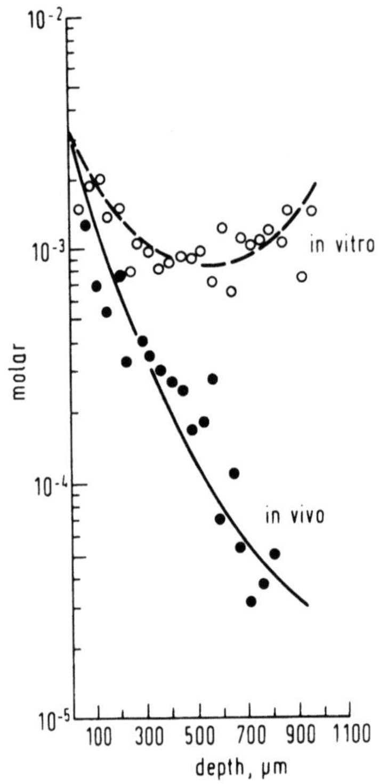

Fig. 1. Molar distribution of hydrocortisone in human skin in vitro ○ – – – ○ and in vivo ● – – – ● (1% hydrocortisone in polyethyleneglycol ointment; 1,000 min penetration period) (according to Zesch and Schaefer [11])

Table 1. Penetration of desoximetasone in vitro

(0.25% OINTMENT)

D_H = THICKNESS OF HORNY LAYER = 20 μM
D_E = THICKNESS OF EPIDERMIS = 160 μM
D_C = THICKNESS OF CORIUM = NUMBER OF SLICES X 40 μM

HORNY LAYER (μG/CM² X D_H)	EPIDERMIS (μG/CM² X D_E)	DERMIS (μG/CM² X D_C)
SKIN SPECIMEN A		
0.44	0.13	0.03
HORNY LAYER REMOVED	1.62	2.54
SKIN SPECIMEN B		
1.12	0.074	0.053
HORNY LAYER REMOVED	1.44	1.69

Radiolabelled/non Labelled Drugs

Experiments in hairless rats revealed a straight forward correlation between the amount of substance entering into the stratum corneum within 30 min and that being absorbed into the body after a single topical application [5]. These findings were confirmed recently for human skin [2] (Fig. 2, 3). In normal skin, the horny layer reservoir is

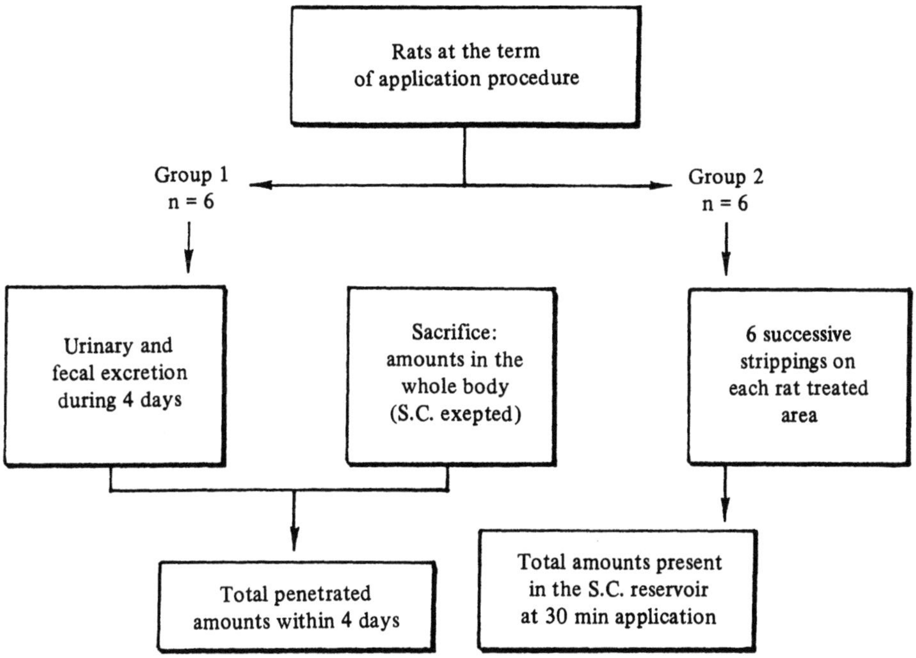

Fig. 2. Determination of the four days penetration and the stratum corneum reservoir on thirty minutes application, for each molecule (According to Rougier et al. [5])

the rate determining step for skin permeability. This allows the quantification of percutaneous absorption by measuring the concentration gradient in the horny layer only.

Similar correlations can be established for diseased skin in animals. Whether this also holds true for excematous human skin and whether epidermal – and dermal – concentrations can be calculated according to this method, has yet to be confirmed.

Vasoconstriction/Penetration

Comparing the curves for ointment concentration versus penetration for hydrocortisone to those for ointment concentration versus vasoconstriction, distinct differences are observed (Fig. 4) [3]: vasoconstriction levels off at about 1% hydrocortisone in the applied ointment whereas the penetration rates increase further with increasing drug concentrations, though not in a linear fashion. This underlines the limited use of such dynamic studies for skin pharmacokinetics as discussed above.

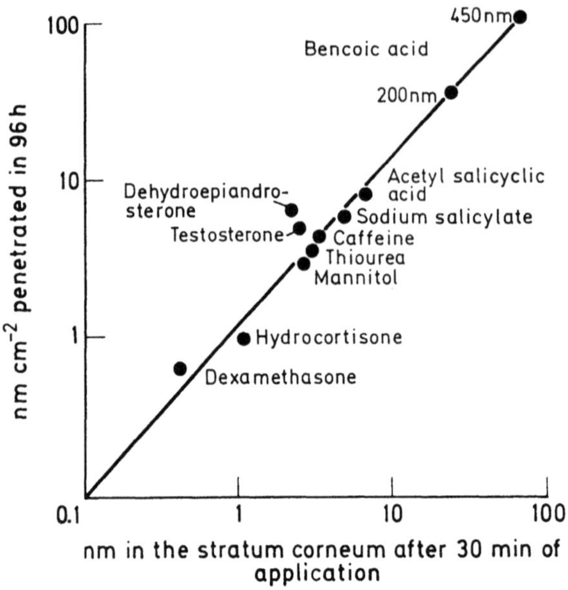

Fig. 3. Correlation between the penetration of the products after 96 h and their concentration in the stratum corneum after 30 min of application. According to Rougier et al. [5]

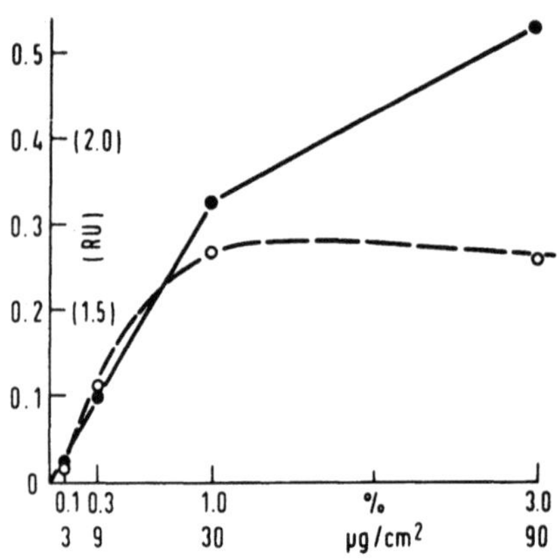

Fig. 4. Vasoconstriction (in relative units RU) after application of increasing concentrations of hydrocortisone in ointment in comparison to the penetration rate into human skin in vivo of radiolabelled hydrocortisone at the same concentrations and from the same ointment. (According to Krantz et al. [3])

Topical Versus Systemic Drug Administration

After topical drug application a typical distribution in the skin is observed with high concentrations in the epidermis and a steep non linear gradient towards the deeper dermal layers (Fig. 1, 5).

It has been literally impossible to follow quantitatively the drug distribution in human skin after systemic administration (100 cm of epidermis are equal to 100 mg, that is about 1/750.000 of the body weight). In animals the distribution pattern of cyproterone acetate an antiandrogen in the skin after systemic administration was compared to that after topical application. As expected an entirely different gradient was obtained, the epidermal concentrations being highest after topical but lowest

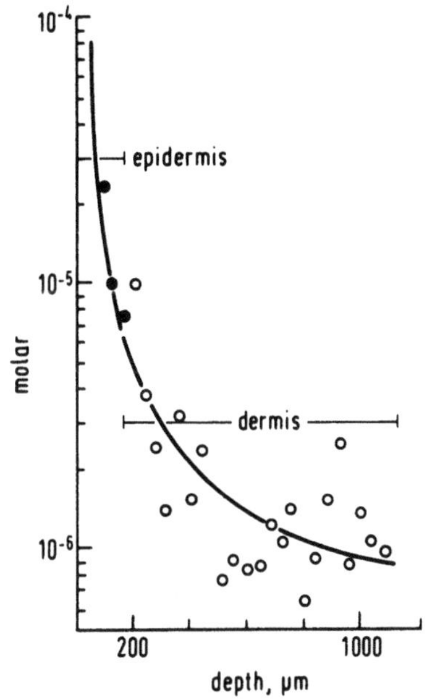

Fig. 5. Molar concentration of vitamin A acid 0.1% in isopropanol after 100 min in vivo. (According to Schaefer and Zesch [7])

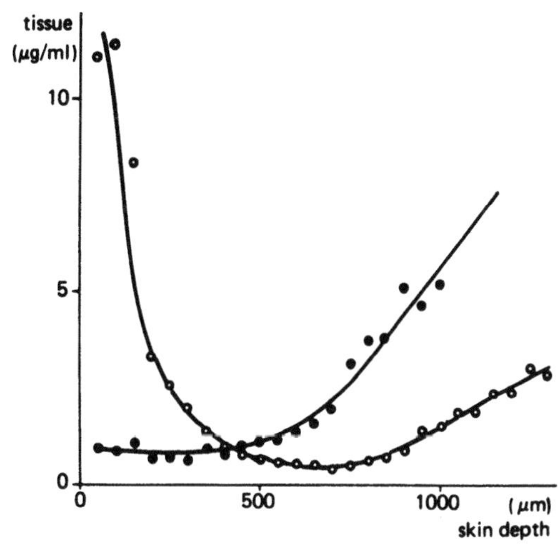

Fig. 6. Cyproterone acetate distribution in the skin of the rat after oral (•) and topical (○) administration of equal doses over 7 days. Oral dose: 7 x 0.4 mg/200 g rat; topical dose: 7 x 0.4 mg/16 cm² skin. According to Täuber [9]

after systemic application (Fig. 6). In general terms this is in accordance with therapeutic experience. Eczema is treated topically whereas dermatoses with a dermal origin (deep vasculitis) are accessible to systemic medication only.

Table 2. Penetration of desoximetasone into normal and diseased skin in vivo

(100-130 MIN PENETRATION PERIOD)

THICKNESS OF HORNY LAYER D_H = 20 μM
THICKNESS OF EPIDERMIS D_E = 160 μM
THICKNESS OF DERMIS D_C = 40 μM X NUMBER OF SLICES

SKIN	CONC. %	HORNY LAYER (μG/CM² X D_H)	EPIDERMIS (μG/CM² X D_E)	DERMIS (μG/CM² X D_C)
NORMAL	0.05	0.31	0.031	0.0094
PSORIATIC	0.05	0.24	0.145	0.12
NORMAL	0.25	0.55	0.035	0.0089
PSORIATIC	0.25	0.90	0.496	0.181

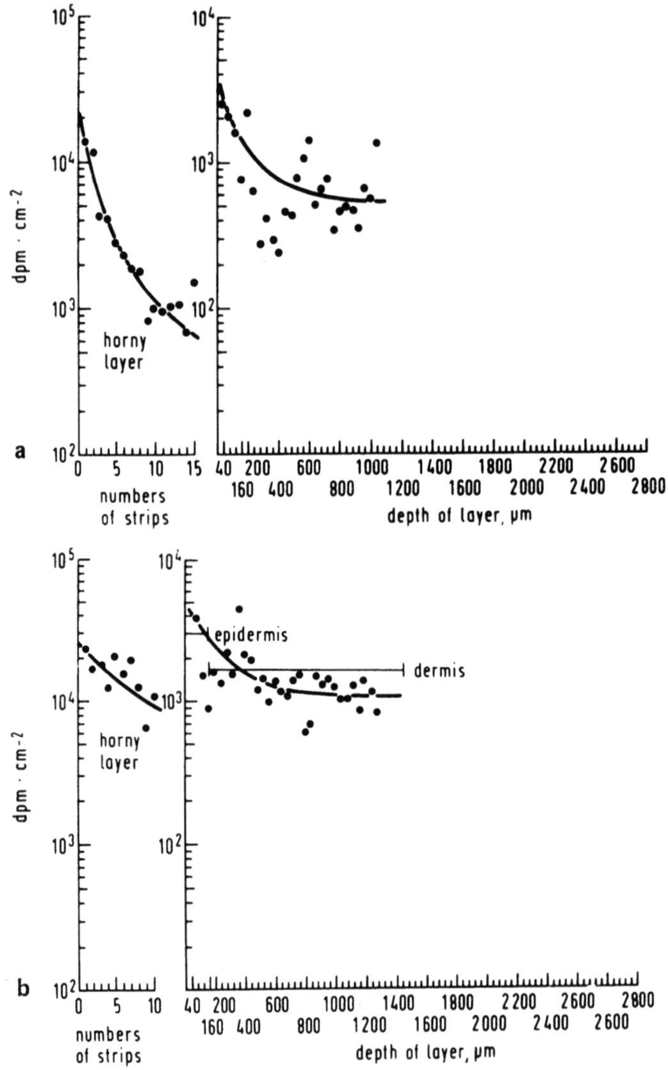

Fig. 7a, b. Distribution profile after topical application of radiolabelled triamcinolone acetonide to uninvolved **a** and to involved **b** skin of a psoriatic patient. According to Schaefer et al. [8]

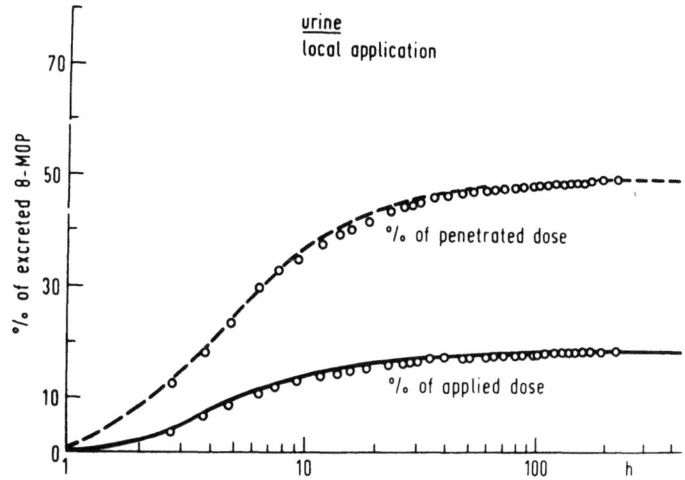

Fig. 8. Time course of urinary excretion after topical application of radiolabelled 8-methoxypsoralene to human skin. Note the slow excretion rate (According to Schaefer et al. [8])

Normal Versus Diseased Skin

Penetration studies are of special interest, whenever investigations can be performed on the involved versus the uninvolved skin in the same patient. In fact, it could be predicted from results like those shown in Table 1 that whenever the horny layer barrier is disturbed – penetration into the diseased area will be facilitated. That this is in fact the case, could be shown by us in several patients [8] (Table 2, Fig. 7a and b).

It should be emphasized that only direct studies on concentrations in the skin itself are of value, since urinary excretion after topical application is a very slow process (Fig. 8) compared to penetration into diseased skin (Table 2). Large differences in concentration in the skin will be minimized whenever the urinary excretion is followed for several days, e.g., allowing a steroid permeating through non-involved skin to catch up with the amount permeating through diseased skin [10].

Conclusions

In skin pharmacokinetics penetration into non-involved skin after topical application is only of interest from the safety point of view, i.e., in comparison to penetration into the involved area; it appears that this problem can be approached by a recently published method [2, 5].

Unless analytical methods employing non-labelled substances are developed which are as sensitive as those used in detection of radiolabelled substances, it seems technically impossible to study the pharmacokinetics of systemically administered drugs in human skin in vivo.

The permeability of diseased skin can be of great interest, whenever the kinetics can be followed properly. Short contact therapy [6] is only one example. One has to keep in mind that these studies are so laborious that they never allow statistical evaluation. In other words, they can only serve for range finding. This means that human cutaneous pharmacokinetics in vivo cannot yet be handled and considered as a true model. Investigations on human skin in vitro and by animal models must serve as substitutes.

References

1. Bartek MJ, Labudde JA, Maibach HI (1972) Skin permeability in vivo: comparison in rat, rabbit, pig and man. J Invest Dermatol 58:114–123
2. Dupuis D, Rougier A, Roguet R, Lotte C, Kalopissis G (1984) In vivo relationship between horny layer reservoir effect and percutaneous absorption in man and rat. J Invest Dermatol 82:353–356
3. Krantz G, Schaefer H, Zesch A (1977) Hydrocortisone (cortisol) concentration and penetration gradient. Acta Dermatol Venereol (Stockh) 57:269–273
4. Reifenrath WG, Chellquist EM, Shipwash EA, Jederberg WW, Krueger GG (1984) Percutaneous penetration in the hairless dog, weanling pig and grafted athymic nude mouse. Br J Dermatol III, Suppl 27:123–135
5. Rougier A, Dupuis D, Lotte C, Roguet R, Schaefer H (1983) In vivo correlation between stratum corneum reservoir function and percutaneous absorption. J Invest Dermatol 81:275–278
6. Schaefer H, Farber EM, Goldberg L, Schalla W (1980) Limited application period for dithranol in psoriasis. Br J Dermatol 102:571
7. Schaefer H, Zesch A (1975) Penetration of vitamin A acid into human skin. Acta Dermatol Venereol (Stockh) Suppl C 74:50–55
8. Schaefer H, Zesch A, Stüttgen G (1977) Penetration, permeation and absorption of triamcinolone acetonide in normal and psoriatic skin. Ach Dermatol Res 258:241–249
9. Täuber U (1982) Metabolism of drugs on and in the skin. In: Brandau R (ed) Dermal and transdermal absorption, pp 133–151
10. Wester RC, Bucks D, Maibach HI (1983) In vivo percutaneous absorption of hydrocortisone in psoriatic patients and normal volunteers. J Am Acad Dermatol 8:645–647
11. Zesch A, Schaefer H (1975) Penetration of radioactive hydrocortisone in human skin from various ointment bases II. In vivo experiments. Arch Dermatol Forsch 252:245–256

The Vasoconstrictor Test as a Model for Developing Topical Formulations

B. W. Barry and R. Woodford

Topical steroid bioassays are the most refined, sophisticated bioassays used to test and develop skin preparations [13]. In particular, the blanching or vasoconstrictor assay may be employed to screen new steroids for clinical efficacy and also to determine the bioavailability of steroids from topical formulations; thus, we can use the test not only to evaluate the intrinsic activity of a steroid for correlation with possible clinical anti-inflammatory action, but also to investigate fundamental bipharmaceutical relationships. In the present context, by bioavailability we mean the relative absorption efficiency for a drug as determined by the release of the steroid from its formulation and its subsequent penetration through the stratum corneum and viable epidermis into the dermis to produce the characteristic blanching effect.

In this paper, we will concentrate only on some examples of our own work in which we have used our modification of the McKenzie-Stoughton assay to screen novel steroids, to help to develop and test commercial formulations, to investigate fundamental aspects of percutaneous absorption and to develop dosage regimes for topical steroids (Fig. 1). Recent reviews which consider such work and other investigators' programmes include those of Barry [3] and Schaefer et al. [16].

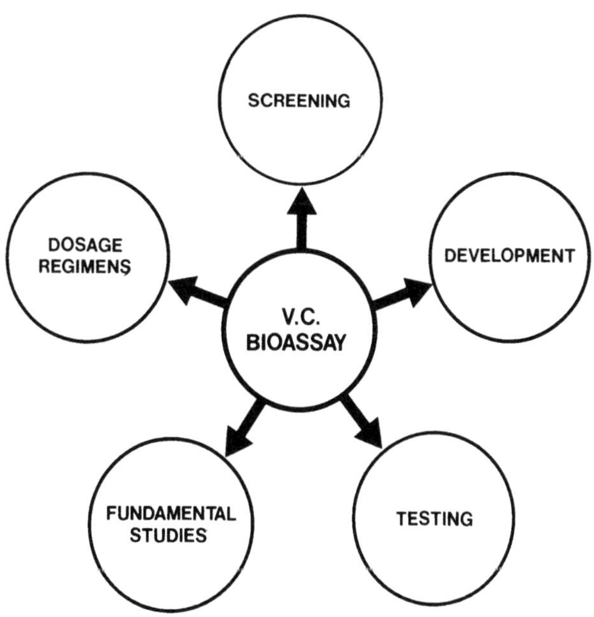

Fig. 1. Possible uses for the vasoconstrictor test

Bioassay Design

For our standard assay, we select volunteers who have not received topical application of corticosteroids for at least two months prior to the investigation. We apply 5 ± 1 mg (or μl) of each formulation to the washed flexor surface of both forearms using small areas (7 x 7 mm) punched out from Blenderm polyethylene tape, according to a randomisation plan. For our occluded assay, the sites are covered with Melinex polyester film for 6 h. In the nonoccluded design, we remove the Blenderm tape after applying the preparations and use a perforated plastic screen instead of the Melinex film. This method removes the occlusive, hydrating effect of the water-impervious film. After the application period, we remove tapes, films and screens, wash the arms with soap and water at skin temperature, gently dry, and estimate the degree of pallor 10 min later (to allow transient erythema to subside) using a 0–4 scale with half-point ratings. Further readings of pallor are taken for up to 5 days after application, in a double blind manner. We construct a full blanching profile from the average results of 10 volunteers (20 replicates for each preparation). To assess the retention of steroid in the stratum corneum (the so-called steroid reservoir effect) we reocclude sites with Melinex film for 6 or 12 h at up to 14 days after the commencement of an experiment; the blanching is then estimated by single or multiple readings.

Provided that experienced investigators adhere to a strict protocol using a controlled panel of volunteers, we conclude that this bioassay is sensitive, accurate and reproducible. The dose-response plots for the vasoconstrictor test follow well defined pharmacological relationships when expressed on Cartesian, semilog, log-probability or log-logistic coordinates. The function of choice for presenting such blanching data is log-logistic because such plots predict a zero response at zero concentration, fully characterise a sigmoid curve over its entire range, are continuous over the entire concentration range and predict a maximum response as the concentration increases indefinitely [10].

We can now look at some examples of how we have used the blanching assay in several areas of biopharmaceutics (Fig. 1).

Screening of Novel Synthetic Steroids

When medicinal chemists synthesise new topical steroids, their intrinsic vasoconstrictor activities, free from complex vehicle effects, may be determined by conducting the bioassay with the molecules dissolved in a simple solvent such as ethanol. This was in fact the way in which blanching test was introduced into dermatology [15, 17]. To illustrate general principles, we can look at some results from a programme which assessed 14 new steroid derivatives, synthesised on the basis of predicted structure-activity relationships, together with a standard steroid, betamethasone 17-valerate [4]. Pure steroids were dissolved in ethanol and applied to volunteers' forearms in the standard manner and blanching profiles were constructed by plotting the percent total possible score as a function of time after application. A typical set of curves is illu-

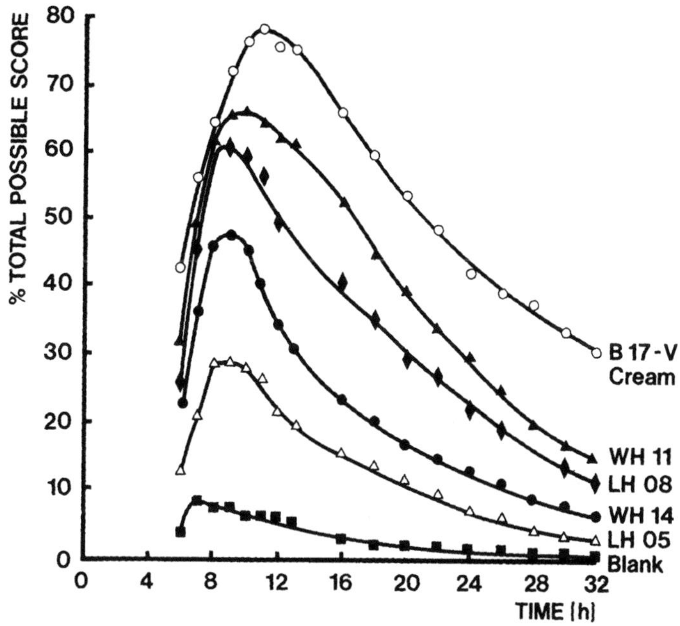

Fig. 2. Vasoconstrictor profiles obtained after occluded application of betamethasone 17-valerate (B17-V Cream), ethanolic solutions of test steroids (coded designation) and ethanol blank (after Barry and Brace [4])

strated in Fig. 2. We see the characteristic shape of blanching curves, with a delay of 6 h until the first reading time, an increase to a maximum in response and then a steady decline.

Examination of the size and shape of the blanching curve can provide information with regard to the activity and rate of penetration through the skin of individual steroids. We can also measure the area under the blanching profile and use this as a measure of bioavailability. Using these methods, it was possible to rank the experimental steroids in order of their vasoconstrictor activities.

Of course, selection of a steroid for clinical use will involve many considerations not dealt with in such a simple vasoconstrictor study. We would need to investigate the toxicology of candidate drugs (e.g. effect of overdosage, excessive effect at normal dosage, side-effects, hypersensitivity reactions and idiosyncracies, if any). An awareness of drug interactions, within or outside the body, and which may have diverse effects, may influence the final selection. Even though there is still no irrefutable evidence that anti-inflammatory efficacy and vasoconstriction are directly related (merely a series of strong analogies), the blanching test is a valuable method for rapid preclinical assessment of potential topically active steroids. After giving due consideration to the physical chemistry of the steroids and their onset of action, duration of action, and maximum intensities from alcoholic solution, we can select the correct vehicle for each drug. Each formulation must be stable on storage and the drug must be compatible with the ingredients of the base, so that a low working concentration of the drug can provide optimum bioavailability. These criteria imply that the steroid should possess maximum thermodynamic activity in the topical vehicle so that the medicament is readily released from the base into the skin. These arguments presuppose that maximum possible effects are required; with potent steroids, such optimisation procedures may increase the incidence of well-known side effects, such as skin-thinning. Thus, many factors affect the final choice of a steroid to be submitted to clinical trial.

We can now turn to work which has utilised the vasoconstrictor test to develop topical formulations.

Development of Topical Formulations

When a manufacturer develops preparations for a new steroid, it is uneconomical in time and expense, as well as being unethical, for the company to submit all possible formulation candidates to a full clinical trial. One of the most valuable features of the blanching bioassay is that it permits the pharmaceutical scientist to select only the promising formulations for progression to the clinic. For our example of the value of the vasoconstrictor test in developmental pharmaceutics, we can review some work on the bioavailability and activity of topical steroids presented in aerosol quick-break foams. As a dermatological dosage form, the quick-break aerosol foam offers the advantages of high activity (the steroid is in solution and concentrates by vehicle evaporation on the skin), ease of application, sterility, controlled dosage from a metering valve, economy in use, suitability for smooth or hairy skin and reduced possibility of inhaling potent steroid compared to aerosol sprays. A logical development involved devising suitable formulations and comparing them with alternative dosage forms [20].

The blanching profiles in Fig. 3 represent the results obtained for 0.025% betamethasone 17-benzoate applied as the foam concentrate, the collapsed foam or as

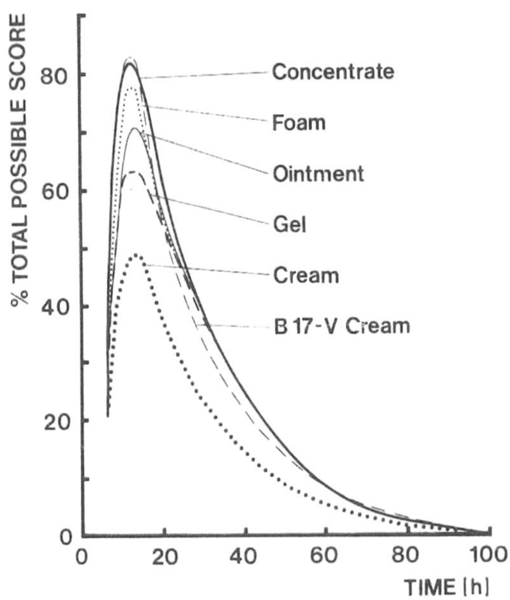

Fig. 3. Blanching responses to betamethasone benzoate formulations containing 0.025% steroid and to betamethasone 17-valerate (B17-V Cream) (after Woodford and Barry [20])

alternative formulations in traditional ointment, gel or cream preparations. As in all our work, we included 0.1% betamethasone 17-valerate cream (Betnovate cream, Glaxo) as a marker preparation to control the bioassay and to allow us to compare, for all our trials, one preparation with another. The superiority of the aerosal foam preparation is readily apparent, as is the inferiority of the cream. The aerosol quick-break foam proved to be a good delivery system for other steroids-clobetasol propionate, flumethasone pivalate, triamcinolone acetonide, desonide, hydrocortisone butyrate and betamethasone valerate. Maybe now that the concern has abated relating halogen-containing propellants to depletion of the ozone layer, manufacturers may show a renewed interest in this novel dosage form. The quick-break aerosol foam should be a suitable vehicle for a wide range of topical corticosteroids and may be a useful delivery system for other medicaments, offering the prospects of a sterile preparation which will remain so during use.

Testing of Marketed Products – Ranking for Clinical Efficacy

Over the last 12 years we have assessed the activities and bioavailabilities of some 100 proprietary formulations by means of our single-application vasoconstrictor test [e.g. 3, 6–8]. The work ranked the preparations in a classification which often demonstrated comparability between the conclusions of clinical trials for anti-inflammatory acitvity and the blanching results. Although we cannot as yet prove a correlation between clinical efficacy and vasoconstriction, in the absence of comprehensive comparative clinical investigations we can base the classification of these preparations on vasoconstrictor studies. In general, there is little evidence for *gross* lack of correlation between clinical anti-inflammatory activities and vasoconstrictor assays competently performed. However, to obtain satisfactory correlation, it is important, among other things, to perform a multipoint vasoconstrictor assay to determine the full blanching profile, as single point readings can sometimes lead to changes in ranking between preparations, depending on the time of assessment. For example, Fig. 4 illustrates

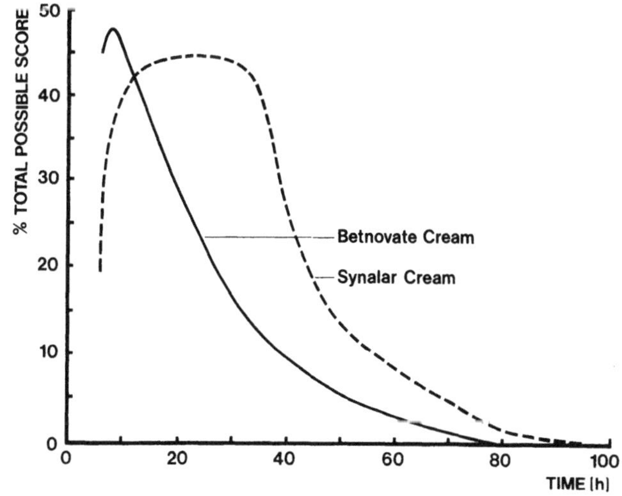

Fig. 4. Vasoconstrictor profiles for betamethasone 17-valerate and fluocinolone acetonide – nonoccluded application (after Barry [2])

blanching curves for betamethasone 17-valerate and fluocinolone acetonide as tested under non-occluded conditions. The betamethasone 17-valerate may be ranked weaker, equal to or stronger than the fluocinolone acetonide, depending on the time of reading [2, 6]. Only if we take the full profile into account may we reasonably compare steroid preparations for situations where the response-time relationships differ.

Fundamental Studies on Percutaneous Absorption

The blanching assay provides a means for testing various aspects of skin permeation theory, *in vivo*, and in humans, the recipients of topical therapy. Examples of our past and current investigations include work on the thermodynamic activity of steroids in topical formulations and the modification of transdermal therapy by penetration enhancers.

Thermodynamic Control

We tested six solutions of mechlorosone dibutyrate formulated to contain 0.2% steroid in different blends of polar solvents (hexylene glycol, propylene glycol, propylene carbonate, PEG 400, water). All systems were at 90% saturation and therefore ideally at the same thermodynamic activity [21]. The results of our occluded bioassay are illustrated in Table 1, which compares the activities of the different steroid solutions using our three standard parameters:

a) area under the curve values,
b) summed percent total possible score, and
c) the square root transformation of the sum of scores divided by the number of volunteers (Tm/10) – this parameter permits statistical ranking of the solutions.

Our results revealed that the solutions were statistically different ($p = 0.05$), probably because solvent components were not inert. Thus, individual solvents could have acted as skin irritants, dehydrating agents, penetration enhancers, accelerants or sorption promoters, or they could have complexed with the steroid and so reduced its thermodynamic activity.

A simpler picture should develop if we were to compare only similar solvent systems – that is, formulations which use the same solvents (but in slightly different ratios) to maintain constant the thermodynamic activity of the steroid while the overall concentration of the drug may differ. Table 1 illustrates the results for gels of type D, E, F in which for each type the steroid concentration was either 0.1% or 0.2% (both still at 90% saturation). The data show that, within any solvent system, there was no significant difference between gels containing different concentrations of steroid. This result agrees with simple thermodynamic predictions on the basis that complicating effects would be approximately the same within any one pair of formulations.

Table 1. Vasoconstrictor data for (*) 0.2% steroid solutions, arranged in rank order of area under the curve values and (**) 0.1% and 0.2% steroid gels (modified from Woodford and Barry 1982)

	Area under the curve (% x h)	Summed % total possible score	Tm/10 mean value[a]	Bioavailability[b]	
				(1)	(2)
Solutions					
E	2,650	413	5.72	1.00	1.00
A	2,140	345	5.23	0.81	0.84
B	2,080	339	5.18	0.79	0.82
D	1,910	305	4.92	0.72	0.74
F	1,780	265	4.59	0.67	0.64
C	1,720	282	4.73	0.65	0.68
Gels					
E 0.2%	2,430	412	5.68	1.00	1.00
E 0.1%	2,240	380	5.48	0.92	0.92
F 0.2%	2,220	331	5.10	1.00	1.00
F 0.1%	1,890	304	4.90	0.85	0.92
D 0.2%	2,020	336	5.14	1.00	1.00
D 0.1%	1,590	281	4.71	0.79	0.84

* If Tm/10 values of two preparations differ by more than the minimum significant range value k, there is a significant difference between the preparations ($p = 0.05$). For solutions, $k = 0.28$; for gels, $k = 0.58$

** Defined by the relationships

(1) $\frac{\text{"area under the curve" for preparation}}{\text{"area under the curve" for most active preparation}}$

(2) $\frac{\text{summed \% total possible score for preparation}}{\text{summed \% total possible score for most active preparation}}$

Penetration Enhancers

For our second example in this section, we can look at an aspect of our work in which we examined the potential for increasing the bioavailability of steroids by using penetration enhancers. In our first trial we used the standard occluded vasoconstrictor assay to assess the bioavailability of 0.1% betamethasone 17-benzoate presented in several potential enhancers – 2-pyrrolidone (2-P), N-methyl-2-pyrrolidone (NMP), 1-ethyl-2-pyrrolidone (EP), dimethylformamide (DMF), diethyl-m-toluamide (DEET), 75% DEET in ethanol (DEET 75%) and propylene glycol (PG). Acetone (AC) represented a volatile solvent and dimethylisosorbide (DMI) served as a baseline vehicle for comparison purposes. Our first trial looked simply for obvious effects, and so we did not adjust the test solutions to the same thermodynamic activity, neither did we allow for different stratum corneum – solvent partition coefficients [5].

The top of Fig. 5 illustrates histograms of the bioavailabilities, calculated as the area under the blanching curve for the steroid in the test solution divided by the area for

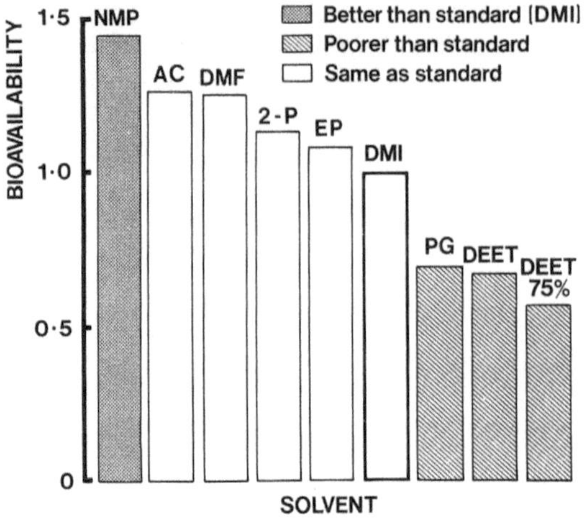

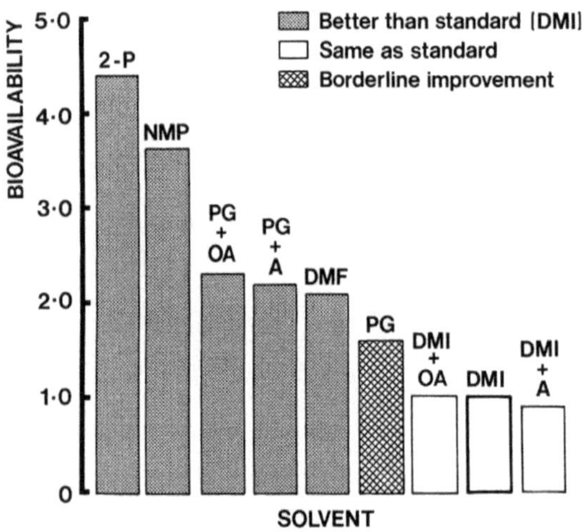

Fig. 5. Histograms of the biovailabilities of steroid solutions; penetration enhancer experiments.Top diagram – uncontrolled thermodynamic activity, occluded test; bottom diagram – controlled thermodynamic activity, nonoccluded test. Key: NMP, N-methyl-2-pyrrolidone; AC, acetone; DMF, dimethylformamide; 2-P, 2-pyrrolidone; EP, N-ethyl-2-pyrrolidone; DMI, dimethylisosorbide; PG, propylene glycol; DEET, diethyl-m-toluamide; DEET 75%, diethyl-m-toluamide in ethanol; OA, oleic acid; A, azone

the steroid in DMI. Only NMP increased steroid bioavailability; acetone, DMF, 2-P and EP were equal to DMI while PG, DEET and DEET 75% were poorer.

This experiment had two main deficiencies – the lack of thermodynamic control and the occluded test procedure; because water is such a good penetration enhancer, it was possible that occlusive hydration obscured the enhancing effects of the test solvents. In our next series of experiments we therefore maintained the initial activity of the steroid approximately constant at 10% saturation (ignoring the effect of differences in the activity coefficients of the steroid in the different solvents on dilution) and we tested the formulations using a nonoccluded vasoconstrictor assay. This avoided the swamping effect of a fully hydrated stratum corneum (Bennett et al. in preparation). We used aqueous solutions of 2-P, NMP, DMF, PG and DMI. Some systems also included 2% azone (A), 1,5% oleic acid (OA) or 5% OA – these concentrations provided saturated solutions of azone or oleic acid and thus maximum thermodynamic activities for these enhancers.

We see from Fig. 5 (bottom) that the bioavailabilities were often increased compared with the DMI standard and that 2-P, NMP, PG+OA, PG+A and DMF were significantly better than the standard. It is interesting to note that oleic acid and azone only showed enhancing activity when combined with PG and not when dissolved in DMI. It may be that azone needs to be combined with a specific solvent such as PG before it will exhibit its enhancer activity; this is so for materials such as oleic acid (Cooper, personal communication).

Development of Dosage Regiments for Topical Steroids

Up until now, we have discussed only single application tests for topical steroid preparations and the resulting profiles. However, an interesting feature of the blanching response is the development of tachyphylaxis to multiple application of topical steroids [1, 11, 12]. Barry and Woodford [9] applied preparations of fluocinonide, fluocinolone acetonide, betamethasone 17-valerate, hydrocortisone 17-butyrate, and hydrocortisone according to a multiple dosage regimen. The blanching response first increased and then waned with continued application over a 5 day period. After 2 days during which no steroid was applied, the application sites recovered considerably, but tachyphylaxis again set in with subsequent repeated applications. Even the weak hydrocortisone preparations elicited acute tolerance similar in nature to that caused by the more potent steroids. Woodford et al. [23] developed this approach further in work which assessed the bioavailabilities and activities of three amcinonide preparations and betamethasone 17-valerate using three multiple dosage regimen blanching assays. This work assessed the clinical implications of the tachyphylactic response to topical steroids and suggested that the most advantageous dosage regimen is a once-daily application with no loading dose. We could achieve considerable savings in medication costs, improved patient compliance and a reduction in side-effects if less frequent applications were generally as effective as the currently recommended regimes of several applications a day [14, 18, 19]. However, it is possible that for the weakest steroid preparations (containing hydrocortisone) twice daily application would be desirable [22].

Conclusions

We can readily see that the vasoconstrictor bioassay is a valuable technique for assessing the combination of intrinsic activity at a receptor site and skin penetrability of a topical steroid, as modified by vehicle effects. It is an *in vivo* test in humans (the recipients of therapy) and it shows excellent correlation with *in vitro* tests and clinical response.

References

1. Altmeyer P, Zaun H (1976) Reflexphotometric determinations of vasoconstriction after topical application of steroids. V. Vasoconstriction phenomenon and tachyphylaxis after repeated steroid application. Arch Derm Res 255:51–56
2. Barry BW (1976) Bioavailability of topical steroids. Dermatologica 152 suppl 1:47–65
3. Barry BW (1983) Dermatological formulations. Percutaneous Absorption. Dekker, New York Basel
4. Barry BW, Brace AR (1975) Vasoconstrictor activities of some novel synthetic steroids in alcoholic solution. J Invest Dermatol 64:418–422
5. Barry BW, Southwell D, Woodford R (1984) Optimisation of bioavailability of topical steroids: penetration enhancers under occlusion. J Invest Dermatol 82:49–52
6. Barry BW, Woodford R (1974) Comparative bioavailability of proprietary topical corticosteroid preparations; vasoconstrictor assays on thirty creams and gels. Br J Dermatol 91:323–338
7. Barry BW, Woodford R (1975) Comparative bioavailability of proprietary topical corticosteroid preparations: vasoconstrictor assays on thirty-one ointments. Br J Dermatol 93:563–571
8. Barry BW, Woodford R (1976) Proprietary hydrocortiscone creams. Vasoconstrictor activities and bioavailabilities of six preparations. Br J Dermatol 95:423–425
9. Barry BW, Woodford R (1977) Vasoconstrictor activities and bioavailabilities of seven proprietary corticosteroid creams assessed using a non-occluded multiple dosage regimen; clinical implications. Br J Dermatol 97:555–560
10. Barry BW, Woodford R (1978) Activity and bioavailability of topical steroids. *In vivo/in vitro* correlations for the vasoconstrictor test. J Clin Pharm 3:43–65
11. DuVivier A, Stoughton RB (1975) Tachyphylaxis to the action of topically applied corticosteroids. Arch Dermatol 111:581–583
12. DuVivier A, Stoughton RB (1976) Acute tolerance to effects of topical glucocorticosteroids. Br J Dermatol 94 suppl 12:25–32
13. Haleblian JK (1976) Bioassays used in development of topical dosage forms. J Pharm Sci 65:1417–1436
14. Maibach HI, Stoughton RB (1973) Topical corticosteroids. Med Clin North Am 57:1253–1264
15. McKenzie AW, Stoughton RB (1962) Method for comparing percutaneous absorption of steroids. Arch Dermatol 86:608–610
16. Schaefer H, Zesch A, Stüttgen G (1982) Skin permeability. Springer, Berlin Heidelberg New York, pp 690–697
17. Schlagel CA (1972) Penetration and action of glucocorticoids. In: Montagna W, Van Scott EJ, Stoughton RB (eds) Advances in biology of skin, vol XII: Pharmacology and the skin. Appeton-Century-Crofts, New York, pp 339–356
18. Senter TP, Stimpson DH, Charles G (1983) Comparison of two therapeutic regimens using the same topical corticoid for stable psoriasis. Western J Med 139:657–662
19. Sudilovsky A, Muir JG, Bocobo FC (1981) A comparison of single and multiple application of halcinonide cream. Intern Soc Trop Dermatol 20:609–613
20. Woodford R, Barry BW (1977) Bioavailability and activity of topical coticosteroids from a novel drug delivery system, the aerosol quick-break foam. J Pharm Sci 66:99–103
21. Woodford R, Barry BW (1982) Optimisation of topical steroids: thermodynamic control. J Invest Dermatol 79:388–391
22. Woodford R, Barry BW (1984) Alphaderm cream (1% hydrocortisone plus 10% urea): investigation of vasoconstrictor activity, bioavailability and application regimens in human volunteers. Curr Ther Res 35:759–767
23. Woodford R, Haigh JM, Barry BW (1983) Possible dosage regimens for topical steroids, assessed by vasoconstrictor assays using multiple applications. Dermatologica 166:136–140

Skin Models in the Understanding of the Physiological and Pathological Properties of Collagen

Ch. M. Lapiere

Skin has contributed much to the understanding of the chemical structure, the processes involved in synthesis, polymerization and degradation as well as the physiological properties of collagen in connective tissues. Significant progress has resulted from the study of abnormal skin (both human and animal) in various heritable and acquired disorders. This review will illustrate the contribution made by skin models in our understanding of the biochemical pathways needed to build a functional connective tissue.

Skin contains the largest mass of collagen (~ 60%) of the body in all the animal kingdom and this protein represents 70% of its dry weight. Collagen is a family of proteins sharing a common basic structure in which glycine occupies every third position in most of the polypeptides. Skin contains 5 of the 6 major types of collagen, two in fibrillar form (type I in thick fibre and type III in thin fibres) and three associated with the cells (type IV and type V associated with the basement membranes and type VI surrounding the fibroblasts and the smooth muscle cells). The chemical structure of these various types of collagen can be found in the review of Bornstein and Sage [7].

Defective Genes and Control Mechanisms

All collagens are made of the association of three polypeptides. Nothing indicates that the primary structure of the defined types of polypeptides differs from tissue to tissue. Furthermore by screening genomic libraries of skin fibroblasts from chicken, calf and human [6] only one single copy of specific genes has been observed. The major difference characterizing a tissue lies in the proportion and architectural organization of the various types of collagen forming its structure and in the type and amount of macromolecules filling the interfibrillar extracellular space (proteoglycans and glycoproteins). Posttranslational enzyme-mediated modelling of the primary gene product, one of the hallmarks of the fibrous proteins of the connective tissue (i.e. hydroxylation of proline and lysine, glycosylation of hydroxylysine, etc. [25]) varies to some extent from tissue to tissue.

Skin is a major source of collagen type I and III that can be extracted, purified and studied *in vitro*. Polymers can be formed by heating at body temperature a cold solution of collagen monomers in physiological saline at neutral pH. Collagen type I forms large bundles of polymers, collagen type III thin bundless and mixtures of the two bundles of intermediate size [27]. The proportion of these two types of collagen is known to vary in different section of the dermis, type III being higher where the bundles are thinner as in the papillary dermis [31], type I predominating in the reticular dermis made of thick bundles.

Collagen type III has been called embryonic since it is the first to occur in skin during the fetal life. It decreases in proportion with advancing development and age [19]. Type III collagen has also been called the vascular type of collagen because of its large proportion in the blood vessels. In the heritable disorder Ehlers-Danlos type IV the gene responsible for the synthesis of collagen type III is missing and/or secretion of this collagen is impaired [10]. In these patients skin is very thin and blood vessels are very fragile. It is called the ecchymotic type of Ehlers-Danlos.

The analysis of the collagen genes coding for α_1 and α_2 type I in skin fibroblasts from patients presenting osteogenesis imperfecta has uncovered a large array of alterations and mainly deletion of exons [12]. The altered size of the polypeptides results in excessive hydroxylation of prolyl and lysyl residues and subsequent hyperglycosylation induced by a reduced rate of helix formation during the assembly of the three α chains ($2\alpha_1$ and $1\alpha_2$) to form the molecule [8]. Although dermal fibroblasts secrete a higher than normal proportion of collagen type III [35] and a reduced proportion of collagen type I the function of skin is little modified but bone is very fragile.

The relationship between the primary structure of the collagen and its function has mostly been derived from analysis of human disorders. One very interesting animal model however demonstrates that the insertion of a viral genome in the type I collagen gene results in a lethal mutation [22].

Defective Hydroxylation and Glycosylation

Hydroxyproline in the collagen molecule is required to insure the thermal stability of the triple helix. It is formed upon the activity of prolylhydroxylase on defined prolyl residues in the nascent collagen polypeptide. This enzyme requires ascorbate, divalent iron, molecular oxygen and α-ketoglutarate as cofactors and cosubstrates.

Skin changes in scurvy include bruising, bleeding gums, perifollicular petechiae and defective healing. Early investigation of skin healing in the scorbutic guinea pig has demonstrated a reduced deposition of fibrillar collagen and the accumulation of what is known at the present time to be underhydroxylated collagen. The defective activity of prolylhydroxylase requiring ascorbic acid as a cofactor is responsible for the defect. When underhydroxylation is extensive the collagen polypeptides accumulate in the cells. If the defect is only partial, the secreted collagen undergoes denaturation at body temperature and proteolytic degradation. Reduced type I and III collagen polymers impair wound healing. Defective basement membranes (collagen type IV) might be responsible for blood vessels fragility. Fibroblast cultures have to be supplemented with ascorbic acid to ensure optimal hydroxylation of peptidyl proline.

Hydroxylysine also results from the activity of the specific lysylhydroxylase requiring the same cofactors and cosubstrates as prolylhydroxylase. Hydroxylysine is the support of a mono (galactose as in bone) or disaccharide (galactose – glucose as in skin) glycosylation of collagen. The physiological property of this enzyme-mediated glycosylation is not yet clear. It has been suggested that it regulates the formation of intermolecular crosslinks in collagen polymers. Ehlers-Danlos type VI is caused by a reduced or missing activity of procollagen lysylhydroxylase. Skin, fascia and occular membranes contain collagen largely (50% or over) or completely (more than 90%) deficient in hydroxylysine. The rheological properties of ligaments and occular membranes are altered [37].

Non enzymatic glycosylation of collagen on sites different to hydroxylysine has been observed in diabetes. It is proposed to represent a mechanism of collagen fibers alteration and increased crosslinking in this disease [42]. It can be experimentally produced in the rat [29].

Defective Oligosaccharide Side Chains in Precursor Extension

Asparagine-linked complex type oligosaccharides are bound to the carboxyl terminal precursor peptide (pC-) of collagen type I and III. It is supposed to be used in the intracellular migration, vesicle packing and addressing of the newly synthesized molecules. An oligosaccharide side chain of the same type is linked to the aminoterminal extension peptide (pN) of collagen type III [43]. An excess of this type of oligosaccharide side chains has been observed in the dermatosparactic calf.

Tunicamycin that is known to inhibit this type of glycosylation depresses procollagen to collagen processing [18]. Monensin reduces the rate of translocation of the molecules in the Golgi apparatus and increases glycosylation of these molecules.

Defective Secretion

Secretion of completed procollagen molecules is an active process mediated by cytoplasmic microtubules and microfilaments. It is specifically defective for collagen type III in some forms of the Ehlers-Danlos type IV [9]. It can be reduced by colchicine and related compounds providing a potential explanation of the activity of such drugs in sclerosing processes.

Defective Removal of Precursor Sequences

Among the biosynthetic products of fibroblast the collagen precursor, i.e., the collagen molecule extended at either extremities by polypeptide sequences, has to be processed to allow polymerization to proceed to completion in the extracellular space.

In vivo the carboxyterminal extension (pC) is cleaved first and the initial polymers are made of aminoterminal (pN) extended collagen precursor. The precursor (pN) peptide is also rapidly removed as fibres grow and mature. Each precursor extension is excised by a specific protease, pN peptidase and pC peptidase. Different peptidases are specific for collagen type I and type III [33].

A heritable disorder in the calf, dermatosparaxis, results from the lack of activity of the pN peptidases [26]. Collagen polymers in skin are made of pN-collagen, poorly organized in fibres and bundles. The skin is very fragile in this disorder. Dermatosparaxis in the calf has been the first demonstration of a defect of one of the post-translational enzyme mediated processing of the collagen molecule. The skin of these calves has provided the first and still now the main source of pN-collagen. It has been extensively used for analysing the sequence of the aminoterminal extension, its tertiary structure and immunological properties. It has been the substrate that has allowed us to determine the existence, the specificity and the mode of action of the pN-procollagen peptidase. A wide variety of biochemical studies have used this collagen as a standard for the biosynthesis of collagen precursors in cell free system, its increased solubility properties and its involvement in the organization of collagen polymers *in vitro* and *in vivo*. Dermatosparaxis in the calf has opened the field of collagen precursors and precursor processing enzymes – now several hundred publications rich.

Dermatosparaxis also exists in sheep [20] and in the cat [13]. In the human an alteration of skeletal tissues and joints makes the Ehlers-Danlos syndrome type VII or arthrocalasis multiplex. It ressembles dermatosparaxis by the existence of collagen precursor in skin. One form of the syndrome is transmitted as an autosomal recessive trait and related to a defective activity of procollagen peptidase while another type is autosomal dominant. In such patients half of the α_2 chains are still extended by pN peptides [44]. This defect has been related to a mutation of the primary structure of the polypeptide affecting the cleavage site specific of the peptidase.

As opposed to fibroblasts isolated from normal calf skin, the shape of the fibroblast-like cells from dermatosparactic (D-) skin is epithelioid at confluence and their density at confluence is higher. The D-cells respond more extensively to stimulation by EGF. They display a higher number of EGF receptors at their cell membrane. They do not retract a collagen lattice in which they are cultivated [16]. Some of the above differences are shared by D-cells from sheep skin but not by fibroblasts isolated from skin of human ED VII [17]. The various alterations characterizing the D-fibroblasts could be explained by a mutation affecting one enzyme involved in the processing of the oligosugar side chains of various glycoproteins.

In bovine dermatosparaxis, abnormal properties of skin fibroblasts are not shared by fibroblasts isolated from other connective tissues such as tendon and blood vessels. The phenotypic expression of dermatosparaxis in the calf also predominates in skin while tendon, blood vessels and bone do not display such clinical or biochemical alterations. A similar difference has also been observed in the Ehlers-Danlos type VI [37]. It indicates that fibroblasts are differentiated cells and do not express an identical posttranslational machinery in the various connective tissues. Some similarity observed between the human disease and the animal disorders seems to justify their use as models. However, obvious differences demonstrate the lack of complete analogy and the complexity of the processes involved in these diseases.

Defective Deamination of Lysyl and Hydroxylysyl Residues

The oxidative deamination of lysyl and hydroxylysyl residues is catalyzed by lysyl oxidase, an enzyme operating on extracellular polymeric collagen to form aldehydes that are precursors of covalent intra- and intermolecular crosslinks in collagen and elastin. This enzyme requires copper as a cofactor.

Lathyrism was known long before discovering that β-aminopropionitrile (β-APN) is a specific and irreversible inhibitor of lysyloxidase [38]. Lathyrism in the chick, the guinea pig and other animal species can be induced by feeding either ground sweet peas (Lathyrus odoratus) or its active compound to produce massive skeletal and skin alterations in the growing animal [30]. Fragility of the connective tissues is the cardinal sign of the defect. The resulting absence of intermolecular crosslinks has been most usefully exploited to extract large amount of native monomeric collagen from skin and other connective tissues. It is used in fibroblast culture to facilitate the collection of newly synthesized collagen and elastin. On the basis of some analogy in the cardiovascular manifestations lathyrism has been compared to the Marfan syndrome. This comparison is hazardous.

The X-linked cutis laxa disorder in human subjects [11] and aneurysm prone mice [40] are caused by mutations affecting some alleles of the X-chromosome which seem to represent heritable disorders related to defective lysyloxidase activity. The phenotypic alterations differ in what is observed in lathyrism in which altered copper metabolism might be responsible for the defect. Copper deficient pigs [41] also present some of these connective tissue disturbances. Various other types of cutis laxa (autosomal dominant, recessive or acquired) depend on alterations of the elastic fibers of skin and other organs. They represent different types of disease.

Altered copper metabolism is the cause of the Menkes' kinky hair syndrome, also inherited in an X-linked pattern [14]. Various enzyme activities are defective and responsible for keratin and melanin alterations, central nervous system dysfunction while the reduced lysyloxidase activity is the cause of connective tissue alterations.

Defective Bundle Formation

The mechanisms involved in aligning the collagen polymers to form the bundles responsible for the mechanical properties of skin depend on the activity of the cells and physical properties of the medium in which polymerization occurs. For example oriented polymers can be created by aligning the molecules or their nuclei during polymerization in a mechanical [28] or in an electromagnetic field [46]. The role of cells in the process is illustrated by the retraction of a three dimensional matrix of collagen fibres as a support for cultured fibroblasts. The cell attach to the fibres, tract and align them. This process is the basis for production of an equivalent of living dermis [4].

Defective packing of the collagen fibres is a common character of several types of the Ehlers-Danlos syndrome. It occurs in calf and sheep dermatosparaxis and seem in these diseases related to the persistance of aminoterminal precursor sequences [36]. In the human Ehlers-Danlos types I and II, defective bundle formation occurs in the absence of defined alteration of the collagen molecule. The mechanism responsible for the defect is still unknown. A similar disorder has been observed in cats as a dominant genetic trait. Colonies of siblings of affected and normal cats have been raised [34]. Dogs and minks [23] present similar defects.

Scleroderma is characterized by the deposition of excessive and compact connective tissue localized first around the deep vessels of the dermis. It might be related to immunological alterations responsible for selection of hyperactive fibroblasts. The tight skin mutant mouse [32] has been claimed to be a better model than others for scleroderma [24]. The defect is autosomal dominant and the excess accumulation of collagen Tsk/+ fibers occurs in the dermis, the hypodermis and the skeletal connective tissue. No visceral involvement except emphysema has been observed in the affected mouse. Interestingly old Tsk/+ mice develop delayed type hypersensitivity to elastase soluble lung peptides [15].

Defective Degradation

The first and required step for polymeric collagen degradation is mediated by collagenase, a specific enzyme produced by various types of cells. The first model for its demonstration has been the rapid regression of the tadpole tail at metamorphosis [21]. Collagenase is under strict local control. It needs to be activated. It is inhibited by various circulating and tissue proteins [45]. Collagenase is largely increased in diseased and uninvolved skin as well as in fibroblasts of patients suffering of recessive dystrophic epidermolysis bullosa [3]. Improvement in the blistering has been observed by treating these patients with diphenylhydantoin [2]. Interestingly this drug produces gingival hyperplasia in some genetically determined epileptic humans perhaps by inhibition of collagenase secretion or activity. A similar condition has been reported in the cat [5].

Defective Association of Collagen with Others Connective Tissue Components

The physiological properties of the various connective tissues depend on an adequate interaction between the polymeric collagen structures and various proteo- and glycosamino-glycans, glycoproteins and elastic fibres. Several examples in pathology demonstrate this interaction as an increased hyaluronate synthesis in some cases of Marfan's syndrome [1] the alteration of the elastic fibers in various types of pseudoxanthoma elasticum and the increased thickness of the vascular basememt membranes in diabetes.

The thickening of vascular basement membrane can be induced in the rat by intoxication using streptozotocin. Detailed analysis of the alteration of this complex structure that is basement membrane has been performed by Rohrbach et al. [39] by rendering diabetic rats bearing a sarcoma producing basement membrane components. An increased concentration of laminin, a basement membrane glycoprotein and a reduced synthesis of proteoheparan sulfate might account for the thickening of the vascular membranes and their increased permeability.

Conclusion

Classical models allowing an overall estimation of the activity of cells involved in connective tissue formation and their biosynthetic activity have not been discussed as granuloma formation, skin or tendon wound healing, skin atrophy, etc. We have focussed here on animal models, heritable or acquired alterations to analyze steps in the complex metabolic pathway responsible for the edification of a connective tissue. The critical use of these models in pharmacology is able to answer very specific questions.

References

1. Appel A, Horwitz AL, Dorfman A (1979) Cell free synthesis of hyaluronic acid in Marfan syndrome. J Biol Chem 254:12199–12203
2. Bauer EA, Cooper TW, Tucker DR, Esterly NB (1980) Phenytoin therapy of recessive dystrophic epidermolysis bullosa: clinical trial and proposed mechanism of action on collagenase. New Engl J Med 303:776–781
3. Bauer EA, Gedde-Dahl T, Eisen AZ (1978) The role of human skin collagenase in epidermolysis bullosa. J Invest Dermatol 68:119–124
4. Bell E, Ivarsson B, Merrill C (1979) Production of a tissue like structure by contraction of collagen lattices by human fibroblasts of different proliferative potential in vitro. Proc Natl Acad Sci USA 76:1274–1278
5. Bergenholtz A, Hanstrom L (1979) The effect of diphenylhydantoin upon the biosynthesis and degradation of collagen in cat palatal mucosa in organ culture. Biochem Pharmacol 28:1–7
6. Boedtker H, Fuller F, Tate V (1983) The structure of collagen genes. In: Hall DA, Jackson DS (eds) International review of connective tissue research, vol 10. Academic Press, pp 1–63
7. Bornstein P, Sage H (1980) Structurally distinct collagen types. Ann Rev Biochem 49:957–1003
8. Byers PH, Bonadio JF, Steinmann B (1984) Invited Editiorial Comment: Osteogenesis imperfecta: update and perspective. Am J Med Genet 17:429–435
9. Byers PH, Holbrook KA, Barsh GS, Smith LT, Bornstein P (1981) Altered secretion of type III procollagen in a form of type IV Ehlers-Danlos syndrome: biochemical studies in cultured fibroblasts. Lab Invest 44:336–341
10. Byers PH, Holbrook KA, McGillivray B, Macleod PM, Lowry RB (1979) Clinical and ultrastructural heterogeneity of type IV Ehlers-Danlos syndrome. Human Genet 47:141–150

11. Byers PH, Siegel RC, Holbrook KA, Narayanan AS, Bornstein P, Hall JG (1980) X linked cutis laxa: defective cross link formation in collagen due to decreased lysyl oxidase activity. New Engl J Med 303:61–65
12. Chu ML, Williams CJ, Pepe G, Hirsch JL, Prockop DJ, Ramirez F (1983) Internal deletion in a collagen gene in a perinatal lethal form of osteogenesis imperfecta. Nature 304:78–80
13. Counts DF, Byers PH, Holbrook KA, Hegreberg GA (1980) Dermatosparaxis in a himalayan cat: biochemical studies of dermal collagen. J Invest Dermatol 74:96–99
14. Danks DM, Cartwright E (1973) Menkes'kinky hair disease: further definition of the defect in copper transport. Science 179:1140–1143
15. DeLustro FA, Mackel AM, LeRoy EC (1983) Delayed-type hypersensitivity to elastase-soluble lung peptides in the tight-skin (Tsk) mouse. Cellular Immunol 81:175–179
16. Delvoye P, Krieg T, Lapière ChM (1983) The lack or reduced capacity of contracting a collagen lattice by dermatosparactic calf and sheep skin fibroblasts is not a common character of fibroblasts from Ehlers-Danlos syndrome. J Invest Dermatol 80:373(A)
17. Delvoye P, Nusgens BV, Lapière ChM (1983) The capacity of retracting a collagen matrix is lost by dermatosparactic skin fibroblasts. J Invest Dermatol 81:267–270
18. Duksin D, Davidson JM, Bornstein P (1978) The role of glycosylation in the enzymatic conversion of procollagen to collagen. Studies using tunicamycin and concanavalin A. Arch Biochem Biophys 185:326–332
19. Epstein EH (1974) α_1 $(III)_3$ human skin collagen. Release by pepsin digestion and preponderance in fetal life. J Biol Chem 249:3225–3231
20. Fjolstad M, Helle O (1974) A hereditary dysplasia of collagen tissues in sheep. J Pathol 112: 183–188
21. Gross J, Lapière ChM (1962) Collagenolytic activity in amphibian tissues: a tissue culture assay. Proc Natl Acad Sci 48:1014–1022
22. Harbers K, Kuehn M, Delius H, Jaenisch R (1984) Insertion of retrovirus into the first intron of α_1 (I) collagen gene leads to embryonic lethal mutation in mice. Proc Natl Acad Sci 81: 1504–1508
23. Hegreberg GA, Padgett GA, Ott RL, Henson JB (1970) A heritable connective tissue disease of dogs and mink resembling Ehlers-Danlos syndrome of man. J Invest Dermatol 54:377–380
24. Jimenez SA, Millan A, Bashey RI (1984) Scleroderma-like alterations in collagen metabolism occurring in the TSK (tight skin) mouse. Arthritis Rheumatol 27:180–185
25. Kivirikko KI, Myllyla R (1982) (10) posttranslational enzymes in the biosynthesis of collagen: intracellular enzymes. In: Cunningham LW, Frederiksen DW (eds) Methods in enzymology, pp 245–304 (Structural and contractile proteins. Part A Extracellular matrix, vol 82)
26. Lapière ChM, Lenaers A, Kohn L (1971) Procollagen peptidase: an enzyme excising the coordination peptides of procollagen. Proc Natl Acad Sci USA 68:3054–3058
27. Lapière ChM, Nusgens B, Pierard GE (1977) Interaction between collagen type I and type III in conditioning bundles organization. Conn Tissue Res 5:21–29
28. Lapière ChM, Nusgens B, Pierard G, Hermanns JF (1975) The involvement of procollagen in spatially orientated fibrogenesis. In: Burleigh M, Poole R (eds) Dynamics of connective tissue macromolecules. North Holland Publ C, pp 33–50
29. LePape A, Guitton JD, Muh JP (1984) Distribution of non-enzymatically bound glucose in vivo and in vitro glycosylated type I collagen molecules. FEBS Letters 170:23–27
30. Levene ChI, Gross J (1959) Alterations in connective tissue properties in the lathyritic chick embryo. Fed Proc 18:355
31. Meigel WN, Gay S, Weber L (1977) Dermal architecture and collagen type distribution. Arch Dermatol Res 259:1–10
32. Menton DN, Hess RA, Lichtenstein JR, Eisen AZ (1978) The structure and tensile properties of the skin of tight skin (Tsk) mutant mice. J Invest Dermatol 70:4–10
33. Nusgens BV, Goebels Y, Shinkai H, Lapière ChM (1980) Procollagen type III N-terminal endopeptidase in fibroblast culture. Biochem J 191:699–706
34. Patterson DF, Minor RR (1977) Hereditary fragility and hyperextensibility of the skin of cats: a defect in collagen fibrilogenesis. Lab Invest 37:170
35. Penttinen RP, Lichtenstein JR, Martin GR, McKusick VA (1975) Abnormal collagen metabolism in cultured cells in osteogenesis imperfecta. Proc Natl Acad Sci 72:586–589

36. Pierard GE, Lapière ChM (1976) Skin in dermatosparaxis. Dermal microarchitecture and biochemical properties. J Invest Dermatol 66:2–7
37. Pinnell SR, Krane SM, Kenzora JE, Glimcher MJ (1972) Heritable disorder of connective tissue. Hydroxylysine-deficient collagen disease. New Engl J Med 286:1013–1020
38. Pinnell SR, Martin GR (1968) The cross-linking of collagen and elastin: enzymatic conversion of lysine in peptide linkage to α-Aminoadipic-δ-Semialdehyde (allysine) by an extract from bone. Proc Natl Acad Sci USA 61:708–714
39. Rohrbach DH, Wagner CW, Star VL, Martin GR, Brown KS, Yoon JW (1983) Reduced synthesis of basement membrane heparan sulfate proteoglycan in streptozotocin induced diabetic mice. J Biol Chem 258:11672
40. Rowe DW, McGoodwin EB, Martin GR, Sussman MD, Grahn D, Faris B, Franzblau C (1974) A sex-linked defect in the cross-linked of collagen and elastin associated with the mottled locus in mice. J Exp Med 139:180
41. Sandberg LB, Zeikus R, Coltrain IM (1971) Tropoelastin purification from copper-deficient swine: a simplified method. Biochim Biophys Acta 236:542
42. Schnider SL, Kohn RR (1980) Glucosylation of human collagen in aging and diabetes mellitus. J Clin Invest 66:1179–1181
43. Shinkai H, Lapière ChM (1983) Characterization of oligosaccharide units of p-N-collagen type III from dermatosparactic bovine skin. Biochim Biophys Acta 758:30–36
44. Steinmann B, Tuderman L, Peltonen L, Martin GR, McKusick VA, Prockop DJ (1980) Evidence for a structural mutation of procollagen type I in a patient with the Ehlers-Danlos syndrome type VII. J Biol Chem 255:8887–8893
45. Stricklin GP, Welgus HG (1983) Human skin fibroblast collagenase inhibitor. Purification and biochemical characterization. J Biol Chem 258:12252–12258
46. Torbet J, Ronzière MC (1984) Magnetic alignment of collagen during self-assembly. Biochem J 219:1057–1059

Applications of the Diazacholesterol Animal Model of Ichthyosis

P. M. Elias, M. L. Williams, M. E. Maloney, P. O. Fritsch and J.-C. Chung

Although scaling skin diseases traditionally have been considered disorders of protein metabolism, i.e., keratinization, there is little support for this hypothesis. Instead, the expanding list of inherited and acquired disorders that are associated with primary lipid abnormalities (Table 1, Fig. 1), implies that it is stratum corneum lipids which regulate stratum corneum cohesion and desquamation (reviewed in [4, 16]). Whereas relevant examples of inherited scaling dermatoses that are associated with abnormal lipid metabolism include recessive x-linked ichthyosis (RCLI), Refsum's disease; multiple sulfatase deficiency, and neutral lipid storage disease, examples of acquired disorders include ichthyosis occurring following administration of hypocholesterolemic agents (reviewed in [16]). Essential fatty acid deficiency (EFAD), both in rodents and in man, represents an additional scaling disorder where stratum corneum barrier function and scaling have been attributed to separate abnormalities in lipid metabolism [14].

Since the EFAD model is characterized both by abnormal barrier function (due to decreased linoleic acid) and by excessive scaling due to epidermal hyperproliferation, it may represent an appropriate model of a high-turnover dermatoses, such as psoriasis and one form of congenital ichthyosiform erythroderma (Table 2). However, the EFAD model can not be expected to approximate retention dermatoses, where epidermal turnover is relatively normal, such as RXLI and ichthyosis vulgaris [16].

Table 1. Importance of epidermal lipids in human scaling disorders: lines of evidence

Inborn errors of lipid metabolism with ichthyosis	
Defined	Recessive x-linked ichthyosis Refsum's syndrome Multiple sulfatase deficiency
Probable	Congenital ichthyosiform erythroderma Neutral lipid storage disease (Chanarin-Dorfman) Harlequin fetus Sjógren-Larssen syndrome
Drug-induced: hypocholesterolemic agents	Triparanol (MER-29) Diazacholesterol Nicotinic acid Butyrophenones
Essential fatty acid deficiency	

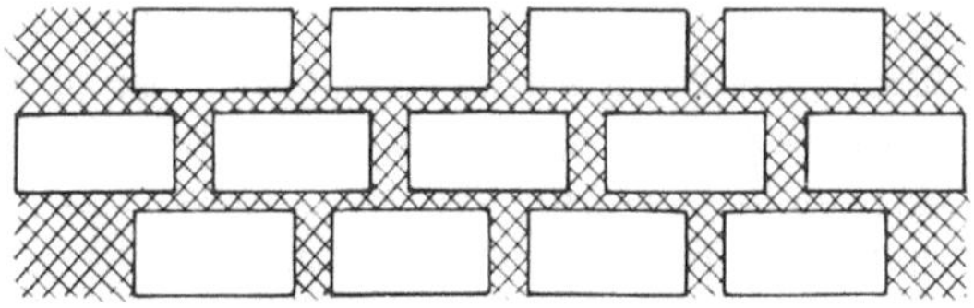

DISORDERS OF CORNIFICATION

BRICKS (PROTEIN)

Disorder	Abnormality
Defined:	
P-P Keratoderma (Richner-Hanhart)	Tyrosinemia
Probable:	
Harlequin fetus (some)	Abn. keratin pattern
I. vulgaris	Abn. ketatohyalin granules
Bullous CIE	Abn. tonofilament attachments; ↑ filaggren; abn. keratin pattern
I. Hystrix	Concentric perinuclear rings of tonofilaments
Trichothiodystrophy	↓ Sulfur-matrix protein

MORTAR (LIPID)

Disorder	Abnormality
Defined:	
RXLI	↑ Cholesterol sulfate ↓ Free sterols
Refsum's	↑ Phytanic acid
EFAD	↓C18:2; C20:4 ↑C20:3
Diazacholesterol	↓ Free sterols ↑ Sphingolipids
Multiple sulfatase deficiency	↑ Cholesterol sulfate
Probable:	
Non-bullous CIE	↑ n-Alkanes
Harlequin fetus (some)	Abn. sterols
Acrodermatitis enteropathica	Same as EFAD
Neutral lipid storage disease	↑ Triglyceride
Sjögren-Larsson Syndrome	Abn. fatty acid metabolism

Fig. 1. Schematic diagram of bricks-and-mortar conceptualization of the stratum corneum. Disorders of cornification have been classified as abnormalities of bricks (protein) and mortar (lipid). Reprinted from [16], with permission

Table 2. Cell turnover in disorders of cornification

High turnover	Normal turnover
Psoriasis	X-linked ichthyosis
Pityriasis rubra pilaris	Ichthyosis vulgaris
Congenital ichthyosiform erythroderma (CIE)	Classic lamellar ichthyosis
Refsum's disease	
Erythrokeratoderma variabilis	
Essential fatty acid deficiency (rodents)	

Availability of a well-defined animal models of ichthyosis would facilitate the development of new forms of therapy for the broad spectrum of economically important scaling dermatoses, e.g., psoriasis, eczema, that are characterized by excessive stratum corneum production and/or retention; and provide new insights into the pathogenesis and molecular basis for abnormal cornification.

In the absence of an ideal animal model, several other systems, in addition to the EFAD rodent, have been advanced as presumed analogues of ichthyosis, including the Rhino mouse [12], the carcinogen-induced mouse papilloma [2], and the mouse tail, which normally demonstrates parakeratosis [20]. Not only do each of these models have deficiencies, but the relationship of each of these models to either retention-type ichthyoses or to hyperproliferative dermatoses have not been defined or even considered.

In attempting to develop more appropriate models of ichthyosis we utilized two approaches: First, systemic administration to hairless mice of drugs that both interfere with cholesterol metabolism and have been associated with ichthyotic abnormalities in man (Table 1). Of the four drugs tested (triparanol, butyrophenones, nicotinic acid, and diazacholesterol, only one of these agents, 20,25-diazacholesterol, reliably produced ichthyosis [9]. The second approach is based upon the observation that certain classes of lipids accumulate in the stratum corneum in two recessive forms of ichthyosis, RCLI [17] and non-bullous congenital ichthyosiform erythroderma (CIE) [18, 19]. Because these lipids are exclusively localized to intercellular domains in the stratum corneum [4], where they are best situated to influence normal desquamation, we applied each lipid topically to hairless mice, and found that both cholesterol sulfate and n-alkanes, the accumulating species in RXLI and CIE, respectively, provoke scaling (Table 4). Since lipids are not strictly drugs, this review will be limited to diazacholesterol-tested rodents and the application of that model.

Ichthyosis in Hairless Mice from Systemically Administered Diazacholesterol

Gross and Clinical Observations

In attempting to develop a drug-induced model of ichthyosis, we administered four agents known to interfere with late stages of cholesterol metabolism (Fig. 2), nicotinic acid, triparanol, the butyrophenone WY-3457, and 20,25-diazacholesterol, 16–60 mg/kg/day (Table 1), to hairless mice that were otherwise given normal laboratory chow and water *ad libitum*. At these doses, only diazacholesterol (30–60 mg/kg/day) consistently produced ichthyosis after 6–8 weeks (Fig. 3), an effect that could be accelerated by utilizing weanlings and by maintenance of animals on a low cholesterol diet [9]. After appearance of scale, the dermatosis could be maintained at a lower daily dose of diazacholesterol (15 mg/kg/day). When animals were maintained on higher doses of diazacholesterol (30–60 mg/kg/day) for prolonged periods (3–6 months) they eventually lost weight and died.

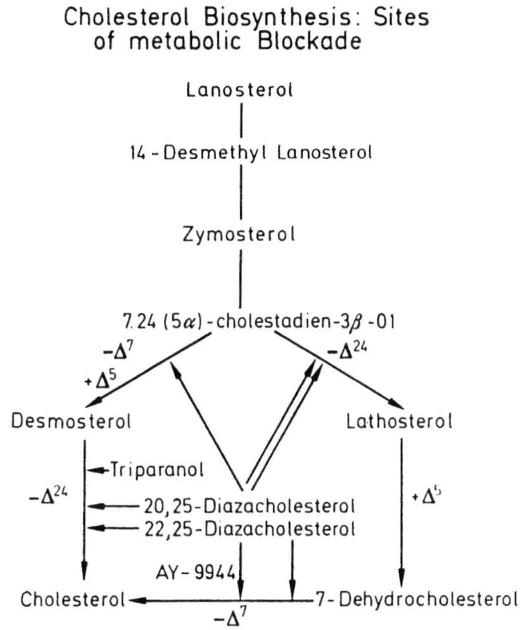

Drug	Enzyme Blocked: Low Dose	Enzyme Blocked: High Dose
Triparanol:	Δ^{24} reductase	Δ^{24} reductase
AY-9944:	Δ^{7} reductase	Δ^{7} reductase & Δ^{5} dehydrogenase
20,25-Diazacholesterol: (SC-12937)	Δ^{24} reductase	Δ^{24} reductase & Δ^{7} reductase
22,25-Diazacholesterol:	Δ^{24} reductase & Δ^{24} reductase	Lethal

Fig. 2. Diagram of final stages of cholesterol synthesis showing sites of blockade by inhibitors of the desmosterol shunt (diazacholesterol, triparanol) which produce ichthyosis (see text) and the 7-dehydrocholesterol shunt (AY-9944), which is not associated with ichtyhosis

Although ichthyosis appeared over the entire cutaneous surface simultaneously, it was most pronounced over the shoulders, which displayed a "shaggy bark" appearance (Fig. 3) and over the tail, which demonstrated prominant, circumferential bands of scale. There was no associated erythema or evidence of inflammation, and none of the animals developed abnormal transepidermal water loss, as occurs in essential fatty acid deficiency [9].

When diazacholesterol-treated animals were returned to a normal diet, the scaling abnormalities gradually disappeared over three to six weeks. Reversion to normal was most dramatic over the tail, where the hyperkeratotic scale often was shed as a whole snake-like encasement in a cephalad-to-caudad direction. Administration of 2% choles-

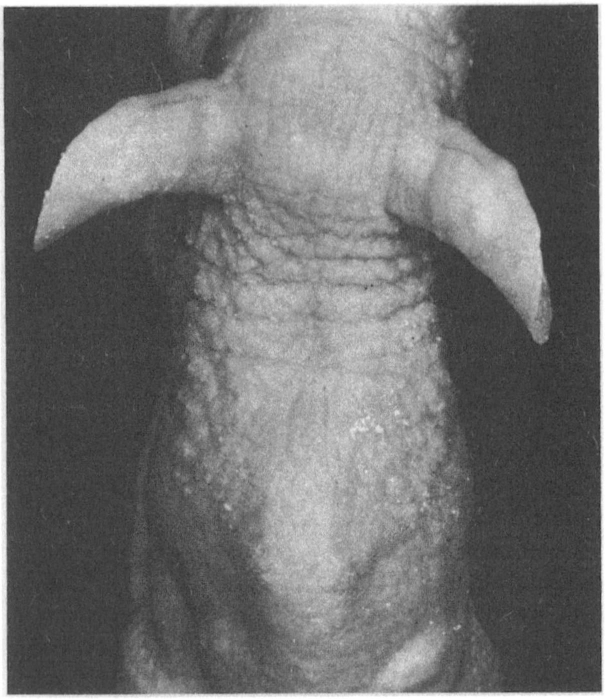

Fig. 3. Gross appearance of hairless mice after 10 weeks treatment with diazacholesterol. Note "shaggy-back" appearance of shoulders and back. Reprinted from Elias PM, et al (1983) Lab Invest 48:565–577, with permission

terol diet to animals, along with the diazacholesterol, prevented the development of ichthyosis, and co-administration of cholesterol with drug after the appearance of ichthyosis, also resulted in clearing in three-to-four weeks. Finally, both topical 7-dehydrocholesterol and topical cholesterol, administered as a 1% solution, but not topical desmosterol, reversed scaling locally, a further indication of the specificity of Δ^{24}-reductase blockade in the induction of the scaling abnormality (Fig. 2).

Quantitation of Stratum Corneum Thickness

In order to obtain an accurate assessment of stratum corneum thickness, we utilized 4–5 μm perpendicular sections of biopsy samples from five separate sites from seven animals that demonstrated ichthyosis versus four age-matched controls stained with 0.1% aqueous 8-anilino-1-napthalene sulfonic acid (ANS). On fluorescence microscopy ANS selectively depicts stratum corneum hydrophobic membrane domains [7], making this a useful technique for measurements not only of stratum corneum thickness, but also for the visualization of cellular organization [7]. In such studies we only assess sections that demonstrate an intact layer of flexible collodion on the epicutaneous surface, and only measure those that demonstrate the narrowest cross-sectional thickness of the underlying viable epidermis, a precaution that minimizes the contribution of tangential sectioning to thickness measurements. Finally, we measure the narrow orthokeratotic region between the hair follicle and the parakeratotic zone because this site corresponds to that used in prior semi-quantitative assays of pharmacologic agents on the stratum corneum [20].

Although the stratum corneum was always thicker in diazacholesterol treated animals (unless co-treated with a high cholesterol diet) than in controls ($p < 0.001$), there was considerable variation from animal-to-animal. Since each animal demonstrated little variation over its own cutaneous surface, each animal served as his own control in subsequent pharmalogic assays of topical agents (see below).

Light Microscopic Observations

On light microscopic examination of either fixed or frozen-sectioned epidermis, the principal finding was hyperkeratosis in the absence of acanthosis or parakeratosis [9]. This finding was most pronounced histologically in tail skin, where diazacholesterol treatment results in exaggerated parakeratosis and orthokeratosis, i.e., there is no disruption of the usual alternating pattern of para- and orthokeratosis [9].

Since abnormal scaling and barrier function have been associated with reduced intercellular lipids in essential fatty acid deficiency [5, 6], we compared the distribution of neutral lipids in diazacholesterol animals with normal tail epidermis, EFAD epidermis and normal non-tail epidermis in frozen sections stained with oil red 0. Since both diazacholesteroltreated and normal tail exhibited very little stratum corneum lipid, depletion of lipids appears to be a general feature of mouse tail stratum corneum [9].

Thin Section Electron Microscopy

Thin sections of Epon-embedded, diazacholesterol-treated epidermis demonstrated massive thickening of the stratum corneum, but no qualitative abnormalities of individual corneocytes [9]. Although some corneocytes in diazacholesterol-treated animals displayed non-membrane bound droplets, comparable structures also are encountered occasionally in control and revertant animals [9]. Moreover, the number of epidermal lamellar bodies, the putative secretory organelle for stratum corneum lipids [4], appears to be normal in diazacholesterol animals [9]. Regardless of treatment, lamellar body contents appear empty in tail epidermis, presumably a further feature of the "delipidized" nature of tail stratum corneum in general [9]. And presumably, as a result, epidermal intercellular domains exhibit a characteristic moth-eaten appearance in both thin sections and freeze-fracture replicas [9], comparable to that described in EFAD epidermis [cf. 5]. Moreover, replicas of freeze-fractured tail stratum corneum demonstrate only fragments of intercellular lamellations, with reversion of the fracture plane to the true plasma membrane [9], again as occurs in EFAD [5].

Table 3. Comparative labelling indices of topical lipid-treated vs. control animals

Treatment	Labelling index (mean + SEM)*
Diazacholesterol (n = 3)	10.8 ± 1.4%
Control (n = 3)	7.2 ± 0.3%

* labelled cells/1,000 basal cells counted. Significance of difference ($p < 0.1$). (Reproduced from Maloney et al. (1984) J Invest Dermatol 83:252–256 with permission)

Cell Cycle Alterations

Although histologic examination of diazacholesterol-treated epidermis did not reveal features of hyperproliferation, i.e., no acanthosis, parakeratosis, or increase in mitotic figures, the mitotic indices from three different diazacholesterol animals were significantly elevated in comparison to normal (Table 3, $p < 0.01$). Thus, the diazacholesterol model, at least according to this parameter, has this feature in common with hyperproliferative scaling dermatoses, such as epidermolytic hyperkeratosis (EHK) and psoriasis rather than to pure retention ichthyoses, such as ichthyosis vulgaris and RXLI [16].

Lipid Biochemistry

In order to determine the underlying lipid biochemical defect in the diazacholesterol model, intact whole stratum corneum sheets were prepared from full-thickness skin samples from severely affected diazacholesterol treated, age-matched controls, revertant animals, utilizing overnight incubations of the tissue on the staphylococcal epidermolytic toxin. As previously described [11], each sheet was further trypsinized to remove residual granular cells, avoiding exposure of the stratum corneum itself to the proteolytic enzyme [11], extracted with Bligh-Dyer solvents, the lipid weight percent/dry weight determined, and the extracts analyzed by sequential, quantitative, one-dimensional thin layer chromatography [11]. Because of the putative importance of free sterols in the pathogenesis of scaling, the free sterol fraction from the neutral lipid separation was analyzed further by gas liquid chromatography (GLC) and high pressure liquid chromatography (HPLC).

In these studies the diazacholesterol and control samples consistently displayed roughly equal stratum corneum lipid weight percentages (3–5%). In contrast to normals, the diazacholesterol-treated animals displayed decreased free sterols with a compensatory elevation in sphingolipids (Table 4 [9]. Further analysis of the non-polar lipid species also revealed several distinctive abnormal bands in the sterol/wax ester region in the diazacholesterol, that were not present in normal animals (Table 4). Finally, the free sterols when analyzed further by HPLC and GLC, revealed an almost total lack of cholesterol, with massive accumulation of desmosterol and of several other cholesterol precursors, including lanosterol, 7-dehydrocholesterol, and zymosterol in diazacholesterol-treated animals (Table 5), in keeping with almost total blockade of the Δ^{24}-reductase (Fig. 2).

Table 4. Stratum corneum lipid composition of diazacholesterol vs. normal mouse* (from Lab Invest (1983) 48:565–577, with permission)

Fraction	Diazacholesterol (n = 3)	Control (n = 3)	Revertant (n = 1)
Lipid weight %	4.6 ± 0.8	3.6 ± 0.05	–
Polar lipids	12.3 ± 1.4	21.4 ± 1.6	17.7
Sphingolipids	43.7 ± 1.7 ($p < 0.01$)*	23.3 ± 2.0	32.8
Glycolipids	5.7	2.2	–
Ceramides	38.0	26.1	–
Neutral lipids	56.2 ± 1.7	75.4 ± 2.0	67.0
Free sterols	14.6 ± 0.7 ($p < 0.01$)**	25.9 ± 1.5	17.2
Diglycerides	7.6 ± 1.0	13.9 ± 1.4	8.1
Free fatty acids	14.6 ± 1.1	12.0 ± 1.8	18.2
Triglycerides	5.0 ± 1.4	5.5 ± 1.0	3.8
Unknown I	1.9	tr	2.2
Unknown II	1.5	tr	1.4
Unknown III	1.4	tr	1.7
Unknown IV	3.4	2.8	7.4
Sterol esters	4.2	7.2	4.8
Hydrocarbons	1.5	6.4	2.2
Total	107.5%	104.7%	117.5%

* Expressed as weight percent of total neutral lipid
** Significant differences by the two-tailed Student's Test

Table 5. Distribution of free sterols in diazacholesterol and control animals (from Lab Invest (1983) 48:565–577, with permission)

Animal	Cholesterol	Desmosterol	7-Dehydrochol	Lanosterol
Diazacholesterol (n = 2)	14.7 ± 5.5	76.7 ± 4.9	ND	8.6 ± 0.53
Control (n = 3)	89.2 ± 1.14	tr	10.1 ± 0.41	ND

ND = Not done; tr = trace

Comments

Pathological scaling can be the end result of either epidermal hyperproliferation, leading to over-production of stratum corneum, or of abnormal stratum corneum retention [4, 16]. Based upon their known cellular kinetics, psoriasis, CIE, and EHK may be considered disease prototypes of an overproduction process [26], while in contrast ichthyosis vulgaris and RXLI presumably result solely from abnormal retention [16]. Over the past decade several animal models have been advanced as potential analogues of human scaling dermatoses including the Rhino mouse [12], the EFAD rodent [14], and chronic scaling produced by either detergent treatment or by topical application of certain fatty acids [13]. With the exception of the EFAD rodent [5, 6], none of these models have been thoroughly defined. The diazacholesterol mouse appears to display several features of a retention hyperkeratosis, i.e., massive hyperkeratosis without acanthosis or parakeratosis, and normal permeability barrier function, features that contrast with those of high-turnover models, such as the EFAD rodent, detergent-treated skin, mouse papillomas, and of course, their human paradigms, psoriasis, EHK, and CIE. Yet, since the diazacholesterol mouse does manifest increased mitotic activity (Table 3), it also cannot be considered a pure model of retention hyperkeratoses. We suspect that diazacholesterol-induced ichthyosis results from hypocholesterolemia, which produces its own constellation of features, comprising increased mitotic activity, abnormal desquamation and normal permeability barrier function.

Although the molecular basis of diazacholesterol-induced ichthyosis is not known, the evidence provided here suggests a primary abnormality of lipid metabolism. Interestingly, the total stratum corneum lipid content, which appears to be related to stratum corneum permeability barrier function in normal human skin [7], is not different from normal. It is noteworthy that the untreated mouse tail is also more scaly than the remainder of the cutaneous surface [20]. Hence, decreased total lipids could contribute to a general tendency for stratum corneum retention by tail epidermis, reflecting a similar tendency for a thickened stratum corneum on the human palm and sole, where the lipid content is under 2% of the total dry weight [7, 11].

In contrast, several lines of data point to a qualitative rather than a quantitative role for abnormal free sterols in the diazacholesterol induced scaling: First, although the lipid weight percentages of the two groups were similar, the diazacholesterol animals demonstrated a significant decrement in total free sterols. Second, several sterol species appeared that normally are present only in small quantities. Although diazacholesterol, as does triparanol, is generally assumed to cause blockade of the Δ^{24} reduction of desmosterol to cholesterol [1, 9], studies in cultured fibroblasts dosed with diazacholesterol suggest that blockade can be more profound, affecting still more proximal steps in the post-lanosterol pathway (Fig. 2). Third, the ichthyosis can be overridden by systemic co-administration of free cholesterol. Finally, since topical cholesterol or 7-dehydrocholesterol, but not desmosterol, reverse diazacholesterol induced scaling [9], this data further indicates that other free sterols can substitute for cholesterol itself in stratum corneum membrane function. These findings largely rule out the possibility that the diazacholesterol effect on scaling is due to an unrelated metabolic consequence of drug treatment unrelated to its impact on sterol biosynthesis.

Despite the clear implications that cholesterol itself is required for normal desquamation, the specific mechanism(s) whereby intercellular sterols might regulate desquamation is not known. It is quite likely, however, that cholesterol is required for normal membrane bilayer formation and fluidity. Thus, abnormal lipid thermal transitions [15] may underlie the pathological scaling in this model

Applications of Drug-Induced Ichthyotic Models

Since the models described above have been well defined by several different morphological, histological, cell kinetic, and in some instances, biochemical parameters, one or more should comprise ideal testing grounds for the assessment of therapeutic agents aimed at scale reduction. Although there are numerous keratolytic agents available to the clinician, none have excited as much attention as the retinoids. Retinoids represent a relatively new class of synthetic vitamin A derivatives with activity against acne, disorders of keratinization, and cancer [2]. The paucity of information about the comparative potency of retinoids in disorders of keratinization can be attributed to the current lack of an appropriate animal model of abnormal stratum corneum desquamation. Although the essential fatty acid deficient (EFAD) rodent is analogous in some respects to hyperkeratosis due to epidermal hyperproliferation, as occurs for example in psoriasis, CIE and EHK [26], it is less useful model as a model for abnormal stratum corneum retention, as occurs in RXLI and ichtyhosis vulgaris [26]. Moreover, the EFAD rodent is too fragile for extensive systemic or topical therapeutic experiments.

As described above, we have recently developed a well-defined model of retention hyperkeratosis by administering the hypocholesterolemic agent 20,25-diazacholesterol to hairless mice [11]. Since the clinical, morphological, histochemical, histometric, cell cycle, and biochemical characteristics of the model are now established, we have utilized the diazacholesterol mouse for comparing the potency of all-trans retinoic acid with several new synthetic retinoids in this system [5].

Assessment of Topical Retinoid Potency in the Diazacholesterol Analogue

Production of Ichthyosis in Hairless Mice. As described above, two-to-three month old male hairless mice given 30–60 mg/kg/day of diazacholesterol develop ichthyotic changes after 8–12 weeks which are most pronounced over the back and tail (Fig. 3). Since the tail manifested the most exaggerated scaling, we chose this site for topical drug applications. Volumes of ~100 μl of each of retinoid arc serially diluted, and applied once daily to circumscribed areas on the tail with each animal serving as his own control [5]. Applications are continued for two weeks, or until a clinical response occurs. At three-or-four day intervals, and at the termination of experiments, the clinical response is graded from 0 to 4+, with 0 indicating no response and 4+ indicating removal of all visible scale to a glistening surface [Table 4].

Obvious concentration-dependent differences in scale reduction can be seen in comparison to the vehicle control with serial dilutions of each of the seven drugs that have been tested to date. The clinical results are summarized in Table 6. Both normal and ichthyotic animals treated with 0.1% or higher concentrations of the arotinoid or its ethyl ester develop generalized erythema, diffuse, fine scaling, and a positive Nikolsky sign after four or five days. If treatment is prolonged beyond five days at this dose level, both normal and diazacholesterol-treated animals inevitably expire. Although animals dosed with 0.01% concentration of etretinate survive, they developed fine scaling over untreated parts of the body that we have ascribed to systemic toxicity

Table 6. Influence of topical retinoids on scaling

Retinoid	Concentration (%)	Time to reduce scaling (days)					
		3–4	5–6	8–9	12–14	18–21	24–28
Arotinoids*	1.0	++	0	0	0	0	0
	0.1	++	0	0	0	0	0
	0.01	++	++++	ND	ND	ND	ND
	0.001	–	±	+++	ND	ND	ND
	0.001	–	–	–	ND	ND	ND
Etretinate	1.0	–	–	+	++++	ND	ND
	0.1	–	–	–	+	+++	ND
	0.01	–	–	–	–	–	ND
Tetrazole-retinamide	0.1	–	–	±	+++	ND	ND
	0.01	–	–	–	–	ND	ND
Tretinoin	1.0	–	–	–	–	+	++
	0.1	–	–	–	–	–	–
	0.01	–	–	–	–	–	–

* Higher doses resulted in redness, generalized scaling, + Nikolsky, and early death (0) shortly after an dramatic reduction in scaling. Doses below 0.01% did not produce systemic toxicity. ND = Not done, – = no changes evident, and + = small amount of scale reduction to ++++ = smooth surface with no scale. Reproduced from Chung et al. [3], with permission

Table 7. Stratum corneum thickness following topical retinoid treatment of diazacholesterol mice. Reprinted from [3], with permission

Retinoid	Concentration (%)	Thickness (μ) (mean ± SEM)			Signifiance (vs. vehicle)
		Pre-Rx (non-vehicle)	Post-Rx	Vehicle*	
Tretinoin (n = 4)	1.0	ND	23.9 ± 0.07	54.6 ± 0.06	$p < 0.001$
	0.1	ND	43.9 ± 0.10	54.6 ± 0.06	$p < 0.05$
	0.01	ND	53.8 ± 0.12	54.6 ± 0.06	NS
Arotinoid (n = 2)	0.001	ND	20.0 ± 0.12	39.6 ± 0.16	$p < 0.001$
	0.0001	ND	35.6 ± 0.26	39.6 ± 0.16	NS
Etretinate (n = 4)	1.0	68.3 ± 6.6	12.2 ± 0.91	41.3 ± /.45	$p < 0.001$
	0.1	68.3 ± 6.6	27.7 ± 1.45	41.3 ± /.45	$p < 0.001$
	0.01	68.3 ± 6.6	37.4 ± 2.15	41.3 ± /.45	NS
Tretrazole-retinamide (n = 1)	0.1	72.0 ± 1.2	16.3 ± 2.0	29.7 ± 2.7	–
	0.01	72.0 ± 1.2	21.1 ± 1.2	29.7 ± 2.7	–

* The vehicle employed with etretinate and all-trans-retinoic acid was the Cremophore vehicle (see Methods), while Eucerin was used in the single tetrazole-retinamide experiment

from absorbed drug, because unusual sites such as the face and paws demonstrated scaling.

Assessment of Stratum Corneum Thickness. Prior to biopsy, the skin surface was coated with a thin film of flexible collodion to prevent fragmentation during frozen sectioning, as described above [9, 11]. Measurements of 8 μm perpendicular sections of biopsy sample sterol with ANS confirmed the clinical observations that the retinoids produce profound thinning of the stratum corneum (Table 7). Thickness measurements provide additional quantitative criteria for comparing the potency of different concentrations of each retinoid. As can be seen in Table 7 even dilute concentrations of some of the synthetic retinoids produce a profound reduction in stratum corneum thickness.

Comments

The development of new synthetic retinoids with activity directed against disorders of keratinization has been facilitated by the availability of a new animal model for comparing the potency and/or toxicity of such compounds [11]. Although several *in vitro* and *in vivo* assays have been advanced previously for comparing the antineoplastic activity of retinoids [12], the only previously available means for evaluating the retinoids antikeratinizing activity has been in cell culture systems [12]. Although *in vitro* systems have the advantages of being both quantitative and of providing information about the possible molecular sites of action of the retinoids, these assays have several disadvantages in comparison to *in vivo* assays: First, they do not necessarily reflected potential activity *in vivo*. Second, they are less likely to predict usefulness for pathological states where scaling may be due to a variety of different metabolic abnormalities. Finally, they are not useful for predicting either systemic or mucocutaneous toxicity [12], which, if sufficiently severe, can prevent clinical application of any drug, regardless of its demonstrated effectiveness in pre-clinical testing.

To the extent that the diazacholesterol mouse can be considered another systemic lipidosis with scaling, this model can be considered analogous to a substantial proportion of retention-type ichthyotic diseases of man. Because the ichthyosis in this model is accompanied by moderate hyperproliferation (Table 3), it is possible that this analogue will have relevance for other scaling skin deseases, such as psoriasis, CIE, EHK, or even warts, where scaling may be largely caused by over-production rather than abnormal retention. On purely clinical grounds it is possible to compare the potency of different retinoids in this system, and these results can be substantiated by quantitation of stratum corneum thickness in frozen cross sections.

Several other *in vivo* bioassays have demonstrated limited usefulness for the assessment of antikeratinizing potency. These include the anti-keratotic [2] and anti-parakeratotic [20] assays in mouse tail, the reduction of parakeratosis and scaling in essential fatty acid deficient mice [5], and the reduction of epidermal utriculi in Rhino mice [12]. The most commonly employed *in vivo* screen of the antikeratinizing potential of retinoids is the mouse papilloma assay [4]. Whereas this assay usefully pinpoints compounds with anticancer activity paralleling their antikeratinizing activity, direct assay, such as the one described here, provides a more desirable test for a strict antikeratinizing effect.

We recently reported that the systemic administration of two commonly employed synthetic retinoids, isotretinoin (13-cis retinoic acid, Accutane) and etretinate (aromatic retinoid, Tigason) produce dose-dependent alterations of the cutaneous barrier in otherwise normal hairless mice [10]. Although this observation corroborated the clinical observations that etretinate is about one order of magnitude more potent than isotretinoin [10], the doses required to produce abnormal barrier function were much higher than those used in clinical practice. Moreover, it could be argued that normal skin may respond differently to retinoids than does diseased skin.

Of all the assays developed to date, only the ichthyotic mouse model [5] and the recently developed barrier disruption assay [10] have demonstrated the capability of providing quantitative comparisons of antikeratinizing activity *in vivo*. A comparison of the results obtained in both these assays reveals close agreement between the two methods [8], suggesting that the ability of systemic retinoids to disrupt barrier integrity may be comparable to the capacity of topical retinoids to reduce abnormal scaling.

Several relevant observations about the comparative potency of the various topically applied retinoids deserve mention: One of the disappointing aspects of the synthetic retinoids has been their apparent lack of activity against either disorders of keratinization or acne when applied topically [12]. It has been widely suggested that the sort of toxicity (irritation) produced by all-trans retinoic acid may be a requirement for topical effectiveness. This study tends to disprove that notion, since scale reduction was accomplished without the cost of irritation, holding considerable hope that synthetic retinoids will ultimately prove to be useful not only as systemic drugs, but also as topical agents.

References

1. Ahrens RA, Dupont J, Thompson MJ (1965) Sterols in brain and liver of young rats fed 20,25-diazacholesterol. Proc Soc Exp Biol Med 118:436–438
2. Bollag W (1979) Retinoids and cancer. Cancer Chemother Pharmacol 3:207–215
3. Chung J-C, Law MYL, Elliott ST, Elias PM (1984) Diazacholesterolinduced ichthyosis in the hairless mouse: Assay for comparative potency of topical retinoids. Arch Dermatol 120:342–347
4. Elias PM (1983) Epidermal lipids, barrier function and desquamation. J Invest Dermatol 80:44–49
5. Elias PM, Brown (1978) The mammalian cutaneous permeability barrier: defective barrier function in essential fatty acid deficiency correlates with abnormal intercellular lipid composition. Lab Invest 39:574–583
6. Elias PM, Brown BE, Ziboh VA (1980) The permeability barrier in essential fatty acid deficiency: evidence for a direct role for linoleic acid in epidermal barrier function. J Invest Dermatol 74:573–578
7. Elias PM, Cooper ER, Kore A, Brown BE (1981) Percutaneous transport in relation to stratum corneum structure and lipid composition. J Invest Dermatol 76:297–301
8. Elias PM, Fritsch P, Lampe MA, Williams ML, Brown BE, Nemanic MK, Grayson S (1981) Retinoid effects on epidermal structure, differentiation and permeability. Lab Invest 44:531–540
9. Elias PM, Lampe MA, Chung J-C, Williams ML (1983) Diazacholesterol-induced ichthyosis

in the hairless mouse. I. Morphologic, histochemical, and lipid biochemical characterization of a new animal model. Lab Invest 48:565–577
10. Elias PM, Williams ML (1981) Retinoids, cancer and the skin. Arch Dermatol 117:160–181
11. Lampe MA, Burlingame AL, Whitney J, Williams ML, Brown BE, Roitman E, Elias PM (1983) Human stratum corneum lipids: Characterization and regional variations. J Lipid Res 24: 120–130
12. Mezick JA, Chabria MC, Thorne EG, Capetola RJ (1982) Topical and oral (P.O.) effects of 13-cis-retinoic acid (13-cis-RA) and all-trans retinoic acid (RA) on horn-filled utriculi size in the Rhino mouse. Clin Res 30:598A
13. Nguyen TT, Ziboh VA, Vermatsu S, McCullough JL, Weinstein G (1981) New model of a scaling dermatosis: Induction of hyper-proliferative in hairless mice with eicosa-5,8,11-trienoic acid. J Invest Dermatol 76:384–387
14. Prottey C (1976) Essential fatty acids and the skin. Br J Dermatol 94:579–587
15. Rehfeld SJ, Elias PM (1982) Mammalian stratum corneum contains physiologic lipid thermal transitions. J Invest Dermatol 79:1–3
16. Williams ML (1983) The ichthyoses-Pathogenesis and prenatal diagnosis: a review of recent advances. Ped Dermatol 1:1–24
17. William ML, Elias PM (1981) Stratum corneum lipids in disorders of cornification I. Increased cholesterol sulfate content of stratum corneum in recessive x-linked ichthyosis. J Clin Invest 68:1404–1410
18. Williams ML, Elias PM (1982) n-Alkanes in normal and pathological human scale. Biochem Biophys Res Comm 107:322–328
19. Williams ML, Elias PM (1984) Elevated n-alkanes in congenital ichthyosiform erythroderma: phenotypic differentiation of two types of autosomal recessive ichthyosis. J Clin Invest 74: 296–300
20. Wrench R (1981) Scale prophylaxis: a new antiparakeratotic assay. Arch Dermatol 117:213–216

The Transcutaneous Oxygen Pressure ($tcPO_2$) as a Non-Invasive Measurement of Aerobic Skin Metabolism and Skin Permeability in Psoriasis

W. Schalla

Transcutaneous partial oxygen pressure ($tcPO_2$) depends upon a number of factors [1] (Table 1). The commercially available devices are designed to control the arterial PO_2 – particularly in anaesthesiology and perinatology. In order to neglect for such purposes most local skin factors, the modified Clark electrode within the probe is heated to 44–45 °C. Using such an apparatus Tronnier et al. [2] found a decreased $tcPO_2$ in psoriasis and acne which returned to normal after effective treatment. We have extended this approach using two different electrode temperatures to confirm those results.

Material and Methods

We used a $tcPO_2$ monitoring system (TCM 1204 Radiometer, Copenhagen) in the usual manner [1] to compare uninvolved with involve abdominal skin in psoriasis, i.e., with an electrode temperature of 44 °C which corresponds to a skin surface temperature of about 42 °C. It is claimed that at this temperature the local skin factors and the ambient factors can be neglected on normal skin, i.e., the $tcPO_2$ values are correlated to the arterial PO_2.

By lowering the temperature the local factors become more important. Therefore, we also measured the same sites at an electrode temperature of 37 °C raising the skin temperature to about 35 °C. To avoid the influence of the higher temperature on these measurements we always started at 37 °C followed by 44 °C.

Table 1

Factors influencing $tcPO_2$	37 °C	44 °C
Arterial blood gases	(–)	(–)
Local skin blood flow	+?	–?
Local skin temperature	–	–
O_2 consumption in the skin	–	+
Diffusional resistance	+++	–
Vaseline	–	–

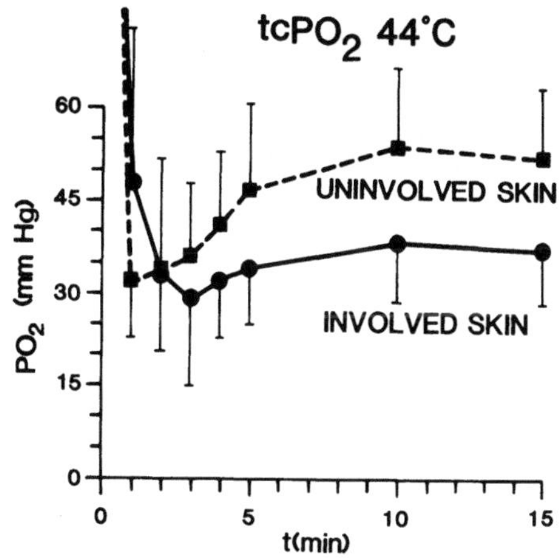

Fig. 1. $tcPO_2$ ($\bar{x} \pm$ S.E.M.) of untreated psoriatic skin at an electrode temperature of 44 °C (42 °C skin temperature) versus time after application of the electrode (n = 8)

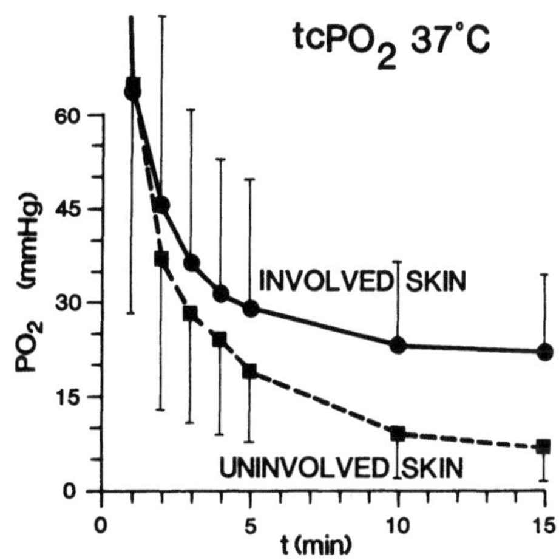

Fig. 2. $tcPO_2$ ($\bar{x} \pm$ S.E.M.) of untreated psoriatic skin at an electrode temperature of 37 °C (35 °C skin temperature) versus time after application of the electrode (n = 9)

Five patients could be followed during treatment. They were given three times weekly a short contact regimen of anthralin followed by UVB irradiation. Anthralin in vaseline was applied in gradually increasing concentrations from 0.3 up to 1 (or 3)%, for 10 up to 30 min, according to the severity of side effects. The UVB dose was slowly augmented from 0.03 up to 0.3 J cm^{-2}.

Results

As shown in Fig. 1 the $tcPO_2$ at 44 °C has been equilibrated some 10 min after application of the calibrated probe. The 15 min values on untreated uninvolved skin (52 ± 12 mmHG) are statistically significantly higher compared to involved skin (37 ± 9 mmHg). At an electrode temperature of 37 °C more time is needed for equilibration (Fig. 2). Fifteen min after application of the probe the $tcPO_2$ on either skin site is

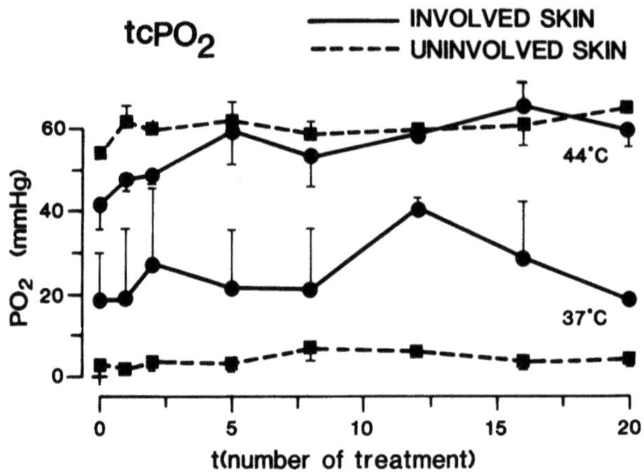

Fig. 3. tcPO$_2$ ($\bar{x} \pm$ S.E.M.) of psoriatic skin 15 min after application of the electrode (37 and 44 °C, respectively) versus the number of treatments with anthralin + UBV (1 patient dropped out after 11 treatments because of clearance)

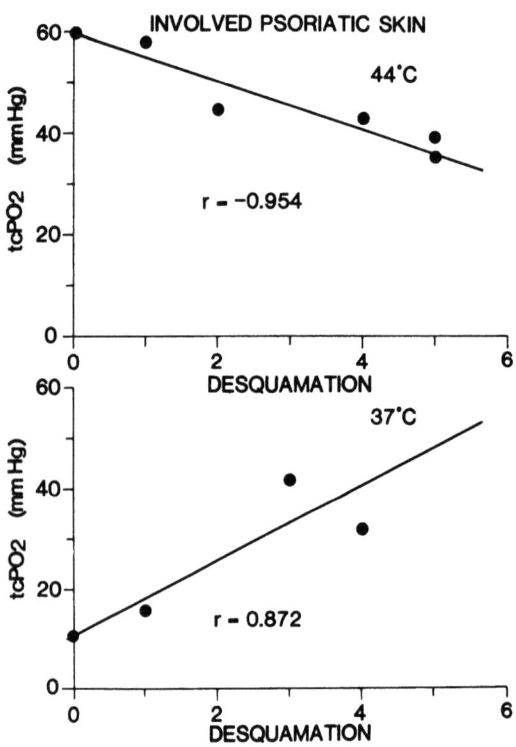

Fig. 4. Correlation between tcPO$_2$ at electrode temperatures of 37 and 44 °C and desquamation in one patient

lower, but in contrast to 44 °C the values on involved skin (22 ± 13 mmHg) are higher compared to uninvolved skin (7 ± 5 mmHg).

During the low dose regimen of short contact therapy with anthralin combined with UV irradiation, the tcPO$_2$ on involved skin at 44 °C increased to that of uninvolved skin within the first two weeks of treatment whereas the decrease at 37 °C to the level of uninvolved skin took longer (3 weeks for 1 patient, 6 weeks and more for the 4 others) (Fig. 3).

Within the results for each individual there is a correlation between $tcPO_2$ and clinical findings (Fig. 4). The desquamation usually, but not always correlated best with the $tcPO_2$, whereas there was no good correlation with infiltration. No general slope for the different clinical findings correlated to the $tcPO_2$ can be given because of greater interindividual variations among patients.

Discussion

Most of the factors which affect the $tcPO_2$ measurements will act in involved and uninvolved skin in the same manner (Table 1) or could mask to some extent the difference observed. We therefore have to concentrate only on those factors which are different in these two types of skin. The skin permeability for water [3, 4] and drugs [5] is increased in psoriatic lesions. For this reason the higher $tcPO_2$ values at 37 °C of involved compared to uninvolved skin are most likely caused by an increased skin permeability for gases such as oxygen.

This explanation does not hold true for the $tcPO_2$ at 44 °C because the values in involved skin are lower and not higher. It is known that the O_2 consumption in psoriatic skin slices is increased in vitro [6]. Our results suggest that this is also the case in vivo. The lack of correlation of $tcPO_2$ to infiltration at 44 °C favours the metabolic activity of keratinocytes being the cause of the difference.

References

1. Huch R, Huch A, Lübbers DW (1981) Transcutaneous PO_2. Thieme, Stuttgart New York
2. Tronnier H, Böttger E-V, Hoffmann E (1979) Transcutane PO_2-Messung unter UV-Therapie von Psoriasis und Akne (vorläufige Mitteilung). Zeitschrift Hautkrankheiten 54:546–550
3. Marks J, Rogers S, Chadkirk B, Shuster S (1981) Clearance of chronic plaque psoriasis by anthralin – subjective and objective assessment and comparison with photochemotherapy. Br J Dermatol 105, Suppl 20:96–99
4. Rajka G, Thune P (1976) The relationship between the course of psoriasis and transepidermal water loss, photoelectric plethysmography and reflex photometry. Br J Dermatol 94:253–261
5. Schaefer H, Zesch A, Stüttgen G (1982) Skin permeability. Springer, Berlin Heidelberg New York
6. Herdenstamm C-G (1962) On the in vitro metabolism of labeled glucose in normal and psoriatic skin slices. Acta Derm Venerol (Stockh) 42, Suppl 47

Experimental Contact Dermatitis Using 2,4-Dinitrochlorobenzene in Humans

J. A. A. Hunter, M. M. Carr, P. A. Botham, D. J. Gawkrodger, E. McVittie,
J. A. Ross and I. C. Stewart

The Clinical Problem

Contact dermatitis, due to both allergenic (sensitising) and irritant (non-sensitising) agents is a major clinical problem in industry and the home. Although its prevalence in a population depends on the nature of employment and industry, and environmental factors, it has been shown that one type of contact dermatitis alone, hand eczema, affects 2% of men and 3% of women [1]. Industrial dermatitis in fact accounts for about 65% of all spells of absence from work in the United Kingdom recorded under prescribed diseases [9].

Previous Animal Studies

Silberberg [11] was the first to note interactions between Langerhans cells and lymphocytes when she reported peripolesis (lymphocyte-Langerhans cell apposition) occurring in cutaneous contact allergic reactions to mercury but not in normal skin and not in mercury-induced irritant reactions. She and her colleagues went on to carry out an elegant series of experiments which were to implicate the Langerhans cell in antigen presentation to lymphocytes [12]. They made most of their observations on experimental contact dermatitis induced in the guinea pig with dinitrochlorobenzene and ferritin, the latter having the advantage of being detectable by electron microscopy. They found that 2–6 h after cutaneous challenge with these substances the number of Langerhans cells increased in the epidermis and dermis and that peripolesis was often noted. They also found that Langerhans cells were increased in dermal lymphatics and in the local lymph nodes. Recently Wolff and Stingl [15] have reviewed the evidence supporting the central role played by the Langerhans cell in antigen presentation, in the afferent limb of the contact hypersensitivity response, and later as a target cell, in the efferent limb.

Choice of Experimental Model

The choice between the mouse and the guinea pig as a model for the study of contact sensitivity is not a difficult one. Polak [10] summarised some of the drawbacks of studying mice. In this animal contact sensitivity is short-lasting as is tolerance. The skin reaction may become apparent macroscopically as soon as 4 h after challenge and histologically the infiltrate at 24 h post-challenge comprises mainly polymorphonuclear granulocytes. These findings contrast sharply with those in the guinea pig and human where contact sensitivity is usually life long and is manifest histologically by a predominantly lymphocytic reaction 24 h after challenge. However, the mouse has found favour as an experimental model for the study of contact sensitivity because animals are cheap, many inbred strains are available and there are numerous commercial monoclonal antibodies against mouse cell surface determinants.

Traditionally the guinea pig has been the animal of choice for experimental studies of contact sensitivity. The use of syngeneic strains facilitates experimental reproducibility and protocols seldom founder on ethical or toxicity gronds. Lympho-reticular tissue can also be readily examined, a point of some importance when investigating the mechanism of contact sensitivity. Nevertheless, in spite of these advantages, the human model must be the most relevant one when considering human disorders which inevitably concern most medical research workers. Naturally there are also worries about extrapolating findings noted in guinea pig skin, where percutaneous absorption and metabolism of chemicals is very different from that in human skin, to the situation in the human. Moreover, skin biopsy can be readily performed in humans and causes minor inconvenience only. Finally another major advantage in studying contact sensitivity in humans is that monoclonal antibodies against human but not guinea pig lymphocyte subsets and Langerhans cells are commercially available allowing phenotypic dissection of the inflammatory infiltrate. Accepting that lymph node examination in the human is not justifiable in any experimental protocol the practicability of investigating the mechanisms of contact dermatitis in this model hinges on toxicological and ethical considerations.

Toxicological and Ethical Considerations in the Human

2,4-Dinitrochlorobenzene (DNCB) is a strong cutaneous sensitiser. Sensitisation almost always occurs in immunologically competent persons after a single application of 1,000 μg and the compound is used as a test of cell mediated immunity [6, 13].

DNCB has also been used in considerably larger amounts therapeutically for the treatment of warts [4] and alopecia areata [8].

Three main criticisms have been raised against the use of DNCB in humans:

- The severity of local reactions following the challenge reactions in sensitised individuals. In the past painful inflammatory lesions have developed into large blisters leaving slowly healing ulcers.

- Its mutagenicity
- The development of an allergic contact dermatitis when sensitised individuals are subsequently exposed to DNCB or a closely related chemical.

We considered these objections to the use of DNCB and our experimental protocol was designed to minimise them:

We chose to use small sensitising (100 μg) and challenge (40 μg) doses of DNCB. Sensitivity has been induced in almost all instances and the challenge reactions have not been painful or severe. Twenty-four hours after challenge an erythematous, indurated plaque with associated microvesicles but without blistering, has been produced.

The prediction of the carcinogenic potential of a chemical is very complex and has been well reviewed recently by Ashby [3]. Much emphasis has been attached to the Ames test [2] which detects the mutagenic potential of the chemical on a strain of bacteria (Salmonella). DNCB, when tested in this way, is mutagenic but the important question is, of course, whether it is carcinogenic in animals. There is considerable evidence, summarised by Ashby [3], indicating that the Salmonella assay is exquisitely sensitive and that it can only define a large pool of potential carcinogens from which only a few real mammalian carcinogens will emerge. For example, a number of natural substances such as broiled fish, hamburgers, oxygen and thymidine are mutagenic in vitro [3] but probably non-carcinogenic in vivo. Animal testing in vivo is therefore important and we are reassured, to some extent, by the work of Weisburger et al. [14] who were unable to demonstrate that DNCB was a carcinogen in long-term studies on the mouse and the rat. We believe that it is highly unlikely that DNCB is carcinogenic in the amount (maximum total dose – 220 μg) which we chose to use in our studies.

One potential longterm complication for sensitised individuals must include the possibility of further contact with DNCB causing contact dermatitis. Fortunately the chemical is used extremely rarely in industry but is involved in the manufacture of some dyes, is present in an algicide of some water cooling systems and is used in the production of rubber chemicals and explosives. It is also employed occasionally in analytical chemistry especially to detect nicotinic acid and its derivates. The possible cross-reactivity of DNCB with chloramphenicol was discounted by Eriksen [7].

We therefore designed our experimental protocol accordingly:

We chose to carry out most of the studies in relatively fit patients recently diagnosed as having an inoperable bronchogenic carcinoma. Only those with a Karnovsky performance status of 8 or over (normal activity with effort, some signs and symptoms of disease) were studied and DNCB sensitisation and challenge constituted part of an overall assessment of their immunological competence before treatment with radiotherapy. Although the prognosis was poor in these patients there was no evidence to suggest that their cell mediated contact hypersensitivity was abnormal at the time of the investigations.

Subjects likely to encounter DNCB because of the above reasons were excluded.

Sterile, pre-packed DNCB impregnated patches, kindly provided by the Institut Merieux (Lyon), France, were used and applied to the forearm with great care. Disposable gloves were also worn by the doctor applying the patch tests so that staff contact with DNCB and the possibility of sensitisation to it were further minimised.

As a result we, and our local Physicians' Advisory Ethical Committee, believe that there are no compelling criticisms of our study at least on ethical or toxicological grounds. Indeed two volunteers from our staff, including the first author of this paper, who were not involved in the laboratory work, have been sensitised and challenged with DNCB. Nevertheless, written consent was obtained from all individuals whom we investigated after careful explanation of the study including discussion of possible complications.

Broad Outline of Methods

As the experimental details of this work have been described elsewhere [5] only a broad description will be given here.

Twenty patients have been studied, 18 with bronchogenic carcinoma undergoing an immunological work-up before definitive treatment and two healthy volunteers from our staff. Each was sensitised with 100 μg DNCB on an occlusive patch test applied to forearm skin, and challenged at least two weeks later with up to three 40 μg DNCB patch tests on the opposite forearm. Ellipse biopsies were taken from unchallenged skin in all cases and from challenged sites 1, 3, and 6 h (no clinical reactions), 12 h (erythema) or 24 h after application (erythema, oedema and occasionally vesiculation). A maximum of 3 biopsies was taken from any individual.

Standard indirect immunofluorescence techniques and indirect immunoperoxidase methods, carried out on fresh 5 μ frozen sections were used to localise the hapten and to determine the nature of the cellular infiltrate. DNCB was detected by using a polyclonal antibody to DNCB raised in the rabbit, lymphocyte subsets by commercially available monoclonal antibodies, Langerhans cells by the monoclonal OKT6 and by monoclonals directed against MHC Class-II (HLA-D) antigens. Routine transmission electronmicroscopy was performed.

Results

DNCB associated with Langerhans cells and keratinocytes within one hour of application of the challenge patch. At 6 h numerous large DNCB-positive dendritic cells appeared in the superficial dermis especially around the appendages (Fig. 1). Simultaneously OKT6-positive cells and Langerhans-like cells (identical ultrastructural features to Langerhans cells but without Birbeck granules) in the same distribution (Fig. 2) were seen in different sections.

An increase in epidermal Langerhans cells occurred at 12 h but their normal dendritic pattern was lost. After 12 h their number decreased and returned to the normal level by 24 h (EM counting). Apposition between Langerhans cells and T-lymphocytes was not seen electronmicroscopically up to 12 h and only rarely between 12 and 24 h.

Fig. 1. 6 h post DNCB challenge. DNCB seen in epidermis and in dermal dendritic cells (Anti-DNCB antibody) (x350)

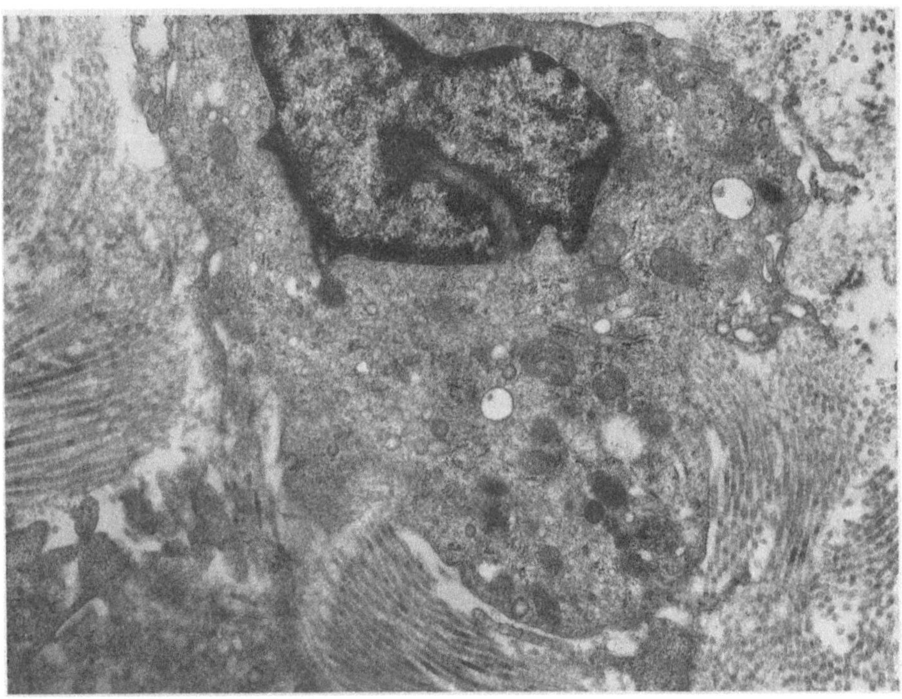

Fig. 2. 6 h post DNCB challenge. Langerhans-like cell (with no Birbeck granules) in papillary dermis (x21,000)

Summary

We chose to use the human model for our investigations on allergic contact sensitivity. We noted that DNCB localised and passed rapidly through the epidermis and was seen in large dermal dendritic cells at 6 h after challenge. At the same time and at the same site, cells staining positively with Langerhans cell markers and appearing ultrastructurally like Langerhans cells were seen. We conclude that DNCB is carried into the dermis by Langerhans cells in the challenge reaction and, as we observed minimal peripolesis up to 24 h post-challenge, we think it probable that most DNCB is presented to T-lymphocytes in the local lymph nodes (rather than in the skin) to stimulate a lymphocyte response.

Acknowledgements

This project was supported by a grant from I.C.I. Central Toxicology Laboratory, Macclesfield. We are also grateful to the Institut Merieux, Lyon, France for the gift of DNCB patches. Dr. M. Sudlow, Professor D. C. Flenley, Dr. R. McHardy and Dr. G. A. Neeaishy kindly allowed us to study patients under their care.

References

1. Agrup G (1969) Hand eczema and other hand dermatoses in Sweden. Acta Derm Venereol (Stockh) 49:1–91
2. Ames BN, McCann J, Yamasaki E (1975) Methods for detecting carcinogens and mutagens with the Salmonella/mammalian-microsome mutagenicity test. Mutation Res 31:347–374
3. Ashby J (1983) The unique role of rodents in the detection of possible human carcinogens and mutagens. Mutation Res 115:177–213
4. Buckner D, Price NM (1978) Immunotherapy of verrucae vulgares with dinitrochlorobenzene. Br. J Dermatol 98:451–454
5. Carr MM, Botham PH, Gawkrodger DJ, McVittie E, Ross JA, Stewart IC, Hunter JAA (1984) Early cellular reactions induced by dinitrochlorobenzene in sensitised human skin. Br J Dermatol 110:631–641
6. Catalona WB, Taylor P, Rabson A, Chretien P (1972) A method for dinitrochlorobenzene contact sensitization. A clinico-pathological study. New Engl J Med 286:399
7. Eriksen K (1978) Cross allergy between paranitro compounds with special reference to DNCB and chloramphenicol. Contact Dermatitis 4:29–32
8. Happle R, Cebulla K, Echternacht-Happle K (1978) Dinitrochlorobenzene therapy for alopecia areata. Arch Dermatol 114:1629–1631
9. Office of Health Economics (1973) Skin Disorders O.H.E. Publication No 46:10
10. Polak L (1980) Monographs in Allergy, vol 15. Immunological Aspects of Contact Sensitivity. Karger
11. Silberberg I (1973) Apposition of mononuclear cells to Langerhans cells in contact allergic reactions. An ultrastructural study. Acta Derm Venerol (Stockh) 53:1–12

12. Silberberg-Sinakin I, Thorbecke GJ (1980) Contact hypersensitivity and Langerhans cells. Acta Derm Venerol (Stockh) 75:61–67
13. Strauss GH, Bridges BA, Greaves M, Hall-Smith P, Price M, Vella-Briffa D (1980) Inhibition of delayed hypersensitivity reaction in skin (DNCB) test by 8-methoxypsoralen photochemotherapy. Lancet ii:556
14. Weisburger EK, Russfield AB, Hamburger F, Weisburger JH, Boger E, Van Dougen CG, Chu KC (1978) Testing of twenty-one environmental aromatic amines or derivatives for long-term toxicity or carcinogenicity. J Environ Path Tox 2:325–356
15. Wolff K, Stingl G (1983) The Langerhans cell. J Invest Dermatol 80:17S–21S (Suppl)

In-Vivo-UVA-Tests: Erythema, Pigmentation, Phototoxicity

R. Rüger, E. Hölzle, G. Plewig and A. Galosi

The time course and dose requirements for the effects of UV-B-irradiation on human skin are well characterized. Since 1931, when Wucherpfennig [13] introduced the "Lichttreppe", this classical test has been used to determine sensitivity to UV-B-erythema. Skin responses to UV-A, however, are less well and only qualitatively defined. Studies aiming at quantification of UV-A-effects are few [2, 4, 6, 8, 11]. They have been hampered by lack of suitable light sources. In 1977 a new apparatus with high radiation energy between 320–460 nm was introduced; a description of the appatus and its dermatological applications was published in 1981 [7]. Advantages of the device are high intensity in the UV-A-range with virtually no additional UV-B and the possibility or irradiation of large areas of the skin surface.

With increasing awareness of the harmful effects of long wave UV-light, sunscreens absorbing UV-A have gained interest. In contrast to the well established sun protection factor of UV-B-absorbing agents appropriate and standardized models for evaluating UV-A-sunscreen efficacy are lacking. Only preliminary test procedures in animals [1, 3] and humans [5] have been proposed. In the present study in-vivo responses of human skin to UV-A were evaluated in a quantitative way. Skin reactions investigated included erythema, immediate pigment darkening and delayed tanning and phototoxic responses to topically applied 8-methoxypsoralen. In further experiments some of these skin responses were used to determine efficacy of UV-A-sunscreens.

Methods

Subjects. A group of 43 patients phototested for suspected light sensitivity disorders and 60 volunteers with no apparent skin disease participated in this study. Informed consent was obtained from all subjects.

Light source. A high intensity UV-A-source equipped with a high pressure metal-halogenide lamp [7] was used (UVASUN 5000 Mutzhas, Munich, FRG).

Immediate pigment darkening (IPD). On the volar forearm or lower back squares measuring 2 x 5 cm were irradiated with 5–30 J/cm^2 using increments of 5 J/cm^2. Threshold doses for IPD were determined 10 min later.

Table 1. UV-A-protective test materials and their ingredients

Product	Ingredients	Concentration in %
Contralum ultra cream (Hermal Chemie, FRG)	4-Isopropyl-dibenzoylmethane	5
	Methylbenzylidene campher	5
	2-Phenylbenzimidazol-5-sulphonic acid	2
Piz Buin (Greiter, Austria)	2-Ethyl-hexyl-p-methoxycinnamate	5
	2-Hydroxy-4-methoxy-benzophenone	3
AV-UV-205[a] (Basotherm, FRG)	Phenylpyridylpropandione	
Contralum ultra Base		

[a] in aqueous lipogel (Neribas cream, Schering, FRG)

Tanning. Minimal tanning doses (MTD) were evaluated by a second grading 24 h after irradiation on the same testsites used for investigation of IPD. Multiple exposures were employed to determine tanning capacity (TC). TC was estimated in patients with skin type II and III after irradiations with 40 J/cm^2 on four successive days. Test sites on the lower back measured 5 x 10 cm. Readings were graded on a four-point scale one hour after the fourth exposure. 0 = no pigmentation, 1 = slight pigmentation, 2 = moderate pigmentation, 3 = dark pigmentation.

Phototoxic reactions. Localized photosensitivity was induced on the mid and lower back in areas measuring 10 x 10 cm by painting 0.15% 8-MOP solution (Meladinine-Lösung, Basotherm, Biberach, FRG) onto the skin as a thin layer using a cotton applicator. Minimal phototoxic doses (MPD) and doses required for a 2+ erythema response were determined 72 h after irradiation with UV-A.

Sunscreen testing. Materials used are listed in Table 1. They were applied to test sites as a thin layer (approximately 0.05 ml/25 cm^2) 30 min prior to each irradiation. UV-A-absorbing properties were evaluated by determination of MTD, TC and MPD on normal skin and on skin pre-treated with the test compounds. From these results protection factors were calculated in accordance with the recommendations for UV-B sunscreen testing [9]. The test areas in selected subjects were biopsied and specimens routinely processed.

Results

Erythema. Erythemal responses to UV-A irradiation were inconsistent and depended largely on the skin type of subjects. In patients with light complexion (type I and II) erythema occurred immediately after irradiation but was difficult to distinguish from

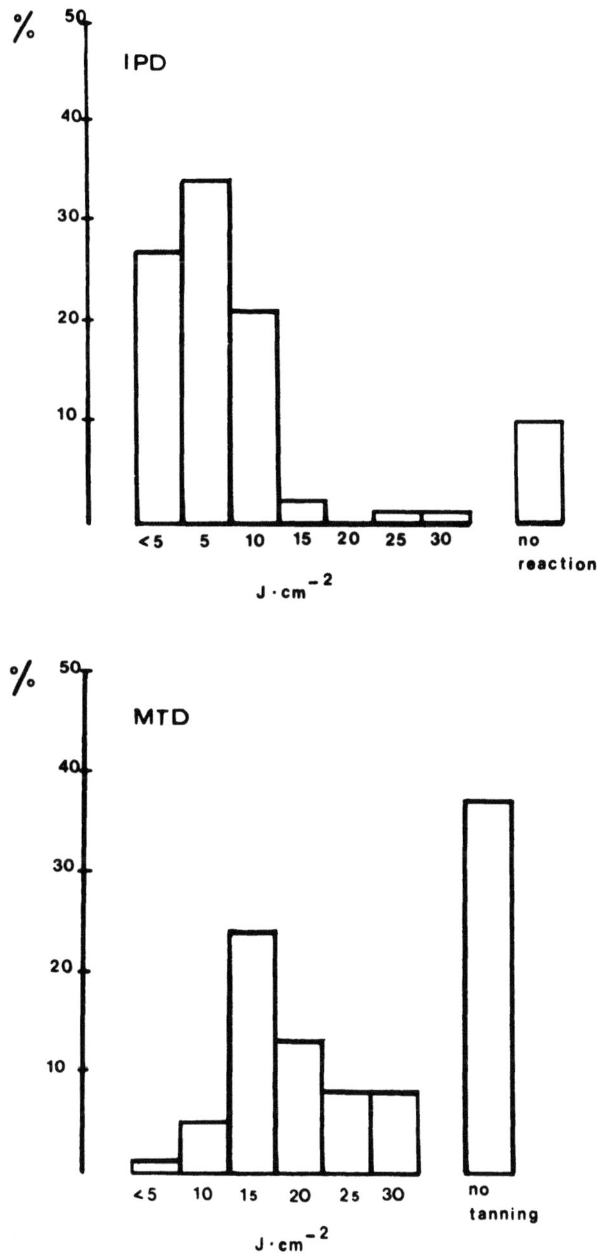

Fig. 1. Threshold doses for immediate pigment darkening (IPD) and minimal tanning dose (MTD) in 69 patients

heat erythema. With doses up to 30 J/cm^2 UV-A erythema was noticed only in a few subjects 24 h after irradiation. In patients with skin type III and IV erythema was obscured by IPD or delayed tanning. Therefore quantitative data on UV-A erythema were not obtained in the present study.

Immediate pigment darkening. IPD was observed in 89% out of 69 patients. It was dose dependent and showed a characteristic greyish hue and correlated with skin

Table 2. Correlation between skin type, immediate pigment darkening (IPD) and minimal tanning dose (MTD) in 74 patients

Skin type	IPD	MTD
	J/cm², M ± SD	
I–II	16.5 ± 13.4	24.2 ± 15.3
II–IV	9.5 ± 4.5	18.7 ± 10.9

Table 3. Minimal tanning dose (MTD) is increased on skin pretreated with Contralum ultra cream ($p < 0.05$)

Control	Contralum[R] ultra cream
J/cm², M ± SD, n = 6	
13.3 ± 5.2	20.0 ± 6.3

Table 4. Tanning capacity is decreased by skin protected with Contralum ultra cream ($p < 0.05$, $n = 6$)

Irradiation	1st	2nd	3rd	4th
Control	1.8 ± 0.8	2.3 ± 0.8	2.3 ± 0.5	2.5 ± 0.6
Contralum ultra cream	1.0 + 0.6	1.2 + 0.8	1.3 + 0.8	1.3 + 0.8

0 = No pigmentation;
1 = Slight pigmentation;
2 = Moderate pigmentation;
3 = Dark pigmentation

type. Patients with a darker complexion reacted more intensely than subjects with fair skin.

Fair skinned people required higher threshold doses or failed to react to doses up to 30 J/cm². This relation is shown in Fig. 1 and Table 2. Within 24 h IPD faded and blended into delayed tanning.

Tanning responses. At 24 h tanning was present in 63% of the patients. The intensity of the tan at 24 h was lower than in the IPD-reaction. Thus minimal tanning doses are higher than threshold doses for IPD; e.g., immediate pigment darkening was achieved in most patients with doses less than 15 J/cm² compared to minimal tanning doses of 20–30 J/cm² or more in the majority of the subjects (Fig. 1). In addition, the same relation of MTD to skin type was seen as in IPD (Table 2).

Table 5. Minimal phototoxic dose (MPD) in relation to different concentrations of 8-MOP-solution

8-MOP (%)	0.01	0.05	0.15
		M ± SD, n = 7	
MPD J/cm^2	1.5	0.9 ± 0.35	0.5 ± 0.25

Table 6. Evaluation of UV-A-protection factor in the topical phototoxic model

	MPD	2+ E	Protection factor	
	J/cm^2, n = 7		MPD	2+ Erythema
Control	0.7	2.0		
Contralum ultra base	0.6	1.3	0.9	0.6
AV-UV-205	0.8	2.1	1.1	1.1
Piz Buin-6	2.3	5.6	3.7	2.8
Contralum ultra cream	8.3	11.6	11.7	5.7

Repeated irradiations with 40 J/cm^2 UV-A each on four successive days induced a moderate to dark pigmentation which resembled the chocolate brown tan following sun exposure. The increase of pigmentation score from day to day as shown in Table 4 reflects multiple IPD-reactions blending into progressive melanization.

Phototoxic reactions. The phototoxic response to 8-MOP and UV-A is dose-related to both irradiation and concentration of the sensitizing solution. By means of a chess-board titration the quantitative relationship between 8-MOP concentration and UV-A-dose was evaluated. With 0.15% 8-MOP the minimal phototoxic dose was found to be 0.5 ± 0.25 J/cm^2 (Table 5). This concentration was used in sunscreen evaluations because of low variability of test reactions and ease of availability of the 8-MOP solution as a proprietary product.

Sunscreen testing. Contralum ultra cream proved effective in reducing pigmentary responses to UV-A. MTD was found to be 13.5 J/cm^2 on normal skin as compared to 20.0 J/cm^2 on skin protected by the sunscreen (Table 3). TC was graded as 2.5 on normal skin and decreased to 1.3 on sites pretreated with Contralum ultra cream (Table 4). The difference in degree of pigmentation is already present one hour after the first irradiation. This indicates protection of IPD-reaction. The ratio between pigmentation score with and without suncreen protection remains at the same magnitude in the three following readings.

Results of sunscreen assessment in the topical phototoxic model are summarized in Table 6. It is shown that the vehicle enhances UV-A-phototoxicity. This may be due to increased transmittance of the lubricated horny layer. Compound AV-UV-205

Table 7. UV-A-protection factors for Contralum ultra cream using different test models and reading-points

Test parameter	n	Protection factor
Minimal tanning dose	6	1.5
Tanning capacity	6	1.9
Minimal phototoxic dose	7	11.7
2+ Phototoxic erythema	7	5.7

a 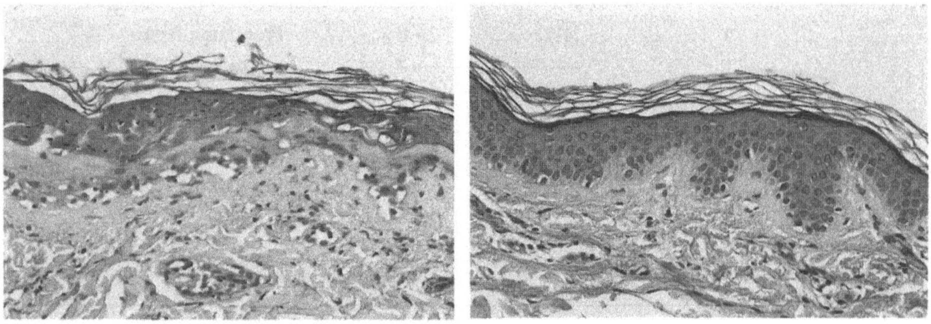 b

Fig. 2a, b. Phototoxic response 72 h after application of 0.15% 8-MOP-solution followed by UV-A-irradiation. **a** Normal skin was irradiated with 2.0 J/cm² UV-A. A heavy phototoxic reaction with necrosis of the upper epidermis, vacuolar alteration of basal zone and perivascular lymphocytic infiltrate in the upper dermis is evident. **b** Skin pretreated with Contralum ultra cream was irradiated with 10.0 J/cm² UV-A. No phototoxic response is seen (HE x32)

is little effective. In contrast, Piz Buin-6 exerts a moderate and Contralum ultra cream a good UV-A-protective effect. Using MPD as reading point is more discriminating than evaluation of a 2+ erythema reaction; UV-A protection factors are higher.

Prevention of phototoxic reaction by a sunscreen is also documented by histologic examination. Unprotected skin shows a high grade phototoxic reaction to 0.5% 8-MOP followed by irradiation with 2 J/cm² UV-A whereas sites pretreated with Contralum ultra cream looked normal even when irradiated with the fivefold UV-A-dose of 10 J/cm² (Fig. 2).

Discussion

Quantitative assessment of UV-A-crythema is difficult. The erythema response is best seen in fair skinned subjects. It is, however, superimposed by unspecific erythema due to heating of the skin. On pigmented skin IPD followed by tanning obscures a specific UV-A-erythema. For that reason test models for quantitative analysis of UV-A-erythema cannot be recommended at present.

Pigmentary responses to UV-A are well described and can be quantitatively evaluated by determining threshold doses for IPD and tanning. In addition, it may be worthwhile to grade the intensity of a tan following multiple exposures. It is noteworthy that UV-A-pigmentation involves true melanogenesis in the absence of a sunburn reaction. The underlying mechanisms are obscure and light- as well as electronmicroscopic findings on UV-A-pigmentation are disputed [10, 12].

It seems appropriate to utilize various skin responses to UV-A as test models for UV-A-sunscreens. Erythema, IPD, tanning and phototoxic reactions elicited by 8-MOP were assessed in the present study, but only pigmentary responses and phototoxic reactions were useful systems. Results, however, differed greatly. Various protection factors were obtained using different models, e.g., IPD, MPD, or different end points, e.g., threshold erythema or 2+ erythema reactions. Table 7 summarizes experimental results obtained with Contralum ultra cream. Protection factors range from 1.5 to 11.7. This variation is most probably due to differing action spectra of these skin reactions. Introducing other sensitizers and adding the oral phototoxic model would further increase variability. More experimental work has to be addressed to this question until specific recommendations can be given for useful and predictive test models.

Acknowledgement

The technical assistance of Mrs. M. H. Habig is greatly appreciated. The study was supported by a grant from Deutsche Forschungsgemeinschaft (Br 147/51).

References

1. Akin FJ, Rose AP, Chamness TW, Marlowe E (1979) Sunscreen protection against drug-induced phototoxicity in animal models. Toxicol Applied Pharmacol 49:219–224
2. Bachem A (1955) Time factors of erythema and pigmentation, produced by ultraviolet rays of different wavelength. J Invest Dermatol 25:215–218
3. Folsom KG, Gange RW, Mendelson IR (1981) The ability of UVA to induce ODC in topical 8-MOP treated mouse epidermis: Use of this effect to measure UVA sunscreen potency (Abstract). J Invest Dermatol 76:330
4. Hausser I (1938) Über spezifische Wirkungen des langwelligen ultravioletten Lichts auf die menschliche Haut. Strahlentherapie 62:315–322
5. Ippen H, Kölmel K (1980) Lichtschutz gegen Ultraviolett A. Ärztliche Kosmetologie 10:219–226
6. Kaidbey KH, Kligman AM (1978) The acute effects of long-wave ultraviolet radiation on human skin. J Invest Dermatol 72:253–256
7. Mutzhas MF, Hölzle E, Hofmann C, Plewig G (1981) A new apparatus with high radiation energy between 320–460 nm: Physical description and dermatological applications. J Invest Dermatol 76:42–47
8. Parrish JH, Ying CY, Pathak MA, Fitzpatrick TB (1974) Erythemogenic properties of long wave ultraviolet light. In: Pathak MA, Harber LC, Seiji M, Kukita A, Fitzpatrick TB (eds) Sunlight and man. University of Tokyo Press, Tokyo, pp 131–141

9. Pathak MA (1982) Sunscreens: Topical and systemic approaches for protection of human skin against harmful effects of solar radiation. J Am Acad Dermatol 7:285–312
10. Pathak MA (1984) Pigmentary responses to solar UVA. In: Urbach F, Gange W (eds) The biologic effects of UVA. Proceedings of a workshop (in press)
11. Pathak MA, Fanselow DL (1983) Photobiology of melanin pigmentation: Dose/response of skin to sunlight and its contents. J Am Acad Dermatol 9:724–733
12. Ryckmanns F, Schmoeckel C, Plewig G, Braun-Falco O (1983) UVA-Pigmentation: Ultrastructural and morphometric analysis. J Cut Pathol 10:399
13. Wucherpfennig V (1931) Biologie und praktische Verwendbarkeit der Erythemschwelle des UV. Strahlentherapie 40:201–244

Thermography as a Model for Cutaneous Vascular Responses

G. Stüttgen

Thermography is the best non invasive contact method to characterize heat radiation from the skin surface of large areas. The thermal resolution is limited to 0.1–0.2 °C and to a size of 1 cm^2. For areas less than this the heat difference has to be more than 0.2 °C.

Calibrated electronic thermometers are more suitable for measurement of small heat differences from very small areas of skin. Thermography and contact thermometry measure different properties. The wave lengths on which thermography depends (2–20 μm) are absorbed by water and humidity of the skin surface. The reference value is 0.1–0.2 °C lower in the case of hydration. Interpretation of the heat pattern of the skin must take into account changes due to conductive heat transport and heat absorption in superficial cooler oedematous areas. Convective heat transport and heat production due to tissue metabolism which are both projected on the skin surface can be hidden by structural changes.

The skin emits heat rays with wavelengths of 2–20 μm to its surroundings. The skin is nearly impermeable to these wavelengths and therefore heat rays measured between 2 and 20 μm are reflected by layers in the epidermis. The amount of emitted heat correlates directly with the temperature of the skin surface according to the Stefan-Boltzmann law. The skin temperature is a result of heat conduction and heat convection, primarily from the blood. The amount of energy liberated from metabolic processes in the skin is minute. Temperatures from depths of up to 2 cm beneath the skin surface are reflected in the surface skin temperature under physiological conditions and in the absence of a direct connection by blood vessels with the skin surface. The temperature of the skin surface results primarily from alterations in the microcirculation either through neural influences or substances affecting vessel diameter. Detectors, sensitive enough to record heat radiation between 2 and 20 μm have been devised and in the medical field two types of detectors are used: Indium-antimonide (InSb) and mercury-cadmium-telluride (HgCdTe). Their sensitivity to the skin's heat radiation differ distinctly. Both detectors must be cooled with liquid nitrogen to a temperature of 77 K or −196 °C. The InSb detector only registers a small portion of the skin's heat radiation, the short wavelengths. The skin does not act as a black emitter within this range. The HgCdTe detector detects most of the skin's heat radiation (8–12.5 μm). The skin acts almost as an ideal black emitter within this range of wavelengths. Dermatological thermograms may be difficult to interpret because of the relative cooling effect of dermal oedema, vesicles or blisters in the upper skin layers, and evaporative water loss in response to sweating. Anatomical and physiological factors must also be considered.

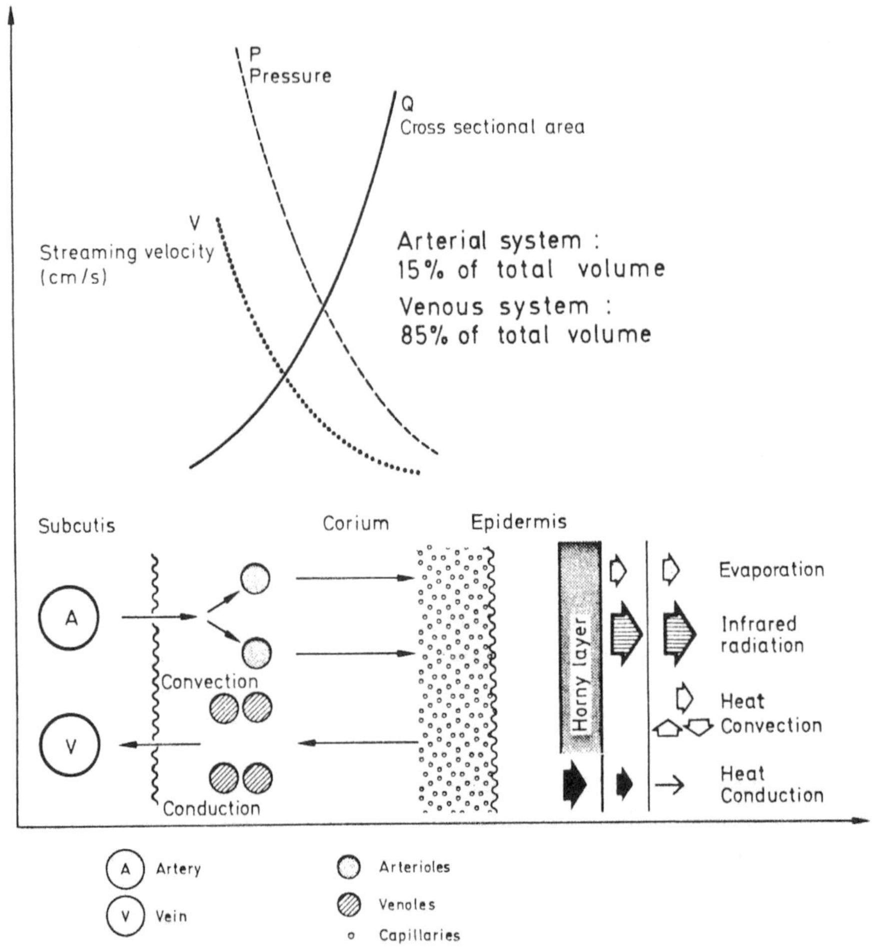

Fig. 1. Heat conduction from the skin to the skin surface

Apart from the well known clinical signs of redness, swelling and surface damage, dermatoses are frequently characterized by changes in blood flow, and thus by alterations in heat transport from the body core by the skin microcirculation. Skin blood flow is one of the most important factors influencing surface temperature. Insensible water loss and sweat gland secretion are most important in skin temperature regulation, but the thickness of the different skin layers influences heat conduction from the level of the microcirculation to the surface (Fig. 1).

Preparing the Patient

Acclimatization in an environmentally controlled room at approximately 20 °C and 40% humidity for at least fifteen minutes before measurements are taken is necessary. Cooling of the skin occurs after clothing has been removed. A dynamic study can be performed at this time examining the degree of cooling in relation to the time intervals following the removal of clothing. Further dynamic tests are possible either with air or water cooling or warming after the acclimatization period has been completed. The rewarming behaviour of small areas of skin with increased temperature can be studied by cooling them with a rubber bag containing water of a known temperature. The emission capacity of human skin can be altered by dermotological and/or cosmetic skin treatment. Only dry skin and skin free of residue, be it cosmetic or dermatologic, acts as a true black emitter for infrared thermography. Otherwise, the surface skin temperature is not a reflection of the emitted radiation. Serious errors occur when moist or wet skin is used. The hydration of skin also plays an important role. A layer of water vapour leads to absorption of the infrared energy radiated by the skin, and less heat radiation reaches the infrared camera so that cooler skin areas are simulated. It is advisable for the patient to bathe on the evening preceding examination. On the day of examination all local teatments should be avoided.

Erythema

Skin erythema is characterized by an increase in the number of erythrocytes within the subpapillary capillary plexus. In the cases of increased blood flow through the capillary network, erythema may be associated with high skin surface temperatures owing to greater heat transport. Skin temperature is dependent more on blood flow than on the amount of blood in the skin. For example, haemangiomas can present "cool" lesions with a low surface temperature because they are not integrated into the normal blood circulation but are by-passed by shunts.

The intensity of erythema does not show a linear correlation with convective and conductive heat transport.

In erythema induced by various means such as dermographism, the application of nictotinic acid benzylester, UV-B irradiation, or dithranol, the mediators released are cleared in the direction of blood and lymph drainage.

As a result heat reflection occurs not only from the erythematous region, but also from the surrounding apparently unaffected skin area. Venous blood vessel texture is increased within areas of drainage. The development of erythema following UV-B radiation correlated with the maximum skin temperature after 24 h. The erythema following antrhalin also appeared after 24 h, but showed a maximum skin temperature 24–28 h later than the UV-B erythema. The erythema correlated with the area of skin in contact with anthralin of UV-B, whereas increased heat radiation is from a much wider area (Fig. 2).

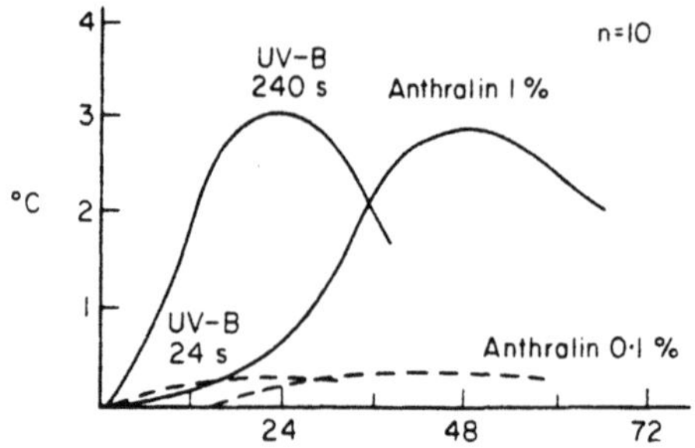

Fig. 2. Single exposure of UV-B and anthralin: time lapse between maximum heat radiation at 24–48 h

The cause of this time difference is the liberation of mediators with different vasodilator activity. The difference can only be assessed with thermography as the erythema formation is similar after both treatments. Thermography is a very instructive method of showing the interaction between direct changes in heat radiation of the surface of the skin and its relation to reflex processes. Short reflex arcs, such as the axon reflex, and long reflex arcs, such as the flush phenomenon provoked by visceral cutaneous reflexes, can be illustrated by thermography. Reflex changes in skin temperature by cooling the skin of patients with cold contact urticaria with the short-term development of large areas of warmed skin and also reflex changes in cholinergic urticaria after prior heating of such skin areas where the urticarial eruption follows are reflected in the picture of telethermography. Reflex arcs between extremities affect the arterial blood flow and show individual variation, as in the case of the cessation of permanent local vasocontriction by indirect cooling. The thermographic illustration of convective heat transport to the skin surface can be obstructed by the cooling effect of sweat secretion, e.g., water vapour exudation and by changes in conductive heat transport in abnormal skin.

Urticaria

The cause for an urticarial reaction is an increased vascular permeability combined with sufficiently high blood pressure in the blood vessels of the microcirculation, predominantly in the venous part. It is useful to distinguish two types of development of oedema in the skin:

1) the water diffusion from the deeper layers of the skin vessels,
2) the increased permeation of water from the superficial net of microvasculature.

In the first case the temperature of the fluid shows the same temperature as the blood inside the vessels of the deeper vascular net. This urticarial weal shows an increased heat radiation from the skin surface. In the second instance the capillary and venous vessels show a diminished temperature in relation to the deeper net, and in this case the development of such an oedema without an increased convective heat transport has a cooling effect on the skin. Taking dermographism as an example, this situation

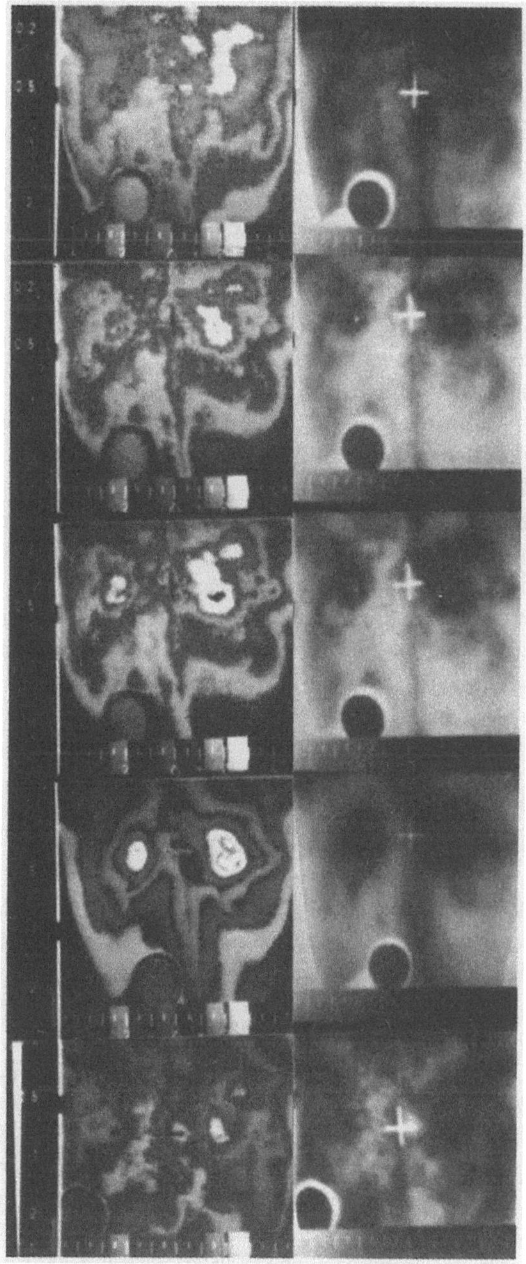

Fig. 3. Type I allergic reaction of the immediate type. Documentation of the reactions before the intracutaneous injection of the allergens after 2, 4, 30 min and 24 h later. *Left side:* injection of cat hair (Bencard 0.05 ml). *Right side:* injection of horse hair (Bencard 0.05 ml); skin temperature is once again related to the center of the reaction

is demonstrated in the area of an erythema which has been provoked by topical application of nicotinic acid benzylester. After the skin has been irritated by pressure, the small area of dermographism is characterized by a decrease of the skin temperature by approximately 0.5 °C. The urticarial reaction provoked by nicotinic acid cannot be diminished by topical application of antihistamines. The pattern of reaction suggests that the mediators involved are kinins which are not inhibited by antihistamines.

In cold-induced urticaria reflex-erythema consequent to cold contact and the subsequent increase of heat radiation (greater than 20 °C) can decrease the heat radiation of the weal itself without affecting its size. The cooling effect from the oedema can be concealed by a simultaneous increase in convective heat transport.

In cholinergic urticaria the increased heat radiation after provocation precedes the urticarial eruption. Nicotinic benzylester (NBE) markedly increased the heat radiation whereas the urticarial eruption in the NBE-field-provoked in the same manner as in dermographism shows a distinct cooling effect.

Histamine seems to be the primary mediator in the cases of urtication following physical skin irritation and elicits not only an increase in arterial blood flow, but also increased vessel permeation.

In an allergic reaction of the immediate type, stepwise increases of the infrared radiation develop from three minutes onwards and can be followed on the AGA Thermovision in the form of isotherm-pictures (isothermograms). The onset of the appearance of skin wheal occurs some minutes earlier than the development of the maximal increase of infrared radiation. This shift with respect to time is probably due to oedema formation, which initially reduces the increased heat flow to the skin surface in regions of allergic reaction. Only during the following minutes does the continuously increasing heat supply take prominence over heat absorption by the oedema (Fig. 3).

Papular Reaction

An allergic reaction of the delayed type (tuberculin type) is accompanied by a circumscribed increase in heat radiation. The same situation is found after the intracutaneous injection of pyrexal (coli endotoxin). Such an increase in temperature can be reduced by pretreating the skin test areas with topical corticosteroids such as clobetasol and betamethasone valerate.

Clinically the papule seems to be less changed by the corticosteroids than the reduction in convective heat transport to the skin. In such a delayed reaction, in contrast to the effect in urticaria, the corticosteroids, which have permeated into the deeper layers of the skin, reduce the heat radiation but influence the erythema and the papules to a lesser degree. The varying mediators of inflammation in the different types of reaction may be responsible for such different inhibitory effects of corticosteroids on heat radiation, permeability of the vessels, and cellular infiltrate (Fig. 4).

Contact allergy is also characterized by increased heat radiation. The formation of vesicles has a distinct cooling effect on small areas of the skin only in the acute phase of the eczema. The increased heat radiation of the skin surface in the area of an allergic contact eczema is still visible even though the eczema is healed morphologically. The areas of former eczematous reactions still demonstrate an increased heat transport to the skin surface. Similar observations have also been made after the disappearance of

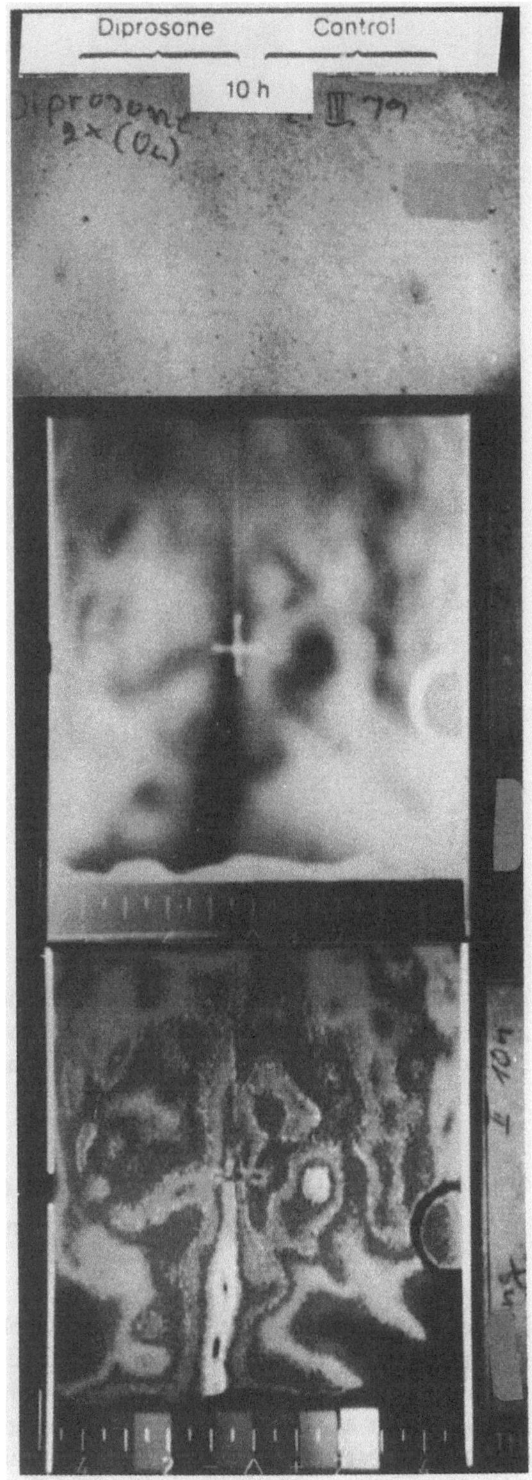

Fig. 4. Pyrexal reaction and corticosteroid effects. Pyrexal reaction (0.1 ml injected in the area pretreated with betamethason-17,21-dipropionat cream), after 10 h. Control pretreated with basis cream without corticosteroid

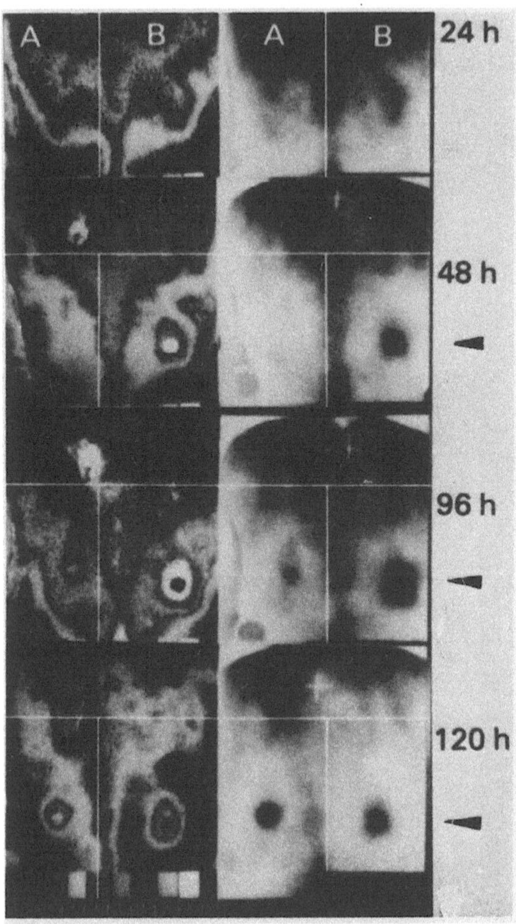

Fig. 5. Pretreatment with corticosteroid ointments delays the development of allergic eczematous reaction against paraphenylendiamine 1%. Demonstration of the course of heat radiation with a shift at 48 h. A = Pretreated area with corticosterioid; B = Pretreated area with ointment basis

positive test reactions of the immediate type. Such reactions are characterized by an increased heat transport to the skin surface. Even when they are no longer visible such locally increased skin temperatures can last between five and seven days.

Telethermographic Analyses of Topical Corticosteroid Effects in Skin Reactions of the Immediate and Late Types

Neither intracutaneous injection nor topical application of corticosteroid (McKenzie test) produces direct vasoconstriction combined with a diminished convective heat transport. The blanching phenomenon with slight decrease of the skin contact temperature occurs as a result of an altered transparency of stratum corneum, increased water vapour emission, and because of a diminished volume of erythrocytes in the upper part of the capillary – venule system. The addition of corticosteroids to intracutaneously injected pollen allergen solution inhibits the immediate type of reaction (type I) and causes a simultaneous decrease in heat radiation, as proven by thermography. Pretreatment with topical corticosteroids inhibits the development of cold-induced urticaria, dermographism, as well as an immediate type reaction, type I, without there being a decrease in heat radiation. Pretreatment with topical corticosteroids also inhibits the development of allergic eczematous contact reaction (type IV) and the toxic papular reaction after injection of E. coli endotoxin (pyrexal test). Both reactions are accompanied by a distinct decrease in heat radiation. Topical pretreatment with corticosteroids delays the normal development of delayed type reaction. It requires an additional one – two days for full development. In general, the topical application of corticosteroids is more effective in preventing urticarial eruptions than in suppressing convective heat transport (Fig. 5). Thermography is useful in demonstrating these effects in quantitative terms and analysing the mode of action.

Alterations of Subcutaneous Fatty Tissue

Fatty tissue is an excellent insulator preventing conductive heat transport. The influence of the subcutaneous fatty tissue on heat radiation of the skin can be demonstrated in the case of the increased fatty tissue of lipomas or the decrease of subcutaneous fatty tissue in scleroderma or anetoderma which includes the subcutaneous fat. In localised scleroderma, where the fatty tissue is diminished, it can also be seen that heat radiation from the skin lesion is markedly increased, because the heat radiation of the deeper layers of the body reaches the skin surface without being inhibited by the fatty tissue.

Varicose Veins and Thermography

The enlargement of the blood volume in varicose veins increases the skin temperature. The temperature of the veins is high enough to increase the heat radiation from the skin around these veins due to AV shunts. Subcutaneous injections of catecholamines and serotonin lead to vasoconstriction not only of the normal but also of varicose veins. The same effect can be shown by iontophoresis of the same substances. The vasoconstriction of veins is followed by reduced heat flow to the skin surface. The vasoconstrictory effect is not restricted to the area of varicose veins, but includes broader skin area in the neighbourhood of the enlarged veins and leads obviously to a decrease in convected heat transport affecting the arterioles.

References

1. Aarts NJM, Gautheri M, Ring EFJ (eds) (1975) Thermography. Proceedings. Proceedings of the 1st European Congress on Thermology, Amsterdam 1974. S. Karger, Basel München Paris London New York Sydney
2. Engel J-M, Flesch U, Stüttgen G (eds) (1985) Thermological method. VCH Publishers
3. Gautherie M, Albert E (eds) (1982) Biomedical thermology. Progress in Clinical and Biological Research, vol 7. Alan R Liss, Inc, New York
4. Ring EFJ, Phillips B (eds) (1984) Recent advances in medical thermology. Proceedings of the 3rd Int. Congress of Thermology, Bath 1982. Plenum Press, New York
5. Stüttgen G, Flesch U (eds) (1985) Dermatological thermography VCH Verlagsgesellschaft, Weinheim Deerfield Beach/Florida Basel
6. van Voss SFC, Thomas P (eds) (1969) Medical thermography. Proceedings of a Boerhaave Course for Postgraduate Medical Education, Leiden. S Karger, Basel New York

The Long-Haired Syrian Hamster: An Animal Model for Androgen Induced Hair Growth

N. Orentreich and J. R. Matias

The mechanism controlling the growth of secondary sexual hair has been the subject of many recent reviews [1, 3, 4]. More recently, Toscano et al. [6] suggested that hirsutism may represent an "evolving syndrome" which progressively involves the peripheral androgen metabolism, the adrenal, and then the ovaries, mediated probably through functional changes in the hypopituitary axis. Further advances in our understanding of this disorder and progress in the development of effective methods of treatment have been hampered by the absence of a convenient animal model. The sebaceous glands may be used as a model to evaluate androgenic and antiandrogenic compounds. However, data derived from sebaceous glands may not apply to the sexual hair since they are structurally and functionally unrelated.

This paper describes the suitability of the long-haired hamster as an animal model for androgen induced hair growth.

Materials and Methods

Animals. The long-haired Syrian (LHS) hamster is an autosomal recessive mutation of the wild-type Syrian (WTS) hamster [5]. This mutation is popularly known in the pet industry of the United States as the "teddy bear hamster." Specimens of this mutant strain were purchased from local pet shops and were randomly bred in our laboratory. At various times, the laboratory population was augmented by the purchase of 4-week-old animals from local breeders. WTS hamsters were purchased from Engle Laboratory Animals, Inc. (Farmersburg, Indiana, USA).

Hamsters were maintained at a photoperiod of 16 h of light and 8 h of darkness. Food (Rodent Laboratory Chow, Ralston Purina Co., St. Louis, Missouri, USA) and tap water were provided ad libitum.

Hair Measurements. Small tufts of hair from different regions of the hamster skin were cut with micro-surgical scissors at the level of the skin surface. The hairs were spread on a microscope slide and the maximum length of hair found in each region of the skin was measured under a magnifying lens with the aid of an electronic graphics calculator (Model 1224, Numonics Corp., Lansdale, Pennsylvania, USA). The diameters of 15 randomly chosen hairs from each region of the skin were determined using a Leitz Ortholux II microscope equipped with a drawing attachment and interfaced with the graphics calculator. All diameter measurements were made at a magnification of 625x.

The total mass of hair was determined by shaving the entire hair coat with an electric clipper using a size 40 blade (Model A5, Oster Corp., Milwaukee, WI, USA). In other experiments, the hair coat was trimmed with scissors at a level of 20 mm from the base of the shaft using a 20 mm metal strip as guide. Hair below and above 20 mm were evaluated separately. Prior to weighing, the hair samples were freed of debris, defatted by extraction with ether and dried overnight in a desiccator over anhydrous $CaSO_4$.

Castration. Male LHS hamsters at 8 weeks or at 16 weeks of age were castrated via the scrotal route under sodium pentobarbital anesthesia (50 mg/kg; i.p.). Control animals were sham-operated.

Androgen Stimulation. Testosterone (Steraloids, Inc., Waltham, Massachusetts, USA) was suspended in an aqueous vehicle composed of 0.2% sodium carboxymethylcellulose, 0.4% Tween 80, 0.8% benzyl alcohol and physiological saline (q.s.). Three-month-old male, female, and gonadectomized male LHS hamsters (castrated at 8 weeks of age) were injected daily in the mid-dorsum with either 0.1 or 1.0 mg of testosterone in 0.1 cc of the vehicle for a period of 7 weeks. Control animals were injected daily with 0.1 cc of the aqueous vehicle.

Statistics. Student's t-test was used to evaluate the significance of the data.

Results

The sexual dimorphism of the pelage in the LHS hamsters is evident (Fig. 1). Hypertrichosis in male LHS hamsters occurred primarily along the flanks and the rump areas. The hair in these regions is characterized by long, wavy strands which vary in length from 30 to 60 mm depending upon the age. Short hairs, ranging in length from 10 to 20 mm were normally present along the head, mid-dorsum, and ventral skin. Occasionally, LHS females may also demonstrate some degree of hypertrichosis along the rump and flanks. However, the magnitude of hair growth was considerably smaller in comparison to male LHS hamsters.

The maximum hair length from various skin regions of LHS and WTS hamsters was compared (Fig. 2). With the exception of the head region, the hair length of male LHS hamsters was longer in comparison to the female LHS hamsters at all regions sampled. The greatest difference between male and female LHS hamsters occurred along the rump and flanks. In general, the hair length was shorter in WTS hamsters than in the long haired mutant strain. Figure 3 shows the hair diameter of the long-haired flank and short-haired mid-dorsal skin of male and female LHS hamsters. The mid-dorsal hair diameter was similar for both sexes while the flank hair diameter in male LHS hamsters was 13% greater than the females ($p < 0.01$).

Table 1 demonstrates the comparison between LHS and WTS hamsters using gravimetric measurement of the hair coat. The weights of the hair coat of male and

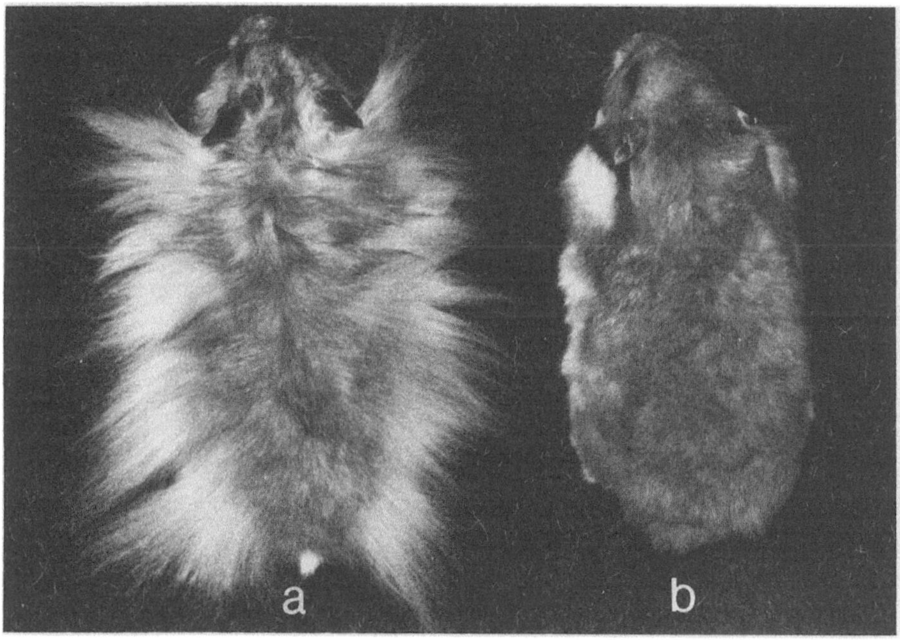

Fig. 1a, b. Comparison of the hair coat between **a** LHS male, and **b** LHS female

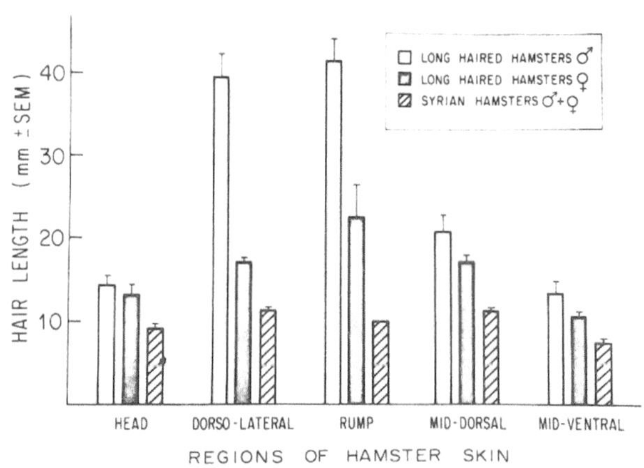

Fig. 2. Differences in the hair length between 12-week-old male and female LHS and wild-type Syrian golden hamsters. Each bar graph represents the mean and SE of 5 to 6 animals. The data for the male and female wild-type hamsters were pooled since differences between the two sexes were not observed

female WTS hamsters were again combined since significant differences were not observed between the two sexes. The total weight of hair in wild type Syrian hamsters was smaller in comparison to LHS hamsters ($p < 0.001$). Differences in the total weight of hair between male and female LHS hamsters were not significant at the 0.05 level of probability.

Hairs were clipped at a level 20 mm above the skin surface. This distance was chosen since it represented the maximum length of hair normally found in the short-haired areas (Fig. 2). Thus, the weight of the hair collected above this level represents a quantitative marker of the sexually dimorphic character of the LHS hamster pelage.

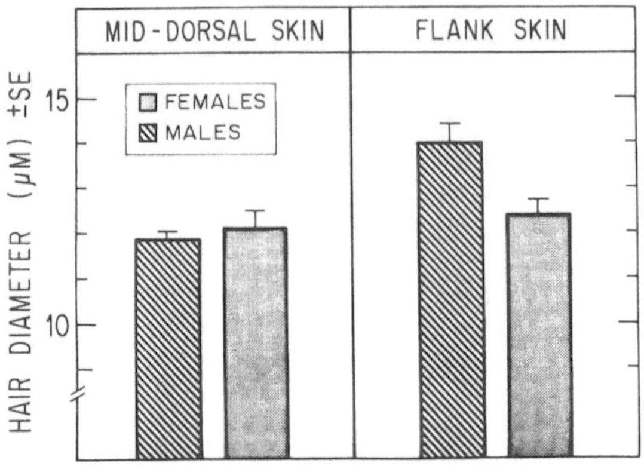

Fig. 3. Diameter of hair from flank and mid-dorsal skin of LHS hamsters. Mean ± SE of 5 animals per data point

Table 1. Weight of hair in 12-week-old LHS and WTS hamsters. Mean ± S.E.

	LHS hamsters		WTS hamsters male & female
	male	female	
(n)	(7)	(7)	(10)
Body weight, g	120 ± 3	117 ± 7	124 ± 2
Total hair, g	1.50 ± 0.05	1.32 ± 0.09	0.88 ± 0.04
Short hair*, g	1.40 ± 0.05	1.31 ± 0.09	–
Long hair**, mg	100 ± 1	5 ± 2	–

* mass of hair up to 200 mm in length
** mass of hair cut above 20 mm from the skin surface

Table 1 shows that the weight of the hair below 20 mm (short hair) was similar for both sexes while the weight of hair above 20 mm was 20-fold greater in males than in females ($p < 0.001$). In the interest of brevity, the hair above 20 mm from the base of the shaft will be referred to by the term "long hair." The subsequent evaluation of this model system utilized the measurement of the long hair weight (LHW).

The relationship between the LHW and sexual maturity was evaluated in LHS hamsters. Sex differences in the amount of long hair may be observed as early as 8 weeks of age (Fig. 4). LHW continued to increase with age in the male. The hypertrichosis observed in 8–12 week old females was transitory and was no longer evident at 20 weeks of age. The effect of castration on the growth of "long hair" was investigated using 4-month-old male LHS hamsters. The LHW declined to values similar to female LHS hamsters at two months post-castration (Fig. 5).

The effect of testosterone administration on the growth of hair in male, female, and gonadectomized males is presented in Fig. 6. In female LHS hamsters exogeneous testosterone at a dose of 0.1 mg/day did not stimulate the production of long hair. Increasing the dosage to 1 mg/day produced a significant increase in the LHW ($p < 0.001$).

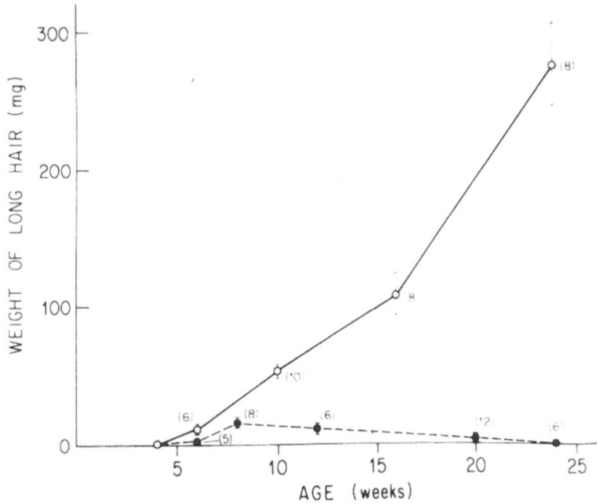

Fig. 4. Relationship between weight of LHW and age in male (○) and female (•) LHS hamsters. The number of animals per data point is shown in the parentheses

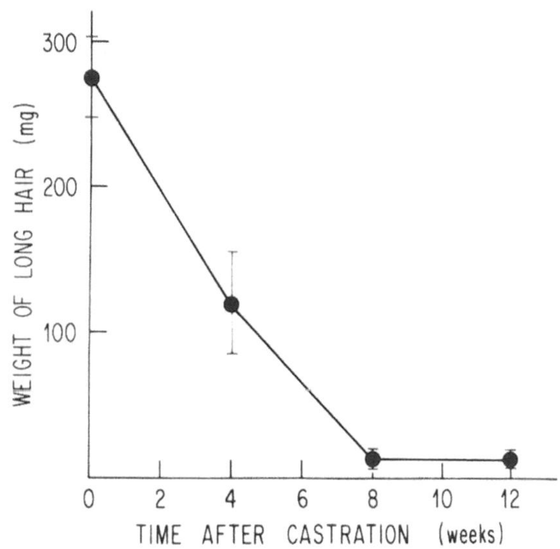

Fig. 5. Influence of castration on LHW of adult male LHS hamsters. Mean ± SE of 6–8 animals

Castration at 8 weeks of age prevented further increase in LHW normally observed in sham-operated and untreated males. In contrast to female LHS hamsters, testosterone injections at a dose of 0.1 mg/day to castrated males produced an increase in LHW compared to the vehicle injected animals. At a dose of 1 mg/day a 6-fold increase in LHW was apparent in the castrated males. An increase of approximately 50% and 150% was observed when normal males were injected with testosterone at doses of 0.1 and 1.0 mg/day, respectively.

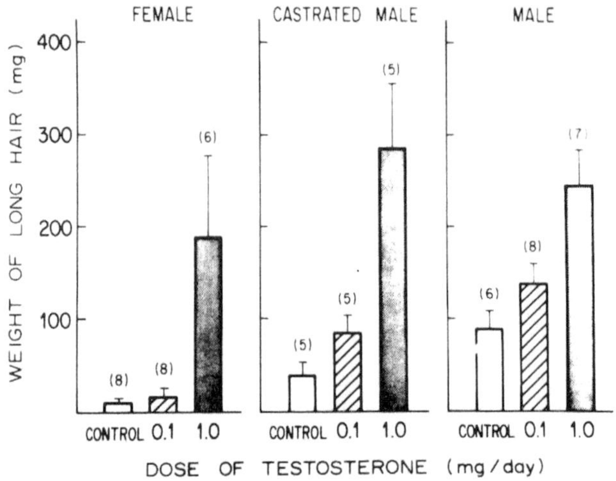

Fig. 6. Effect of the subcutaneous administration of testosterone on LHW of male, female, and castrated male LHS hamsters

Discussion

Much of our knowledge concerning the mechanism of androgen action comes from the studies of gene defects in human beings and animal model systems. These mutations are manifested by abnormalities either in the enzymes for testosterone synthesis, the 5α-reductase enzyme, the androgen receptor protein, or the post-receptor responses [2]. To further understand the mechanism of androgen-induced hypertrichosis, there is a need for in vivo models in which androgens turn on the hair follicle cells to produce longer, thicker hairs.

Mutations for long hair occur in a wide variety of mammals, such as selected breeds of sheep, cats or dogs. However, the mutation observed in the LHS hamster is unique because its expression is dependent upon the presence of androgens. Hair growth in the LHS hamster has a striking similarity to that observed in the facial hair of man. Castration of LHS hamsters causes loss of long hair while testosterone replacement stimulates the growth of long hair in females and castrated males. It is also of interest to note that testosterone administration to the intact male LHS hamster increased the mass of long hair, suggesting that not all hair follicles are stimulated at physiologic levels of androgens.

Although LHS hamsters have existed in the pet industry for decades, the exact origin of these animals remains unknown. Schiemke et al. [5] were the first to introduce this mutation in the scientific literature by describing the mode of inheritance, coat texture and hair length. However, the sexual dimorphism of the pelage was not noted. Although the sex difference in the hair coat was quite obvious, a more quantitative method of measuring these differences was important.

The present study utilized morphometric and gravimetric methods to describe the hormonal regulation of hair growth in this model. Measurement of the LHW is the most convenient and appropriate technique. Because lengths of hairs at any given region are variable, the determination of maximum length may also be used as an indicator of the growth potential of the hair coat in each region. However, hair length measurements produce inaccurate results since females may also occasionally produce some long hairs. The measurement of hair diameter is time consuming, particularly when measuring large samples. Furthermore, the differences between males and females are smaller when diameter and length measurements are utilized as markers of the sexual dimorphism.

The LHS hamster is small and easy to handle. The relatively inexpensive cost of procurement and maintenance make them available for use by most research laboratories. For these reasons the LHS hamster is an ideal and convenient model for the evaluation of androgens, antiandrogens, and other compounds which may affect hair growth.

References

1. Ebling FJ, Hale P (1983) Hormones and hair growth. In: Goldsmith LA (ed) Biochemistry and physiology of the skin. Oxford University Press, pp 522–522
2. Leshin M, Wilson JD (1981) Mechanisms of androgen-mediated hair growth. In: Orfanos CE, Montagna W, Stüttgen G (eds) Hair research, Springer, Berlin, pp 205–209
3. Pochi PE (1980) Hormonal control of hair growth with special references to steroids applied topically. In: Mauvais-Jarvis P, Vickers CFH, Wepierre J (eds) Percutaneous absorption of steroids. Academic Press, New York, pp 155–165
4. Rentoul JR (1983) Management of the hirsute woman. Int J Dermatol 22:265–272
5. Schimke DJ, Nixon CW, Connelly Me (1974) Long-hair: a new mutation in the Syrian hamster. J Hered 64:236–237
6. Toscano V, Adano MV, Caiola S, Foli S, Petrangeli, Casille D, Sciarra F (1983) Is hirsutism an evolving syndrome? J Endocrinol 97:379–387

Biochemical Markers in the Mouse Tail Model of Psoriasis

P. T. Bladon, N. F. Cooper, E. J. Wood and W. J. Cunliffe

Psoriasis is characterised by an abnormal hyperproliferative stratum corneum [27]. In recent years the major and characteristic proteins of epidermis, the keratins, have been the subject of intense research effort and there have been major advances in knowledge of the biosynthesis and structure of these proteins [10–12, 14, 15, 18, 30, 31, 36] and several investigations have shown that the expression of keratin polypeptides in psoriatic lesions is abnormal [2, 7, 20, 23, 25, 28, 34, 39, 40]. However it is by no means clear to what extent, if any, the presence of an abnormal set of keratin polypeptides is responsible for the manifestations of the disease. Many components of the epidermis are observed to be abnormal in psoriasis, but in most cases this abnormality is thought to be a secondary phenomenon. The restoration of a normal set of keratins when psoriatic lesions resolve has been reported [26, 35].

A number of models for psoriasis have been proposed including the induction of hyperplasia by adhesive tape stripping [19] and the buccal epithelium in normal humans [17], but no pathological animal model is known. The normal adult mouse tail has alternating rings of ortho- and para-keratotic stratum corneum characterised by the presence and absence of a granular layer respectively and this has been used as a model [6, 32, 42–44]. In the ortho-keratotic interscale regions round the hair follicles a granular layer is present and the stratum corneum is histologically similar to that of normal human epidermis. In contrast, in the scale regions the granular layer is thin or absent and nucleated cells are visible in the stratum corneum. These regions of parakeratotic stratum corneum are histologically similar to those found in the stratum corneum of psoriatic plaques in humans.

In this study the effects of coal tar, an established treatment for psoriasis, and of an aromatic retinoid (Ro11-1430, Motretinide), were compared in the mouse tail model by both histological and biochemical parameters. The possible use of this model for screening treatments for psoriasiform diseases is discussed.

Methods

Male or female haired (CFLP) or hairless (hrhr) mice were allocated to the following treatment groups.

Group 1: control group (no treatment)
Group 2: 0.3% commercial coal tar solution (Polytar – Stiefel Laboratories)

Group 3: 100% crude coal tar
Group 4: 0.3% Motretinide in an inactive base
Group 5: placebo group (base for group 4)

Mice were caged separately and the treatment described was applied once daily for four weeks. Treatment was applied only to the 2.5 cm of tail nearest the body. At the end of each treatment period the animals were sacrificed by dislocation of the neck, tail skin was stripped from the underlying cartilage and samples were subjected to histology (haematoxylin and eosin staining). Epidermis was separated from underlying dermis by soaking in 1% (v/v) aqueous glacial acetic acid followed by microdissection. Prekeratin was extracted from tail epidermis using 0.1 M citric acid-trisodium citrate buffer, pH 2.65 [8] and epidermal keratin was extracted by the procedure of Winter, Schweizer and Goerttler [41]. One- and two-dimensional electrophoretic analysis of epidermal prekeratin and keratin from treated and untreated mice was performed. Two-dimensional analysis was by Non-Equilibrium pH Gradient Electrophoresis (NEpHGE) [29].

After high resolution one-dimensional SDS/polyacrylamide gel electrophoresis and staining with Coomassie blue, gel tracks were scanned by densitometry. This was carried out by using a spectrodensitometer (Model SD 300) and density computer (Model SDC 300, supplied by Schoeffel Instrument Corp. and processed by a Hewlett Packard Integrator Model 3390A). Gels supported between two glass plates were scanned by transmission. In our hands optimal conditions for scanning were: wavelength 605 nm, slit width 0.4 mm, scan speed 25 mm/min. After scanning the relative concentration of each polypeptide was determined (as a percentage of the total protein concentration). The data are presented as histograms.

Results

Mice from each group were examined after treatment for the effect of each compound on the physical appearance of the tail. After treatment mouse tails from groups 2, 4 and 5 appeared to have the same macroscopic appearance as those in group 1. However, the tails of mice from group 3 were erythematous and there was a reduction in the number of hairs in the treated area. The photograph (Fig. 1) shows typical examples from groups 1 and 3 (marked C and CT respectively).

Histology (Fig. 2) showed that treatment with coal tar, whether in a crude state or as a commercial solution (0.3%), induced the formation of a continuous granular layer which gave rise to a continuous orthokeratotic stratum corneum instead of alternating ortho- and parakeratotic regions. The same effect was produced by treatment with a 0.3% Motretinide solution.

Figure 3 shows a comparison between prekeratin from normal and from coal tar – treated mouse tails. Normal tail prekeratin is shown in track 1 and prekeratin from crude coal tar (100%) treated mouse tail is shown in track 2.

Prekeratin extracted from mouse tails which had first been treated with crude coal tar (100%) was found to be deficient in a polypeptide of molecular weight 73,000.

Fig. 1. Macroscopic appearance of normal and coal tar-treated mouse tails. Tails of male, CFLP haired mice were treated with crude coal tar for a period of four weeks. Crude coal tar (CT) was applied to the region of each tail nearest the body with a paint brush. The control group of mice (C) was also kept under the same environmental conditions. Group size: 16 animals

Densitometric tracings of tracks 1 and 2 from (Fig. 3) are shown in Fig. 4 below. As can be seen both from the gel picture (Fig. 3) and the densitometric tracings (Fig. 4), the polypeptide of mol. wt. 73,000 was reduced, in prekeratin from the crude coal tar-treated tail whilst the mol. wt. 73,000 polypeptide is relatively increased. Densitometric tracings (scans) of polyacrylamide gels containing prekeratin from treatment groups 2 and 4 showed a similar profile to those obtained after crude coal tar treatment.

Prekeratin extracted from placebo, Motretinide and 0.3% commercial coal tar-treated tails was analysed by one-dimensional SDS/polyacrylamide gel electrophoresis. Scans of the gel tracks containing these proteins were made and the percentage of each polypeptide as a fraction of the total polypeptide composition is expressed as a histogram in Fig. 5. A scan of the gel of untreated mouse tail prekeratin was also made, i.e., control group (Fig. 4a). As can be seen from comparing the histograms for active treatment, i.e., Motretinide and 0.3% commercial coal tar, with the control group, the mol. wt. 73,000 polypeptide decreases with respect to the mol. wt. 57,000 band. The result is even more striking when a histogram of the densitometric tracing of a gel containing prekeratin from crude coal tar treated mouse tails is plotted (not shown).

Mouse tail keratin extracted from Motretinide, placebo and commercial coal tar (0.3%)-treated mouse tails was analysed by one dimensional SDS/polyacrylamide gel

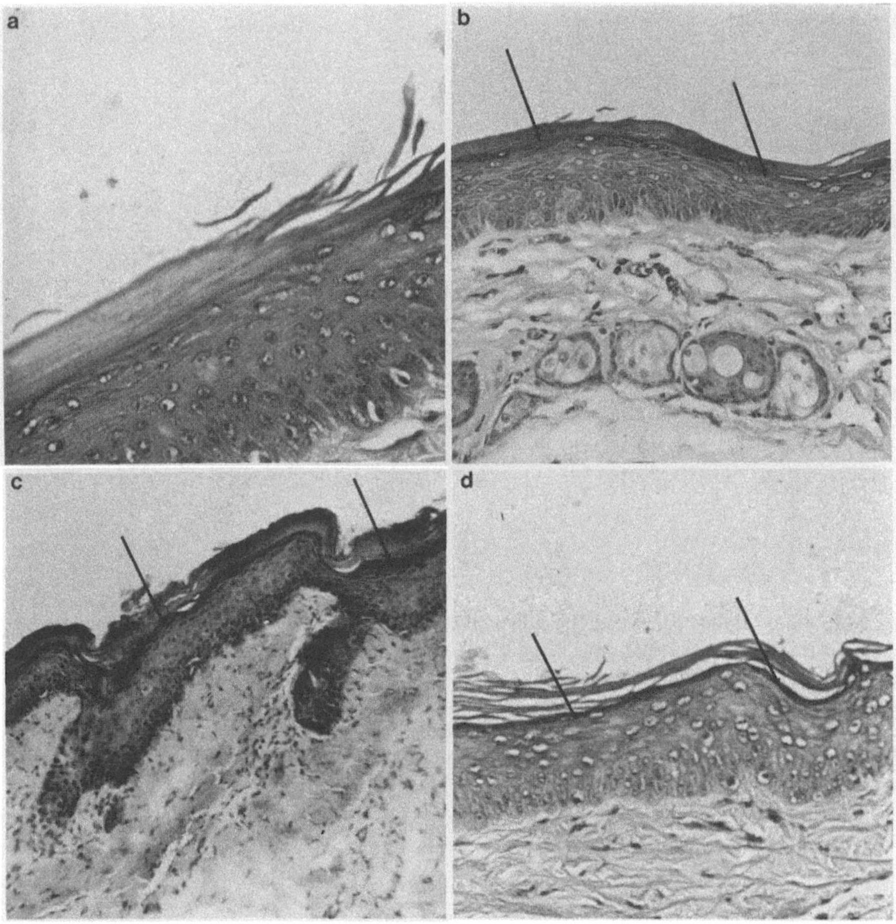

Fig. 2a–d. Histological analysis of untreated and treated mouse tails. **a** Section of a normal mouse tail through the parakeratotic region. The stratum corneum is scaly and no granular layer is visible. **b** Low-power magnification of the epidermis of a 0.3% commercial coal tar treated mouse tail. A continuous granular layer (arrowed) is visible through the epidermis. **c** Low-power magnification of the epidermis of crude coal tar treated tail. A granular layer (arrowed) is easily visible **d** 0.3% Motretinide treated mouse tail epidermis. A continuous granular layer (arrowed) can again be seen

electrophoresis. The gel tracks were then scanned using a densitometer and the results are presented in Fig. 6. By eye the gel patterns were somewhat similar to each other but careful quantitative analysis by scanning densitometry revealed that distinct changes had occurred with respect to the proportions of each of the keratin polypeptides present.

In the Motretinide and coal tar treated groups the proportion of mol. wt. 73,000 to mol. wt. 67,000 polypeptide had changed. The mol. wt. 67,000 polypeptide had become more abundant than the mol. wt. 73,000 polypeptide in these groups when compared with prekeratin from the control and placebo groups. Previous work by ourselves [3] and others [11, 13] indicates that this effect is likely to be due to a

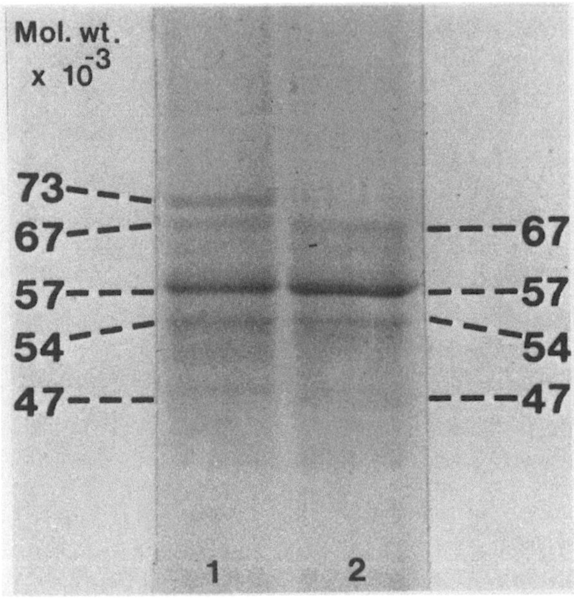

Fig. 3. Effect of crude coal tar on the polypeptide profile of mouse tail epidermis. Prekeratin was extracted from mouse tail epidermis using 0.1 M citric acid-trisodium citrate buffer, pH 2.65, and analysed by SDS/polyacrylamide gradient (7.5–17.5%) gel electrophoresis (track 1). Prekeratin extracted from the epidermis of mouse tail which had been treated with crude coal tar (track 2) was deficient in the polypeptide of highest molecular weight (73,000). During the electrophoresis [24] the positive electrode was at the bottom of the gel. Molecular weights of mouse tail prekeratin polypeptides are shown ($x10^{-3}$)

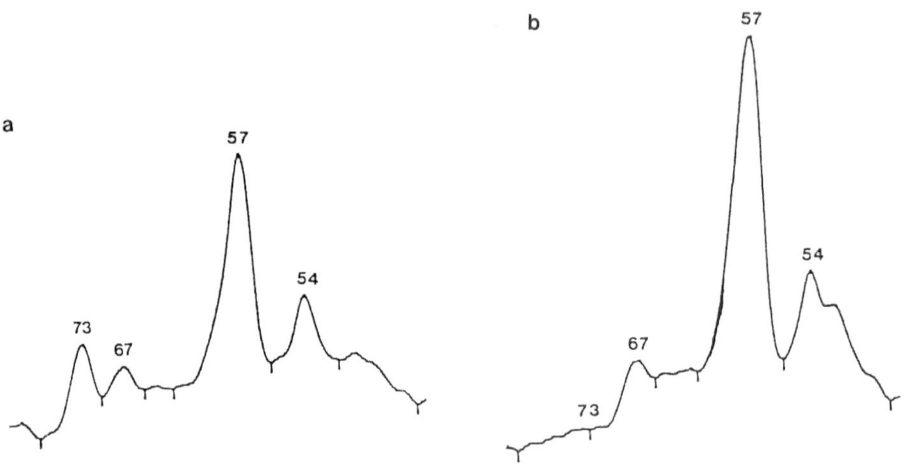

Fig. 4a, b. Scanning densitometry of prekeratin from control and crude coal tar treated mouse tails. Scanning densitometry was performed on SDS/polyacrylamide gels of prekeratin (one-dimensional) which had been extracted from normal **a** and from crude coal tar treated **b** mouse tails. The numbers show the molecular weights ($x10^{-3}$) of mouse tail prekeratin polypeptides. As can be seen by comparing the scans, the 73,000-dalton polypeptide of mouse tail prekeratin is depleted after crude coal tar treatment and relatively more 57,000-dalton polypeptide is expressed when compared with the control prekeratin

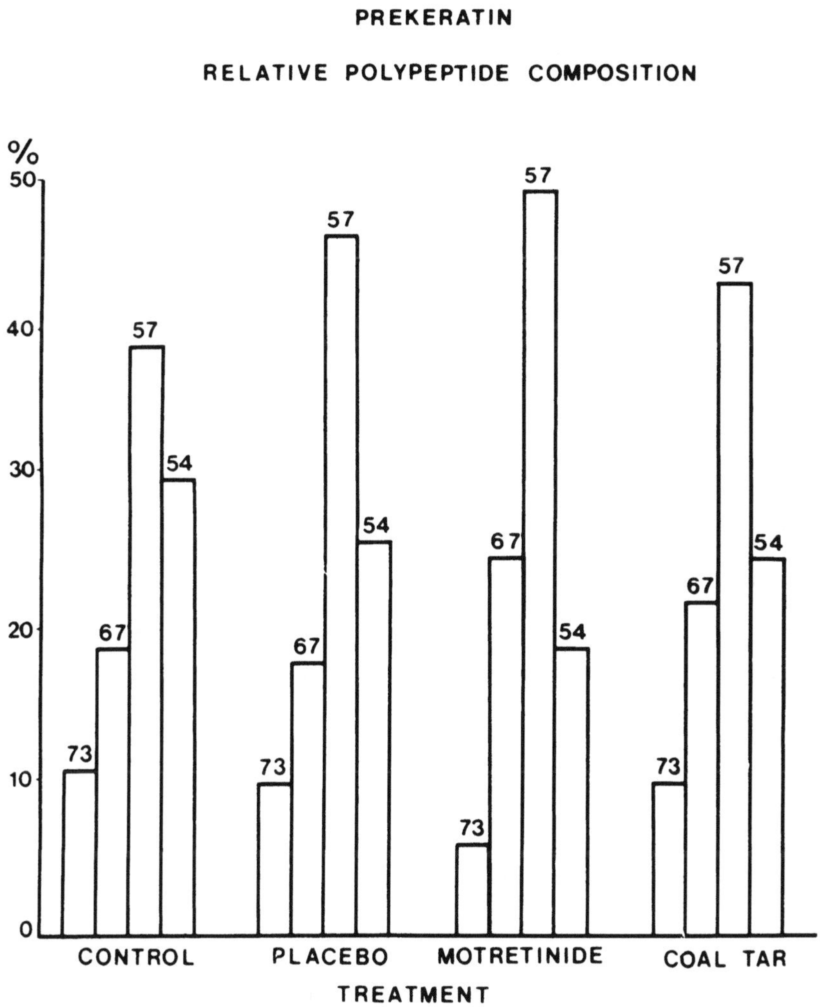

Fig. 5. Histograms of scanning densitometry traces (prekeratin). Quantitative analysis of the traces of SDS/polyacrylamide gels which contained prekeratin from mice on four different treatment regimes: control, placebo, Motretinide and 0.3% commercial coal tar. The histograms show the percentage abundance of each polypeptide relative to the total quantity of polypeptides in any particular prekeratin extract. The numbers give the molecular weights of the prekeratin polypeptides ($\times 10^{-3}$). As can be seen, when treated with Motretinide or coal tar the 73,000-dalton polypeptide decreases, particularly with respect to the 57,000-dalton band

change in the expression of keratin polypeptides and not to a product-precursor relationship between them.

Two-dimensional SDS/polyacrylamide gels of prekeratin extracted from normal and Motretinide-treated mouse tails are shown in Fig. 7. Prekeratin taken from Motretinide-treated mouse tails was relatively depleted in the mol. wt. 73,000 polypeptide as described above. No significant change in the isoelectric variants of prekeratin from Motretinide-treated tails was observed when compared with normal mouse tail pre-

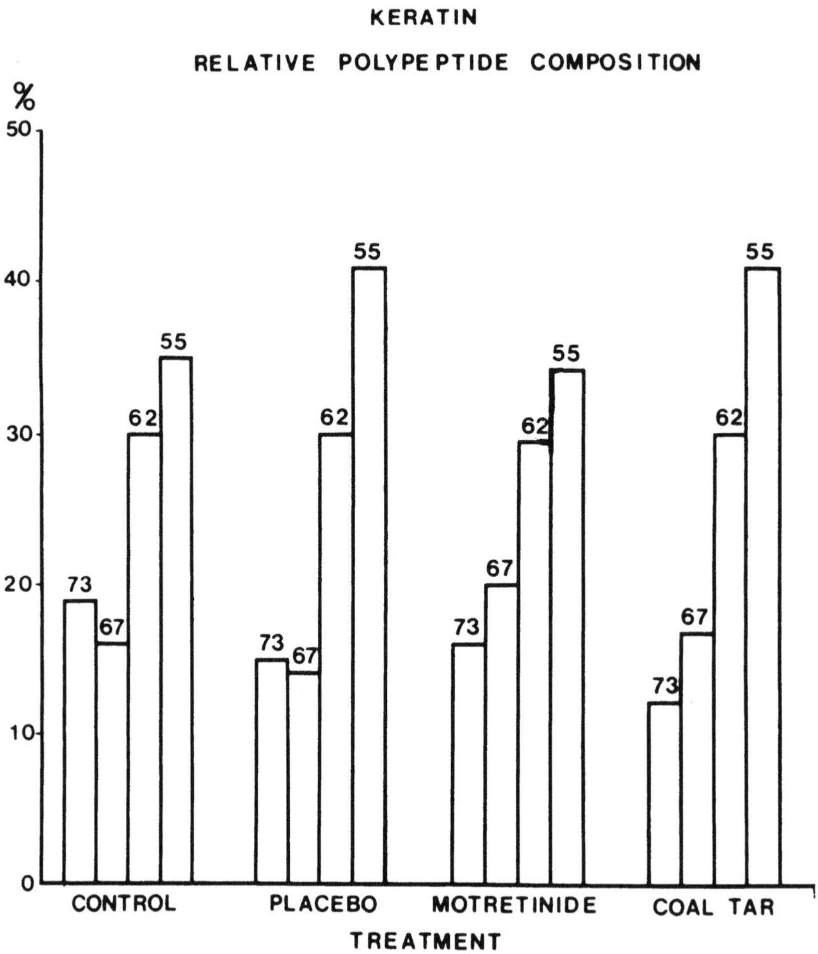

Fig. 6. Histograms of scanning densitometry traces (keratin). Quantitative analysis of SDS/polyacrylamide gels which contained keratins from mice on four different treatment regimes: control, "placebo", Motretinide and 0.3% commercial coal tar. The histograms show the percentage abundance of each polypeptide relative to the total amount of polypeptides in any particular keratin extract. The numbers give the molecular weights of the keratin polypeptides ($x10^{-3}$). As can be seen, when treated with Motretinide or coal tar the 67,000-dalton polypeptide increases particularly with respect to the 73,000-dalton band

keratin. Similar results were found with crude coal tar, and commercial coal tar treated mouse tail prekeratin (i.e., these compounds produced no change in the isoelectric variants of prekeratin).

Discussion

The mouse tail model of psoriasis has been used for some time by a number of groups [22, 42–44]. Principally, histology has been used to monitor changes although some immunological aspects have been examined too [9]. The present work demonstrates

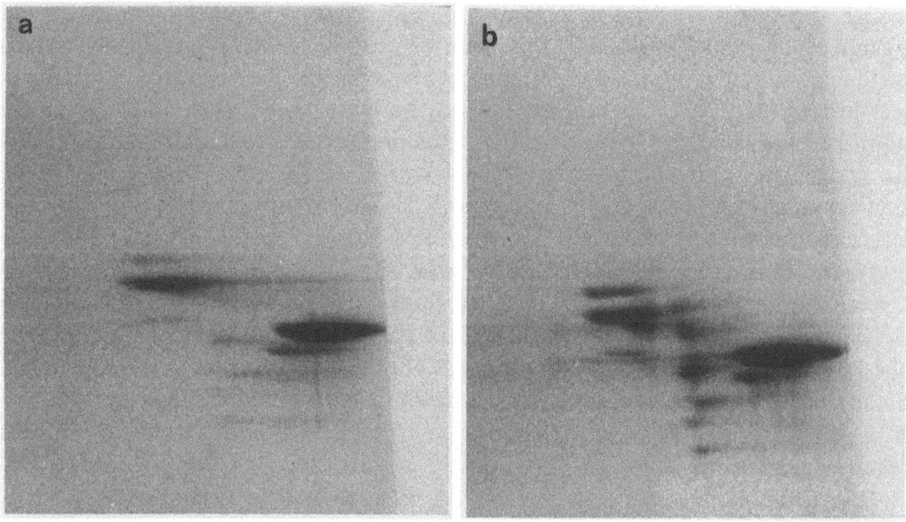

Fig. 7a, b. Two-dimensional gel electrophoresis of prekeratin from normal and Motretinide-treated mouse tails. Two-dimensional gel electrophoresis (NEpHGE followed by SDS-PAGE) on prekeratin extracted from normal and Motretinide-treated mouse tails. The prekeratin from Motretinide-treated tails was relatively depleted in the polypeptide of mol. wt. 73,000 but there was no change in the isoelectric variants of any prekeratin polypeptide **a** compared with normal mouse tail prekeratin. **b** Similar results were obtained with prekeratin from crude coal tar and commercial coal-tar-treated mouse tails

the effect of a known therapeutic substance (coal tar), both in its crude form and in a commercially available form (0.3%) in the mouse tail model of psoriasis. Histology reveals parakeratotic regions are induced to convert to orthokeratotic regions by coal tar and, in addition, by the aromatic retinoid Motretinide. Essentially this involves stimulation of the formation of a granular layer in areas of previously parakeratotic stratum corneum.

In the present study we have sought to extend the histological findings to the molecular level. Like human epidermal keratins, animal epidermal keratins have been investigated extensively [5], and we have studied the effect of topically applied therapeutic compounds on the expression of keratin polypeptides in the mouse tail model. The prekeratin and keratin polypeptide profiles were found to show an altered pattern, indicative of a change in their expression under the influence of these compounds. However, there was no change in the isoelectric variants, which is taken to imply that their degree of phosphorylation was unchanged. It has been known for some time that the keratins are phosphorylated proteins [1, 3, 16, 21, 37, 38], and the lack of any change after treatment with coal tar or Motretinide suggests that the major effects of these substances are at a translational or a transcriptional level rather than at a post-translational level.

Other workers [13] have shown that vitamin A at least has its effect at the level of transcription and this is therefore likely to be the level at which the aromatic retinoid, Motretinide, exerts its action. However, it remains that the effect of coal tar may be at the translational level; we have no information on this. In any case, the mouse-tail model of psoriasis is useful in looking at potentially therapeutic compounds for the treatment of psoriasis whatever the class of compound being investigated. Thus, preliminary investigations at the histological level may later be supported by biochemical investigation. It should be emphasised, however, that prekeratin found in the normal mouse tail is not the same as that found in the human psoriatic lesion. Perhaps for comparison with the mouse tail model studies on the effect of treating normal skin or uninvolved psoriatic skin with crude coal tar or Motretinide would be useful. Other studies in our laboratory have indicated that the mouse tail model may be useful for looking at the prophylactic effects of anti-psoriatic compounds. For example, the application of crude coal tar to the newborn mouse tail prevents the development of a parakeratotic stratum corneum. Alternating regions of parakeratotic and orthokeratotic epidermis would normally become apparent on day 9 after birth. Small quantities of keratin protein from newborn mouse tails could be analysed by SDS/polyacrylamide gel electrophoresis and visualised by the extremely sensitive silver staining procedure [4].

In conclusion, the mouse tail model of psoriasis is a versatile model, which is readily available (much more so, say, than rat or mouse vaginal epithelium or human buccal epithelium) and which also displays a true, though normal, parakeratosis. It has the disadvantage that it is not a pathological condition. Rather one is looking at an altered state of normal differentiation. Despite these limitations, however, it remains a convenient and comparatively inexpensive tool for investigations in an area where a real, pathological model is totally lacking.

Acknowledgements

The authors thank Mr. I. Hazari (Vick International) and Mrs. S. Gunning (Organon Laboratories Ltd.) for their technical assistance. We are grateful for the financial support of Organon Laboratories Ltd., (P.T.B.) and Vick International (N.F.C.). We thank the Department of Medical Photography at Leeds Infirmary for help with illustrations. This work was supported in part by an equipment grant from the Psoriasis Association.

References

1. Aoyagi T, Umeda K, Kato N, Fukaya T, Kobayashi H, Koizumi H, Miura Y (1983) Epidermal growth factor stimulates phosphorylation of pig epidermal keratin protein. J Invest Dermatol 81:49–53
2. Baden HP, McGilvray N, Cheng CK, Lee LD, Kubilus J (1978) The keratin polypeptides of psoriatic epidermis. J Invest Dermatol 70:294–297
3. Bladon PT, Bowden PE, Cunliffe WJ, Wood EJ (1982) Prekeratin biosynthesis in human scalp epidermis. Biochem J 208:179–187
4. Bladon PT, Cooper NF, Wright RM, Wood EJ, Cunliffe WJ (1983) Selection of a silver stain-

ing procedure to detect low quantities of human epidermal "keratin" protein following electrophoresis in polyacrylamide gels. Clin Chim Acta 127:403–406
5. Bladon PT, Taylor M, Wood EJ, Cunliffe WJ (1983) Isolation characterisation and comparison of mammalian epidermal prekeratins. Comparat Biochem Physiol 748:653–660
6. Bladon PT, Wood EJ, Cunliffe WJ (1984) Intracellular epidermal fibrous proteins. Clin Exp Dermatol 9:18–27
7. Bowden PE, Wood EJ, Cunliffe WJ (1983) Comparison of prekeratin and keratin polypeptides in normal and psoriatic human epidermis. Biochim Biophys Acta 743:172–179
8. Bowden PE, Cunliffe WJ (1982) Modification of human prekeratin during epidermal differentiation. Biochem J 199:145–154
9. Didierjean L, Wrench R, Saurat JH (1983) Expression of cytoplasmic antigens linked to orthokeratosis during the development of parakeratosis in the newborn mouse-tail epidermis. Differentiation 23:250–255
10. Fuchs E (1983) Evolution and complexicity of the genes encoding the keratins of human epidermal cells. J Invest Dermatol 81:141–144
11. Fuchs E, Green H (1979) Multiple keratins of cultured human epidermal cells are translated from different mRNA molecules. Cell 17:573–582
12. Fuchs E, Green H (1980) Changes in keratin gene expression during terminal differentiation of the keratinocyte. Cell 19:1033–1042
13. Fuchs E, Green H (1981) Regulation of terminal differentiation of cultured human keratinocytes by vitamin A. Cell 25:617–625
14. Fuchs E, Coppock SM, Green H, Cleveland DW (1981) Two distinct classes of keratin genes and their evolutionary significance. Cell 27:75–84
15. Gibbs PEM, Freedberg IM (1982) Epidermal keratin messenger RNAs. A heterogenous family. Biochim Biophys Acta 696:124–133
16. Gilmartin ME, Culbertson VB, Freedberg IM (1980) Phosphorylation of epidermal keratins. J Invest Dermatol 75:211–216
17. Harrison PV, Skerrow D (1982) A comparative study of psoriatic and non-psoriatric buccal mucosa. Br J Dermatol 106:637–642
18. Hanukoglu I, Fuchs EV (1982) The cDNA sequence of a human epidermal keratin: Divergence of sequence but conservation of structure among intermediate filament proteins. Cell 31: 243–252
19. Hunter I, Skerrow D (1981) The effect of increased tissue turnover on the keratinization of human epidermis. Biochim Biophys Acta 674:155–159
20. Hunter I, Skerrow D (1982) The proteins of living psoriatic epidermis. Biochim Biophys Acta 714:164–169
21. Ikai K, McGuire JS (1983) Phosphorylation of keratin polypeptides. Biochim Biophys Acta 760:317–376
22. Jarrett A, Spearman RIC (1964) In: Tavener D, Trousce J (eds) Histochemistry of the skin: Psoriasis. Universities Press PLC, London
23. Kitajima Y, Tsureda Y, Mori S, Okaro Y, Nozawa Y (1982) A simplified method of studying fibrous proteins in psoriatic scales obtained by tape stripping. Br J Dermatol 106:629–636
24. Laemmli UK (1970) Cleavage of structural proteins during assembly of the head of bacteriophage T4. Nature 227:680–683
25. Levine M, Mcleod A (1979) Fibrous proteins of normal and abnormal human epidermis. Br J Dermatol 100:401–408
26. LeVine MJ, McGilvary N, Baden HP (1980) Effect of therapy on keratin polypeptide profiles of psoriatic epidermis. Arch Dermatol 116:1028–1030
27. Marks R (1981) Psoriasis. Martin Dunitz Limited, London
28. Matoltsy AG, Matoltsy MN, Cliffel PJ (1983) Characterisation of keratin polypeptides of normal and psoriatic horny cells. J Invest Dermatol 85:185–188
29. O'Farrell PZ, Goodman HM, O'Farrell PH (1977) High resolution two-dimensional electrophoresis of basic as well as acidic proteins. Cell 12:1133–1142
30. Roop DR, Hawley-Nelson P, Cheng CK, Yuspa SH (1983) Keratin gene expression in mouse epidermis and cultured epidermal cells. Proc Natl Acad Sci 80:716–720

31. Schweizer J, Goerttler K (1980) Synthesis in vitro of keratin polypeptides directed by mRNA isolated from newborn and adult mouse epidermis. Europ J Biochem 112:243–249
32. Schweizer J, Marks F (1977) A developmental study of the distribution and frequency of Langerhans cells in relation to formation of patterning in mouse tail epidermis. J Invest Dermatol 69:198–204
33. Schweizer J, Winter H (1982) Keratin polypeptide analysis in foetal and in terminally differentiating newborn mouse epidermis. Differentiation 22:19–24
34. Skerrow D, Hunter I (1978) Protein modifications during the keratinization of normal and psoriatic human epidermis. Biochim Biophys Acta 537:474–478
35. Staquet MJ, Faure MR, Reano A, Viac J, Thivolet J (1983) Keratin polypeptide profile in psoriatic epidermis normalized by treatment with etretinate (aromatic retinoid Ro10-9359). Arch Dermatol Res 275:124–129
36. Steinert PM, Rice RH, Roop CR, Trus BL, Steven AC (1983) Complete amino acid sequence of a mouse epidermal keratin subunit and implications for the structure of intermediate filaments. Nature 302:784–800
37. Steinert PM, Wantz ML, Idler WW (1982) o-phosphoserine content of intermediate filament subunits. Biochem 21:177–183
38. Sun TT, Green H (1978) Keratin filaments of cultured human epidermal cells. J Biol Chem 253:2053–2060
39. Thaler MP, Fukuyama K, Inoue N, Cram DL, Epstein WL (1978) Two tris-urea-mercaptoethanol extractable polypeptides found uniquely in scales of patients with psoriasis. J Invest Dermatol 70:38–41
40. Thaler M, Fukuyama K, Epstein WL, Fisher KA (1980) Comparative studies of keratins isolated from psoriasis and atopic dermatitis. J Invest Dermatol 75:156–158
41. Winter H, Schweizer J, Goerttler K (1980) Keratins as markers of malignancy in mouse epidermal tumours. Carcinogenesis 1:391–398
42. Wrench R, Britten AZ (1975) Evaluation of coal tar fractions for use in psoriasiform diseases using the mouse tail test (I). High and low temperature tars and their constituents. Br J Dermatol 92:569–574
43. Wrench R, Britten AZ (1975) Evaluation of coal tar fractions for uses in psoriasiform diseases using mouse tail test (II). Tar oil acids. Br J Dermatol 82:575–579
44. Wrench R, Britten AZ (1975) Evaluation of coal tar fractions for use in psoriasiform diseases using the mouse tail test (III). High boiling tar acids. Br J Dermatol 93:67–74

Corneocyte Exfoliative Cytology: A Model to Study Normal and Diseased Stratum Corneum

E. Hölzle, G. Plewig and A. Ledolter

The stratum corneum serves as a barrier between viable tissue and the external environment. Its various functions include protection against toxic hazards, trauma, microbes, UV-light and thermal stimuli as well as conservation of water balance. Corneocytes are the building blocks of the stratum corneum. Their structure and biochemistry can be studied with non-invasive methods in the desquamating portion of the stratum corneum. This may help to gain some insight into the biology of the stratum corneum as a whole.

The original methods to remove corneocytes from the epidermis or nail plate with tape [4, 21], sticky slides [6] or cyanoacrylate [14] were modified by McGinley et al. introducing the detergent scrub method [16]. This procedure permits the study of qualitative and quantitative parameters characterizing the desquamating portion of the horny layer. Measurements include cell counts, cell surface area, cell thickness, morphology of corneocytes, biochemical and immunochemical analysis of corneocyte structural proteins, and materials within the intercorneal space.

In this paper data on some of these parameters are given in relation to biological variables, such as age, body region and season. In addition, morphological changes of corneocytes in inflammatory skin diseases are described. Further experimental studies involving irritation models and the effect of topical or systemic drugs are also presented.

Methods

Removal Technique. Corneocytes were harvested from the epidermis using the detergent scrub method [16]. Sheets of adhering horny cells were removed from nail plates with cellophane tape (Tesafilm D, Baiersdorf, Hamburg FRG) [4].

Staining. A 3:1 mixture of Löffler's methylene blue (Merck, Darmstadt, FRG) and rhodamine B (Chroma, Schmidt & Co, Stuttgart, FRG) was used for staining cell suspensions [12] or cells adhering to tape [4].

Cell Counts. Cell suspensions were counted in a Fuchs-Rosenthal haemocytometer.

Cell Surface Area. Projections of cell outlines were measured in μm^2 by planimetry [12] or with the aid of image-analyzing systems [1, 19] (Videoplan, Kontron, Eching, FRG).

Cell Thickness. Unstained corneocytes adherent to glass slides and protected by a cover glass were studied with a X-100 oil immersion lense with the oil present only between lense and cover glass. Using interference microscopy and polarized light according to the technique of Jamin-Lebedeff [3], thickness could be measured with an accuracy of 0.010 μm [18].

Cell Morphology. Shape of corneocytes, structure of intracellular trabeculae and nuclear remnants were observed with a light microscope [12]; with a scanning electron microscope shape and surface patterns were evaluated [9].

Immunochemical Analysis. Intercorneocyte space substances can be analyzed with immunochemical methods [13]. For results of these studies the reader may refer to the publication of King et al. [13].

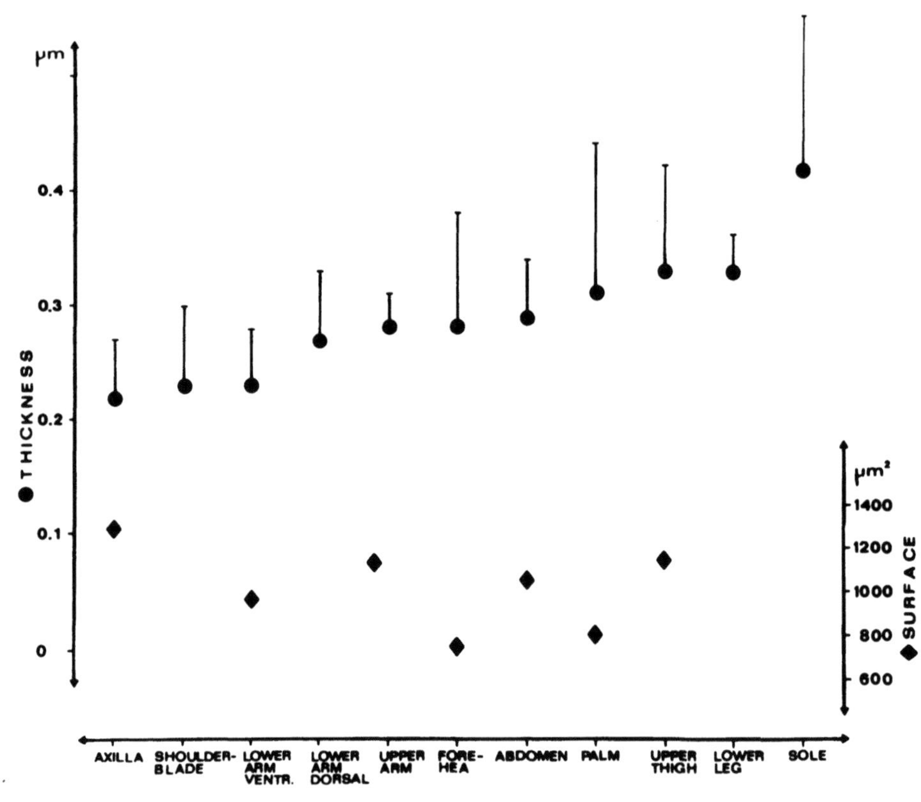

Fig. 1. Thickness and cell surface area of corneocytes from different body regions. Adapted from Plewig and Marples [17] and Plewig et al. [18]

Results

Physiological Variations

Regional Differences. Corneocytes vary in surface area and thickness with body region. Large cells are found in axillae, on trunk, upper arm and thigh. These cells also tend to be thin. Corneocytes on lower leg, sole, palm and forehead are small and thick (Fig. 1). It is known that in many cases epidermal turnover leads to small corneocyte cell surface areas [8, 12]. Small cells also tend to be thicker. This is the generally accepted explanation for variation of corneocyte size and thickness due to body region, age and inflammatory disorders of the epidermis.

Effects of Age. Cell surface area of corneocytes increase with age. This is illustrated in Fig. 2 where corneocytes were taken from the chest of male subjects. An age related increase in size is also seen in most other regions including nail plates [4], but with the exception of the forehead [17, 19]. Corneocytes from the mucous membrane of the tongue also fail to show age-dependent changes in surface area [19].

With increasing age the variability of cell size increases. This is shown by larger standard deviations (Fig. 3, 4).

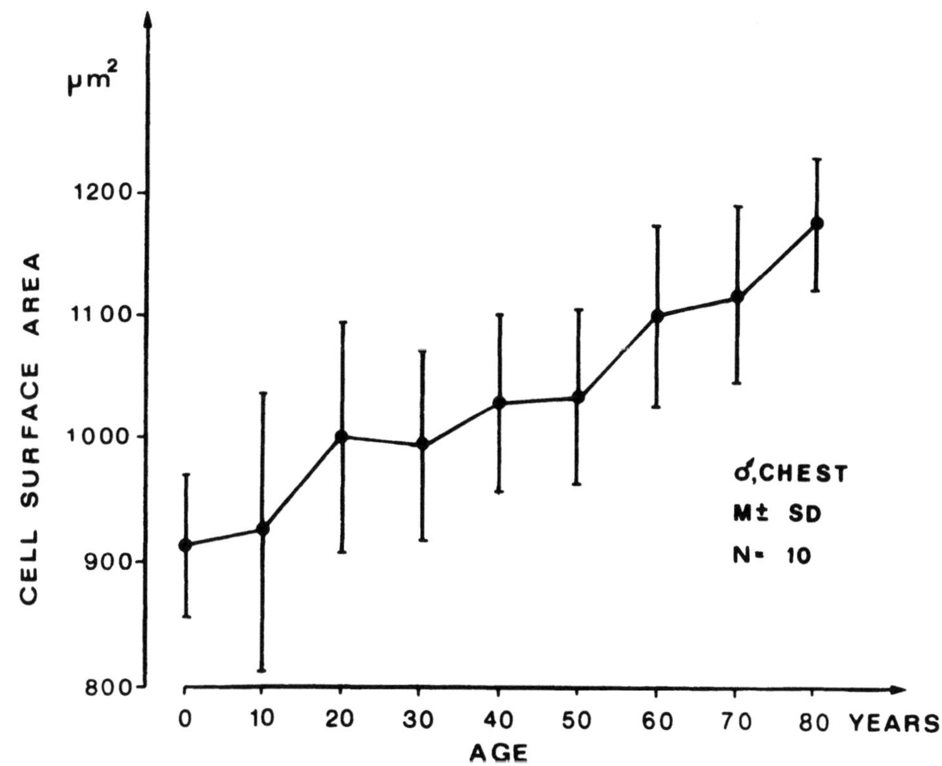

Fig. 2. Relationship between cell surface area and age [19]

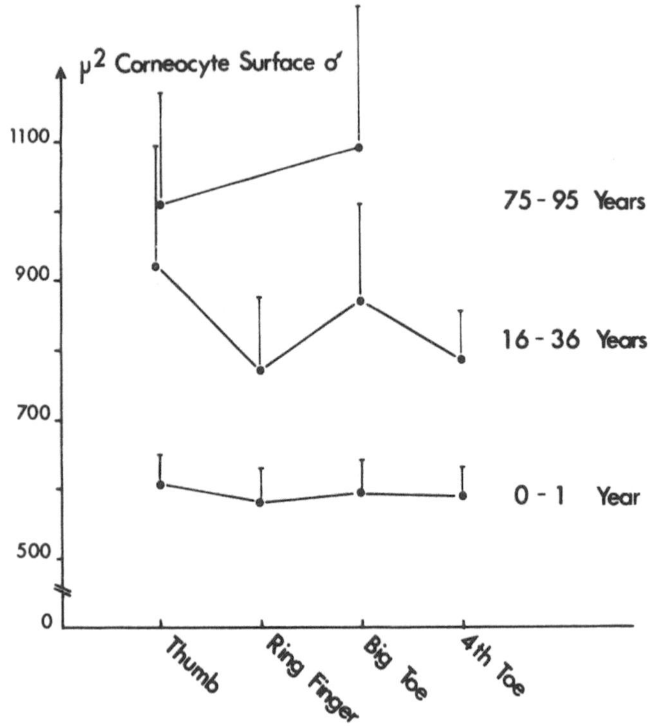

Fig. 3. Corneocyte surface area on nail plates is related to age [4]. Note increased variability in older subjects (n = 20)

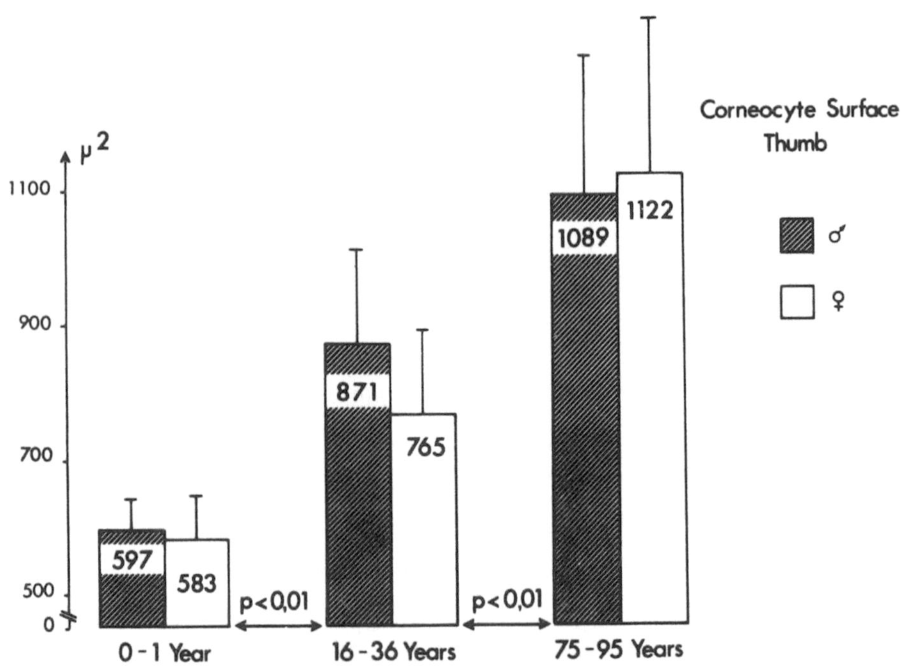

Fig. 4. Surface area and variability of corneocytes from nail plates of the thumb increase with age [4] (n = 20)

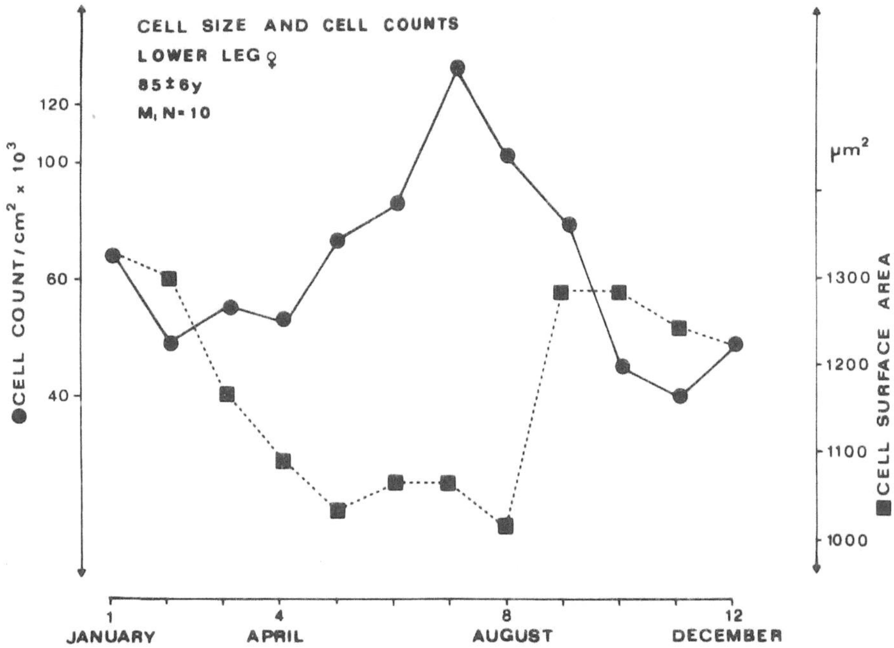

Fig. 5. Seasonal variations of cell surface area and corneocyte counts from lower leg of older females

Seasonal Changes. Measuring corneocyte surface area and cell counts on lower leg and paraumbilical region in the same panel of subjects monthly throughout the year some reproducible fluctuations were found [10]. In the warm season during the months from March to September, the cell surface area decreased and cell counts increased correspondingly (Fig. 5). In addition, smaller cells were found to be thicker. These seasonal changes were most pronounced on lower legs in both older and younger females. A trend was also seen in older males whereas in younger males changes were inconsistent. In the paraumbilical region only little fluctuation was found. As to the underlying mechanism we assume that smaller and thicker corneocytes as well as increased cell counts are caused by increased epidermal turnover in the warm season (personal communication, G. Grove Ph.D., Philadelphia).

Inflammatory Skin Diseases

In normal skin from many body regions cells are of regular shape and show regular trabeculae [19]. Irritation or inflammation leads to irregularly shaped horny cells and loss of trabecular network. Nuclear particles are retained (Fig. 6). This has been shown in allergic contact dermatitis [12], psoriasis [6], and in experimental irritant dermatitis [12].

In inflammatory skin diseases associated with increased epidermal turnover, corneocytes tend to be smaller [5, 12, 20]. Effective treatment reduces corneocytes to normal

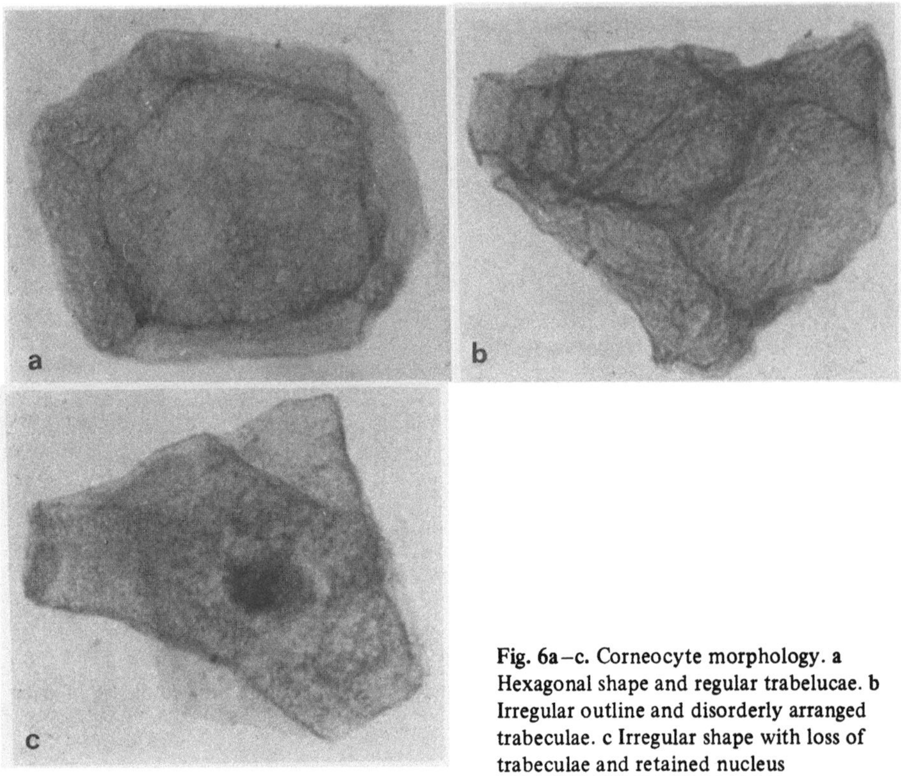

Fig. 6a–c. Corneocyte morphology. **a** Hexagonal shape and regular trabelucae. **b** Irregular outline and disorderly arranged trabeculae. **c** Irregular shape with loss of trabeculae and retained nucleus

size in parallel with the clinical improvement. Relative differences of cell surface area before and after treatment in different skin disorders are depicted in Fig. 7.

Variations in cell size are also detectable in corneocytes from diseased nail plates [4]. In psoriasis nail growth is accelerated and corneocytes are significantly smaller than in lichen planus in which growth rates are decreased (Fig. 8). Thus corneocyte size and turnover rate seem to be correlated in nail plates as well.

Experimental Studies

Topical application of tretinoin and stripping off of the horny layer were used to induce irritant dermatitis [12]. In this experiment corneocytes were found to be smaller and of irregular shape with asymmetric trabeculae. Many cells retained nuclear particles and cell counts per square centimeter skin surface rose considerably. Topical treatment with steroids improved all parameters. The effects of stripping on nuclear remnants within corneocytes are demonstrated in Fig. 9. There is a sharp increase of parakeratotic cells three to four days after stripping followed by an exponential decline during steroid treatment.

Systemic treatment with isotretinoin in acne patients induces dermatitis particularly on the face with visible desquamation. On the upper back of these patients an increase in corneocyte counts and a decrease in mean surface area were observed while the skin

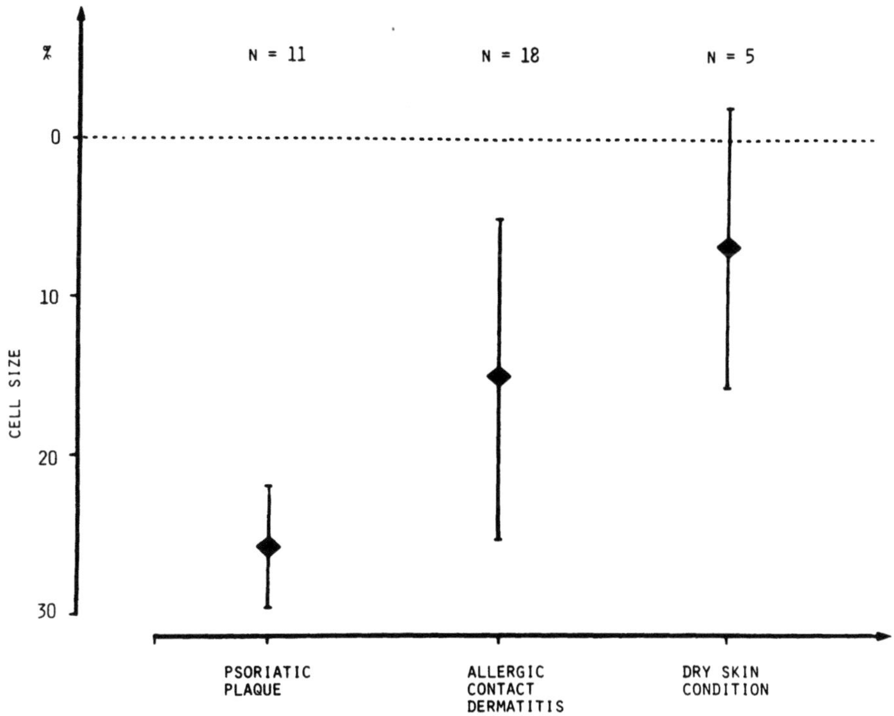

Fig. 7. Relative differences of cell size before and after treatment in psoriatic plaques, contact dermatitis and dry skin condition. Cells are much smaller (8 to 27%) in these three examples of increased epidermopoeisis and get larger with treatment

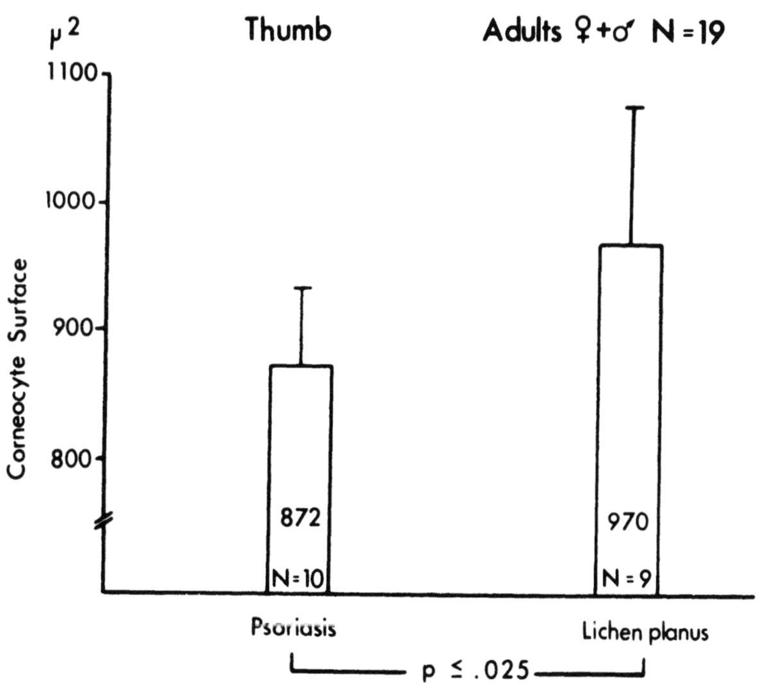

Fig. 8. Corneocytes from nail plates are significantly smaller in psoriasis than in lichen planus [4]

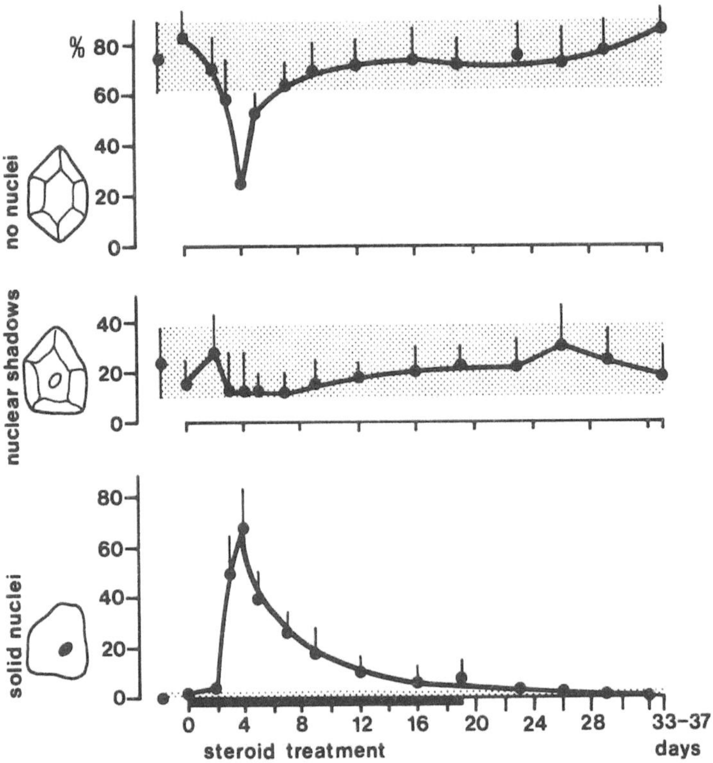

Fig. 9. Effect of stripping on nuclear remnants. Sharp increase 3 to 4 days after stripping followed by exponential decline during steroid treatment [12] (n = 11)

was clinically unchanged (Fig. 10). This coincides with accelerated epidermopoiesis as shown by H^3-thymidine autoradiography [1]. Grove and Kligman demonstrated with the dansyl chloride method that topical application of all-trans retinoic acid decreased stratum corneum transit time and corneocyte surface area [7]. These findings fit the general concept of cell growth rate and its relation to cell dimension.

Application of corticosteroids to normal skin decreases corneocyte surface area (Fig. 11) [11, 15].

This is in contrast to the effect of steroids on inflamed or irritated skin [5, 12, 20] in which initially small cells increase to normal dimensions under efficient treatment. It was shown by others that topical steroids reduce the size of viable epidermal cells [2]. It is conceivable that these smaller keratinocytes finally turn into smaller corneocytes.

Conclusions

With the technique described qualitative and quantitative parameters of exfoliating corneocytes are available to monitor subtle changes in the stratum corneum. The method is applicable to the entire integument including nail plates.

Physiological conditions of the skin are characterized by distinctive corneocyte parameters. This can be shown for body site, age, and season. In addition, changes in

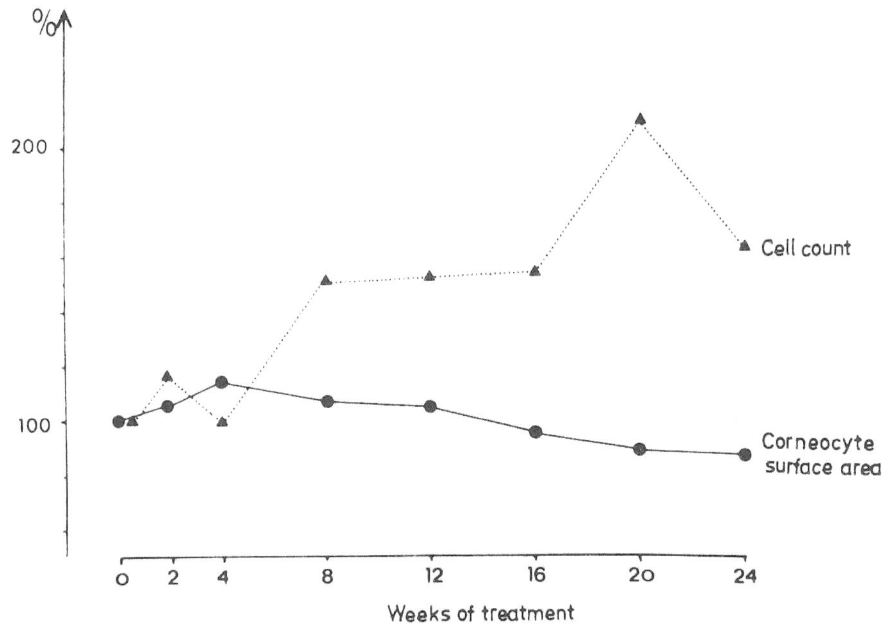

Fig. 10. Mean per cent change in corneocyte cell count and in corneocyte surface area during systemic treatment with isotretinoin [1] (n = 13)

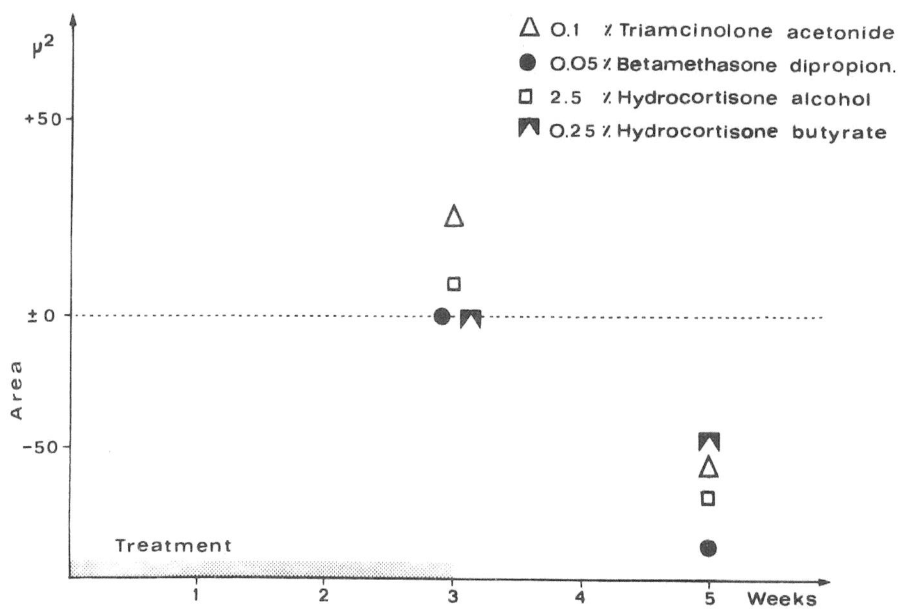

Fig. 11. On normal skin corneocyte surface area decreases after topical application of steroids. The effect is most pronounced two weeks after treatment [11] (n = 11)

corneocyte dimension due to inflammatory skin disorders as well as drug induced changes of epidermopoiesis can be detected.

This skin model offers a simple, non-invasive method for the study of pharmacological effects on human skin.

Acknowledgement

The technical assistance of Mrs. E. Scheuber is gratefully acknowledged. The study was supported by Vick International.

References

1. Breiner W, Scheuber E, Plewig G (1983) Effects of isotretinoin (13-cis retinoic acid, Ro 4-3780) treatment on exfoliative cytology. In: Marks R, Plewig G (eds) Stratum corneum. Springer, Berlin Heidelberg New York, pp 222–226
2. Delforno C, Holt PJA, Marks R (1978) Corticosteroid effect on epidermal cell size. Br J Dermatol 98:619–623
3. Gahm J (1963) Durchlicht-Interferenz-Einrichtung nach Jamin-Lebedeff. Zeiss-Mitteilungen 2/10:389
4. German H, Barran W, Plewig G (1980) Morphology of corneocytes from human nail plates. J Invest Dermatol 74:115–118
5. Goldschmidt H (1979) Surface area measurements of psoriatic corneocytes: Effects of intralesional steroid therapy. J Invest Dermatol 73:558–560
6. Goldschmidt H, Kligman AM (1967) Exfoliative cytology of human horny layer. Arch Dermatol 96:572–576
7. Grove GL, Kligman AM (1983) Corneocyte size as an indirect measure of epidermal proliferative activity. In: Marks R, Plewig G (eds) Stratum corneum. Springer, Berlin Heidelberg New York, pp 191–195
8. Grove GL, Lavker RM, Hölzle E, Kligman AM (1980) Use of nonintrusive tests to monitor age-associated changes in human skin. J Soc Cosm Chem 32:15–26
9. Heilmann BB, Ryckmanns F, Plewig G (1983) Scanning electron microscopy of human corneocytes. In: Marks R, Plewig G (eds) Stratum corneum. Springer, Heidelberg New York, pp 186–190
10. Herrmann S, Scheuber E, Plewig G (1983) Exfoliative cytology: Effects of the seasons. In: Marks R, Plewig G (eds) Stratum corneum. Springer, Berlin Heidelberg New York, pp 181–185
11. Hölzle E, Park J, Plewig G (1980) Einfluß verschiedener Glukokortikoide und ihrer Grundlagen auf die Korneozyten der normalen Epidermis. Aktuel Dermatol 6:75–81
12. Hölzle E, Plewig G (1977) Effects of dermatitis, stripping and steroids on the morphology of corneocytes: A new bioassay. J Invest Dermatol 68:350–356
13. King CS, Dykes PJ, Marks R (1983) Preparation and immunochemical analysis of non-ionic detergent-soluble protein from human stratum corneum. In: Marks R, Plewig G (eds) Stratum corneum. Springer, Berlin Heidelberg New York, pp 68–72
14. Marks R. Dawber RPR (1971) Skin surface biopsy: An improved technique for the examination of the horny layer. Br J Dermatol 84:117–123
15. Marks R, Nicholls S, King CS (1981) Studies on isolated corneocytes. Internat J Cosmet Science 3:251–258

16. McGinley KJ, Marples RR, Plewig G (1969) A method for visualizing and quantitating the desquamating portion of the human stratum corneum. J Invest Dermatol 53:107–111
17. Plewig G, Marples RR (1970) Regional differences of cell sizes in the human stratum corneum, part I. J Invest Dermatol 54:13–18
18. Plewig G, Scheuber E, Reuter B, Waidelich W (1983) Thickness of corneocytes. In: Marks R, Plewig G (eds) Stratum corneum. Springer, Berlin Heidelberg New York, pp 171–174
19. Rogall P (1976) Größe von Hornzellen und Schleimhautzellen in Abhängigkeit vom Alter. Dissertation, Dept Dermatol, University of Munich, FRG
20. Stein W (1978) Hornzellmorphologie exsikkierter Altershaut. Verlaufsmessungen. Dissertation, Dept Dermatol, University of Munich, Munich, FRG
21. Wolf J (1939) Die innere Struktur der Zellen des Stratum desquamans der menschlichen Epidermis. Zeitschr Mikroskop-Anatom Forschung 46:170–202

Modulation of EGF Receptors during Epidermal Hyperplasia

D. A. Basketter

Epidermal growth factor (EGF), first described by Cohen [2], may be an important endogenous mediator of the control of growth and differentiation of skin [1, 4]. As determined by ^{125}I-EGF autoradiography, EGF receptors are localised in skin to cells which have growth potential [6]. In addition, evidence which suggests that there is a close relationship between mitotic rate and EGF receptor density in epidermal basal cells of the neonatal rat has been described by Green et al. [6]. The results presented here demonstrate that concomitantly with the epidermal hyperplasia induced by a chemical stimulus, EGF receptor density is dramatically reduced. An almost total disappearance of EGF receptor sites occurs some 24 h after a sufficient hyperplastic stimulus. Following a period of similar duration, the EGF receptor complement is restored to normal.

Materials and Methods

Colworth Wistar rats weighing about 60 g at the start of the experiment and of either sex were used in this study. Highly purified EGF isolated by the method of Savage and Cohen [8] was labelled with iodine-125 to high specific activity (120 μCi/μg) using iodogen (Pierce).

0.2 ml hexadecane (BDH) was applied to the clipped rear dorsum of test rats for 30 s, left for 15 min, and a repeat application made. 4 mm skin punches were taken from 0.2 mm keratotome slices of both treated and control skin.

Additionally, samples were taken from weight matched rats which were clipped immediately prior to sacrifice. The skin samples were incubated for 2 h at room temperature in medium 199 containing 5 nM ^{125}I-EGF, washed 3x in medium 199 (4 °C) for a total of 30 min, fixed in Bouin's solution overnight and then processed for autoradiography.

Non-specific background labelling was estimated by incubating replicate skin punches in the presence of a 300 fold excess of cold EGF.

Slides were exposed using K2 nuclear research emulsion (Ilford), and when developed were stained lightly with eosin. Autoradiograms were taken on a Leitz Orthoplan microscope using darkground illumination.

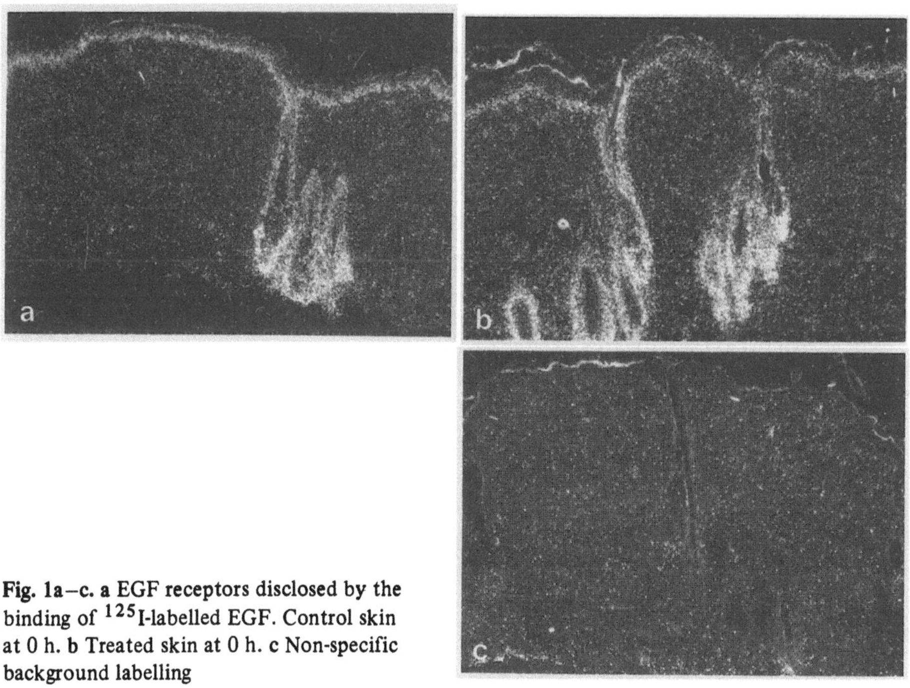

Fig. 1a–c. a EGF receptors disclosed by the binding of ^{125}I-labelled EGF. Control skin at 0 h. **b** Treated skin at 0 h. **c** Non-specific background labelling

Results

The three autoradiograms in Fig. 1 show the location and density of EGF receptors in rat dorsal skin at the start of the experiment. In the autoradiogram which forms Fig. 1c, the skin was incubated in the presence of excess unlabelled EGF, and thus shows only background grains. In contrast, specific labelling of EGF receptors in control skin (Fig. 1a) and in skin taken immediately after hexadecane treatment (Fig. 1b) occurs only on those epithelial cells with growth potential. Particularly striking is the line of silver grains overlying the epidermal basal cell layer and the outer root sheath of the hair follicles. At 4 h and 8 h post-treatment, no significant change in this pattern is apparent (data not shown). However by 24 h, the EGF receptors in hexadecane treated skin have almost completely disappeared from the epidermal basal layer (Fig. 2b), but are largely unchanged in the outer root sheath of the hair follicle (Fig. 2b). Skin which was clipped at the start of the experiment but not treated with hexadecane is unchanged (Fig. 2a). After a further 24 h, control and hexadecane treated skin look similar once again (Fig. 3a, b) in terms of their EGF receptor density. At this time, histologically, the hexadecane treated skin shows typical signs of acute mild inflammation, with acanthosis, slight oedema and parakeratosis. At 96 h whilst the inflammatory hyperplasia is still quite evident by histology, labelling of EGF receptors in control and treated skin is apparently identical (Fig. 4).

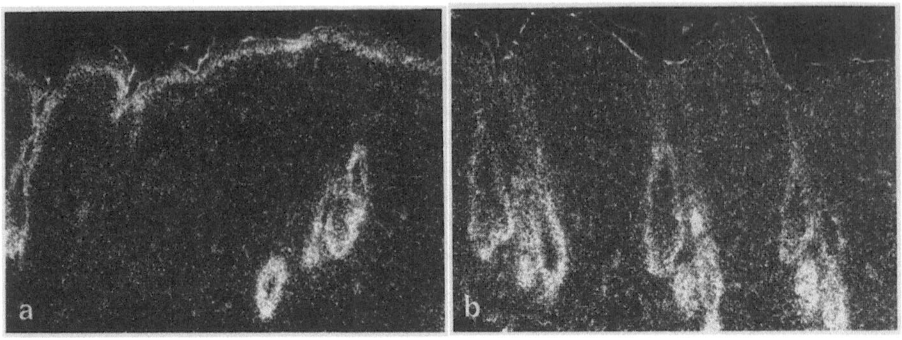

Fig. 2a, b. a EGF receptors disclosed by the binding of ^{125}I-labelling EGF. Control skin at 24 h. **b** Treated skin at 24 h

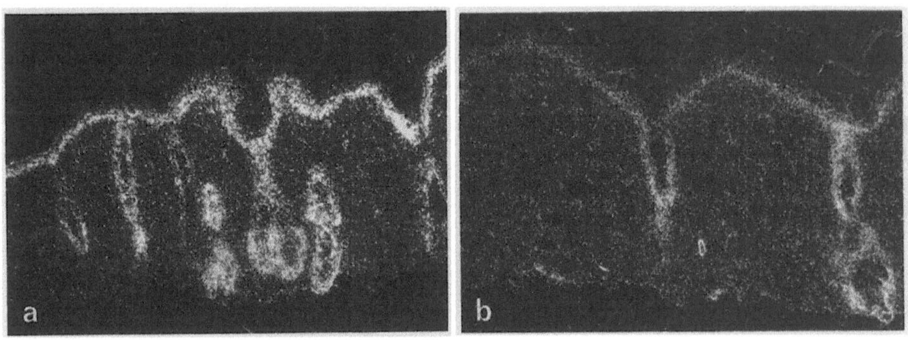

Fig. 3a, b. a EGF receptors disclosed by the binding of ^{125}I-labelled EGF. Control skin at 48 h. **b** Treated skin at 48 h

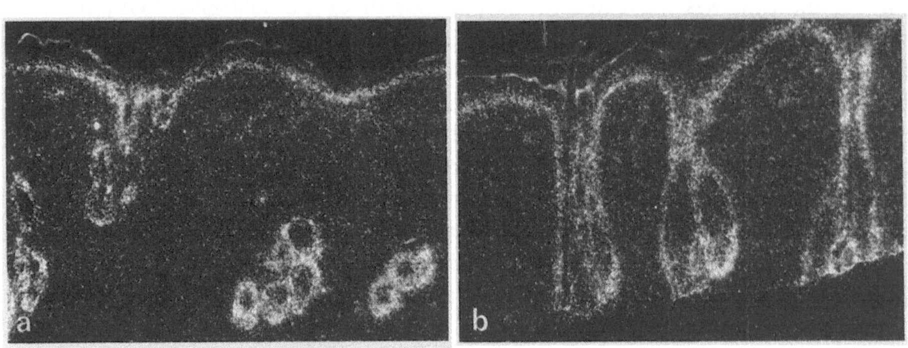

Fig. 4a, b. a EGF receptors disclosed by the binding of ^{125}I-labelled EGF. Control skin at 96 h. **b** Treated skin at 96 h

Discussion

Although EGF receptors may now be readily identified in epidermal tissues [6], their physiological function is not yet proven. The evidence available in the literature strongly suggests that EGF has a role in development of the neonate [5, 10].

In particular, various authors have shown that EGF can promote neonatal epidermal growth and differentiation [3, 7, 9]. In the neonatal rat, Green et al. have extended these observations. They demonstrated that not only are EGF receptors found almost exclusively on the basal cells of the epidermis, but that changes in the number of EGF receptors on each cell is directly related to the rate of epidermal basal cell mitosis. The immediate postnatal decrease in mitotic index was closely correlated with a fall in EGF receptor density.

The observations described in this report are that EGF receptors disappear briefly during the onset of epidermal hyperplasia. It seems unlikely that EGF receptors are masked by hexadecane 24 h after its application for two reasons. Firstly the receptors in the hair follicles remain unaffected by the treatment, and secondly in some experiments the fall in receptor number has occurred (usually to a lesser degree) in control animals in which shaving at the start of the experiment induced a significant epidermal response. So the change in receptor number is apparently connected with the inflammatory hyperplasia. A different mechanism must be operating to that in the neonatal rat however. In this case (inflammatory hyperplasia), an increase in mitotic rate of epidermal basal cells is associated with a fall in receptor density. The most reasonable explanation of such a change might be a local increase in the tissue concentration of EGF, which would then stimulate basal cell division and concomitantly down-regulate EGF receptors. An alternative explanation, which would have a similar result, is that the hexadecane treatment increases the sensitivity of the EGF receptors. This latter possibility might be evaluated using cultured rat keratinocytes. Whatever explanation is evinced, it is likely at present to be quite speculative. Currently studies are underway to define accurately both the onset of mitosis (as measured by the uptake of tritiated thymidine) and the fall in EGF receptor number. If these events are physiologically related, then it would not be unreasonable to expect the latter to precede the former. Additionally, attempts are being made to determine which, if any, cells in skin are particularly important in supplying EGF.

References

1. Carpenter G, Cohen S (1981) EGF: Receptor interaction and the stimulation of cell growth. In: Lefkowitz RJ (ed) Receptor regulation. Chapman and Hall, London, pp 41–66 (Receptors and recognition, series B, vol 13)
2. Cohen S (1962) Isolation of a mouse submaxillary gland protein accelerating incisor eruption and eyelid opening in the newborn animal. J Biol Chem 237:1555–1562
3. Cohen S (1965) The stimulation of epidermal proliferation by a specific protein (EGF). Dev Biol 12:394–407
4. Cohen S (1983) The epidermal growth factor (EGF). Cancer 51:1787–1791

5. Gospadarowicz D (1981) Epidermal and nerve growth factors in mammalian development. Ann Rev Physiol 43:251–263
6. Green MR, Basketter DA, Couchman JR, Rees DA (1983) Distribution and number of epidermal growth factor receptors in skin is related to epithelial growth. Dev Biol 100:506–512
7. Moore GPM, Panaretto BA, Robertson D (1983) Epidermal growth factor delays the development of the epidermis and hair follicles of mice during growth of the first coat. Anat Res 205:47–55
8. Savage CR, Cohen S (1972) Epidermal growth factor and a new derivative. J Biol Chem 247: 7609–7611
9. Steidler NE, Reade PC (1980) Histomorphological effects of EGF on skin and oral mucosa of mice. Archs Oral Biol 25:37–43
10. Thorburn GD, Waters MJ, Young IR, Buntine DD, Hopkins PS (1981) Epidermal growth factor: A critical factor in fetal maturation. In: The Fetus and independent life. Pitman, London, pp 172–198 (Ciba Foundation symposium 86)

Comparative Study of the Keratin Polypeptide Profiles in Human and in Pig Hyperkeratotic Scabies

D. Van Neste, M. J. Staquet, G. P. Martineau and J. P. Ortonne

Increased numbers of patients with hyperkeratotic (or Norwegian) scabies have been reported in the literature in recent years. Various environmental and endogeneous factors are known to be associated with the disease [4]. An animal model could be useful to evaluate the relative importance of each of these factors by modifying them one by one and could provide deeper insight into the disturbed keratinization process which is the clinical characteristic feature of this parasitic disease. As there is strong evidence now that fibrous proteins of different physicochemical properties are intimately associated with the epidermal cell differentiation process [2, 19, 20] and that polypeptide composition of the fibrous protein is subjected to environmental influences, we were interested in evaluating the electrophoretic patterns of the fibrous proteins in scales of human and pig hyperkeratotic scabies.

Material and Methods

Patients

Six patients (five females and one male) aged 37 to 87 years had typical hyperkeratotic scabies. In the elderly patients, the condition was associated with senility (four cases); the younger patients were being treated with systemic steroids for bullous pemphigoid (one case) and steroids were given with cyclophosphamide for a systemic vasculitis syndrome (one case). No topical therapy had been given during the last weeks. The clinical diagnosis was confirmed by direct microscopic observation of numerous scabies mites, larvae and eggs in scales taken from the hands of each patient.

Hyperkeratotic Scabies in Pigs

A case of naturally occurring hyperkeratotic scabies in a Landrace domestic pig and a control pig have also been investigated.

Skin Samples

Scales were obtained from active lesions and from control skin from the back of the hands (humans), and on the inner and outer aspects of the ears (pigs). Scraping was stopped as soon as the underlaying oozing epidermis was uncovered. The samples were stored at –20 °C and transferred to the laboratory for further analysis.

Extraction and Separation of Keratin Polypeptides

The extraction procedure was performed according to the method of Winter et al. [23]. Briefly, after repeated homogenization of scales in high salt buffer (1.5 M KCl, 10 mM NaCl, 10 mM Tris-HCl pH 8, 2 mM dithioerythritol, 0.5% Triton x 100), keratins were dissolved in 10 mM Tris-HCl pH 8, 5 % 2-mercaptoethanol, 5% sodium dodecylsulfate and heated for 3 h at 60 °C. The protein concentration was determined by the Bradford method [3]. Keratins were analyzed on 7.5–17.5% acrylamide gradient and 8.5% acrylamide slab gels respectively according to Laemmli [11]. Gels were scanned with an Isco slab gel scanner (Model 1310).

Immunoblotting

Proteins separated by gel electrophoresis were transferred to nitrocellulose sheets [17]. Sheets were incubated for 18 h at 4 °C with rabbit antikeratin antibodies (5 μg of

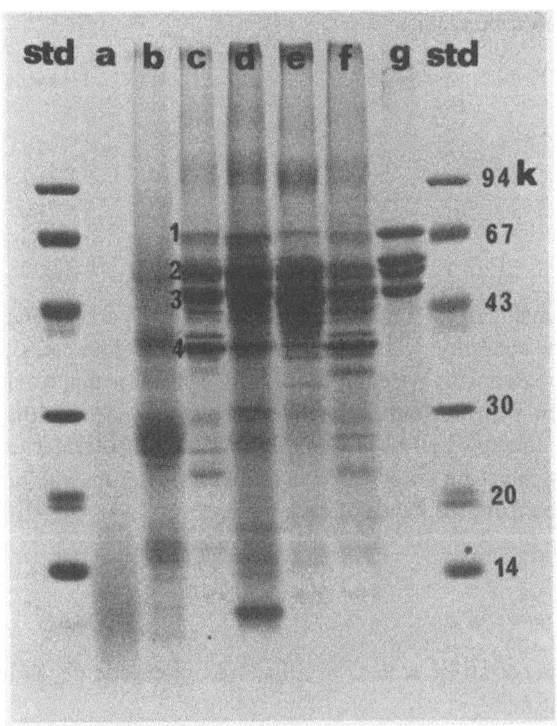

Fig. 1. Keratin polypeptide patterns of 6 different samples of human hyperkeratotic scabies (a to f) and from control human stratum corneum (g). No significant amounts of proteins were obtained from samples a and b. Abbreviations: Std gel standards including: phosphorylase b 94,000, albumin 67,000, ovalbumin 43,000, carbonic anhydrase 30,000, trypsin inhibitor 20,000, and lactalbumin 14,000. The estimated molecular weight of the main keratin polypeptides are 67 [1], 57 [2], 52 [3] and 40 [4] kilodaltons

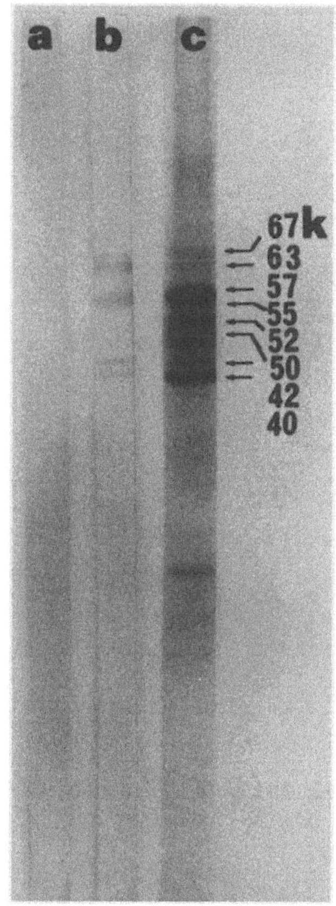

Fig. 2. Immunological identification of human keratin after electrophoretic transfer of polypeptides extracted from hyperkeratotic scabies scales. **a** normal rabbit serum, **b** rabbit antikeratin antiserum (5 μg of purified IgG fraction), and **c** Coomassie blue stained gel

purified IgG fraction) diluted in saline (10 mM Tris-HCl pH 7.4, 0.9% NaCl) containing 3% BSA. After five washes with buffer, the sheets were incubated for two hours at room temperature with peroxidase-conjugated F (ab') 2 fragments of goat antirabbit immunoglobulins. Finally, after five rinses with buffer, antibody fixation was revealed in the presence of 3-amino 9 ethylcarbazol (AEC) [10].

Results

Keratins were extracted from the scales of six patients who had typical hyperkeratotic scabies. For unknown reasons, we failed to extract significant amounts of keratins from two of the patients. The electrophoretic patterns obtained from four different scales were reproducible with only slight differences in the expression of small polypeptides (Fig. 1). When compared to normal human skin, there were changes in the relative amounts of the keratin polypeptides. Densitometric readings of the gels showed that the amount of the 67K polypeptide was decreased since it accounted for 10.4% of the total protein amount in scales in comparison to 18% in normal

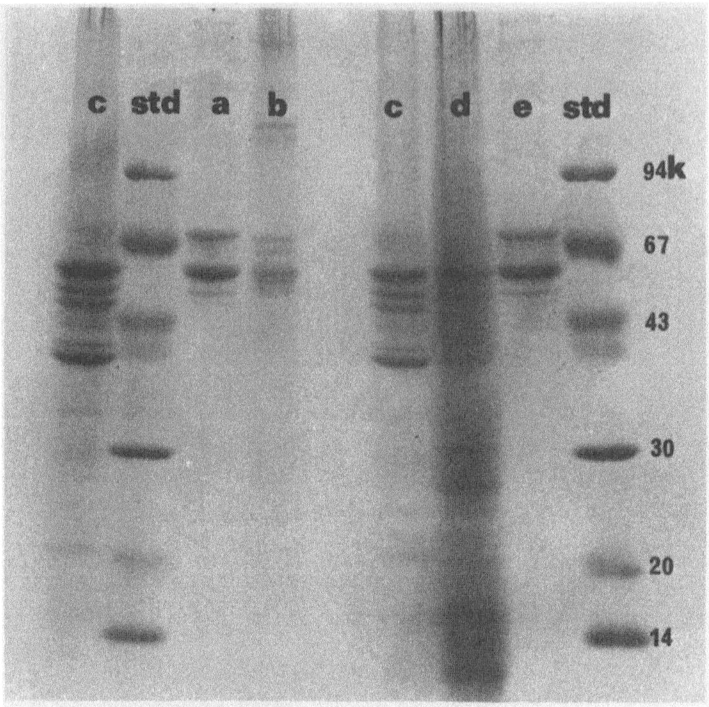

Fig. 3. Keratin polypeptide patterns of human and pig hyperkeratotic scabies. Human normal epidermis (a and e), normal pig ear epidermis (b) and hyperkeratotic scabies in human (c) and pig (d)

stratum corneum. The amounts of the 52K and 50K polypeptides were increased and a polypeptide of molecular weight 40K, not detected in normal skin, was present in large quantitity (18%) in hyperkeratotic scales.

Using immunoblot analysis, the keratin nature has been shown for all these polypeptides by their binding to specific antikeratin antibodies (Fig. 2). Similarly when the keratins of the hyperkeratotic lesions of the parasited pig were compared to the control profile, the 67K band almost completely disappeared and low molecular weight polypeptides were increased (Fig. 3).

Discussion

The keratin pattern obtained from hyperkeratotic scabies scales is different from that of normal human skin. The amount of high molecular weight polypeptides decreased and was compensated for by a relative increase of the amount of the smaller polypeptides. The abnormal pattern was also found in hyperkeratotic scabies of the pig. As we know from previous studies in hyperkeratotic scabies [18] as well as in common scabies [6, 14] the mites penetrate the stratum corneum and destroy underlying epidermal cell layers at the blind end of the burrow. In hyperkeratotic scabies there is psoriasiform hyperplasia of the epidermis with massive inflammatory cell infiltration in the upper dermis [1]. Morphologically, the epidermal changes occurring in hyperkeratotic scabies of the pig are quite similar [13] with collapsing epidermal cells, hyperkeratosis, parakeratosis and secondary epithelial hyperplasia. The decreased amount of the 67K polypeptide found in hyperkeratotic scabies could be explained by the destruction of the cells of the upper layers of epidermis.

Indeed, these cells are known to mainly synthesize the high molecular weight polypeptides [5, 20]. However, similar findings are observed in other pathological conditions where the upper epidermal cells are not destroyed. These conditions including psoriasis [7, 16] warts [15], basal and squamous cell carcinomas [12, 21, 24] are characterized by epidermal hyperplasia. Thus whatever the cause of the disorder, its expression at the epidermal level is reflected by a disturbed differentiation process. For this reason it is impossible to determine whether the reported changes are directly related to the epidermal cell destruction or indirectly reflecting the epidermal reactive process.

Animal models and in vitro experiments will certainly be welcomed to investigate further epidermal changes occurring in scabies. Advantageously pig epidermis has morphological [8, 9] and dynamic [22] similarities to human epidermis.

References

1. Ackerman AB (1977) Histopathology of human scabies. In: Orkin O, Maibach HI, Parrish LC, Schwartzman RM (eds) Scabies and pediculosis, Lippincott, Philadelphia, pp 88–95
2. Baden HP, Kubilus J, Argyris TS (1980) Modification of polypeptide composition in keratinocyte fibrous protein. J Invest Dermatol 75:383–387
3. Bradford MM (1976) A rapid and sensitive method for the quantitation of microgram quantities of protein utilizing the principle of protein dye binding. Anal Biochem 72:248–254
4. Chouvet B, Ortonne JP, Perrot H, Thivolet J (1979) La gale norvégienne: circonstances etiologiques. Ann Dermatol Vénéréol 106:569–574
5. Fuchs E, Green H (1980) Changes in keratin gene expression during terminal differentiation of the keratinocyte. Cell 19:1033–1042
6. Heilesen B (1946) Studies on acarus scabiei and scabies. Histological examinations of the scabies burrow and the eruption in ordinary scabies. Acta Dermatovenereol (Stockh) 26: suppl 14:223–240
7. Hunter L, Skerrow D (1982) The proteins of living psoriatic epidermis. Biochim Biophys Acta 714:164–169
8. Karasek J, Oehlert W (1968) Die Ultrastruktur der Epidermis des Schweines. I. Stratum basale und stratum spinosum. Z Mikrosk Anat Forsch 78:133–144
9. Karasek J, Oehlert W (1968) Die Ultrastruktur der Schweineepidermis. II. Stratum granulosum und corneum. Z Mikrosk Anat Forsch 79:157–169
10. Karcher D, Lowenthal A, Thormar H, Noppe M (1981) Serological identification of viral antigens after electrophoretic transfer. J Immunol Meth 43:175–179
11. Laemmli UK (1970) Cleavage of structural proteins during the assembly of the head of bacteriophage T4. Nature 277:680–685
12. Löning T, Staquet MK, Thivolet J, Seifert G (1980) Keratin polypeptides distribution in normal and diseased human epidermis and oral mucosa. Virchows Arch A 388:273–288

13. Sheahan BJ (1975) Pathology of sarcoptes scabiei infection in pigs. I. Naturally occurring and experimentally induced lesions. J Comp Path 85:87–95
14. Shelley WB, Shelley ED (1983) Scanning electron microscopy of the scabies burrow and its contents, with special reference to the sarcoptes scabiei egg. J Am Acad Dermatol 9:673–679
15. Staquet MJ, Viac J, Thivolet J (1981) Keratin polypeptide modifications induced by human papilloma viruses (HPV). Arch Dermatol Res 271:83–90
16. Staquet MJ, Faure M, Reano A, Viac J, Thivolet J (1983) Keratin polypeptide profile in psoriatic epidermis normalized by treatment with etretinate (aromatic retinoid Ro 10-9359). Arch Dermatol Res 275:124–129
17. Towbin H, Staehelin T, Gordon J (1979) Electrophoretic transfer of proteins from polyacrylamide gels to nitrocellulose sheets: procedure and some applications. Proc Natl Acad Sci 76:4350–4354
18. Van Neste D, Lachapelle JM (1981) Host-parasite relationships in hyperkeratotic (Norwegian) scabies: pathological and immunological findings. Br J Dermatol 105:667–678
19. Viac J, Schmitt D, Staquet MJ, Thivolet J, Ortonne JP, Bustamante R (1980) Binding specificity of guinea pig anti-α-keratin polypeptide sera on human keratinocytes: comparison of their receptors with those of human epidermal cytoplasmic antibodies. Acta Dermatovenereol (Stockh) 60:189–196
20. Viac J, Staquet MJ, Thivolet J, Goujon C (1980) Experimental production of antibodies against stratum corneum keratin polypeptides. Arch Dermatol Res 267:179–188
21. Viac J, Reano A, Thivolet J (1982) Cytokeratins in human basal and squamous cell carcinomas: biochemical, immunohistological findings and comparisons with normal epithelia. J Cut Path 9:377–390
22. Weinstein GD (1965) Comparison of turnover time and of keratinous protein fractions in swine and human epidermis. In: Bustad LK, McClellan RO (eds) Swine in biomedical research. Frayn Printing Co, Seattle, pp 287–297
23. Winter H, Schweizer J, Goerttler K (1980) Keratins as markers of malignancy in mouse epidermal tumors. Carcinogenesis 1:391–398
24. Winter H, Schweizer J, Goerttler K (1983) Keratin polypeptide composition as a biochemical tool for the discrimination of benign and malignant epithelial lesions in man. Arch Dermatol Res 275:27–34

In Vitro Models

Role of Extracellular Matrix in Skin Morphogenesis, Analysed by Dermal Cell Cultures

P. Sengel and M. Kieny

The formation of skin and cutaneous appendages during embryonic development results from precisely-timed and precisely-located dermal-epidermal interactions [33–36]. In this interplay between the two skin tissues, the dermis plays a predominant role. Indeed size, shape, distribution pattern, and growth rate of cutaneous appendages are controlled by the dermal mesenchyme. The dermis thus exerts a constant influence on the morphogenetic and differentiative features of the epidermis.

In most species of amniotes studied so far, the cutaneous appendages – e.g. reptilian scales, avian scales, feathers, hairs – are distributed on the surface of the body according to a more or less rigourously defined pattern. Reptilian and avian scales are commonly arranged in regular rows, feathers are usually organized in a regular hexagonal pattern, hairs are frequently grouped in triads or other geometric configurations.

In many skin diseases likewise, the morphologic expression of the epidermal disorder is characterized by a more or less well defined distribution pattern, where certain areas of skin are affected while adjacent zones retain a normal appearance. Papules, macules, pustules, comedones, warts, naevi, ichthyotic scales, psoriatic lesions, blisters, etc. are common examples.

It appears from recent studies on the development of various epithelial-mesenchymal organs, such as the salivary gland [7]. the cornea [15], the tooth [27, 42, 43], the kidney [11–31], as well as the skin [17, 20] that the extracellular matrix (ECM) of the connective tissue and the micro-architecture of the epithelial-mesenchymal junction [8, 14–41] might play an important role in the transmission of morphogenetic messages from one organ constituent to the other. Similarly, although little is yet known with accuracy on dermal-epidermal interactions in skin pathology, it is probable that the dermis, its ECM and the dermal-epidermal junction are involved in the expression of many epidermal disorders.

The Microheterogeneous Distribution of Extracellular Matrix Components

ECM macromolecules, as revealed by indirect immunofluorescence or other histochemical methods, can be classified into two categories [19, 22, 24]. In the first one are those which, like laminin or type IV collagen, are evenly distributed along the dermal-epidermal junction, throughout the development of feathers, scales or hairs in appendage-forming skin as well as in glabrous regions. These components are probably not directly involved in the morphogenesis of cutaneous appendages. In the second category are those which, like interstitial collagen types I and III, fibronectin [21, 22] and several glycosaminoglycans (GAGs) [16, 39] exhibit a heterogeneous distribution, which changes in space and time during the development of cutaneous appendages. The synthesis, deposition and degradation of these components thus appear to be meaningfully related to skin morphogenesis and to the formation of cutaneous appendages, and these metabolic events might thus play an important role in the transmission of morphogenetic signals from the dermis to the overlying epidermis [23, 37].

In the chick embryo for instance, in zones of high morphogenetic activity like the outbulging feather or scale bud, or the ingrowing feather follicle, the density of interstitial collagens and of sulfated GAGs is low, whereas that of fibronectin is high. Conversely, in zones of histological stability, like the base of appendage buds, interplumar, or interscale regions, the density of interstitial collagens and sulfated GAGs is high and increases steadily with age, while that of fibronectin is low and decreases during development. Likewise, Streptomyces hyaluronidase sensitive Alcian blue stain along the dermal-epidermal junction is at first uniform, then becomes more intense inside feather rudiments than in surrounding interplumar skin; later, when the feather bud bulges out, label is first maximal at the apex and posterior slope of the dermal-epidermal junction, then, as the bud elongates, becomes more intense at the cranial than at the caudal slope. Interstitial collagens also exhibit an asymmetric distribution in outgrowing buds: type I collagen is denser under the cranial than under the caudal slope, while type III collagen shows a maximal density under the caudal slope and around the posterior hinge region of the bud.

At later stages of feather development, when the base of the elongating feather filament grows inward to form the feather follicle, interstitial collagens are seen to increase in density in interplumar skin, while their density diminishes around the feather follicle and inside the feather filament. By contrast, fibronectin is more abundant inside the feather filament and around the feather follicle than in interfollicular skin. Alcian blue stained GAGs are also heterogeneously distributed along the dermal-epidermal junction of the feather filament base, being denser at the cranial than at the caudal face.

In normal glabrous skin, such as that of the midventral apterium, or in appendageless skin of mutant (scaleless) [22] or hydrocortisone-treated embryos [9], the distribution of ECM components is homogeneous. Interstitial collagen types I and III are abundant and deposited in increasing amounts with age, notably along the dermal-epidermal junction. By contrast, fibronectin, which is present throughout the depth of the dermis at early stages, gradually disappears from the dermal mesenchyme and is finally completely absent from the dermal-epidermal junction at later stages.

In the mouse embryo, similar observations have been made regarding the distribution of fibronectin [38] and bullous pemphigoid antigen (BPA) [2]. In undifferentiated skin, before the onset of hair development, both components are uniformly distributed, the former throughout the thickness of the dermis, the latter as a continuous underlining of the dermal-epidermal junction. By the time the hair placodes, and later the hair buds, form, fibronectin accumulates around the ingrowing hair peg, notably along the dermal-epidermal junction, and also underneath it in the dermal region into which the epidermal bud is going to penetrate. By contrast, BPA becomes attenuated and disappears from around the lower part of the ingrowing hair bud, and remains absent, at later stages, from the lower part of the hair follicle and from around the dermal papilla, while it persists in the upper part and in interfollicular skin.

These observations lead to the idea that interstitial collagens and several still unidentified GAGs might constitute a firm or semi-solid framework on which to found cutaneous appendages, while fibronectin might promote cell and tissue mobility. In inter-appendage skin, the abundance of interstitial collagens and the scarcity of fibronectin might cause a stabilization of skin tissues, while inside developing appendages the absence or sparseness of collagen and the accumulation of fibronectin might facilitate or promote morphogenetic movements [23]. Likewise, in the skin of mouse embryos, BPA might block the transmission of morphogenetic messages and, through increased adhesiveness between epidermal basal cells and the basement membrane, consolidate developmentally non active regions of skin and appendages.

In Vitro Culture of Dermal Cells

In order to experimentally test the hypothesis that the morphogenetic performance of skin cells might be influenced by their environment, the effect of several substrates was tested on in vitro cultured cells. In a first study, the investigation was carried out on 7-day embryonic chick dermal cells in primary cultures on homogeneous or heterogeneous substrates of type I collagen and of fibronectin, as compared to culture dish polystyrene [40].

Rate of cell patterning was quantified by the definition of an arbitrary scale of 10 stages, progressing from isolated cells (stage 0), through first cell contacts (stage 1), initial alignment and coordinated grouping of neighbouring cells (stage 2), formation of a loose network where cells cover less than half the available surface (stage 3), formation of a dense network where cells cover more than half the available surface of substratum (stage 4), formation of a subconfluent network where cells occupy almost all the available surface (stage 5), initial confluence where cells occupy all the available surface, but still maintain their initial width (stage 6), crowded confluence where cells become narrow and crammed together on the available surface (stage 7), formation of a criss-cross pattern where cells pile up on top of each other in several layers (stage 8), to the conglomeration of cells into bulging clusters (stage 9).

Skin was obtained from the back of 7-day chick embryos, at a stage when the first feather rudiments are being laid down. After treatment of the skin pieces with cold (4 °C) trypsin (1% in calcium- and magnesium-free saline, approximately 30 min), the epidermis was mechanically peeled off the dermis and discarded. The dermis was minced, further trypsinized (same conditions) and dissociated into single cells. The washed cell suspension was used to seed the plastic (Falcon or Corning) culture dishes (35 mm diameter) at a density of 3.5×10^5 cells per dish in 1.7 ml of medium. Culture medium was either Eagle's Minimum Essential Medium (MEM) or a 1 : 1 mixture of MEM and F12 (MEM/F12), supplemented with foetal calf serum (usually 5%) and 1% penicillin. Cultures were maintained at 38.5 °C, in a gas phase of 5% CO_2 in air and saturating humidity, for a period of 7 or 10 days, routinely without change of medium.

In order to precisely localize selected zones of the culture for day after day observation and photographic recording, landmarks were scratched on the bottom of the culture dish; these consisted of two orthogonally oriented sets of three parallel lines, thus delimiting 16 roughly square areas of approximately 9 mm^2.

Necessity of Foetal Calf Serum

Various concentrations of foetal calf serum (FCS) were tested: 20%, 10%, 5%, and 0%. It was found that variations of the concentration of FCS did not significantly influence the rate of cell patterning, but that FCS could not be omitted or replaced by commercially available serum substitutes such as NU-serum (Collaborative Research) (Fig. 1, 2). For further studies the minimal concentration of 5% was chosen as standard.

Removal of Fibronectin from FCS

Since one of the substrate material to be tested was fibronectin, an attempt was made to use FCS depleted of fibronectin by immunoabsorption. The rate of cell patterning was significantly lowered when the fibronectin-depleted serum was used at a concentration of 5%. When the concentration of the depleted serum was raised to 10%, the cultures progressed at approximately the same rate as with 5% whole FCS, with a notable decline however beyond 8 days of culture.

Influence of Substrate on the Rate of Cell Patterning

Collagen Substrate

The bottom of the plastic culture dishes was coated with collagen using the following procedure. Acid soluble type I collagen extracted from calf skin was sprayed as a methanol solution (Cerad, Lyon). As methanol evaporates, collagen molecules precipitate on the bottom of the culture dish and form a uniform coat of so-called molecular non fibrillar collagen.

When dermal cells were cultured on this collagen substrate, the rate of cell patterning was significantly lower than on polystyrene substrate. While on plastic, cultures reached a mean stage of 7.5, on collagen they did not go beyond 4.6 in 7 days (Fig. 3). This inhibitory effect of collagen substrate was evident in MEM as well as in MEM/F12.

The influence of fibrous collagen was tested in the following way. A drop (approximately 6 mm in diameter) of acid soluble bovine type I collagen (1 mg/ml at pH 7.4) was deposited on the bottom of collagen coated dishes. The drop was left in place for 45 min at room temperature, after which part of it was removed by suction, and the remainder was rinsed twice locally with culture medium. This procedure resulted in the formation of a disc-shaped gel of fibrous collagen surrounded by molecular collagen of the same type. The behaviour of the dermal cells on the two kinds of collagen substrate was thus compared within the same culture dish. It was found that on the spot cell patterning progressed at a slower rate than elsewhere (Fig. 4), indicating that fibrous collagen had a more pronounced inhibitory effect than molecular collagen.

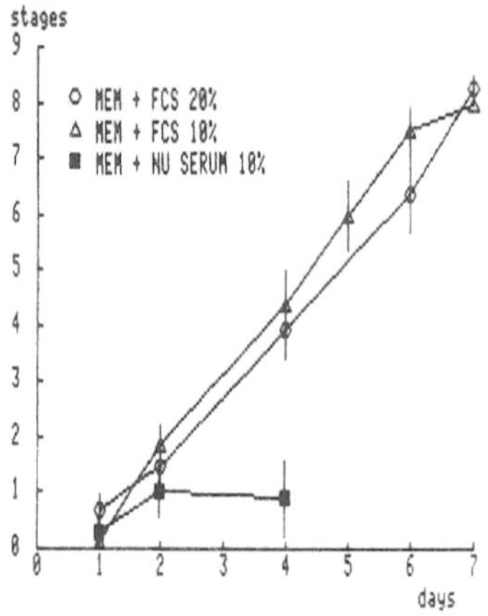

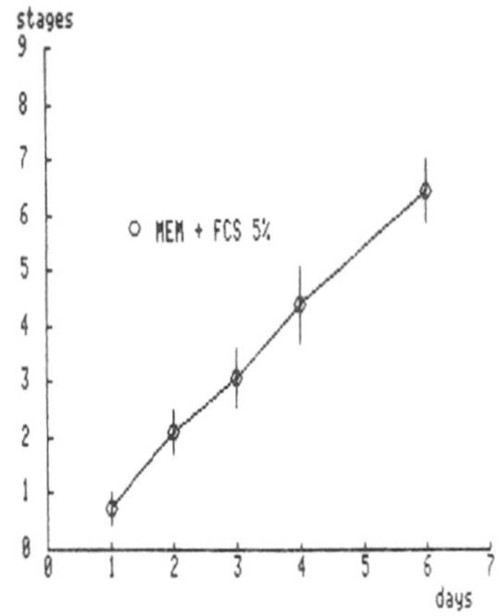

Fig. 1–7. Rate of cell patterning (stages reached by the cultures) as a function of time (days). See text for the explanation of the stages

Fig. 1. Effect of foetal calf serum (FCS) concentration in Eagle's Minimum Essential Medium (MEM). Comparison with NU serum (Collaborative Research)

Fig. 2. Standard evolution of cultures in MEM with 5% FCS on culture dish polystyrene substrate

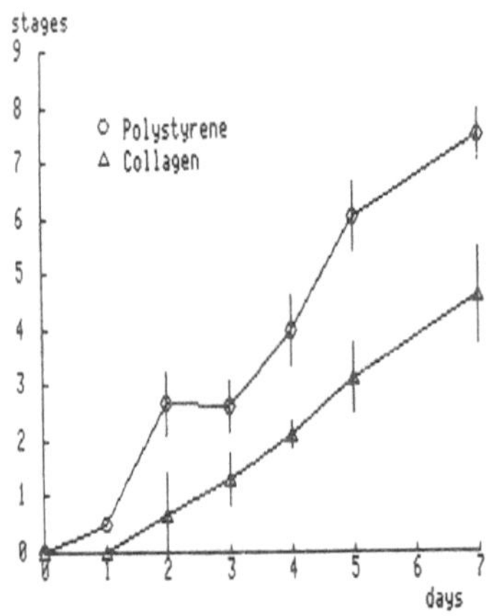

Fig. 3. Effect of bovine acid soluble type I collagen substrate, deposited on the bottom of the culture dishes in non fibrous "molecular" form, compared to polystyrene

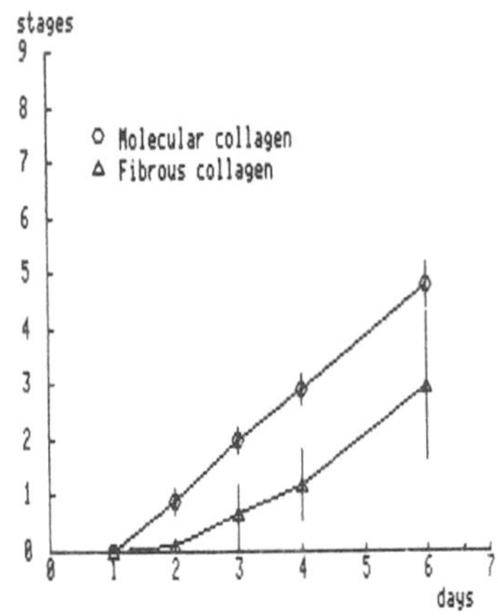

Fig. 4. Comparison between the effect of bovine acid soluble type I collagen in "molecular" and gelled fibrous form. The two types of substrate were offered to the cells within the same dish

Fibronectin Substrate

The culture dishes were coated with fibronectin, using a solution of human serum fibronectin (40 µg/ml) in MEM, which was left in place for 45 min at room temperature. The dishes were then rinsed twice with MEM prior to seeding of the cells. In several other series of experiment, a heterogeneous substrate was offered to the cells

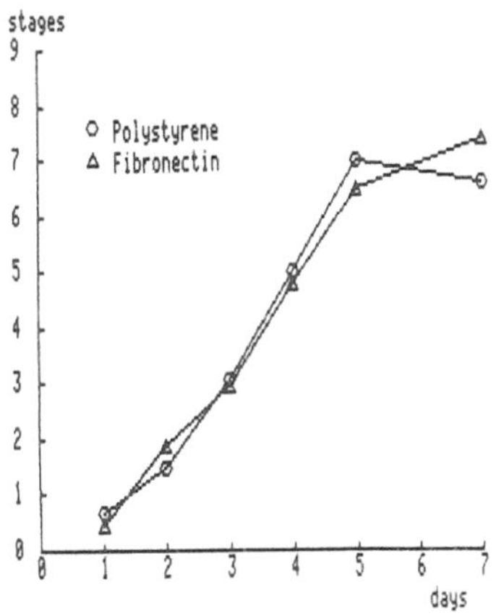

Fig. 5. Effect of human serum fibronectin, compared to polystyrene. The two types of substrate were offered to the cells within the same culture dish

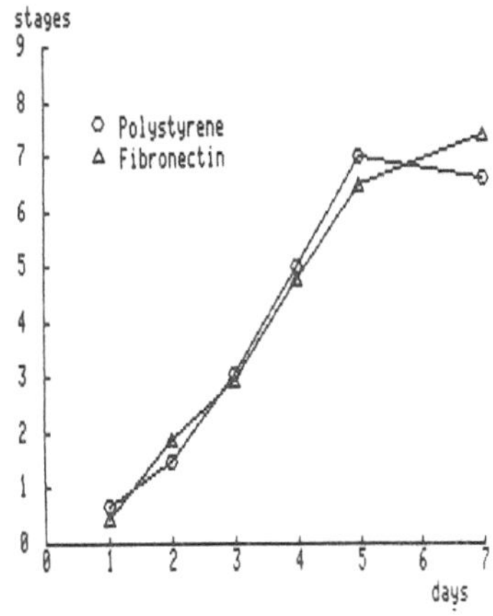

Fig. 6. Comparison between the effect of rinsed and dried human serum fibronectin spots on the bottom of the culture dish

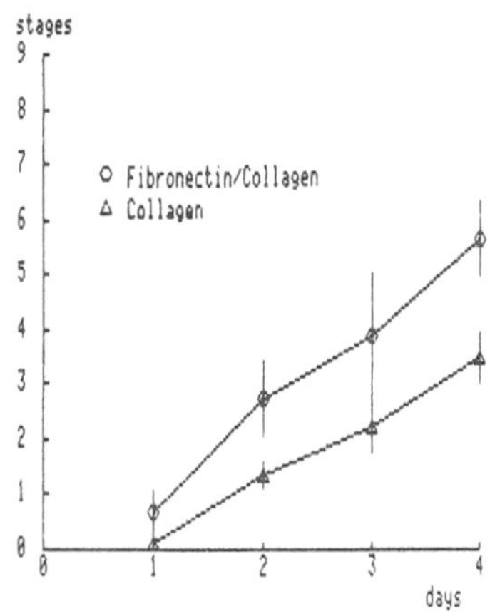

Fig. 7. Effect of human serum fibronectin coat deposited on top of bovine acid soluble type I collagen. The two types of substrate were offered to the cells within the same culture dish

by depositing a roughly circular 5–6 mm diameter drop of fibronectin solution in randomly located spots of the culture dish, in much the same way as with the previously mentioned drops of collagen solution. The drop was removed by suction after 45 min or left to dry out completely. The wet or dry spot was rinsed twice locally, so as to avoid dispersal of fibronectin outside the coated area. The position of the spots with respect to the scratched landmarks was carefully recorded. Thus again the rate of cell patterning on the two kinds of substrate could be compared within the same dish.

There was no difference in the rate of cell patterning between cells growing on the fibronectin coated substrate or on culture dish polystyrene (Fig. 5). There was no significant difference either between cells growing on the rinsed or dried spots of fibronectin (Fig. 6).

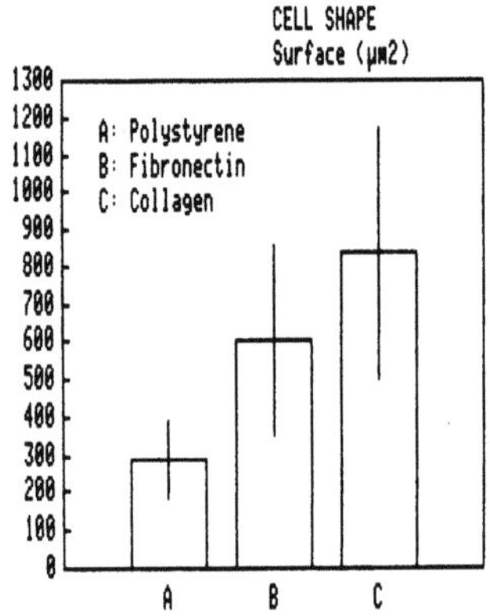

Fig. 8. Effect of substrate (culture dish polystyrene, human serum fibronectin, and bovine acid soluble type I collagen) on the area (surface) occupied by the cells on the bottom of the culture dish

When fibronectin substrate was compared to collagen substrate by depositing a drop of fibronectin solution on the bottom of a dish coated with "molecular" collagen, it was found that the rate of cell patterning on fibronectin was significantly higher than on the surrounding collagen (Fig. 7). Thus fibronectin on top of collagen appeared to completely suppress the inhibitory effect of collagen.

Influence of Substrate on Cell Shape and Locomotion

Selected areas of the cultures were photographically recorded at 15 min intervals for periods of 9–19 h during days 1 to 3 of culture. Areas to be thus analysed were chosen among those where cells had not yet reached stage 1, so that all recorded cells were isolated and seemingly not yet influenced by their neighbours.

One parameter of cell shape under study was the area occupied by a given cell on the substrate. This so-called "cell area" was calculated with the aid of an image analysis computer program, using projection of transparencies onto a digitalizing tablet. The "cell area" of randomly selected cells is of course highly variable, resulting in rather large statistical confidence limits. However, as a rule, on all three types of substrate (polystyrene, fibronectin, collagen), cells tended to progressively spread out with time, so that the area of substrate that they occupied increased as the culture progressed. This was particularly marked on the collagen substrate, where cells were significantly "larger" than on polystyrene. On fibronectin, the "size" of the cells was intermediate between that on polystyrene and that on collagen, although the differences were not statistically significant (Fig. 8).

The large "size" of cells on collagen can be interpreted as resulting from a high adhesivity of the cell surface to collagen. As several studies indicate that cells attach to collagen through interposed fibronectin molecules [1, 18, 19, 28], the greater extension of the cells on collagen than on any of the other two substrates might be caused by a collagen stimulated production and secretion of fibronectin. The strong adhesiveness of cells to the collagen substrate might thus explain the purportedly stabilizing role of collagen in skin morphogenesis.

Another feature of cell behaviour is locomotion. Isolated cells in culture do not move extensively, unless chemoattracted or otherwise stimulated [6, 25, 26]. Rather

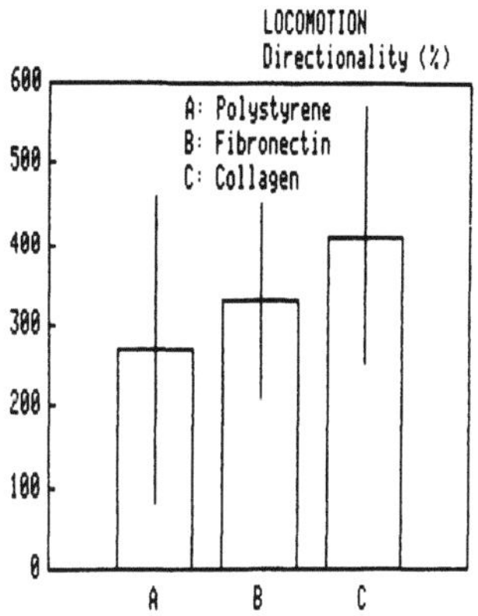

Fig. 9

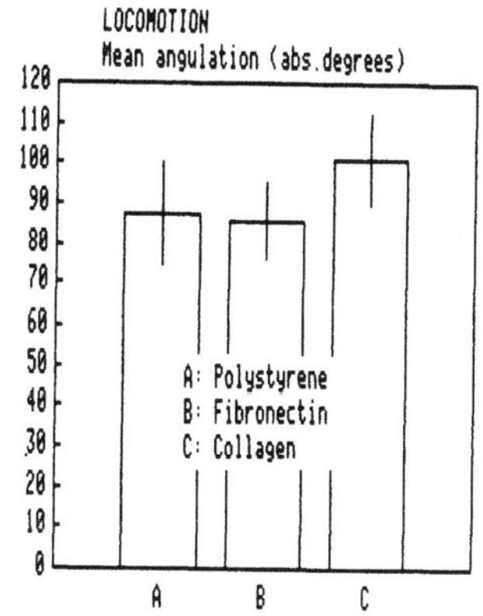

Fig. 10

Fig. 9–11. Effect of substrate (culture dish polystyrene, human serum fibronectin, and bovine acid soluble type I collagen) on cell locomotion

Fig. 9. Directionality, expressed as the percent ratio of the distance between starting and end position to the total length of path

Fig. 10. Angulation, expressed as the absolute value of the mean angle between two successive tracks, recorded at 15 min intervals

Fig. 11. Speed, calculated over a period of 9–19 h of culture

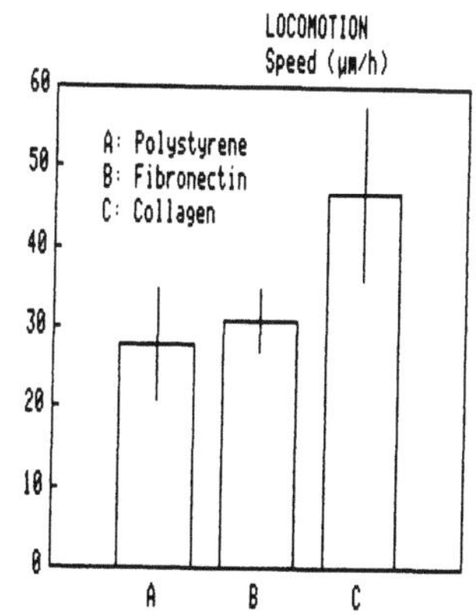

Fig. 11

they fidget about at random, remaining on the same spot for extended periods of time, only exceptionally heading off in one or another direction. This kind of "sur place" locomotion was analysed in a number of randomly chosen and isolated cells, by recording the position of their geometric centre of gravity in a rectangular co-ordinate system, at 15 min intervals. As soon as a given cell under study would establish a contact with a neighbouring cell, analysis was discontinued.

Three parameters were studied:

1) Directionality, which is the percent ratio of the distance between starting and end positions to the total length of the track;
2) Angulation, which is expressed as the absolute value (in degrees) of the mean angle between two successive track portions (at 15 min intervals); and
3) the speed (in micrometres/hour).

The two former parameters were not significantly different on the three types of substrate. Directionality was found to be highly variable from cell to cell and slightly higher on collagen than on fibronectin, and slightly higher on fibronectin than on polystyrene (Fig. 9). Angulation was near 90° on all three substrates (Fig. 10), indicating that cells fidget about completely at random, probably influenced by multiple minor and unknown causes. The third parameter, namely speed, was significantly higher on collagen than on fibronectin or polystyrene (Fig. 11). Whether this somewhat surprising observation is meaningful remains to be investigated by further experimentation. In view of the fact that cells appear to be more spread out on collagen than on the other two substrates, and therefore plausibly to be more adherent to collagen than to fibronectin or polystyrene, it is difficult to understand their higher velocity on collagen. Also, if collagen plays a stabilizing role in morphogenesis, it is unexpected to observe that it can also promote cell locomotion, at least in "molecular" form, which admittedly is not the common form under which it is present in living tissues.

Conclusion

This preliminary analysis of embryonic skin fibroblasts cultured in vitro indicates that the substrate on which the cells grow may indeed influence several features of cell behaviour, such as rate of cell patterning, of spreading and of locomotion.

Thus, while fibronectin was found to have no influence on the rate of cell patterning or of locomotion, collagen had a marked retarding effect on the rate of cell patterning, when compared to polystyrene substrate. Collagen also apparently promoted cell locomotion. Fibronectin, when coated on top of collagen, abolished the inhibitory effect of collagen on the rate of cell patterning. In addition gelled fibrous collagen exerted an even stronger retardation of cell patterning than "molecular" collagen substrate [40].

These results should, for the time being, be interpreted with caution. They are in rough agreement with the view that collagen and fibronectin play a role in the morphogenesis of skin and cutaneous appendages. Indeed immunofluorescence detection of interstitial collagens and of fibronectin in embryonic chick skin showed that collagen was accumulated in high density in zones of morphogenetic stability, while it was removed from and remained scarce in zones of high morphogenetic activity. Conversely, fibronectin was observed to be present in high concentration in morphogenetically active sites, whereas it was scarce or absent in regions that were histologically stabilized [21–23, 37, 38].

The analysis of the behaviour of cultured embryonic fibroblasts therefore sustains the idea that the microheterogeneous distribution of several ECM components in embryonic skin might constitute part of the morphogenetic message that the dermis is known to transmit to the epidermis during the development of skin and cutaneous appendages.

This type of investigation may thus serve as a model for the study of dermal-epidermal interactions in skin diseases. An analysis of the influence of various extracellular matrix macromolecules, and of various combinations of them, can easily be performed using skin cells from human source.

Cell patterning of course reflects various aspects of cellular activity, primarily proliferation and the establishment of mutual contacts, leading to collaborative movements, which may be important for normal and pathological tissue morphogenesis. Cell attachment and spreading on the substratum is but a crude image of the relationships that cells might establish with their three-dimensional environment in situ. They may nevertheless reveal certain meaningful properties of cell surfaces. Locomotory activity on a two-dimensional substrate likewise is but a coarse simplification of cell movements as they occur in the living organism. It is clear that further studies are needed, using more complex environments. Investigations of this kind with three-dimensional gels are under way in many laboratories [3–5, 10, 12, 13, 29, 30, 32, 44] and the techniques that are being developed can serve as models for the study of the properties of human fibroblasts or keratinocytes from normal or diseased skin.

References

1. Akiyama SK, Yamada KM, Hayashi M (1981) The structure of fibronectin and its role in cellular adhesion. J Supramol Struct Cell Biochem 16:345–358
2. Bard S, Micouin C, Thivolet J, Sengel P (1981) Heterogeneous distribution of bullous pemphigoid antigen during hair development in the mouse. Arch Anat Micr Morphol Exp 70:141–148
3. Bell E, Ivarsson B, Merrill C (1979) Production of a tissue-like structure by contraction of collagen lattices by human fibroblasts of different proliferative potential in vitro. Proc Natl Acad Sci USA 76:1274–1278
4. Bellows CG, Melcher AH, Bhargava U, Aubin JE (1982) Fibroblasts contracting three-dimensional collagen gels exhibit ultrastructure consistent with either contraction or protein secretion. J Ultrastr Res 78:178–192
5. Bellows CG, Melcher AH, Aubin JE (1982) Association between tension and orientation of periodontal ligament fibroblasts and exogenous collagen fibres in collagen gels in vitro. J Cell Sci 58:125–138
6. Ben Slimane S, Houllier F, Tucker G, Thiéry JP (1983) In vitro migration of avian hemopoietic cells to the thymus: preliminary characterization of a chemotactic mechanism. Cell Differ 13:1–24
7. Bernfield MR (1981) Organization and remodeling of the extracellular matrix in morphogenesis. In: Connelly TG, Brinkley LL, Carlson BM (eds) Morphogenesis and pattern formation. Raven Press, New York, pp 139–162
8. Démarchez M, Mauger A, Sengel P (1981) The dermal-epidermal junction during the development of skin and cutaneous appendages in the chick embryo. Arch Anat Micr Morphol Exp 70:205–218
9. Démarchez M, Mauger A, Herbage D, Sengel P (1984) Effect of hydrocortisone on skin development in the chick embryo: ultra-structural, immunohistological, and biochemical analysis. Devel Biol 106:15–25
10. Dodd NJF, Schor SL, Rushton G (1982) The effects of a collagenous extracellular matrix on fibroblast membrane organization. An ESR spin label study. Exp Cell Res 141:421–431
11. Ekblom P, Saxén L, Timpl R (1982) The extracellular matrix and kidney differentiation. In: Membranes in growth and development. Alan Liss Inc, New York, pp 429–442
12. England MA, Wakely J (1979) Evidence for changes in cell shape from a 2-dimensional to a 3-dimensional substrate. Experientia 35:664–665
13. Grinnell F (1982) Migration of human neutrophils in hydrated collagen lattices. J Cell Sci 58:95–108
14. Hardy MH, Van Exan RJ, Sonstegard KS, Sweeny PR (1983) Basal lamina changes during tissue interactions in hair follicles. An in vitro study of normal dermal papillae and vitamin A-induced glandular morphogenesis. J Invest Dermatol 80:27–34

15. Hay ED (1981) Collagen and embryonic development. In: Hay ED (ed) Cell biology of extracellular matrix. Academic Press, New York, pp 379–409
16. Jahoda C, Mauger A, Sengel P (1986) Distribution of glycosaminoglycans in the developing feather. In preparation
17. Kitamura K (1981) Distribution of endogenous beta-galactoside-specific lectin, fibronectin and type I and III collagens during dermal condensation in chick embryos. J Embryol Exp Morphol 65:41–56
18. Klebe RJ, Hall JR, Rosenberger P, Dickey WD (1977) Cell attachment to collagen: the ionic requirements. Exp Cell Res 110:419–425
19. Kleinman HK, Klebe RJ, Martin GR (1981) Role of collagen matrices in the adhesion and growth of cells. J Cell Biol 88:473–485
20. Mauger A, Démarchez M, Georges D, Herbage D, Grimaud JA, Druguet M, Hartmann DJ, Sengel P (1982) Répartition du collagène, de la fibronectine et de la laminine au cours de la morphogenèse de la peau et des phanères chez l'embryon de poulet. C R Acad Sci Paris Série III 294:475–480
21. Mauger A, Démarchez M, Herbage D, Grimaud JA, Druguet M, Hartmann DJ, Sengel P (1982) Immunofluorescent localization of collagen types I and III, and of fibronectin during feather morphogenesis in the chick embryo. Devel Biol 94:93–105
22. Mauger A, Démarchez M, Herbage D, Grimaud JA, Druguet M, Hartmann DJ, Foidart JM, Sengel P (1983) Immunofluorescent localization of collagen types I, III, IV, fibronectin and laminin during morphogenesis of scales and scaleless skin in the chick embryo. Wilhelm Roux's Arch Devel Biol 192:205–215
23. Mauger A, Démarchez M, Sengel P (1983) Matrice extracellulaire et morphogenèse de la peau. J Med esthet Chirurg Dermatol 10:193–199
24. Mauger A, Démarchez M, Sengel P (1984) Role of extracellular matrix and of dermal-epidermal junction architecture in skin development. Prog Clin Biol Res 151:115–128
25. Mensing H, Pontz BF, Muller PK, Gauss-Muller V (1983) A study on fibroblast chemotaxis using fibronectin and conditioned medium as chemoattractants. Eur J Cell Biol 29:268–273
26. Rovasio RA, Delouvée A, Yamada KM, Timpl R, Thiéry JP (1983) Neural crest cell migration: Requirements for exogenous fibronectin and high cell density. J Cell Biol 96:462–473
27. Ruch JV, Lesot H, Karcher-Djuricic V, Meyer JM, Mark M (1983) Epithelial-mesenchymal interactions in tooth germs: mechanisms of differentiation. J Biol Buccale 11:173–193
28. Ruoslahti E, Engvall E, Hayman EG (1981) Fibronectin: current concepts of its structure and functions. Collagen Res 1:95–128
29. Sanders EJ, Prasad S (1983) The culture of chick embryo mesoderm cells in hydrated collagen gels. J Exp Zool 226:81–92
30. Sarber R, Hull B, Merrill C, Soranno T, Bell E (1981) Regulation of proliferation of fibroblasts of low and high population doubling levels grown in collagen lattices. Mechanisms Ageing Development 17:107–117
31. Saxén L, Ekblom P, Thesleff I (1982) Cell-matrix interaction in organogenesis. In: Kuehn K, Schoene H, Timpl R (eds) New trends in basement membrane research. Raven Press, New York, pp 257–264
32. Schor SL, Schor AM, Bazill GW (1981) The effects of fibronectin on the migration of human foreskin fibroblasts and Syrian hamster melanoma cells into three-dimensional gels of native collagen fibres. J Cell Sci 48:301–314
33. Sengel P (1958) Recherches expérimentales sur la différenciation des germes plumaires et du pigment de la peau de l'embryon de poulet en culture in vitro. Ann Sc Nat Zool 20:431–514
34. Sengel P (1964) The determinism of the differentiation of the skin and the cutaneous appendages of the chick embryo. In: Montagna W, Lobitz WC (eds) The epidermis. Academic Press, New York, pp 15–34
35. Sengel P (1971) The organogenesis and arrangement of cutaneous appendages in birds. Adv Morphog 9:181–230
36. Sengel P (1976) Morphogenesis of skin. In: Abercrombie M, Newth DR, Torrey JG (eds) Developmental and cell biology series. Cambridge University Press, Cambridge London New York Melbourne

37. Sengel P (1985) Role of extracellular matrix in the development of skin and cutaneous appendages. In: Lash JW, Saxén L (eds) Developmental mechanisms: normal and abnormal. Progr Clin Biol Res 171:123–135
38. Sengel P (1986) Epidermal-dermal interaction. In: Bereiter-Hahn J, Matoltsy AG, Richards KS (eds) Biology of the integument. Vertebrates. Springer, Berlin Heidelberg New York Tokyo, pp 374–408
39. Sengel P, Bescol-Liversac J, Guillam C (1962) Les mucopolysaccharides-sulfates au cours de la morphogenèse des germes plumaires de l'embryon de poulet. Devel Biol 4:274–288
40. Sengel P, Kieny M (1984) Influence of collagen and fibronectin substrates on the behaviour of cultured embryonic dermal cells. Brit J Dermatol 111, suppl 27:88–97
41. Slavkin HC, Bringas P (1976) Epithelial-mesenchymal interactions during odontogenesis. IV. Morphological evidence for direct heterotypic cell-cell contact. Devel Biol 50:428–442
42. Slavkin HC, Cummings E, Bringas P, Honig LS (1982) Epithelial derived basal lamina regulation of mesenchymal cell differentiation. In: Weber R, Burger M (eds) Proc Int Soc Devel Biol IX Congress, Alan Liss Inc, New York
43. Thesleff I, Barrach HJ, Foidart JM, Vaheri A, Pratt RM, Martin GR (1980) Changes in the distribution of type IV collagen, laminin, proteoglycan and fibronectin during mouse tooth development. Devel Biol 81:182–193
44. Yasui N, Osawa S, Ochi T, Nakashima H, Ono K (1982) Primary culture of chondrocytes embedded in collagen gels. Exp Cell Biol 50:92–100

The Development of Epidermis In Vitro. Electronmicroscopic and Autoradiographic Studies in the Limb Bud Culture

U. Schultz-Ehrenburg

A problem of epidermal culture systems is to obtain organotypical differentiation in vitro. For that purpose, the presence of dermal factors is of great importance. The epithelio-mesenchymal interaction is best presented in organ cultures of skin, but adult skin explants show short survival times and a progressive loss of maturation. The most successful culture condition is the embryonal organ culture because it is able not only to survive and to grow, but also to differentiate and to mature.

In the following, the epithelial changes in the limb bud culture of mouse embryos are reported. In this culture system, it is possible to observe an undifferentiated two-layered ectoderm developing into a multilayered squamous epithelium within a few days in vitro [18, 19]. We studied the culture with regard to the following aspects:

- which fine structures of keratinization can be observed and what is their histotopographical organization?
- which similarities and differences in vitro result in comparison with the development in vivo?
- to what extent is it possible to do basic research on epidermogenesis or experimental studies of epidermal differentiation in vitro?

Material and Methods

Upper limb buds of mouse embryos were cultured in an organ culture system according to Aydelotte and Kochhar [2]. Details of the limb bud culture system have been described elsewhere [13, 17, 19]. The limb buds were removed from the embryos on day 11 + 3 h of gestation (= 42 somite pairs) and incubated in the culture for 1 up to 10 days (Fig. 1). The first 24 h of the culture were designated as day 0 in vitro.

For comparison, in vivo developing limb buds were observed at daily intervals from day 11 of gestation up to birth.

For further preparation, the limb buds were fixed and dehydrated in an acctone series and embedded in Mikropal (Ferak, Berlin, FRG). 1 μm semi-thin sections were prepared and inspected light microscopically. For electron microscopic examination, thin sections were cut and contrasted with uranyl acetate and lead citrate.

In the special experimental studies, drugs or radioisotopes were given into the culture medium at different days of culture development for various time intervals. For the autoradiographic investigations, semi-thin sections were prepared as usual and

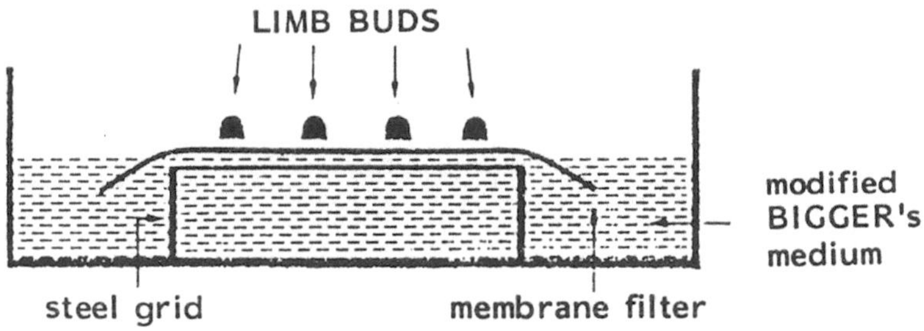

Fig. 1. Cultivation of limb buds from mouse embryos

coated with Ilford K5 photoemulsion by dipping. After exposure, the sections were developed, fixed, stained with alkaline Giemsa solution and embedded in Eukitt.

The Development of the Epidermis in the Culture

The culture started with fore-limb buds from day 11 of gestation (Fig. 2). A special ectodermal structure was present at the apex of the limb buds, the apical ectodermal ridge, which should be referred to in this paper only by mentioning it. Its development in vitro had been described elsewhere [18].

The circumference of the limb buds was mostly covered by a two-layered ectoderm (Fig. 2, 3). The basal cell layer consisted of cuboid or polygonal cells. Only this layer represents the prospective epidermis. The outer cell layer consisted of very flat periderm cells. It is a specific embryonal cell layer without equivalent in the adult epidermis. All cells were completely undifferentiated. Desmosomes or structures of keratinization were detectable nowhere.

Stratification

In the course of 5–6 culture days, stratification and cellular differentiation could be observed step by step in vitro. From the basal cell layer, a stratum intermedium, a stratum granulosum and a stratum corneum developed, and the periderm was desquamated with the formation of a stratum corneum (Fig. 4–6).

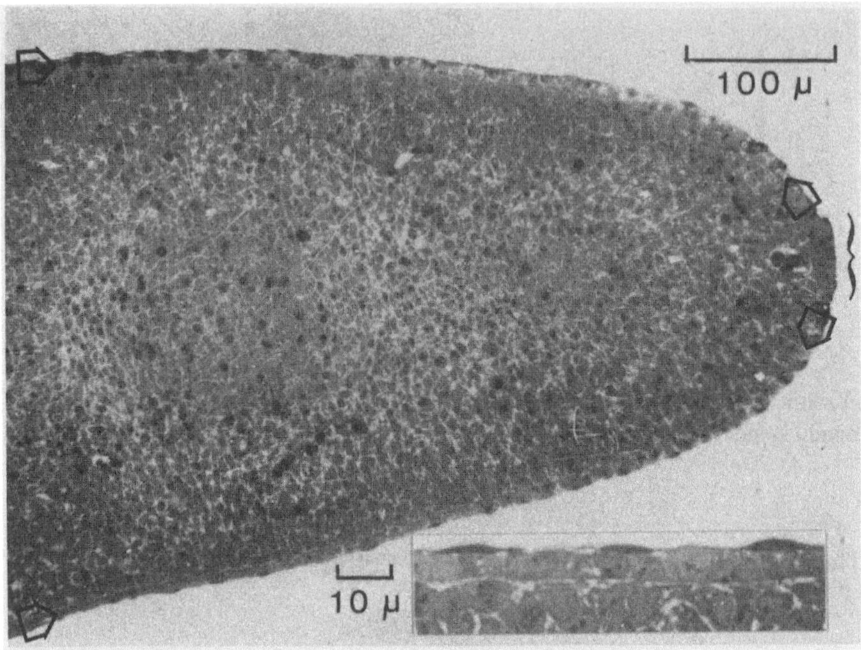

Fig. 2. Fore-limb bud from day 11 of gestation (= 42 somite stage), when put into the culture (day 0 in vitro). With the exception of the apical ectodermal ridge (bracket), the mesenchyme is surrounded by a two-layered ectoderm (arrows and inset)

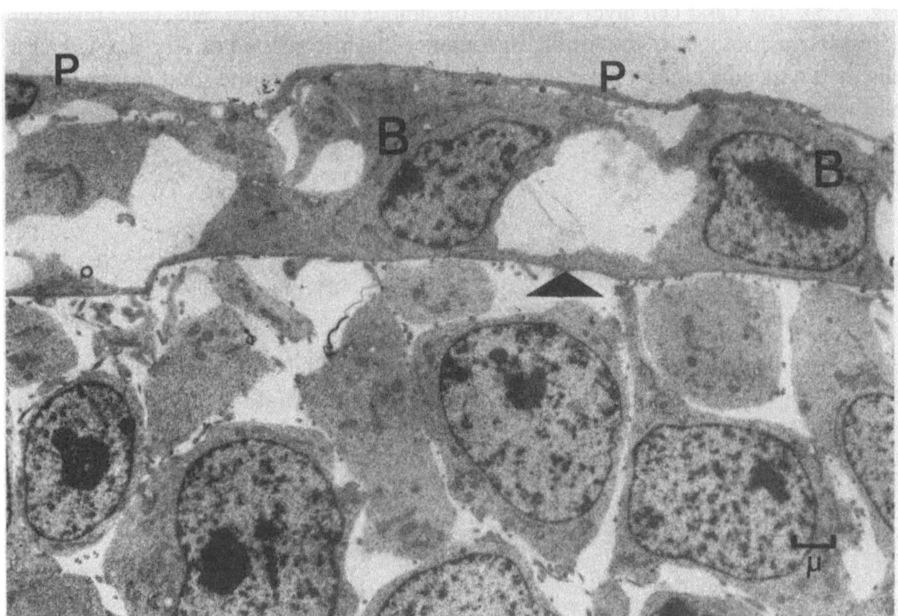

Fig. 3. Limb bud as before on day 0 in vitro. Two-layered ectoderm with periderm (P), basal cell layer (B) and basal lamina (triangle)

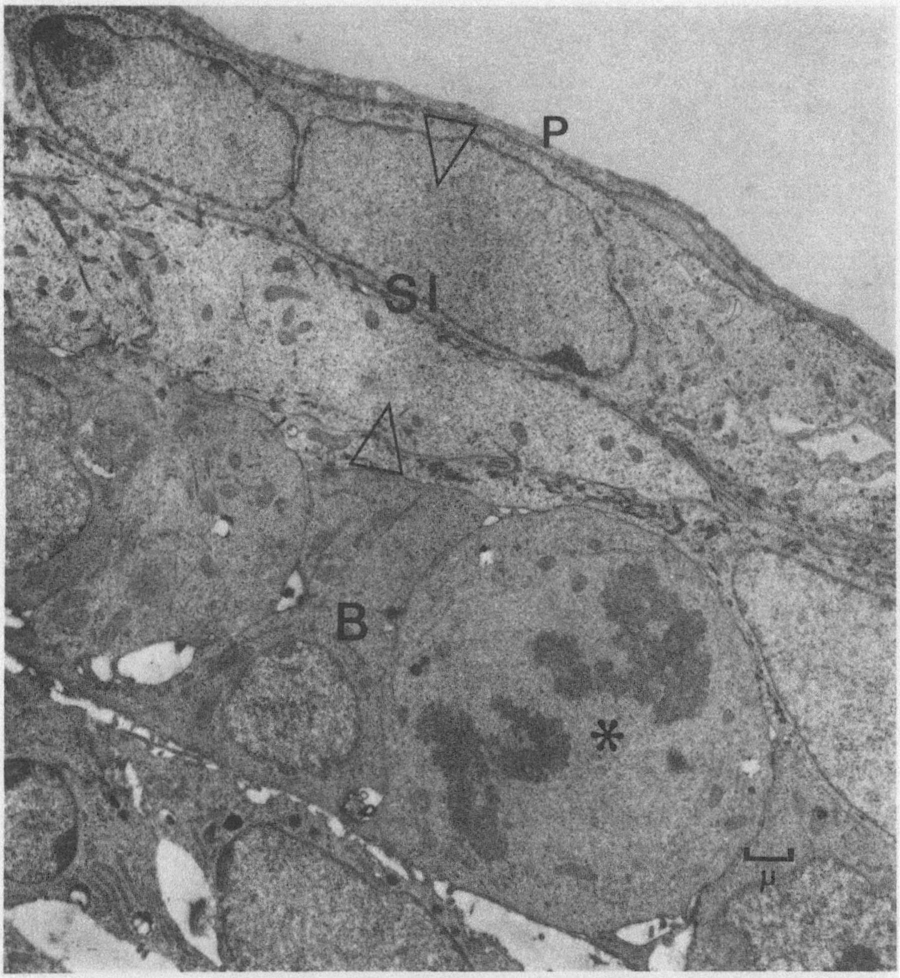

Fig. 4. Limb bud culture on day 3 in vitro. Multilayered epithelium with stratum intermedium (SI). Many desmosomes and some free tonofilaments have developed (compare Fig. 7). * = Mitosis of a basal cell

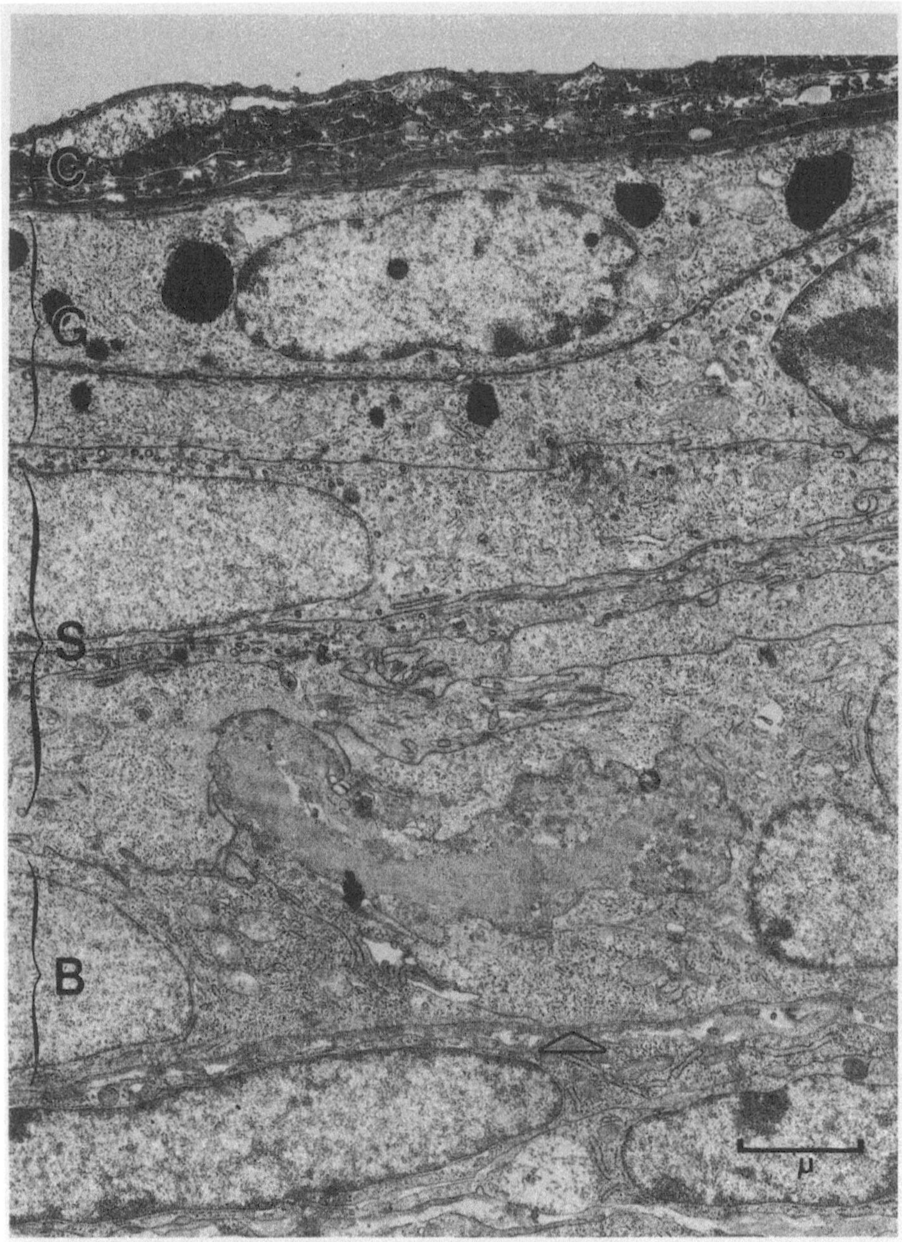

Fig. 5. Day 6 in vitro. Multilayered cornified squamous epithelium. C = stratum corneum, G = stratum granulosum, S = stratum spinosum, B = stratum basale, triangle = basal lamina

	days in vitro				
day 11 in utero →	0	periderm	s.basale	(present)	
	1	↓	↓		
	2				
	3		s.intermedium	(starting)	
	4		s.granulosum	(starting)	strata not all around
	5	desquamation	s.corneum	(starting)	
	6		↓	(not finished)	

Cellular Differentiation

The stratification went hand in hand with fine structural differentiation. At first, desmosomes and tonofilaments appeared, next keratinosomes, keratohyaline granules and dense bodies, and finally cornified cells, intercellular cement substance, composite desmosomes, desmosomal disks and hemidesmosomes (Fig. 4, 5, 7, 8) [18, 19].

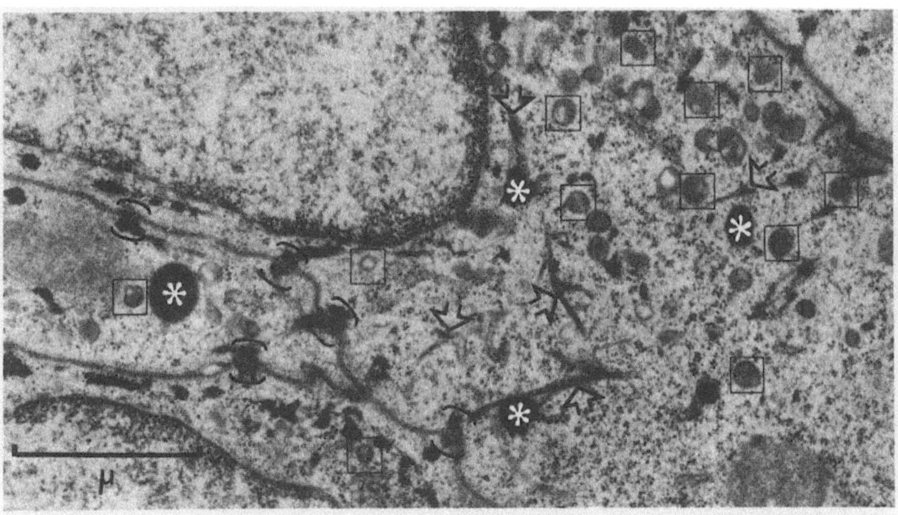

Fig. 7. Cellular epidermal differentiation (day 5 in vitro). Desmosomes (brackets), free tonofilaments (arrows), keratinosomes (rectangles) and keratinohyaline granules (asterisks)

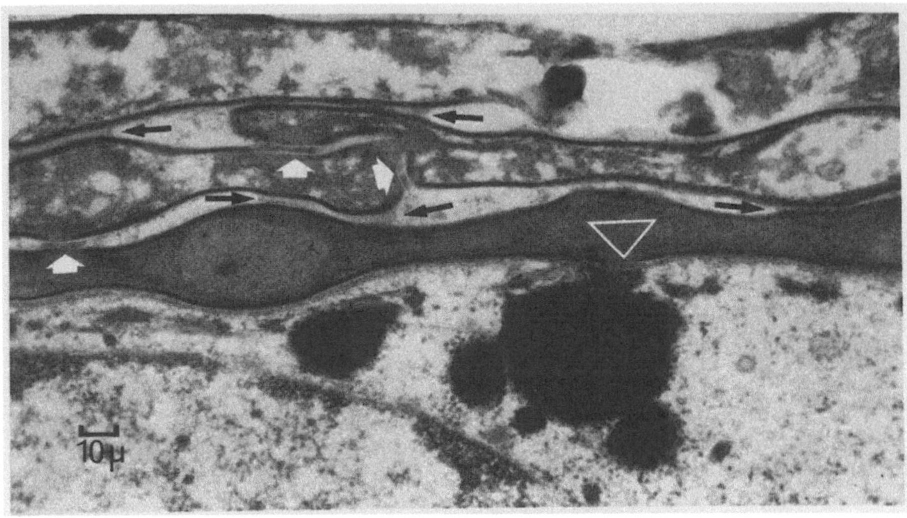

Fig. 8. Cell contacts of horny layer (day 6 in vitro): Composite desmosome (triangle), desmosomal disks (white arrows) and intercellular cement substance (black arrows)

Comparison with the Development in Vivo

The steps of stratification in vivo were similar to the development in vitro, but in vivo the formation of a new stratum was finished within two days (Fig. 9–11). The stratification in vitro proceeded less simultaneously, and in the later stages of development, it was not complete on all sides (Fig. 6).

The cellular differentiation in vitro was quite similar to the changes in vivo. All structures of keratinization and all cellular junctions specific for the in vivo development could be observed in vitro as well. Quantitatively, the cultures showed a deficit of tonofilaments, keratofilaments, horny cell matrix and hemidesmosomes.

The degree of maturation after 6 days of cultivation (Fig. 5) was comparable to the adult interfollicular epidermis corresponding to day 17 in utero (Fig. 11), i.e., two days before birth [19].

days in utero			
11	periderm	s.basale	(present)
12	↓	↓	
13		s.intermedium	(starting)
14		↓	(finished)
15			
16		s.granulosum	
17	desquamation	s.corneum	(starting)
18	loss	↓	(finished)

Fig. 9. Stratification of the limb bud epithelium in vivo

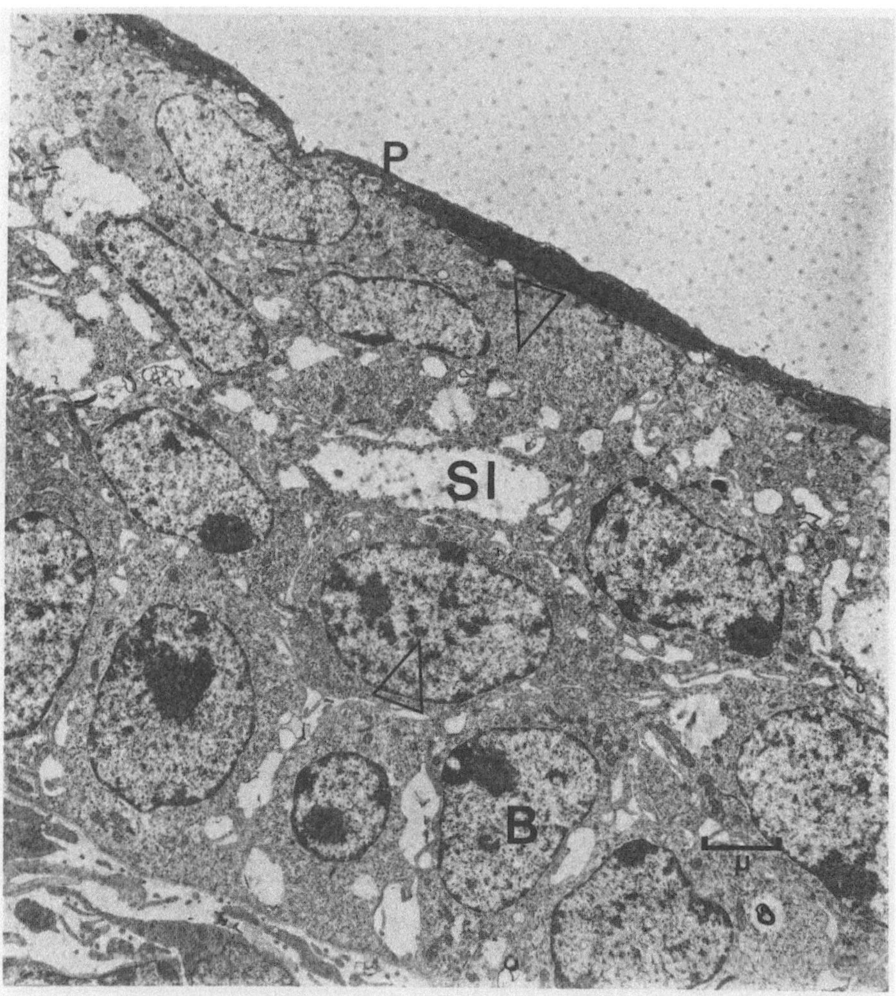

Fig. 10. Limb bud epithelium on day 15 in utero with periderm (P), stratum intermedium (SI), basal cell layer (B) and basal lamina (triangle)

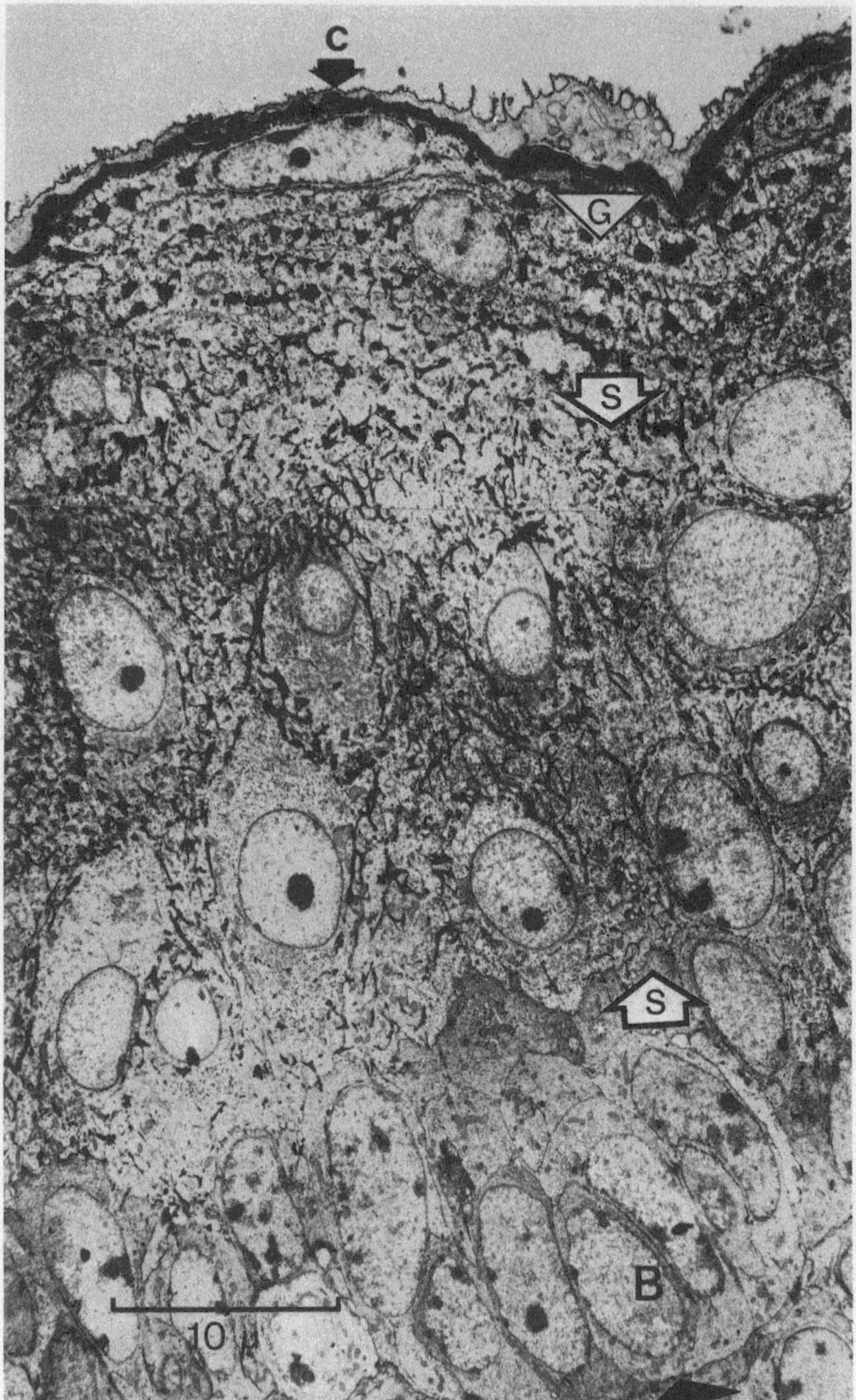
C
G
S
S
B
10 μ

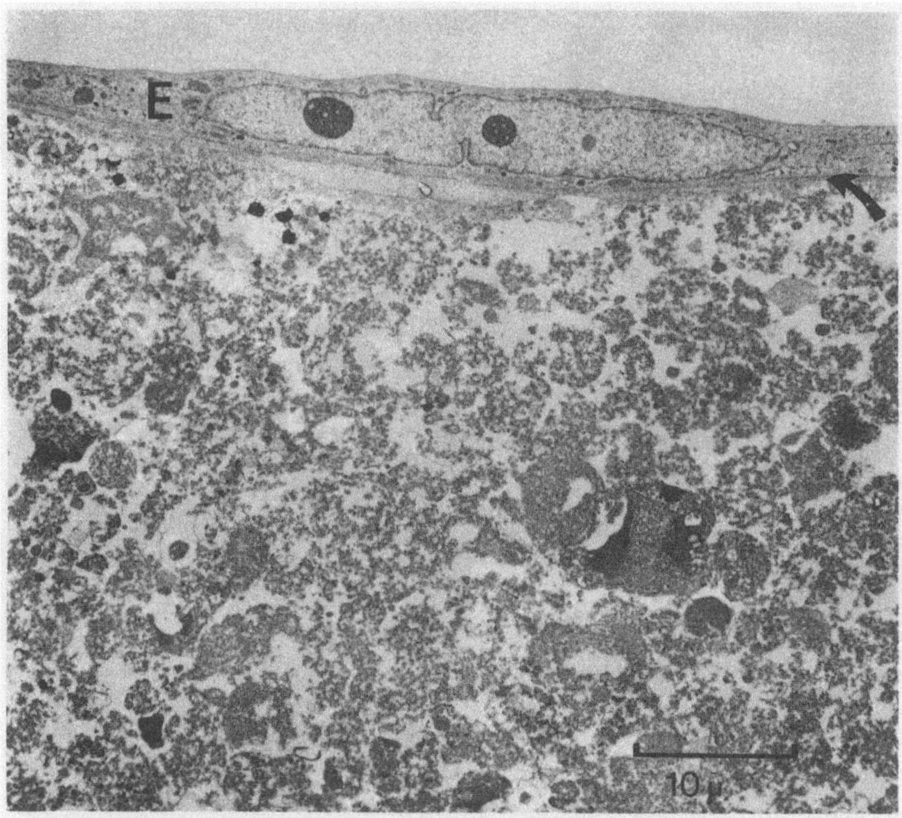

Fig. 12. Treatment with hydroxyurea (1 g/l). Limb bud culture on day 3 in vitro. Single-layered ectoderm (E) with basal lamina (arrow), below which is only cellular debris

Experimental Changes of the Culture

It was further decided to study if the limb bud culture can be used to study experimental influences on epidermal differentiation. For this purpose, antimetabolites and drugs were given into the culture medium, i.e., hydroxyurea, actinomycin-D and cAMP.

Hydroxyurea

With hydroxyurea (0.1 and 1 g/l), an inhibitor of DNA-synthesis [22] growth and differentiation were stopped by both concentrations, but the resistance of ectoderm and

◄ **Fig. 11.** Day 17 in utero. Multilayered cornified squamous epithelium. C = stratum corneum, G = stratum granulosum, S = stratum spinosum, B = stratum basale, triangle = basal lamina

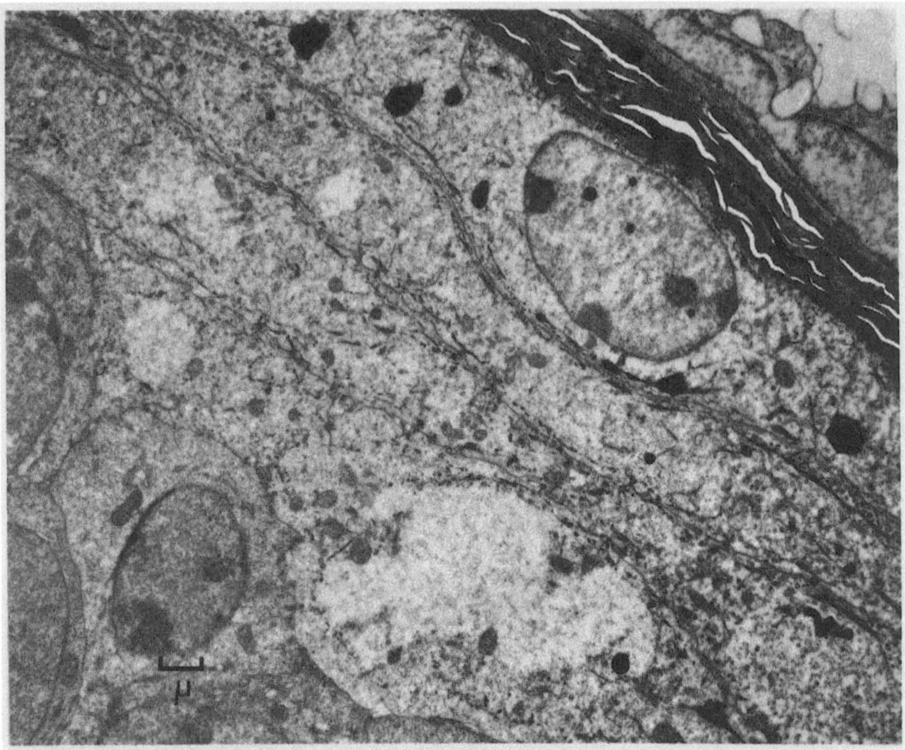

Fig. 13. Treatment with actinomycin-D (0.01 g/l for 3 days). Limb bud culture on day 6 in vitro. No disturbance of stratification and differentiation is recognizable

mesoderm was quite different (Fig. 12). Instead of mesenchyme, the limb buds showed only cell debris with scattered macrophages, containing many lysosomes.

The ectoderm after three and six days of culture was reduced to an one- or sometimes two-layered envelope, but it was alive and contained a few necrotic cells only.

Actinomycin-D

Actinomycin-D (0.01 g/l, an inhibitor of RNA-synthesis [8] was added for different time periods (1, 2 and 3 days). The maturation of the ectoderm was not disturbed. Figure 13 shows a limb bud after 6 days in vitro with well developed epidermal strata. Contrary to the epidermal findings, skeletal malformations were obtained under these culture conditions [29].

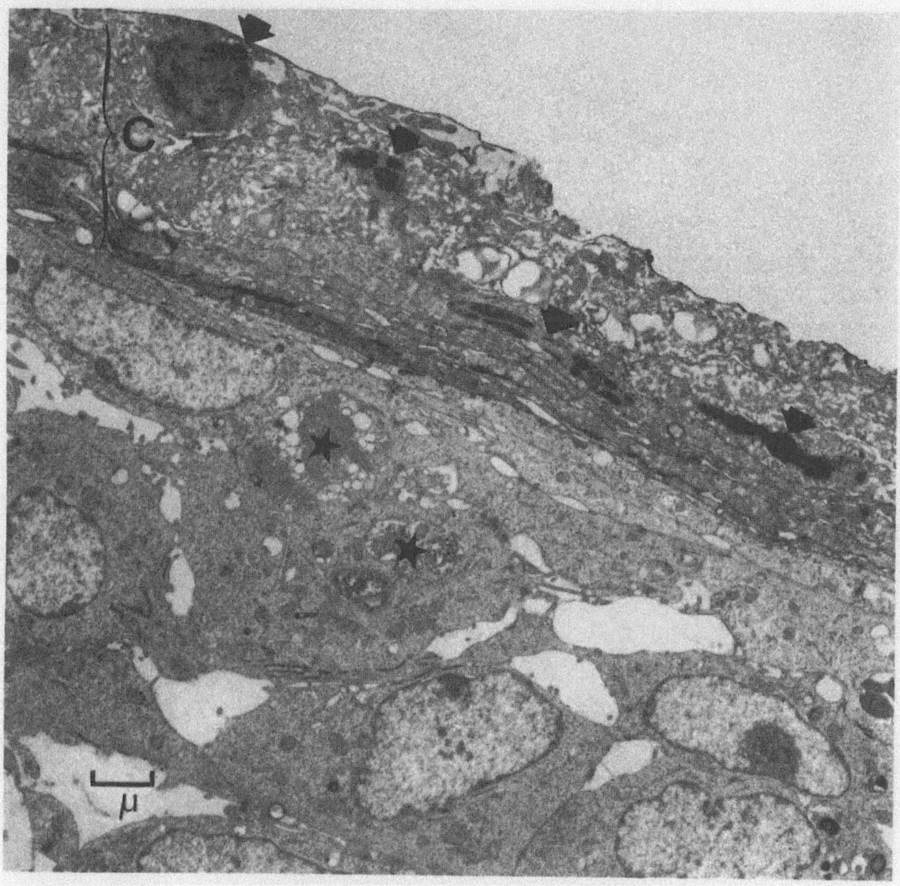

Fig. 14. Treatment with dibutyryl-cAMP (1.0 x 10^{-3} M). Limb bud culture on day 6 in vitro. Thick parakeratotic horny layer (C) with persistent pyknotic nuclei (arrows). No stratum granulosum. Some necrotic figures in the Malpighian layers (asterisks)

cAMP

In this experiment, dibutyryl-cAMP (0.5 x 10^{-3} M and 1.0 x 10^{-3} M) was given into the culture, which is said to be an inhibitor of proliferation and a promoter of cellular differentiation for epidermal cells [4, 5, 23]. The reaction pattern of the limb bud epithelium was studied after three and six days in vitro.

The findings were similar in both concentrations [19]. The proliferation did not change, but a deviation of cellular differentiation was recognizable. The limb buds showed a diminution of tonofilaments and keratohyaline granules, leading to a lack of stratum granulosum and parakeratotic cornification (Fig. 14, 15). The horny cells presented persistent pyknotic nuclei, incomplete thickening of the cell membranes, and a reduction of desmosomal disks.

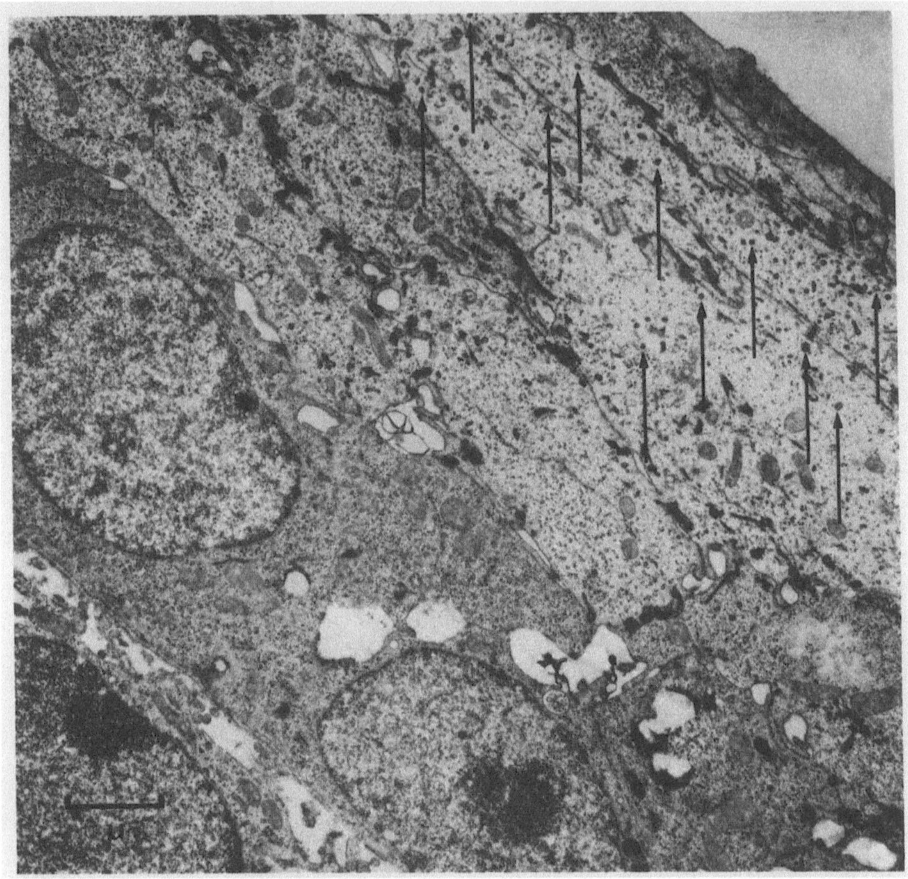

Fig. 15. Treatment with dibutyryl-cAMP (0.5×10^{-3} M). Limb bud culture on day 6 in vitro. Several layers with keratinosomes (arrows), but without keratohyaline granules. C = stratum corneum

According to Friedman [7], cAMP should be considered to be a promoter of the expression of differentiated properties only. Our findings support the possibility that it may also play a role in the process of determination, i.e., of developing on one of several potential paths.

The results might be of interest for researchers in psoriasis because similar structural changes are found in psoriatic epidermis. The reports on the significance of cAMP for psoriasis are controversial. The data by Vorhees' group [16, 24, 25] would speak in favour of a decrease of the cAMP-level in psoriatic lesions, but other authors [1, 11, 28] have found increased or normal levels, which would better correspond to our findings.

Parakeratosis is not specific for psoriasis and appears in many inflammatory diseases of the skin. Parakeratotic differentiation developed without inflammatory influences, since inflammation is not possible in the culture system. Summing up, it may be said that the cAMP treated limb bud culture represents a model of parakeratotic differentiation in vitro.

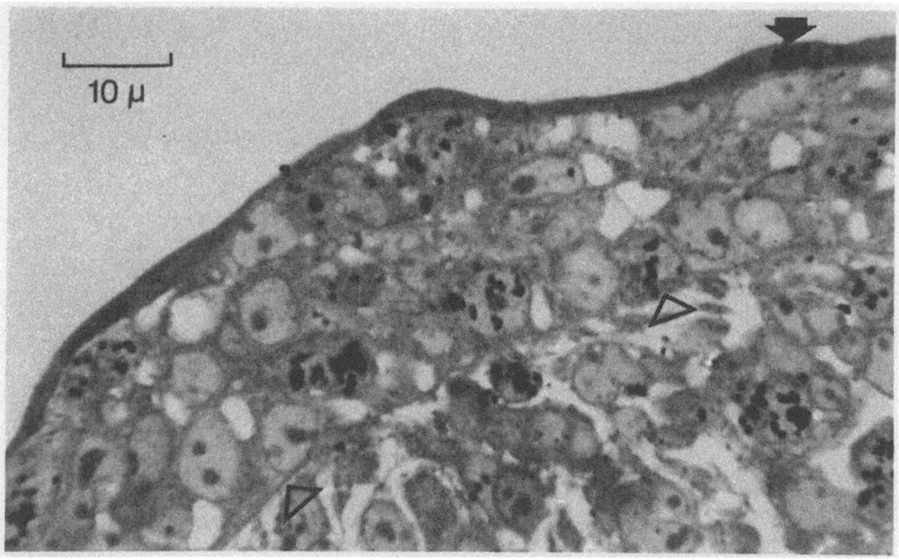

Fig. 16. Autoradiography with ^{3}H-thymidine on day 3 in vitro (labelling for 1 h). Labelled cells in all epidermal layers, there is also one in the periderm (arrow). Triangles = basal lamina

Autoradiographic Investigations on Epidermogenesis in Vitro

In the last two experiments we were interested in the question how far the limb bud culture can be used for autoradiographical investigations. The culture system seemed to be particularly suited with regard to the following aspects:

- The whole epidermogenesis takes place in a very short time, i.e., 5–6 days in vitro.
- The radioactive substance cannot be metabolised by the maternal organism or the placenta. For example, ^{3}H-thymidine passes the placenta, but the labelling of the embryo is up to 8 times lower than that of the mother animal [26].
- The intact epitheliomesenchymal connection is a great advantage for autoradiography in vitro, because the ability to incorporate thymidine is nearly completely lost within 12 h in isolated embryonic epidermal explants [27].

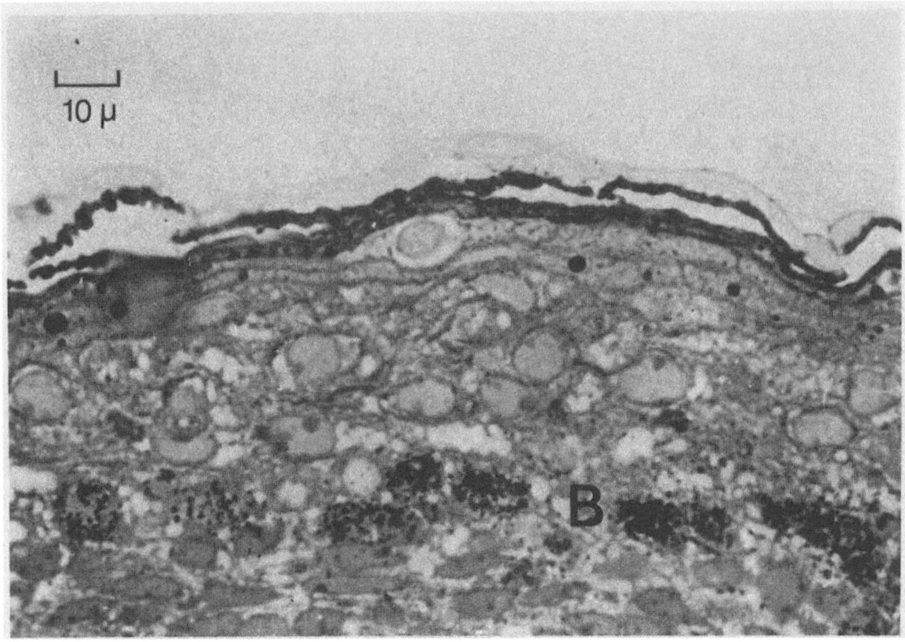

Fig. 17. Autoradiography with ^{3}H-thymidine on day 6 in vitro (labelling as before). Multilayered cornified squamous epithelium. Only the basal cell layer (B) is still labelled

Cell Kinetics

To investigate the cell kinetics of the epidermogenesis in vitro limb bud cultures were labelled with 5 µCi/ml ^{3}H-thymidine (sp.act. 5 Ci/mmol, Amersham-Buchler). Long-term labelling for 24 h and pulse labelling for 1 h were performed at different stages of development, and the migration of labelled cells was studied on day 3 in vitro up to 10 h after pulse labelling [14].

Two successive types of proliferation could be observed. In the early phase of epidermogenesis up to the stage of a stratum intermedium each individual cell layer grew horizontally and had its own cell cycle, including the periderm (Fig. 16).

The change to vertical proliferation started on day 4 in vitro with formation of the stratum granulosum and was finished on day 5 in vitro with the appearance of the stratum corneum. Then the limb buds showed the typical cell kinetics of the adult epidermis, which is regenerated from the basal layer only (Fig. 17) [14].

Previous reports on the cell kinetics of embryonal epidermogenesis using autoradiography are very rare and refer to special stages of development only [9, 27]. Our findings support the conclusions of Smart [20], which were drawn from histological investigations on the place of cell cleavage in mouse embryos in vivo.

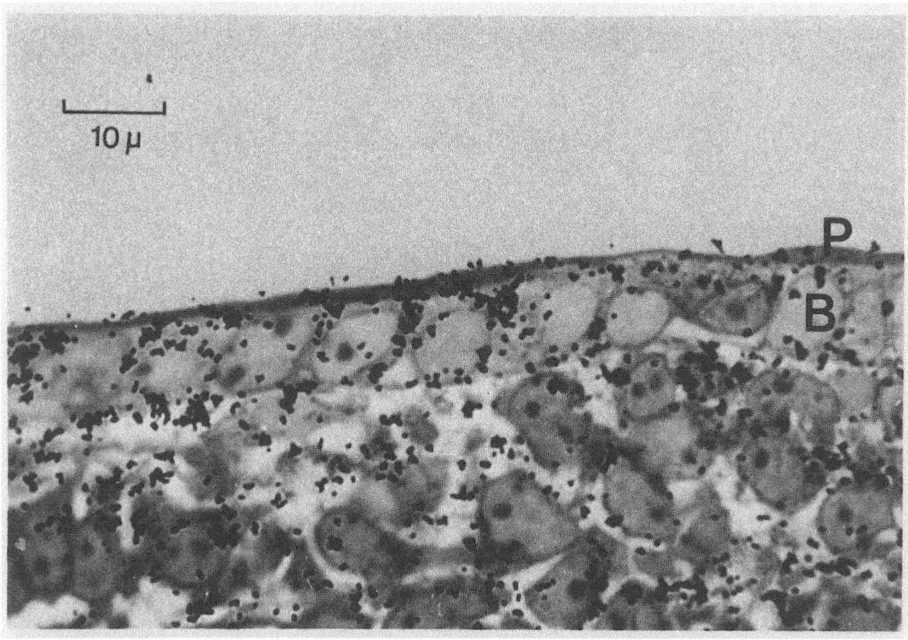

Fig. 18. Autoradiography with ^{3}H-fucose on day 2 in vitro (labelling for 4 h). Labelling of epidermal and peridermal cells, especially in the region of the cell surface. P = periderm, B = basal cell layer

The findings indicate that the reaction of the limb bud epithelium is organotypical in our culture system.

Glycoprotein Synthesis

Finally, the suitability of the limb bud culture for histotopographical metabolic observations was studied. The described changes in cell kinetics during the epidermogenesis might be connected with changes in the contact and adhesive behaviour of the epidermal layers. Since the glycoproteins of the cell surface coat are of special importance in this respect, we performed an autoradiographical study with ^{3}H-fucose [15].

L-fucose is a carbohydrate that is specifically incorporated into carbohydrate side chains. Therefore ^{3}H-fucose is a suitable radioactive precursor to observe glycoprotein synthesis and glycoprotein turnover [3].

The cultures were labelled with 25 µCi/ml ^{3}H-fucose (sp. act. 15 Ci/mmol, Amersham-Buchler). Labelling was performed for 30 min or 4 h on days 2 to 6 in vitro. After incubation for 30 min, the labelling was weak and restricted to the cytoplasm, indicating glycoprotein synthesis. After labelling for 4 h, the labelling was visible also in the region of the cell surface, indicating glycoprotein synthesis and glycoprotein turnover.

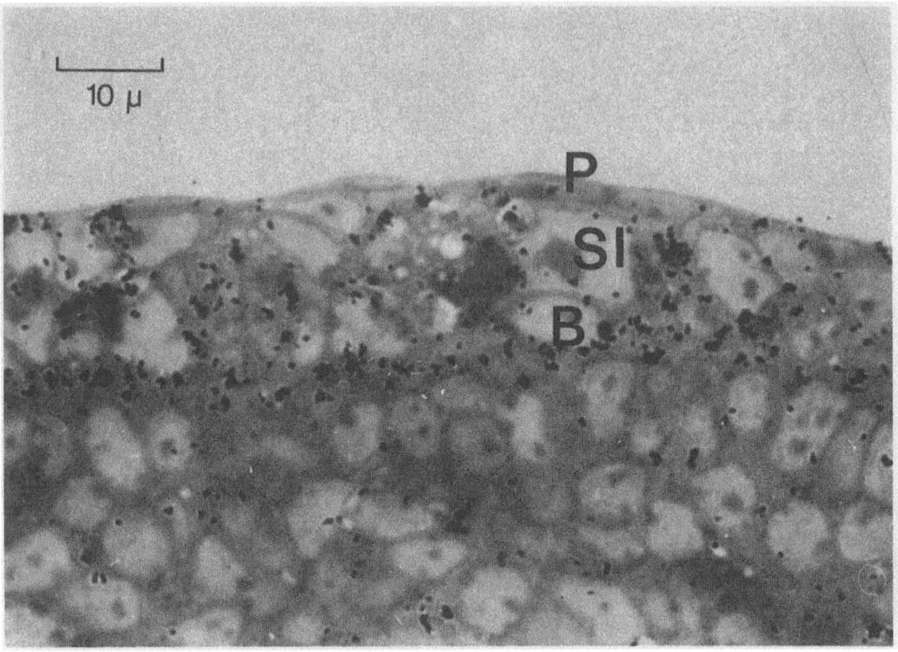

Fig. 19. Autoradiography with ^{3}H-fucose on day 3 in vitro (labelling as before). No change of labelling pattern. Labelling in all epithelial cell layers. SI = stratum intermedium

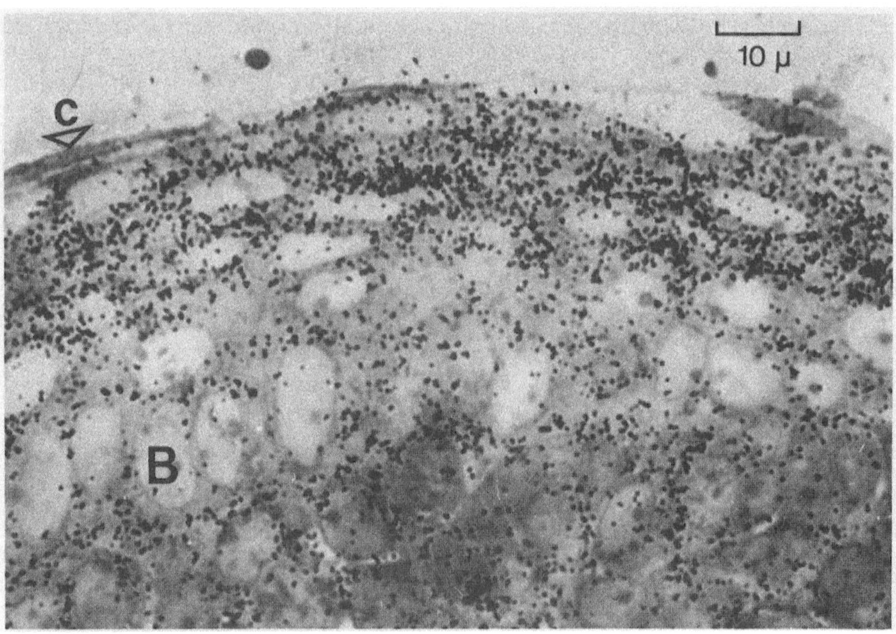

Fig. 20. Autoradiography with ^{3}H-fucose on day 6 in vitro (labelling as before). Strong labelling in the stratum granulosum and in the upper stratum intermedium. C = Lamella of stratum corneum, B = basal cell layer

In the two-layered ectoderm and in the early stratum intermedium, no difference in the labelling of individual layers was found (Fig. 18, 19). Shortly before the appearance of the stratum granulosum, strong labelling with ^{3}H-fucose became visible in the outer intermediate cells and later on in the granular cell layer, too (Fig. 20). No labelling was detectable in the newly formed stratum corneum. Labelling of periderm cells ceased with the occurence of the stratum granulosum, indicating cessation of glycoprotein synthesis in these cells before desquamation.

The findings indicate that the development of the stratum granulosum is connected with a strong increase of glycoprotein synthesis. But why do the other cell layers show no such strong activity during their stratification? The only cellular structures which are exclusively present in the upper intermediate and in the granular layers are the keratinosomes.

The keratinosomes of adult or juvenile epidermis shall mainly consist of lipids [6, 10]. The content of glycoproteins [12, 27] seems to be of minor importance. At least during epidermogenesis, the formation of keratinosomes is associated with marked glycoprotein synthesis.

Our findings show that metabolic processes maintain their histotopographical differences in the limb bud epithelium in the culture system. Thus the culture system is suitable for basic research of biological functions in vitro.

References

1. Adachi K, Iizuka H, Halprin KM, Levine V (1980) Epidermal cyclic AMP is not decreased in psoriasis lesions. J Invest Dermatol 74:74–76
2. Aydelotte MB, Kochhar DM (1972) Development of mouse limb buds in organ culture: chondrogenesis in the presence of a proline analog, L-azetidine-2-carboxylic acid. Dev Biol 28:191–201
3. Benett G, Leblond CP, Haddad A (1974) Migration of glycoprotein from Golgi apparatus to the surface of various cell types as shown by radioautography after labeled fucose injections into rats. J Cell Biol 60:258–284
4. Delescluse C, Colburn NH, Duell EA, Voorhees JJ (1974) Cyclic AMP-elevating agents inhibit proliferation of keratinizing guinea pig epidermal cells. Differentiation 2:343–350
5. Delescluse C, Fukuyama K, Epstein W (1976) Dibutyryl cyclic AMP-induced differentiation of epidermal cells in tissue culture. J Invest Dermatol 66:8–13
6. Elias PM (1981) Membranes, lipids, and the epidermal permeability barrier. In: Marks R, Christophers E (eds) The epidermis in disease. MTP Press, Lancaster, pp 1–30
7. Friedman DL (1976) Role of cyclic nucleotides in cell growth and differentitation. Physiol Reviews 56:652–708
8. Fukuyama K, Epstein WL (1971) Inhibition of RNA and protein synthesis in granular cells by actinomycin-D and puromycin. J Invest Dermatol 56:211–222
9. Gerstein W (1971) Cell proliferation in human fetal epidermis. J Invest Dermatol 57:262–265
10. Grayson S, Elias PM (1982) Isolation and lipid biochemical characterization of stratum corneum membrane complexes: Implications for the cutaneous permeability barrier. J Invest Dermatol 78:128–135

11. Härkönen M, Hopsu-Havu VK, Raji K (1974) Cyclic adenosine monophosphate, adenyl cyclase and cyclic nucleotide phosphodiesterase in psoriatic epidermis. Acta Dermatovenereol (Stockh) 54:13–18
12. Hashimoto K (1971) Cementosome, a new interpretation of the membranecoating granule. Arch Dermatol Forsch 240:349–364
13. Herken R (1977) The influence of D-penicillamine on the proliferation rate of cells from the upper limb bud of mouse embryos in vitro. Teratology 15:159–161
14. Herken R, Schultz-Ehrenburg U (1981) Autoradiographic investigations on the cell kinetics of epidermis and periderm of limb buds from mouse embryos in vitro. Br J Dermatol 104: 277–284
15. Herken R, Schultz-Ehrenburg U (1983) Autoradiographic study of glycoprotein synthesis during embryonic epidermogenesis of the mouse in vitro. Arch Dermatol Res 275:109–113
16. Marcelo CL, Duell EA, Stawiski MA, Anderson TF, Voorhees JJ (1979) Cyclic nucleotide levels in psoriatic and normal keratomed epidermis. J Invest Dermatol 72:20–24
17. Neubert D, Merker H-J, Tapken S (1974) Comparative studies on the prenatal development of mouse extremities in vivo and in organ culture. Naunyn-Schmiedeberg's Arch Pharmacol 286:251–270
18. Schultz-Ehrenburg U (1975) Differentiation of the epidermis in the limb bud culture. In: Neubert D, Merker H-J (eds) New approaches to the evaluation of abnormal embryonic development. Georg Thieme, Stuttgart, pp 213–226
19. Schultz-Ehrenburg U (1980) Die Epidermisentwicklung im In-vitro-Modell. Vergleichende licht-, elektronenmikroskopische und autoradiographische Studien an der Extremitätenknospenkultur. Habilitationsschrift, Freie Universität Berlin
20. Smart IHM (1970) Variation in the place of cell cleavage during the process of stratification in the epidermis. Br J Dermatol 82:276–282
21. Staak WJBM van de, Stadhouders AM, Gilsing H (1969) A comparative electron microscopic and histochemical investigation of membranecoating granules in normal human skin and in the skin of psoriasis vulgaris patients. Dermatologica 138:341–345
22. Timson J (1975) Hydroxyurea. Mutat Res 32:115–132
23. Voorhees JJ, Duell EA, Kelsey WH (1972) Dibutyryl cyclic AMP inhibition of epidermal cell division. Arch Dermatol 105:384–386
24. Voorhees JJ, Duell EA, Bass LJ, Powell JA, Harrell ER (1972) Decreased cyclic AMp in the epidermis of lesion of psoriasis. Arch Dermatol 105:695–701
25. Voorhees JJ, Stawiski M, Duell EA, Haddox MK, Goldberg ND (1973) Increased cyclic GMP and decreased cyclic AMP levels in the hyperplastic, abnormally differentiated epidermis of psoriasis. Life Sci 13:639–653
26. Wegener K, Hollweg S, Maurer W (1964) Autoradiographische Bestimmung der DNS-Verdopplung und anderer Teil-Phasen des Zell-Zyklus bei fetalen Zellarten der Ratte. Zeitschr Zellf 63:309–326
27. Wessels NK (1963) Effects of extra-epithelial factors on the incorporation of thymidine by embryonic epidermis. Exp Cell Res 30:36–55
28. Yoshikawa K, Adachi K, Halprin KM, Levine V (1975) Is the cyclic AMP in psoriatic epidermis low? Br J Dermatol 93:253–258
29. Zimmermann B, Bachmann D, Neubert D, Merker HJ (1975) Induction of skeletal malformations in organ cultures of mouse limb buds. Experientia (Basel) 31:227–228

In Vitro Perfusion of Human Skin

H. Hiernickel

Percutaneous penetration is an essential part of the dermatological toxicology and pharmacology. Only after being absorbed is a topically applied substance able to cause systemic reactions. Until now experiments on these questions could only be performed using dead human skin or with living animals. However with dead skin it is not possible to detect energy-dependent phenomena such as active transport mechanisms or metabolic changes. The second model has the disadvantage that percutaneous absorption shows great interspecies differences.

For this reason we developed a procedure for establishing a long surviving isolated human skin preparation. Our model made it possible to perfuse human cutis, fatty tissue and lymphnodes in vitro. In the following we describe the method of preparation, the technical details and the first results of our experiments.

Material

There are four regions of the human integument, which are suitable for our experiments. The first flap is located in the axilliary region. The vascular supply is highly variable, one being one supplied by multiple musculo-cutaneous arteries, another by direct cutaneous arteries from the subscapular or lateral thoracic artery.

The second donor site is the deltopectoral area. This relatively thin flap is supplied by a complex arteriovenous system. However the diameters of these vessels are very small.

A neurovascular free flap is located on the dorsum of the foot. The blood supply is provided by the dorsalis pedis artery and the veins of the dorsum pedis [2, 8].

The most interesting preparation for our experiments originates from the groin [7]. It has the following advantages:

This iliofemoral flap is supplied by vessels with a large diameter (0.8 to 1.5 mm). These are the superficial circumflex iliac and the superficial epigastric arteries and veins. The arteriovenous system of the groin flap is multiple.

This groin flap has been used very successfully by microsurgeons for free island flap transplantations for many years. The donor site has a large size of about 9.0 to 21.0 cm.

This skin preparation is often available as a "waste product", when we carry out dissection of the inguinal lymphnodes of patients suffering from malignant melanoma (vertical tumor thickness over 0.75 mm). We process according to the operation tech-

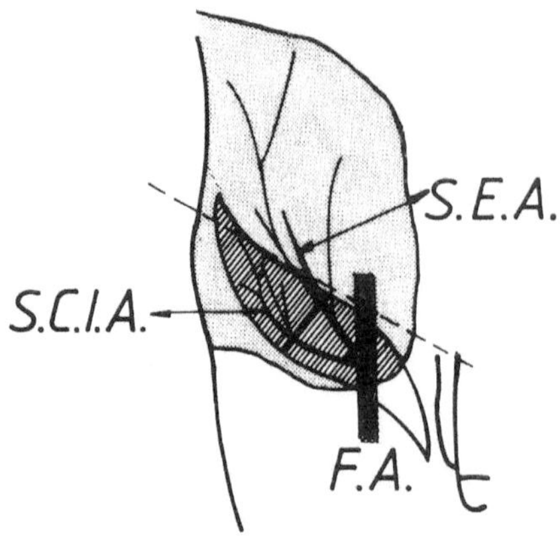

Fig. 1. Design of the groin flap

nique first described by Tritsch [10]. It allows the most caretul dissection of the vertical and horizontal inguinal lymphnodes and shows the best results in wound healing.

Design and Dissection of the Flap

The flap axis line is marked. It extends from the anterior superior iliac spine to a point above the femoral artery 5 cm below the inguinal ligament. Then the flap borders are lined. They have a double-S-spindle form (Fig. 1).

After the lateral to medial dissection of the flap we elevate the preparation carefully from the sartorius muscle fascia. The entrances of the superficial circumflex iliac and superficial epigastric arteries and veins into the femoral artery into the saphenous bulb are carefully isolated. Finally the medial border of the flap is dissected.

The preparation is stored in an isotonic solution of glucose (120 mg per cent). The temperature is 5 °C. To prevent thrombosis 50 IU heparin per ml solution are added [4].

The arteries are dissected carefully and cannulated with Cavafix MT catheters. All the catheters flow into a central Luer-lock-ramp. To enable this connection we reconstructed the ramp with tetrahydrofuran and male respectively female Luer-locks.

Construction of the Perfusion Apparatus

With the knowledge we have gained from more than 7 experiments with our first perfusion machine [5], we have constructed a completely new gas-tight tempered apparatus (Fig. 2 [2]). The atmosphere in this plastic tank consists of 95% oxygen and 5% carbon dioxide, the temperature amounts 37 °C and the relative atmospheric humidity 100%.

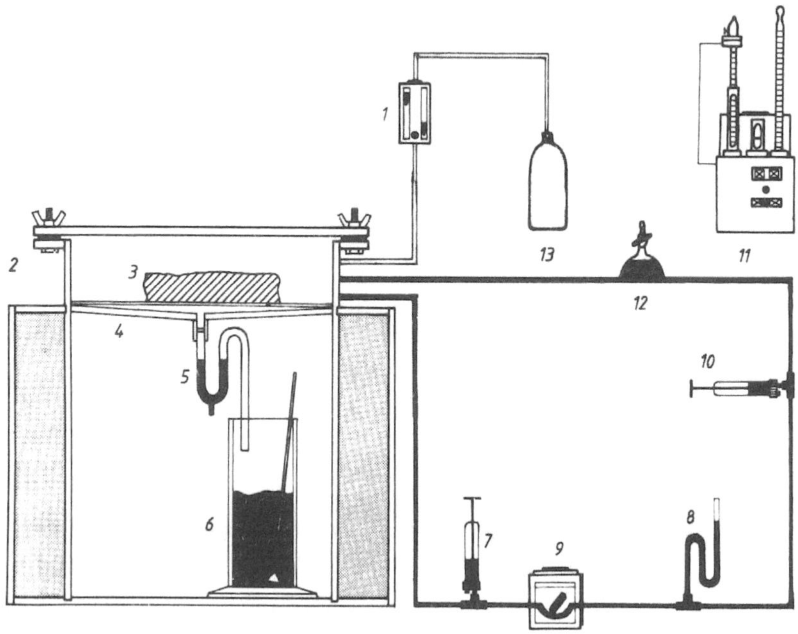

Fig. 2. Construction of the perfusion apparatus

The groin flap is placed on a net (Fig. 2 [3]). The venous fluid is collected by a funnel and led through a syphon into a measuring cylinder. On the lower loop of the syphon there is a device for taking samples (Fig. 2 [5]). In the measuring cylinder there is a glass filter to enrich the perfusion fluid with the gas mixture.

The fluid is kept in motion by a roller pump (Fig. 2 [9]). The perfusion pressure is measured by means of a medifix set. We are working with non pulsatile pressures of 4.5 to 5.5 kPa and with a flow rate of 90 to 120 ml/h. Furthermore there is a bubble trap arranged in the circulation [6] to protect the groin flap from air emboli (Fig. 2 [12]).

Perfusion Fluid: Constituents

We used a non-cellular perfusion fluid. To obtain physiological conditions it consists of the reagents listed in Table 1 [3]. The analysis of the perfusion fluid by the SMAC-technicon system is shown in Table 2, and the organic salts, amino acids, vitamins and other components in Table 3.

Table 1. Perfusion fluid: Constituents

Econazole 20,000 mg; Gibco SMEM suspension (10x) 7,000 ml; glucose 5% 1,600 ml; heparin 1,000,000 IE; $NaHCO_3$ 1,800 ml; penicillin/streptomycin 1,000 ml; rhodalbumin 20% 33,000 ml; prednisolone hemisuccinat 2,000 mg; aprotin 2,500,000 IE; aqua dest ad 100,000 ml

Table 2. Perfusion fluid: SMAC-technicon analysis

Na 164.0 mmol/1; K 3.8 mmol/l; Ca trace mmol/l; Cl 121.0 mmol/1; inorganic phos. 7.4 mmol/l; urea trace mg/dl; creatinine 0.7 mg/dl; protein 6.1 g/l; iron 12.0 µg/dl; glucose 155.0 mg/dl; uric acid trace mg/dl; cholesterol trace mg/dl; glyceride trace mg/dl; bilirubin 0.0 mg/dl; LDH 5.0 U/l; SGOT 0.0 U/l; SGPT 3.0 U/l; gamma-GT 4.0 U/l; alkaline phosphatase trace U/l; pH 7.4; osmolarity 430.0 mosm/kg

Table 3. Perfusion fluid; inorganic salts, amino acids, vitamins, other components

Inorganic Salts

KCL 280.0 mg/l; $MgSO_4$ x H_2O 140.0 mg/l; NaCl 4,760.0 mg/l; $NaHCO_3$ 1,540.0 mg/l; NaH_2PO_4 x H_2O 98.0 mg/l

Amino acids

L-arginine HCL 88.2 mg/l; L-cystine 16.8 mg/l; L-glutamine 204.4 mg/l; L-histidine x H_2O 29.4 mg/l; L-isoleucine 36.4 mg/l; L-leucine 36.4 mg/l; L-lysin HCl 50.8 mg/l; L-methionine 10.5 mg/l; L-phenylalanine 22.4 mg/l; L-threonine 33.6 mg/l; L-trotophan 7.0 mg/l; L-thyrosin 25.2 mg/l; L-valine 32.2 mg/l.

Vitamins

D-Ca-pantothenic acid 0.7 mg/l; choline chloride 0.7 mg/l; folic acid 0.7 mg/l; inosite 1.4 mg/l; nicotinamid 0.7 mg/l; pyridoxin HCl 0.7 mg/l; riboflavin 0.1 mg/l; thiamine HCl 0.7 mg/l.

Other Components

Phenolsulphonphthalein 7.0 mg/l.

Tests of Viability

To find out which skin areas are perfused, we inject 1 ml of patent blue solution, diluted 1:100 into the circulation. It is very important to avoid a reperfusion of this solution, because triphenylmethane-dyes develop a trypsin like activity, which results in the epidermis becoming detached. Furthermore patent blue easily forms crystals, which obstruct the millipore filters [9].

We perfused the preparation for more than 50 h either by a closed or by an open circulatory system. In the former case the perfusion fluid is analysed and then discarded. In the latter case the whole perfusion fluid is exchanged regularly to support the skin with new substrates and to remove the decomposition products. In this period of time we perform numerous analysis with the SMAC-technicon system.

For the last six hours of the perfusions we inject a solution containing 0.4% black ink into the superficial circumflex iliac and superficial epigastric arteries. In this way we label those vessels, which could be perfused. These arteries and veins are detected by examination of the histological slides.

In earlier experiments we have determined the tritium-labelled-thymidine index of the basal layer of the epidermis after 50 h of perfusion using autoradiography [1].

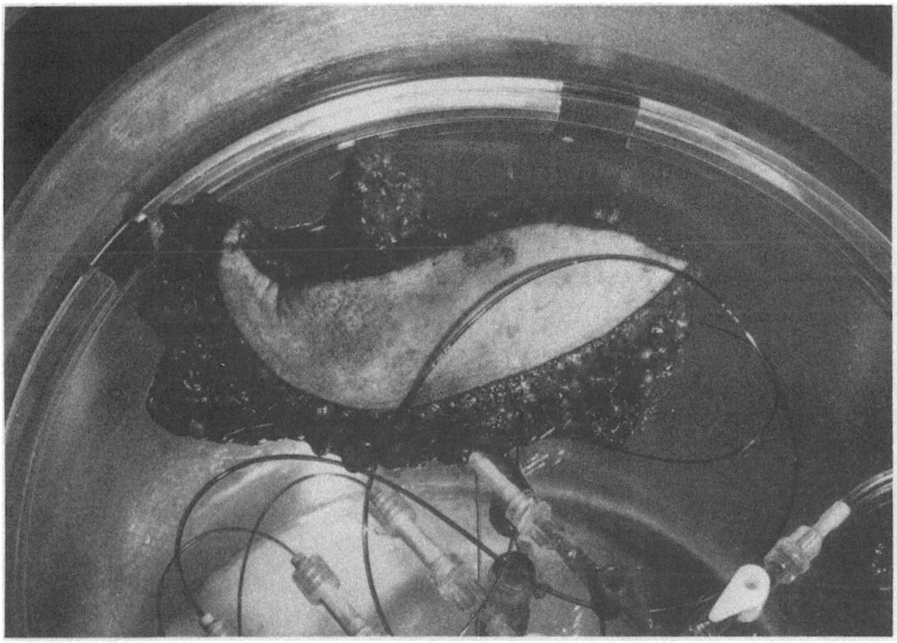

Fig. 3. Groin flap 20 s after the i.a. injection of patent-blue

Results

After the injection of the patent blue solution the skin becomes blue within a few seconds (Fig. 3). By the way of the two cannulated arteries some 90% of the total preparation are perfused in our experiments [11].

After 50 h of perfusion macroscopically and microscopically the skin flap looks like freshly removed tissue.

Among the 21 parameters analysed by the SMAC-technicon system the following ones have not or only minimally changed: sodium, calcium, chloride, inorganic phosphate, urea, creatinine, protein, iron, uric acid, cholesterol, glyceride, SGOT, SGPT, y-GT, alkaline phosphatase, pH and potassium.

Significant changes take place in the concentrations of glucose (Fig. 5), LDH (Fig. 6) and lactate (Fig. 7), most marked in the first four hours of perfusion. In this period of time the groin flaps recover from the hypoxia, which occurs between the end of the operation and the beginning of the perfusion. In these four hours the decomposition products were washed out and the substrate storage sites are filled up. Afterwards there is a linear decrease or increase of these parameters. Generally the substrate consumption and the ion and enzyme loss is within normal limits.

Glucose was the only source of energy. As no or only a slight lipogenesis took place, glucose consumption equals the consumption of energy. The groin flap consumed 9.3 mmol of glucose and produced 1.15 mmol of lactate between the 4th and 20th hour of perfusion. Therefore the calculated share of aerobic oxidation was ap-

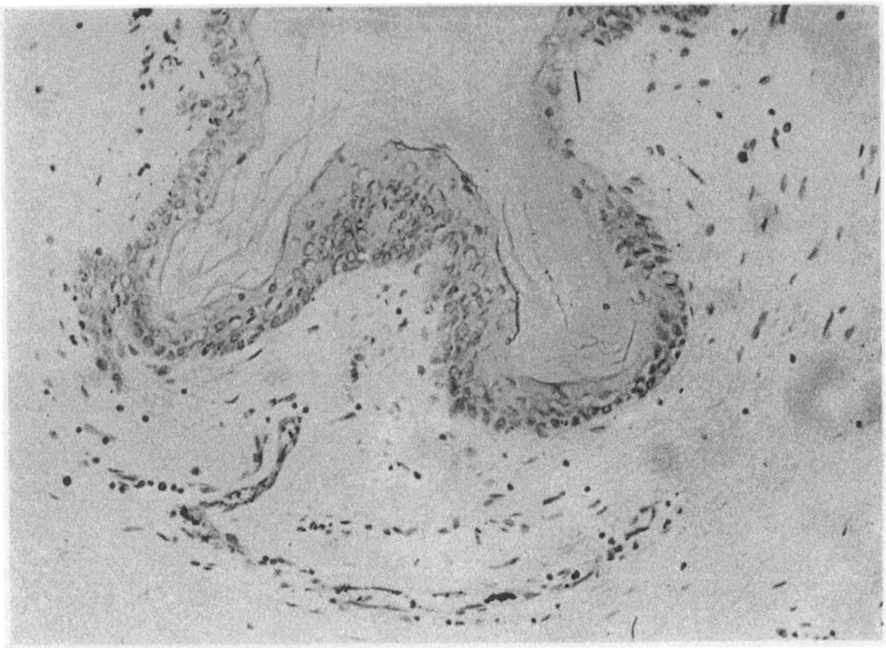

Fig. 4. Perfused black ink in the subepidermal vessles

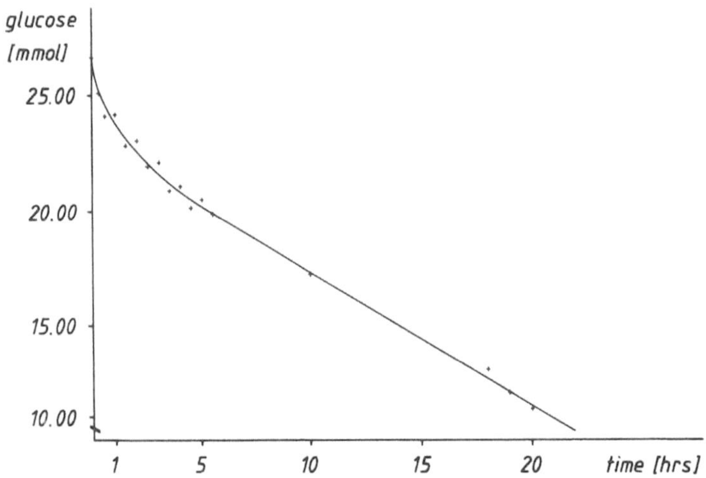

Fig. 5. Glucose consumption

proximately 94% and for anaerobic oxidation approximately 6%. The calculated amount of ATP produced in 1 h by our flap was 22 mmol. The exact values must be determined by further investigations.

The injected black ink could be found in the vessels of the stratum subpapillare (Fig. 4), the sweat glands, the vasa vasorum and the lymphnodes. Thus all essential

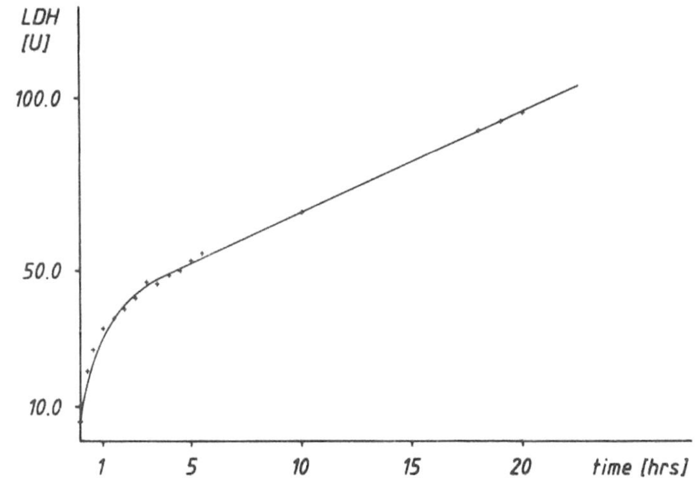

Fig. 6. LDH wash out

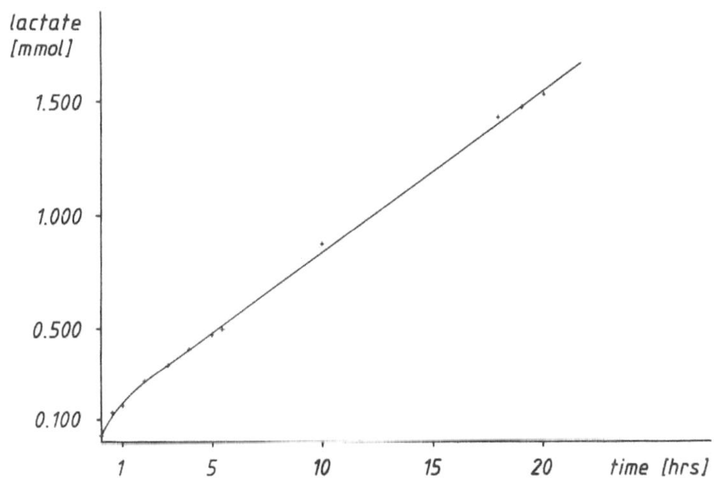

Fig. 7. Lactate production

skin structures can be perfused via the superficial circumflex iliac and the superficial epigastric arteries.

The tritium-thymidine labelling index of the basal layer of the epidermis was approximately 2.5%, which equals the value in normal skin.

Conclusion

The long surviving human groin flaps can be perfused for more than 50 h under conditions of constant temperature and humidity. In this period of time there is steady state biochemically. The perfused skin is not influenced by other organs. We can work with all substances, some of which cannot be used for experiments with man in vivo.

With our human skin model it is possible for the first time to perform pharmacokinetic and biochemical experiments with isolated, living human skin under in vitro conditions. It is particularly important to note that energy dependent metabolism and transport mechanisms can now be studied.

References

1. Baserga R, Malamud D (1969) Autoradiography, techniques and application. Hoeber Medical Division, Harper and Row Publishers, New York Evanston London
2. Biemer E, Duspiva W (1980) Rekonstruktive Mikrogefäßchirurgie. Springer, Berlin Heidelberg New York
3. Collins GM, Green RD, Halasz NA (1979) Importance of anion content and osmolarity in flush solutions for 48 to 72 h hypothermic kidney storage. Cryobiology 16(3):217–220
4. Harii K, Ohmori K, Ohmori S (1974) Successful clinical transfer of ten free flaps by microvascular anastomoses. Plast & Reconstr Surg 53:259–270
5. Hiernickel H (1983) Methode zur Gewinnung langlebiger supravitaler Hautpräparate. Z Hautkr 58:820–833
6. Lowry M, Ross B (1980) Apparatus for organ perfusion. Br J Clin Equip 5/3:104–111
7. McGregor IA, Jackson IT (1972) The groin flap. Brit J Plast Surg 25:3–16
8. Rollin KD, Terzis JK (1981) Reconstructive microsurgery. Little, Brown and Company, Boston
9. Simson MB, Harden W, Barlow C, Harken AH (1979) Visualization of the distance between perfusion and anoxia along an ischemic border. Circulation 60(5):1151–1155
10. Tritsch H (1978) Hautschnittführung zur Ektomie der Leistenlymphknoten. Hautarzt 29: 531–535
11. Ward LC, Buttery PJ (1979) The patho-physiological basis for tests of variability in isolated perfused organs. Biomedicine 30(4):181–186

The Use of Human Scalp and Abdominal Skin as In Vitro Models for Percutaneous Absorption

S. L. Bennett and B. W. Barry

It is generally accepted that human abdominal skin used in vitro is a good model for in vivo percutaneous absorption [2]. However, such skin is often not readily available, especially for many workers within for example, the pharmaceutical industry, so instead they use animal models [5, 6, 27] or synthetic membranes [15, 20]. These systems are liable to provide inferior models as they are not from a human source. Hair transplant clinics provide a readily available supply of scalp skin; although this skin has some limitations we believe it is superior to animal models or synthetic membranes. However the major drawback of the scalp skin is that it is atypical being from patients with male pattern baldness.

Regional differences in percutaneous absorption exist although data vary on the magnitude of the differences and regions of greatest penetration [2, 8, 13]. There are few quantitative results on regional variation for a structurally related series of compounds as most work has been performed using individual materials of interest at the time and therefore the results are difficult to extrapolate to other compounds [17, 24]. Although very little quantitative information exists, it is commonly accepted that scalp skin is more permeable than abdominal skin.

Before scalp skin could be used routinely as a model, its barrier property had to be compared with that of a well established system and abdominal skin was therefore chosen. For this programme scalp and abdominal skin were used to compare a test series of compounds of decreasing polarity; water and the aqueous alcohols were chosen because they are fairly simple, readily available radiolabelled and they provide an ideal homologous series of compounds.

Materials and Methods

Preparation of Materials

Chemicals

[^{14}C] methanol, specific activity 56.9 mCi mmol^{-1}, [^{14}C] ethanol, 61.6 mCi mmol^{-1}, [^{3}H] water, 5 mCi ml^{-1} (Radiochemical Centre, Amersham), [^{14}C] butanol, 10 mCi mmol^{-1}, [^{14}C] hexanol, 5.4 mCi mmol^{-1} (ICN Pharmaceuticals Inc., California), non-radioactive alcohols, methanol, ethanol, butanol and hexanol (BDH Chemicals Ltd.), FisoFluor-1 scintillation fluid (Fisons Scientific Apparatus), Soluene-350 tissue solubiliser (Packard) were used.

Skin Source and Storage
Caucasian abdominal and scalp skin samples, the former obtained as strips from autopsy and the latter as disks from hair transplant operations, were stored in sealed evacuated polythene bags at –24 °C. For convenience the strips were stored flat and the disks on aluminium foil.

Preparation of Skin Membranes

Dermatomed Abdominal Skin for Diffusion Experiments
Excess fat was trimmed from abdominal skin specimens which were then frozen at –24 °C for a minimum of one hour. Immediately prior to dermatoming, the stratum corneum surface of the skin was thawed slightly so that it was just mobile to the touch. A layer 43 μ (±5%) thick was then removed using a Davies dermatome 7 (Duplex Electro Dermatome). This dermatomed layer consisted of the epidermis and some of the dermal tissue.

Sliced Scalp Skin for Diffusion Experiments
The specimens could not be effectively dermatomed as they were only 4 mm in diameter. Therefore they were allowed to thaw and a layer (approximately 600 μ) removed using an open cut-throat razor.

Heat-separated Epidermal Membranes
Abdominal or scalp epidermal layers were removed from the skin samples by a heat-separation technique [17]. After removing the fat, the skin was immersed in water at 60 °C for 45 s. The epidermal layer was gently teased off taking care to ensure it was not stretched. This membrane was used for diffusion experiments.

For partition coefficient work, an additional step was required. The membrane was incubated overnight at 37 °C dermal side down on filter paper saturated with a solution of trypsin (0.0001% w/v) containing 0.5% w/v sodium bicarbonate. Any mushy epidermis was removed by rubbing with cotton wool before drying to constant weight in a dessicator.

Solubility of Hexanol in Water
The solubility of hexanol in water was determined at 25 °C by allowing excess hexanol to equilibrate with distilled water for 24 h in a shaking water bath (Grant Instruments). The concentration of hexanol in the aqueous layer was determined by gas liquid chromatography using a Pye 104 chromatogram with a single flame ionization detector and conditions indicated below:

Column	10% PEG 400 on chromosorb W-AW, 80–100 mesh
Injection port temperature	160 °C
Oven temperature	105 °C
Nitrogen flow	60 ml min^{-1}
Hydrogen flow	50 ml min^{-1}
Air flow	500 ml min^{-1}

The aqueous solubility of hexanol at 25 °C was determined as this was the approximate temperature of the solutions in the diffusion run. All alcohol solutions were made up to this concentration.

Experimental Methods

Diffusion Experiments

Skin sections were hydrated for 4 h and mounted in stainless steel diffusion cells (diffusional area 0.05 cm^2) in an automatic system at 30 ± 1 °C [1, 7]. Water was pumped through the lower receptor compartment at 2.5 ml h^{-1} via a multichannel peristaltic pump and collected in scintillation vials on a rotating table. After 3 days equilibration, to fully hydrate the skin, 10 μl 0.75% v/v aqueous ^{14}C hexanol was applied to the epidermal surface and penetration was monitored every 30 min for 6 h, the donor solution being replaced hourly. After a washing procedure 0.75% v/v aqueous ^{14}C butanol, ^{14}C ethanol and ^{14}C methanol were applied in turn. 10 ml FisoFluor-1 was added to each sample collected, the vials stored overnight and the radioactivity was analysed (Packard Liquid Scintillation Counter, model Tri-Carb 460C). The penetration of ^3H water was determined and skin damage investigated by determining water permeability coefficients before and after alcohol runs.

Partition Coefficient Determinations

Disks of trypsinated stratum corneum were weighed dry on a Stanton CL3 5 decimal place balance and allowed to equilibrate at 30 °C for 24 h with 1 ml 0.75% v/v ^{14}C alcohol solution (methanol, ethanol, butanol or hexanol) or 1 ml ^3H water. These skin samples were blotted on filter paper, weighed wet and dissolved in 1 ml Soluene-350 in a glass scintillation vial. 10 ml FisoFluor-1 containing 0.1 ml glacial acetic acid was added and samples were stored at room temperature overnight to allow chemiluminescence to subside. Acidifying the mixture is an effective way of quickly reducing non-radiation events such as chemiluminescence. 10 ml of the same scintillant was added to duplicate 100 μl volumes of the vehicle at equilibrium and the radioactivity was analysed as before.

Results and Discussion

Diffusion Experiments

Figure 1 illustrates a typical cumulative penetration curve obtained for the diffusion experiments. The results derived all followed this standard profile, with a lag period when the material was initially crossing the skin followed by a linear portion when the amount penetrating was proportional to time. No tailing of the linear portion was seen as a classical skin experiment was carried out, with sink conditions and a solution

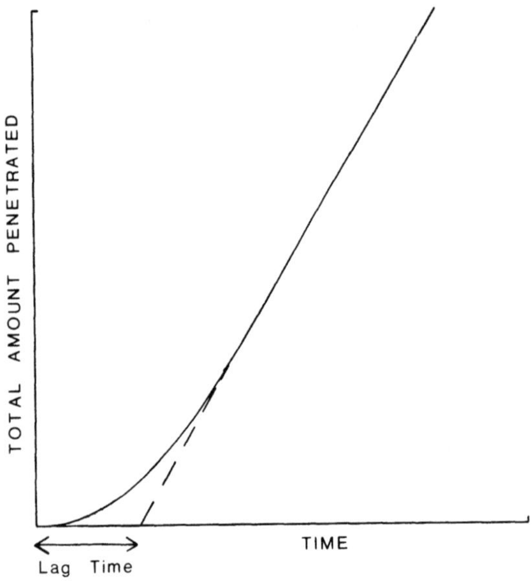

Fig. 1. Standard curve showing approach to steady state penetration

Table 1. Permeability coefficients (Kp) for abdominal and scalp skin at 30 °C following application of water or aqueous alcohol

Compound	Full thickness membrane		Epidermal membrane	
	Abdominal skin $Kp \times 10^3$	Scalp skin $Kp \times 10^3$	Abdominal skin $Kp \times 10^3$	Scalp skin $Kp \times 10^3$
Water	1.1 ± 0.3*	9.5 ± 1.2	1.2 ± 0.2	9.8 ± 1.8
Methanol	0.6 ± 0.1	10.3 ± 3.5	0.8 ± 0.1	10.4 ± 2.0
Ethanol	0.9 ± 0.1	11.3 ± 4.4	0.8 ± 0.1	11.2 ± 2.3
Butanol	3.1 ± 1.1	16.7 ± 5.2	3.3 ± 0.8	19.0 ± 3.1
Hexanol	18.8 ± 1.9	23.9 ± 7.9	20.0 ± 1.5	28.1 ± 9.0

* Standard deviations, n = 4

above the membrane which was not allowed to deplete. By using Fick's first law, permeability coefficients were derived by dividing the slope by the area available for diffusion and the concentration of alcohol or water applied; the results are shown in Table 1. This table shows the permeability coefficients for each compound through the epidermal layer and full thickness abdominal and scalp skin. It can be seen that particularly for the more polar compounds the permeability coefficients were higher for scalp than abdominal skin with the ratio between the two being approximately ten or greater. This was probably due to the differences in the structure of the two skins with scalp having many more hair follicles and sebaceous glands. As the polarity of alcohols decreased, the permeability coefficients obtained became more similar and the difference for hexanol was only significant at the 10% level [3]. The intra-

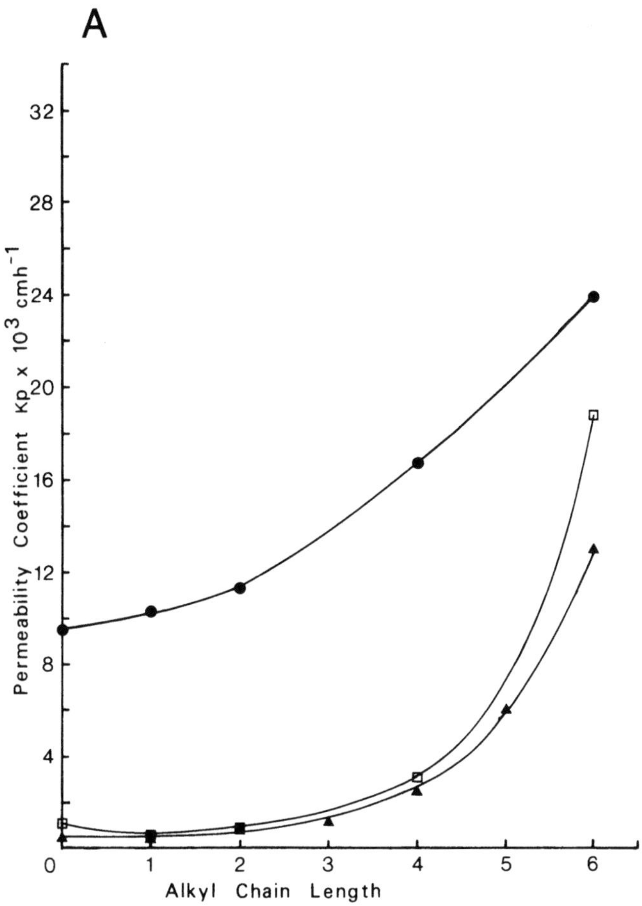

Fig. 2A, B. Permeability of aqueous alcohols through human skin. Standard deviation bars, n = 4. **A** Full thickness skin. ● Scalp skin 30 °C, ▫ Abdominal skin 30 °C, ▲ Scheuplein and Blank (1973) – abdominal skin 25 °C

specimen variation for scalp skin was much higher than for abdominal skin as shown by the larger standard deviations and is further illustrated in Fig. 2.

Figure 2 shows the relationship which exists between permeability coefficient and alkyl chain length. As can be seen from Fig. 2A the trend is not the same for the two different skin types, so that differences at the polar end of the penetrant spectrum are not matched in the non polar region. This further indicates that care would have to be taken when interpreting results obtained using scalp skin as an in vitro model. The abdominal permeability coefficients compare favourably with those of Scheuplein and Blank [22] especially considering that their experiments were conducted at 25 °C compared with 30 °C used in this study. Figure 2B illustrates that within a particular skin type the trend is remarkably similar regardless of whether full thickness skin or just the epidermal layer is used in the investigation, indicating that the differences observed were not due to the dermal tissue, i.e., the stratum corneum provided the rate-limiting barrier in all instances. These differences could not be even partially

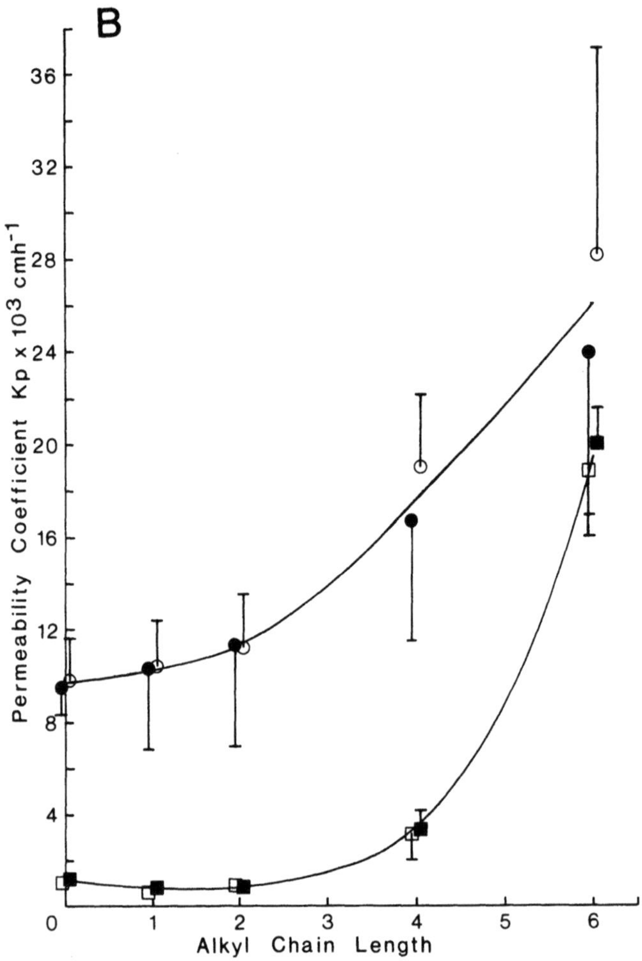

Fig. 2B. ▫ Full thickness abdominal skin, ▪ Epidermal layer abdominal skin, • Full thickness scalp skin, ○ Epidermal layer scalp skin

explained by a greater damaging effect of the alcohols on one skin type then other as water damage ratios indicated that no damage occurred with 0.75% aqueous alcohols used.

The difference in the trends observed for the two skin types could have important consequences if similar trends can be shown for other series of compounds. A literature search has failed to reveal any studies where a series of related compounds has been used to compare scalp skin with that from another site. However, Elias et al. [12] have compared the penetration of water and salicylic acid through leg and abdominal stratum corneum. They found that although leg stratum corneum was approximately twice as permeable to water ($p < 0.025$) as abdominal stratum corneum the difference in the permeability of salicylic acid across these two sites was not as great and was not statistically significant ($p < 0.1$). This suggests that the trends in permeability for leg skin may not be the same as abdominal skin.

Table 2. Stratum corneum: water partition coefficients for aqueous alcohols

	Abdominal skin		Scalp skin	
	Km	log Km	Km	log Km
Water	2.29 ± 0.68*	0.34 ± 0.13	1.68 ± 0.82	0.16 ± 0.29
Methanol	1.53 ± 0.50	0.16 ± 0.16	1.64 ± 0.99	0.14 ± 0.27
Ethanol	2.07 ± 0.75	0.29 ± 0.16	2.63 ± 1.20	0.37 ± 0.23
Butanol	4.57 ± 1.09	0.65 ± 0.10	5.10 ± 1.05	0.70 ± 0.10
Hexanol	12.12 ± 1.34	1.08 ± 0.05	10.84 ± 1.56	1.03 ± 0.06

* Standard deviations, n = 8

Partition Coefficient Determinations

Listed in Table 2 are the stratum corneum water partition coefficients for both skin types, which were found to be similar. The standard deviations are low for all compounds showing the technique was reproducible.

In Fig. 4 the partition coefficient data are compared with those found by other workers [21, 26]. The trends seen are similar with a constant value for the partition coefficient up to a chain length of C_2 or C_3. Above this the value increased up to C_6 in this study and C_8 in the other studies quoted. These partition coefficient values show that there is no difference in the ability of water or n-alcohols to partition into abdominal or scalp skin.

The partition coefficient values obtained illustrate that there is no simple linear relationship between log partition coefficient and chain length for either skin. Both types of skin show the same biphasic trend suggesting that the lower alcohols partition into a different region in the stratum corneum compared with the higher alcohols. If the skin provided a homogeneous membrane as was originally assumed for subtances crossing the membrane [4, 21] then the slope of a graph of log partition coefficient (log Km) vs chain length (n) would be linear. This would be the situation because the partitioning of alkyl homologs between two immiscible phases obey the relationship:

$$\log Km_n = \log Km_o + \pi n$$

where Km_n is the partition coefficient of the homolog of chain length n, Km_o is the partition coefficient of the reference homolog of chain length zero, and π is the group contribution. In view of the biphasic character observed in this study and by other workers [22, 26] the conclusion drawn is that two different solution areas exist, one watery and the other lipoidal. Thus it would seem that water and the lower alcohols, methanol and ethanol, preferentially partition into the water areas and the higher alcohols partition into the lipoidal areas of the stratum corneum. The same pattern has also been observed for mouse epidermis [14].

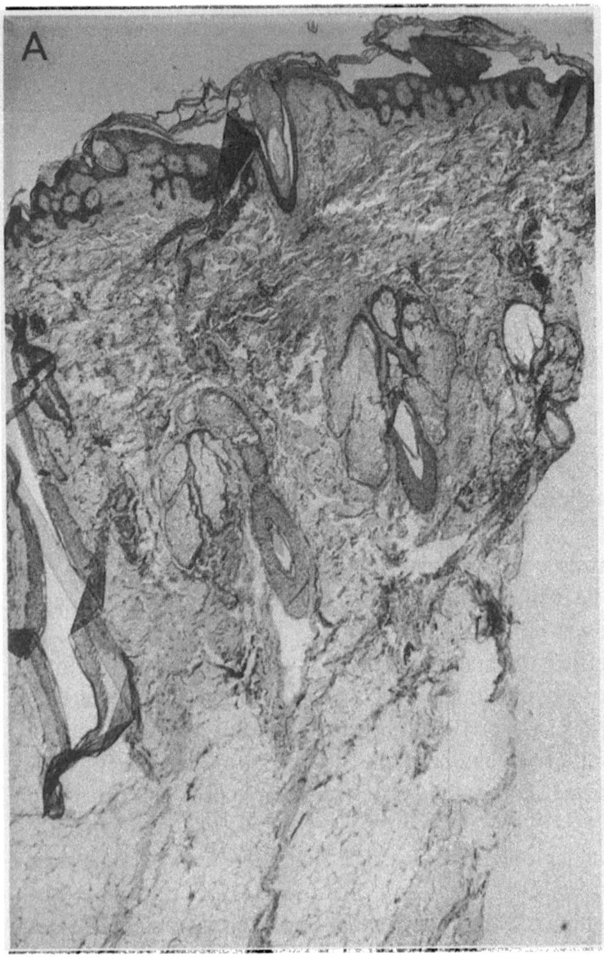

Fig. 3A, B. Sections of male pattern baldness and scalp skin stained with haematoxylin and eosin. **A** Scalp skin, **B** Abdominal skin

Histology

Sections of each skin were stained with haematoxylin and eosin to see whether there were gross differences in the structure. Figure 3 shows stained sections of scalp and abdominal skin at magnification x28. The stratum corneum, just above the dark epidermal layer, is of uniform thickness in abdominal skin, but variable thickness in the scalp. This variable thickness may explain the large variations in permeability coefficients observed using scalp tissue. Abdominal skin has very few hair follicles and sebaceous glands, but abundant eccrine sweat glands. Scalp skin has many more hair follicles and sebaceous glands but fewer eccrine sweat glands. These structural differences may partially explain the different trends seen in permeability coefficients.

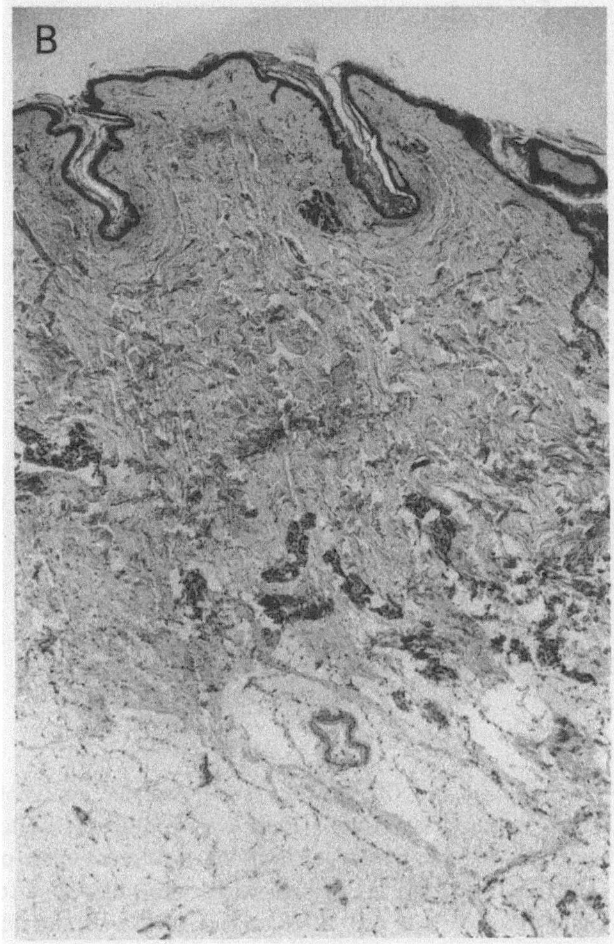

Fig. 3B. Abdominal skin

Diffusional Process

The skin appendages act as diffusion shunts and will have a part to play in the initial penetration of materials across the skin. This route will be more important in scalp than abdominal skin because of the greater number of appendages. Szabó [28] has shown that there are on average 980 appendages per cm^2 in scalp skin compared with 270 in abdominal skin. The fractional area of the skin covered by appendages over the body is quite small, of the order of 10^{-3} cm^2/cm^2 [21]. For larger molecules the route may be important as in the diffusion of the more polar steroids [23]. For alcohols, which penetrate the skin quickly, after the initial diffusion period, the effect of these shunt routes will be negligible even for the scalp skin because of the small diffusional volumes involved [21].

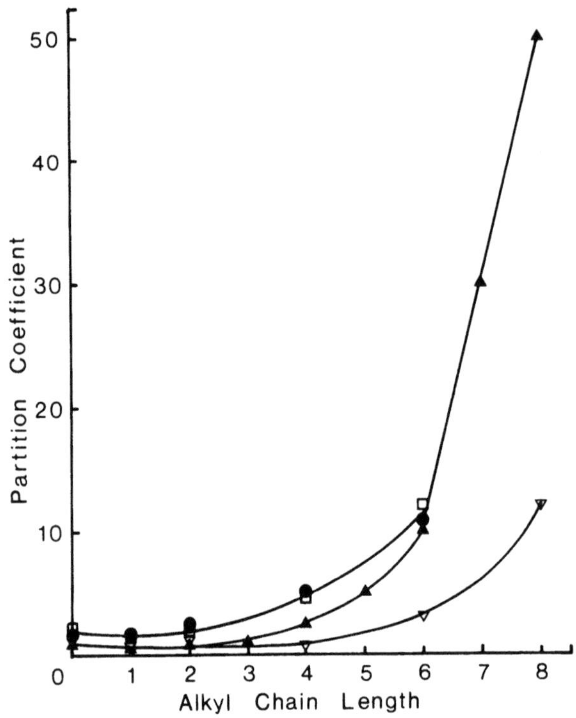

Fig. 4. Stratum corneum water partition coefficients of water at 30 °C Standard deviation bars, n = 8. □ Abdominal skin, ● Scalp skin, ▲ Scheuplein and Blank [22], ▽ Southwell and Barry [26]

Scheuplein and Blank [21] have postulated that polar and non-polar molecules diffuse through the stratum corneum by different routes and that the intra-cellular material which comprises 85% of the membrane mass provides the main diffusional resistance. Water and highly polar molecules would tend to diffuse through the aqueous regions located near the outer surface of the keratin filaments and non-polar lipid-soluble molecules would diffuse through the lipid network between the filaments. The results obtained in the present study agree with two separate pathways being responsible for controlling the penetration rate and suggest that there is a larger volume available for polar diffusion in scalp than abdominal skin, or that the resistance of this route is lower in scalp than in abdominal skin, or that both factors operate. The characteristics of the route available for non-polar diffusion in both skins would appear to be similar.

Historically, the intercellular route of penetration has generally been discounted because of the small volumes involved, i.e., 5% in the dry membrane and 1% in the fully hydrated tissue [4, 21]. More recently however, Elias and Leventhal [10] using stereological measurements have shown that the intercellular spaces comprise 10–30% of the stratum corneum volume. This in turn suggests that these spaces should not be ignored as a route of penetration as the volumes involved are significant. Also there is now evidence that much of the lipid is contained in membrane complexes. Grayson and Elias [16], using a freeze-fracture technique, calculated that the stratum corneum cytoplasm contains less than 3% lipid and membrane preparations account for approximately 80% of the total stratum corneum lipid. It has also been shown that stratum corneum sheets can be dispersed into individual cells by treatment with lipid solvents [9, 25]. As lipids are preferentially extracted from the intercellular spaces, lipophilic compounds would preferentially penetrate the skin using these regions. Butanol has been shown preferentially to traverse the intercellular spaces as detected via a precipitation technique [11]. The results obtained in the present study tend to confirm those of Elias and co-workers suggesting that the intercellular spaces provide the main route of penetration of lipophilic compounds.

Roberts et al. [19], however, suggest that the main resistance to penetration is the lipid barrier in the stratum corneum and that less polar solutes are retarded by aqueous boundary layers as well. Therefore further work needs to be conducted to test whether the diffusion rate of the more lipophilic alcohols is controlled by a boundary layer.

Conclusions

There are four main conclusions that can be drawn from this study:

For model compounds, water and n-alcohols, male pattern baldness scalp skin is more permeable than abdominal skin especially for the more polar compounds.

The partitioning of n-alcohols and water into male pattern baldness scalp and abdominal skin is similar.

The permeability differences between scalp and abdominal skin probably arise from the greater volume of the polar route in scalp skin, or the lower diffusional resistance of this route, or both.

Bearing these differences in mind we conclude that human scalp skin is a better model for percutaneous absorption than animal tissue or synthetic membranes.

Acknowledgement

The authors thank the Science and Engineering Research Council for providing a studentship for SLB and Miss M. Thorley.

References

1. Akhter SA, Bennett SL, Waller IL, Barry BW (1984) An automated diffusion apparatus for studying skin penetration. Int J Pharm 21:17–26
2. Barry BW (1983) Dermatological formulations; percutaneous absorption. Marcel Dekker, New York
3. Bennett SL, Barry BW (1983) A comparison of the permeability of abdominal and male pattern baldness scalp skin. J Pharm Pharmacol 35:suppl 32P

4. Blank IH (1969) Transport across the stratum corneum. Toxicol Appl Pharmacol suppl 3:23–29
5. Bronaugh RL, Stewart RF, Congdon ER (1982) Methods for in vitro percutaneous studies II. animal models for human skin. Tox Appl Pharmacol 62:481–488
6. Chien YW, Keshary PR, Huang YC, Sarpotdar PP (1983) Comparative controlled skin permeation of nitroglycerin from marketed transdermal delivery systems. J Pharm Sci 72:968–970
7. Cooper E (1984) Increased skin permeability for lipophilic molecules. J Pharm Sci 73:1153–1156
8. Cronin E, Stoughton RB (1962) Percutaneous absorption: regional variations and the effect of hydration and epidermal stripping. Br J Dermatol 74:265–272
9. Elias PM (1981) Lipids and the epidermal permeability barrier. Arch Dermatol Res 270:95–117
10. Elias PM, Leventhal ME (1980) Intercellular volume changes and cell surface area expansion during cornification. Eur J Cell Biol 22:439a
11. Elias PM, Nemanic MK (1980) In situ precipitation – a novel cytochemical technique for visualization of permeability pathways in mammalian stratum corneum. J Histochem Cytochem 28:573–578
12. Elias PM, Cooper ER, Korc A, Brown BE (1981) Percutaneous transport in relation to stratum corneum structure and lipid composition. J Invest Dermatol 76:297–301
13. Feldmann RJ, Maibach HI (1967) Regional variation in percutaneous penetration of ^{14}C cortisol in man. J Invest Dermatol 48:181–183
14. Flynn GL, Dürrheim H, Higuchi WI (1981) Permeation of hairless mouse skin II: membrane sectioning techniques and influence on alkanol permeabilities. J Pharm Sci 70:52–56
15. Gary-Bobo CM, Di Polo R, Solomon AK (1969) Role of hydrogen-bonding in nonelectrolyte diffusion through dense artificial membranes. J Gen Physiol 54:369–382
16. Grayson S, Elias PM (1982) Isolation and lipid biochemical characterization of stratum corneum membrane complexes: implications for the cutaneous permeability barrier. J Invest Dermatol 78:128–135
17. Kligman AM, Christophers E (1963) Preparation and isolation of sheets of human stratum corneum. Arch Dermatol 88:70–73
18. Maibach HI, Feldmann RJ, Milby TH, Serat WF (1971) Regional variation in percutaneous penetration in man. Arch Environ Health 23:208–211
19. Roberts MS, Anderson RA, Swarbrick J, Moore DE (1978) The percutaneous absorption of phenolic compounds: mechanism of diffusion across the stratum corneum. J Pharm Pharmacol 30:486–490
20. Roseman TJ, Higuchi WI (1970) Release of methoxyprogesterone acetate from a silicone polymer. J Pharm Sci 59:353–357
21. Scheuplein RJ, Blank IH (1971) Permeability of the skin. Physiological Reviews 51:702–747
22. Scheuplein RJ, Blank IH (1973) Mechanism of percutaneous absorption IV. penetration of nonelectrolytes (alcohols) from aqueous solutions and from pure liquids. J Invest Dermatol 60:286–296
23. Scheuplein RJ, Blank IH, Brauner GJ, MacFarlane DJ (1969) Percutaneous absorption of steroids. J Invest Dermatol 52:63–70
24. Smith JG, Fischer RW, Blank H (1961) The epidermal barrier: a comparison between scrotal and abdominal skin. J Invest Dermatol 36:337–343
25. Smith WP, Christensen MS, Nacht S, Gans EH (1980) Effect of lipids on the barrier function of the stratum corneum. Fed Proc 39:286a
26. Southwell D, Barry BW (1983) Penetration enhancers for human skin: mode of action of 2-pyrrolidone and dimethylformamide on partition of model compounds water, n-alcohols and caffeine. J Invest Dermatol 80:507–514
27. Stoughton RB (1973) Animal models for *in vitro* percutaneous absorption. In: Maibach H (ed) Animal models in dermatology. Churchill Livingstone, Edinburgh London New York, pp 121–132
28. Szabó G (1962) The number of eccrine sweat glands in human skin. In: Montague W, Ellis RA, Silver AF (eds) Eccrine sweat glands and eccrine sweating. Pergamon, Oxford London New York, pp 1–5 (Advances in biology of skin, vol 3)

An In Vitro Human Skin Model for Assaying Topical Drugs against Dermatophytic Fungi

A. M. Kligman, K. J. McGinley and A. Foglia

Many hundreds of chemicals are capable of inhibiting a wide variety of plant and animal parasitic fungi. Only a handful are in use for the treatment of human ringworm infections. While many are rejected on grounds of toxicity, most are disqualified because they lack the capacity to diffuse into the depths of the horny layer where dermatophytic fungi reside. Efficacy is absolutely dependent on achievement of inhibitory concentrations at the base of the stratum corneum. Muskatblit's assessment has been repeatedly confirmed: "chemicals which give striking results of fungus cultures in test tubes prove entirely useless when applied to human skin" [7].

Determining minimum inhibitory concentrations in growth media is but a first step in the process of selecting fungistatic topicals for clinical use. Inactive materials are eliminated.

Pharmaceutical companies often resort to ringworm infections of guinea pigs to obtain more dependable data regarding clinical efficacy. However, the rapid development of intense inflammation with swift spontaneous healing is a serious limitation. This model does not mimic chronic ringworm infection. Treatment must begin within a few days after inoculation else, as in tinea capitis, fungi become inaccessible by penetrating deeply into the hair roots. Guinea pig assays mainly estimate prophylactic capability and have limited relevance to human disease.

Ringworm infections can be established in human volunteers. Wallace et al. have assayed various antifungal agents after inoculating human skin with *T. mentagrophytes* [11]. This model can never become routine. Besides its reliability has not been established.

Since dermatophytes are confined to the dead horny layer, one theoretically should be able to avoid the use of living animals, substituting samples of excised human skin. Indeed, attempts of this kind have been made.

Stoughton's method depends upon the capacity of the anti-fungal compound to attain an inhibitory level in the dermis [9]. Surgical specimens from breast or leg amputations were treated for 20 to 48 h by applying the test agents to the surface within plastic cylinders. After heat separation of the epidermis, six mm punches of dermis were excised and placed surfaceside down on Miami Mycosel Medium inoculated with spores of *T. mentagrophytes.* A clear zone around the disc unequivocally demonstrates that the agent has penetrated through the epidermis into the dermis. A zone of inhibition validates efficacy. A negative result, however, is uninterpretable. The compound may remain trapped in the dermis or be so insoluble in the aqueous medium as to exhibit no outward diffusion from the dermal disc.

Knight's procedure is ingenious and simpler [5]. He applies the test agents to the upper arms for 15 min, followed by rinsing with running water. Using translucent adhesive tape, strippings of monolayers of corneocytes are removed at various intervals, inoculated with *T. mentagrophytes* spores and incubated in moist chambers. The amount of growth is estimated on a 0–3 scale at the end of 7 days. The clinical relevance of this model is moot since only the desquamating portion of the horny layer is sampled. This model relates to prophylaxis, not treatment of established infection.

Dittmar's well-researched method overcomes these objections and comes close to mimicking in vivo realities [4]. He applied the test substances on pieces of human skin. After a suitable interval, a sheet of epidermis is heat separated and the under-surface inoculated with spores of *T. mentagrophytes*. Absence of growth shows that a fungistatic concentration has been established on the internal side of the horny layer barrier.

Our experience with Dittmar's procedure confirms his judgement that the results are highly relevant to clinical efficacy [3]. Dittmar's specifications are rigorous and require careful attention to detail. We have found that very little is lost by stripping the procedure down to bare essentials. It is the purpose of this communication to describe a simplified, short version of Dittmar's technique. We claim that the method reliably predicts clinical effectiveness of antifungal topicals.

Materials and Methods

Preparation of Skin

Abdominal cadaver skin was wiped with 70% ethanol, cut into 4 x 5 cm sections, wrapped individually in plastic film and stored at 20 °C.

After thawing, excess fat was trimmed away, and the sections cleansed thoroughly with gauze pads wetted with 0.1% Triton X-100, followed by rinsing with sterile water and a final defatting with hexane. The skin was placed, surface side up, in a plastic petri dish lined with water-moistened filter paper.

Applications of Test Agents

50 μl was dispensed from a tuberculin syringe and spread evenly over the skin surface with a glass rod. A control sample from the same skin specimen was left untreated. Each sample. lying on the wet filter paper in a closed Petri dish, was kept at room temperature for either 4 or 24 h.

Preparation of Epidermal Sheet

The test material was removed by thorough cleansing with 0.1% Triton X-100, a water rinse and a final wash in hexane. Three 1 x 1 cm squares were cut from each piece of skin and placed in water at 60 °C for one minute. The epidermis was then gently peeled off with a spatula and placed surface side down in a petri dish containing 2% non-nutrient agar with 0.005% choramphenicol. Three triplicate pieces and one control were placed in each dish.

Inoculation and Incubation

A spore suspension of *T. mentagrophytes* was prepared as follows. Spores were grown on Saboraud dextrose agar and incubated for 2 weeks at room temperature. The plate

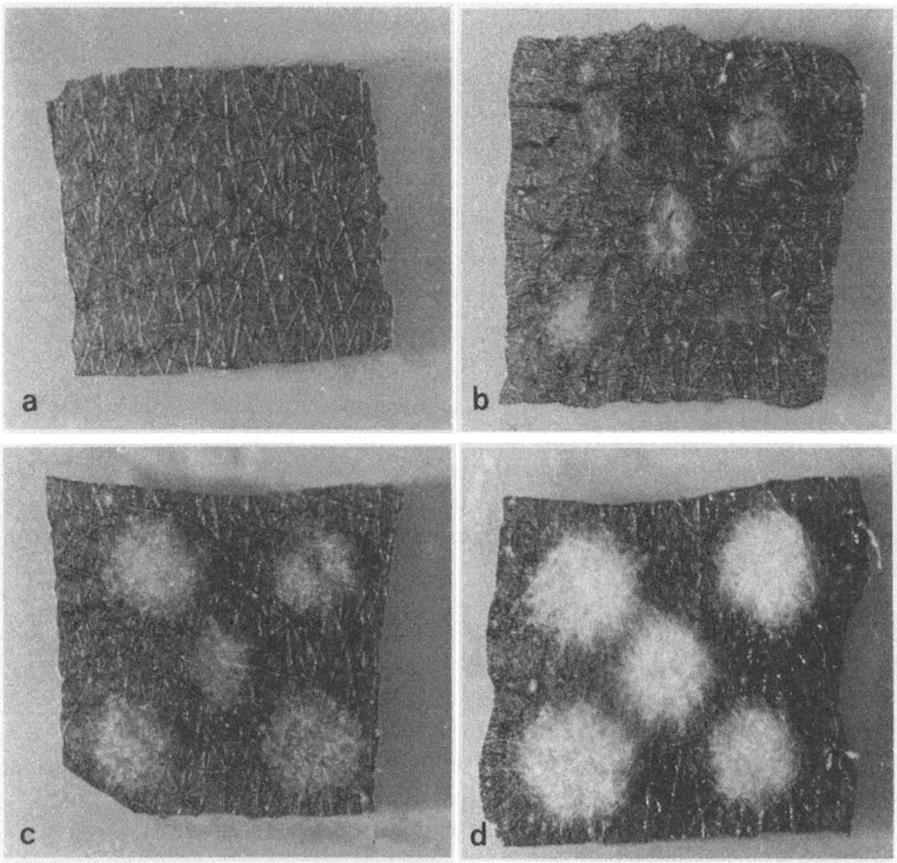

Fig. 1a–d. Grading system. **a** complete inhibition. No growth at any of 5 inoculation points. **b** Moderate inhibition, grade I. **c** Slight inhibition, grade II. **d** no inhibition, indistinguishable from control

was then flooded with 0.01% Tween-80, scraped with a teflon spatula and the suspension filtered through several layers of gauze. After centrifugation, the pellet was resuspended in sterile distilled water at a concentration of 25–50 spores/ml, by counting in a hemocytometer, verified for viability by quantitative plating. The suspension was stored at 5 °C. A new suspension was prepared each month, although we found no loss of viability after several months of storage.

1 μl of the suspension was inoculated by micropipette onto the undersurface of each 1 x 1 cm square at 5 equidistant sites. The plates were incubated at room temperature for either 4 or 5 days, depending upon the appearance of 3 mm fluffy colonies on untreated skin. If that degree of growth was not noted in the control, the study was repeated with a new specimen.

Reading

Growth on treated skin was graded as follows: 0 = complete inhibition, 1 = strong inhibition, barely perceptible colonies, 2 = moderate inhibition, colonies 1 to 2 mm, 3 = no inhibition, similar to control.

Suction Blisters in Human Volunteers

One cm circular blisters were raised up on volar forearm skin using a standard suction apparatus at 2 atmospheres of negative pressure. These are analagous to heat separated epidermal specimens, with the level of detachment at the dermo-epidermal junction. These were used in antifungal assays as a check in the results obtained on abdominal skin.

Results

Concentrations of familiar antifungal chemicals, ranging from 0.001% to 2% in ethanol for 24 h periods were tested with the results shown in Table 1. Wide differences in potency came to light, even within the same class. For example, clotrimazole and ketoconazole were completely inhibitory down to 0.1% but miconazole was incompletely suppressive at 2.0%. Tolnaftate ranked along with pyridinethiones. Ciclopirox was no better than miconazole.

Table 2 depicts the results with proprietary medicaments commonly used in the treatment of ringworm infections, for 4 and 24 h exposure periods. Only Loprox (Hoechst) and Spectazole (Ortho) were active at 4 h. Vioform (Ciba), Halotex (Westwood) and Whitfields Ointment were completely inactive.

Simultaneous comparisons were made on suction blisters and abdominal skin with Tinactin, Lotrimin, Micatin and Spectazole, for 24 h exposures. Agreement was complete, the degree of inhibition for each agent corresponding to that shown in Table 2.

For these same agents we showed that inhibition was not altered by overlaying the inoculated skin with serum or sebum.

Discussion

The arrival of the imidazoles marked a new era in antifungal therapeutics. The old standby's Whitfield's ointment, the undecylenates and the iodinated and chlorinated quinolines had to give way. Unlike their predecessors, the imidazoles have greater efficacy and are less irritating to skin. Still the search for more efficacious topical antifungals in ringworm of the nails, palms and soles goes on.

We believe that Dittmar's original method, and its modification by us, furnishes a realistic and reliable way to rapidly screen and optimally formulate antifungals. The results confirm to clinical experience. We found Whitfield's ointment, undecylenate and haloprogin to be ineffective with Vioform showing marginal activity. These have virtually been crowded out of the market place for treating ringworm infections. There is a curious, but ill-advised, rebirth of interest in the undecylenates which became popular during and after the 2nd World War [6]. Starting with Smith et al's "new look at undecylenic acid" in 1977 [8] two reports have been published comparing Desenex ointment (zinc undecylenate and undecylenic acid to 1% Tinactin cream (tolnaftate)

Table 1. Antifungal chemicals (Ethanolic solutions for 24 h exposures)

Percent Surface Concentration ($\mu g/cm^2$)	2% 50	1% 25	0.5% 12.5	0.2% 5	0.1% 2.5	0.01% 1.0	0.001% 0.1
Sodium pyridinethione	0	0	0	0	0	1	3
Ketoconazole	0	0	0	0	0	1	2
Clotrimazole	0	0	0	0	0	2	3
Tolnaftate	0	0	0	0	0	2	3
Zinc pyridinethione (in DMSO)	0	0	0	1	1	2	3
Sulconazole nitrate	0	0	1	1	1	2	3
Ciclopirox olamine	0	1	2	3	3	–	–
Miconazole nitrate	1	1	1	2	2	3	3
Thiabendazole	2	2	2	2	3	–	–

Table 2. Proprietary antifungal agents

	Surface concentration ($\mu g/cm^2$)	Time	
		4 h	24 h
Loprox[1] (1% ciclopirox olamine)	25	0	0
Spectazole[2] (1% econazole nitrate)	25	0	0
Tinactin[3] (1% tolnaftate)	25	2	1
Lotrimin[3] (1% clotrimazole)	25	2	1
Monistat-Derm[2] (2% miconazole nitrate)	50	2	2
Vioform[4] (1% iodochlorhydroxyquin)	25	3	2
Halotex[5] (1% haloprogin)	25	3	3
Aktrinol[3] (0.2% acrisorcin)	5	3	3
Whitfield's ointment[6] (6% benzoic acid)	150	3	3
Hydrophilic ointment U.S.P.	–	3	3

[1] Hoechst AG
[2] Pharmaceutical Corporation
[3] Schering Corporation
[4] CIBA Pharmaceutical Inc.
[5] Westwood Pharmaceutical Inc.
[6] Eli Lilly & Co.

[1, 10]. Therapeutic equivalence was found. This is at varaince with our findings. Tinactin cream is a highly effective topical antifungal while Desenex has marginal clinical value. The FDA Dermatology Panel III has expressed an opinion which is shared by most dermatologists. "The panel feels that this product (Desenex) is at best weakly fungistatic and recommends adequate documentation beyond that of the base alone." Among the newest introductions Loprox (ciclopirox) and Spectazole (econazole) appear to be more effective than Micatin (miconazole) and Lotrimin (clotrimazole). These former two were the only ones which gave complete inhibition at 4 h, indicating rapid diffusion into the horny layer. Likewise, they were alone in

completely suppressing growth after a 24 h exposure. Clinicans seem to prefer 1% Lotrimin over 2% Miconazole; we found the former superior.

The decisive importance of the vehicle was evident in the comparisons of the same agents in ethanol and in their proprietary cream form. At a 0.2% concentration in ethanol ciclopirox gave no inhibition while clotrimazole was partially suppressive down to 0.01%. The ethanolic tests imply high potential efficacy for ketoconazole and sodium pyridinethione when properly formulated. In ethanol miconazole was not particularly impressive and fared only moderately well in its 2% commercial form. It was barely suppressive (grade 2) at 24 h. Tolnaftate on the other hand registered strongly in ethanol and likewise in its 1% cream base.

For screening new compounds a range of concentrations in ethanol, starting which 2%, will almost certainly identify effective fungistats. Complete inactivity in ethanol warrants elimination from further study. The next step is formulation in various bases.

When these studies began we evaluated the methods described by Knight and by Stoughton. We found too many discrepancies between in vitro and in vivo efficacy. Both lack discriminating power and do not match the reliability of Dittmar's method. As an example, Knight found complete inhibition with Vioform and Haloprogin. Neither were effective in our model. Knight's stripping model mainly showed whether a tropical drug can resist a 10 s rinse.

It is astonishing that Knight failed to show inhibition with 1% Tinactin, a material which was highly inhibitory in our model and whose clinical efficacy is beyond doubt.

Likewise, in Stoughton's dermal disc technique, neither 1% tolnaftate in ethanol nor 1% Tinactin cream showed the slightest inhibition. We found 2% Micatin cream completely inactive and Lotrimin only marginally so with the disc model. This probably reflects the fact that these substances do not diffuse well through the agar medium and so do not give zones of inhibition.

In contrast to these incongruities, we have found complete agreement between the results obtained with Dittmar's model and clinical experiences recorded in many publications.

We are persuaded that this model has greater reliability than double blind clinical studies for discriminating among the now numerous proprietary antifungal topicals. Actually, the procedure is an in-vivo-in-vitro blend which eliminates the numerous variables which afflict clinical trials. Moreover, much time and money can be saved by determining the most efficacious vehicle and concentration before embarking numerous clinical trials. As an example of optimization, we found 2.0% ethanolic thiabendazole only weakly fungistatic. However, a 5.0% concentration in a polyethylene glycol base was as effective as any of the proprietary drugs. Battistine et al. have demonstrated that thiabendazole, when properly formulated in sufficient concentration (10%) is highly effective in various forms of tinea [2]. Our unpublished studies are in agreement with this conclusion.

References

1. Amsel LP, Cravitz L, Vanderwyk R, Zahry S (1979) Comparison of in vitro activity of undecylenic acid and tolnaftate against athlete's foot fungi. J Pharm Sci 68:384
2. Battistini F, Zaias N, Sierra R, Bolivas C, Rubell G (1974) Clinical antifungal activity of thiabendazole. Arch Dermatol 109:695–699
3. Dittmar W (1980) Pyridones. Med Mycology Zbl Bakt Suppl 8. Gustav Fischer, New York, pp 266–268
4. Dittmar W (1981) Zur Penetration und antimyzetischen Wirksamkeit von Ciclopiroxolamine in verhorntem Körpergewebe. Arzneim Forsch Drug Res 31:1353–1559
5. Knight AG (1973) Human models for in vivo and in vitro assessment of topical antifungal agents. Br J Dermatol 89:509–514
6. Lyddon FE, Gundersen K, Maibach HI (1980) Short chain fatty acids in the treatment of dermatophytoses. Int J Dermatol 19:24–28
7. Muskatblit A (1947) Clinical evaluation of undecylenic acid as a fungicide. Arch Dermatol Syph 56:256–262
8. Smith EB, Powell RF, Graham JL, Ulrich JA (1977) Topical undecylenic acid in tinea pedis, a new look. Int J Dermatol 16:52–56
9. Stoughton RB (1970) Bioassay of antimicrobials. Arch Dermatol 101:160–166
10. Tschin EH, Becker LE, Ulrich JA (1979) Comparison of over the counter agents for tinea pedis. Cutis 23:696–704
11. Wallace SM, Shah UP, Epstein WL, Greenberg J, Riegelman S (1977) Topically applied antifungal agents. Arch Dermatol 113:1539–1542

In Vitro Models for Cutaneous Phototoxicity

B. E. Johnson, E. M. Walker and A. M. Hetherington

The term phototoxicity is used for any photosensitized reaction of the skin in which ultraviolet or visible radiation is absorbed by a specific chemical compound (the wavelength range is determined by the molecular structure of the photosensitizing chemical) and the absorbed energy is passed on in some way to cause damage directly to cell and tissue components. These reactions are differentiated from those of photoallergy in which immune system sensitization and elicitation processes are required. It is generally accepted that a phototoxic agent will produce a reaction in any subject so long as sufficient of the chemical is present in the skin which is then exposed to a high enough dose of radiation of appropriate wavelength. Cutaneous phototoxicity appears to take four major forms (Table 1) although it is possible that a single photosensitizer may produce more than one of these reaction patterns.

In vitro models of phototoxicity serve two functions. First, they may provide a screening method for the phototoxic potential of foreign substances such as drugs, plant materials, fragrances, perfume and cosmetic constituents, antiseptics, dyestuffs and other environmental chemicals and also abnormal metabolites or normal metabolites which might occur in higher than normal concentrations. Secondly, they may provide some insight into the cellular mechanisms involved in the skin reactions themselves.

The first description of phototoxicity in which an understanding of the basic photosensitization process appeared, literally, the transfer of energy of fluorescence to chemical energy resulting in cell death, was Raab's study of the acridine photosensiti-

Table 1. Cutaneous phototoxicity

Skin reactions	Associated substances and conditions
Exaggerated sunburn	Drugs such a thiazides, demethylchlorotetracycline and chlorpromazine
Prickling and burning; immediate erythema, oedema or urticaria with higher doses. Sometimes delayed erythema and hyperpigmentation	Coal tar, pitch, antraquinone based dyestuffs, benoxaprofen, erythropoietic protoporhyria
Delayed onset erythema, blistering with slightly higher exposure doses. Low exposure doses may lead to delayed hyper pigmentation only	Psoralens, phytophotodermatitis, berloque dermatitis
Increased skin fragility with blistering in extremities	Nalidixic acid, frusemide. Similar in porphyria cutanea tarda. High dose tetracycline

zed killing of Paramecia [26]. Numerous model systems have been used since this but it is still useful, forming the basis of a comprehensive survey of polycyclic hydrocarbons [6] and a more recent comparison of the effects of different psoralens by Young and Barth [35]. In this paper, we shall describe the five relatively simple model systems used in Dundee with specific reference to the results obtained with well established and also lesser known photosensitizing substances.

For each test system, the radiation source was made up from fluorescent tubes arranged to provide an even irradiance for a number of specimens. In this way dose response relationships and photosensitizer concentration effects may be most easily studied. Westinghouse FS-20 or Philips TL20W/12 type lamps were used to provide UV-B (280–315 nm) while the longer wavelength UV-A (315–400 nm) was obtained with Philips TL20W/80 BLB Thorn/Atlas UV or Sylvania FR74T12/PUVA lamps. Other spectral bands may be studied using the various special fluorescent tubes produced by manufacturers such as Philips and Thorn. The irradiance should be measured with linear thermopiles, PUVA meters or other calibrated photocells, or for the UV wavelengths, ideally by the chemical actinometry method of Hatchard and Parker [10].

The first and simples test is that developed by Daniels [4] using Candida albicans. For screening, the suspected photosensitizing material, for instance pieces of plant, solid drugs, dyestuffs and other chemicals, or filter paper discs impregnated with solutions and then dried, is placed on Sabouraud's dextrose agar plates, freshly seeded with Candida. A non-pathogenic form, C. utilis, may be preferred [15] but may not be so easily available. For best results, a standard thickness of agar, 4 mm, should be used and seeding from a 5 ml sterile water suspension of yeast made up from a standard bacterial loop from a 48 h culture, should produce a damp but not wet surface to the agar. Duplicate plates are set up, one serving as a dark control, the other being exposed to UV-A or visible radiation at an irradiance around 0.2 mW/cm^2 for 48 h. Phototoxicity is indicated by a clear zone around the photosensitizer in the irradiated culture while growth in the dark control is complete (Fig. 1). A result may be detected with

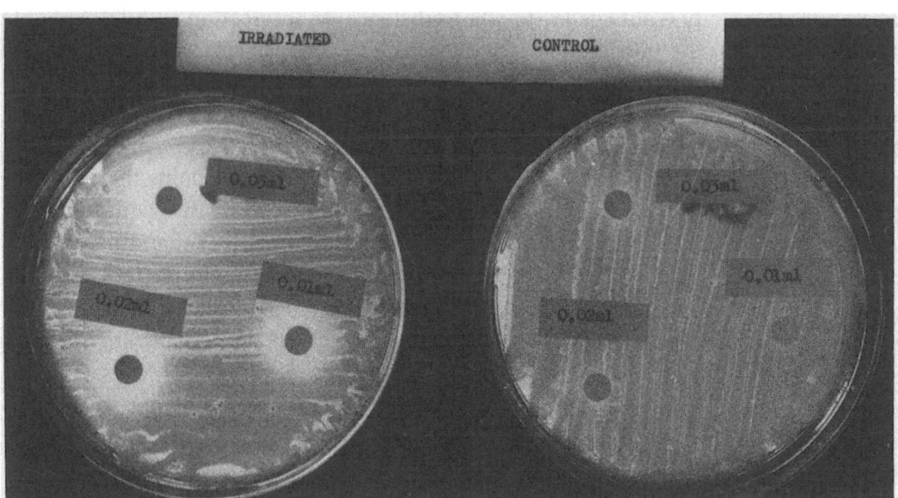

Fig. 1. The Candida albicans test with different amounts of a perfumed product on filter paper discs

Table 2. A simple microbiological method for demonstrating phototoxic compounds. According to Daniels F Jr (4)

	Positive	Negative
Plants	Parsnip, celery, fig, parsley, skimmia japonica, marigold seed, lemon oil, orange oil, tangerine oil	Tansy, zinnia, coreopsis, chrysanthemum, cornflower, dahlia
Drugs	Quinacrine, chlordiazepoxide	Sulfanilamide, quinidine, acetazolamide, griseofulvin, trimeprazine
Others	Crude coal tar, 8-methoxypsoralen, 5-methoxypsoralen, trimethylpsoralen	Carmine, rose bengal, eosin, phyloxine

only 24 h exposure but 48 h ensures good yeast growth and a definitive test. Where the material is toxic in the dark, phototoxicity may still be demonstrated by a larger clear zone in the irradiated culture. The majority of positive tests obtained by Daniels (Table 2) were with plants containing psoralens and the psoralens themselves with recognised preferential effects against DNA. The crude coal tar may also photosensitize through DNA effects but this action is not known for the drugs found to be positive. Our own screening (Table 3) confirmed the efficiency of the psoralens in the test and its use to confirm the phototoxicity of plants known to be causative agents in phytophotodermatitis. However, a number of dyes, shown here to be phototoxic by using visible rather than UV-A irradiation, probably act mainly through cell membrane damage and the non-psoralen fragrance materials have no known interactions with DNA. Nonetheless, the psoralens are so effective in this test that where a similar degree of phototoxicity is obtained with other substance, an interaction with DNA should be considered and the possibilities of photosensitized mutagenesis examined, perhaps by sister chromatid exchange studies with leucozyte preparations [24].

Although a large number of materials are shown to be phototoxic in this model, an equally large number, particularly drugs with recorded photosensitizing side effects, have proved negative. It is not clear why this is so but the finding limits the use of the test on its own as a screen. Because the test is simple and very effective with psoralens, it has proved valuable in investigations of phytophotodermatitis [9] and perfume induced skin changes such as berloque dermatitis [36]. We have found the test to be sensitive down to 0.1–0.5 microgrammes of 8-methoxypsoralen (8-MOP). In quantitative studies of the psoralen content of Heracleum mantegazzianum (Giant hogweed) we ensured that we had extracted most of the contained photosensitizer by testing the residue. Similarly, thin layer chromatogram fractions were scraped from the backing and used in the test to determine the major phototoxic components of the extract. Similar studies have been reported by Möller [22] and Kavli et al. [18].

With carefully controlled conditions of use, a quantitative estimate of psoralen content of a solution may be obtained from measurements of the diameter of the clear

Table 3. Phototoxicity in vitro: Candida

	Positive	Negative
Plant extracts	Umbelliferae: Giant hogweed, cow parsnip	Compositae: Chrysanthemum, tansy, yarrow, sagebrush, ragweed, dahlia, cornflower, coreopsis, aster, wild feverfew, burweed, lettuce, chicory, golden rod, black eyed susan, sunflower, burdock, cocklebur, sneezeweed, dog fennel
Chemicals	8-Methoxypsoralen, 5-methoxypsoralen trimethylpsoralen	Alantolactone, isoalantolactone,
Fragrance materials	Oil of bergamot 6-methyl coumarin, costus root oil, cinnamaldehyde, cinnamyl alcohol, amylcinnamaldehyde, jasmines, laurel leaf oil, colophony	Musk ambrette, balsam of Peru, oak moss, atranorin, usnic acid, hydroxycitronella, eugenol, isoeugenol, geraniol, benzyl salicylate, benzyl benzoate, benzyl alcohol, methyl salicylate
Dyestuffs	Benzanthrone, eosin, methylene blue toluidine blue	Anthraquinone
Drugs	Chlorpromazine, benoxaprofen, minocycline	Amiodarone, azapropazone, carbamazepine, cimetidine, demethylchlorotetracycline, diflunisal, doxycycline, griseofulvin, hydrochlorthiazide, imipramine, methyl dopa, nalidixic acid, oxytetracycline, piroxicam, propanolol, protriptyline, sulphamethoxazole, sulphapyridine, tetracycline, trimethoprim

zone around a filter paper disc. A 6–8 mm disc ideally accepts 2 μl of solution. Zaynoun [37] established a standard curve for varying amounts of 5-methoxypsoralen (5-MOP) and found the technique as accurate as the more complex, solvent extraction/spectroscopic determination method normally used for the estimation of 5-MOP in bergamot oils and perfume preparations.

The second test, photohaemolysis [3, 13], depends on the photosensitized damage to red blood cell (RBC) membranes, phototoxicity being quantified by determination of the haemoglobin released (Fig. 2). The method used in Dundee is derived from those described by Harber et al. [11] and Kahn & Fleischaker [16]. A dilute suspension of

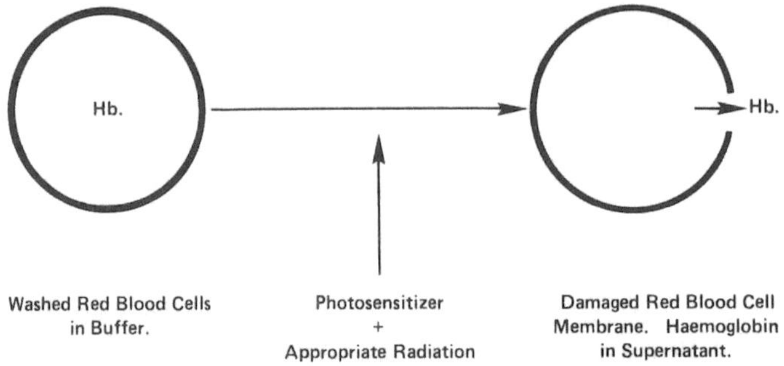

Fig. 2. The process of photohaemolysis

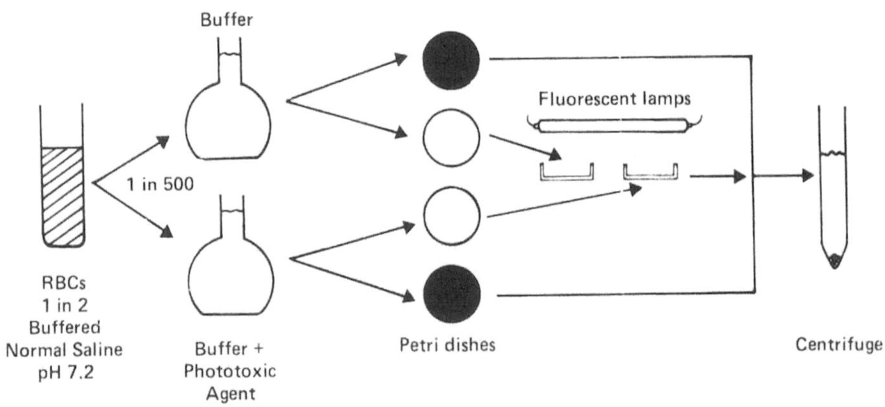

Fig. 3. Schema 1 for photohaemolysis method. From: Hetherington and Johnson [13] with permission

RBCs is made in barbitone buffered saline at pH 7.4 after washing three times. For irradiation, 5 ml aliquots of RBC suspension at a final dilution of 1 in 1,000 are placed in 5 cm diameter tissue culture dishes (Sterilin Ltd.) forming a monolayer. The phototoxic agent is then added in microlitre amounts (Fig. 3). Alternatively, the final dilution step of the RBCs may be made in buffered saline containing the photosensitizer (Fig. 4). Controls for the effects of photosensitizer alone and radiation alone are essential. After 30 min in the dark to allow for equilibration, the RBCs are exposed to a standard 14 J/cm^2 from UV-A or visible sources. Where UV-B is used, the exposure dose is lower, around 4 J/cm^2 to avoid the damaging effects of UV-B alone. With the fluorescent lamp arrangements used, the irradiation times are from 1–2 h. At the end of the exposure, a further 30 min dark equilibration is required before the suspensions are centrifuged for 10 min at 3,000 rpm in a standard benchtop centrifuge.

The haemoglobin in the supernatant is determined by taking 2 ml aliquots, adding 2 ml of Drabkin's solution (KCN 0.05 g · $K_3Fe(CN)_6$ 0.2 g · H_2O 1 l) leaving the

PHOTOHAEMOLYSIS METHOD

1. **Fresh human red blood cells (RBCs) washed and diluted to 500x original volume, in buffered saline pH 7.4. (Michaelis - Barbitone buffer)**

2. **Phototoxic compound added to diluted blood. Control samples receive vehicle only.**

3. **Treated blood is decanted into culture dishes and left for the RBCs to settle as monolayer.**

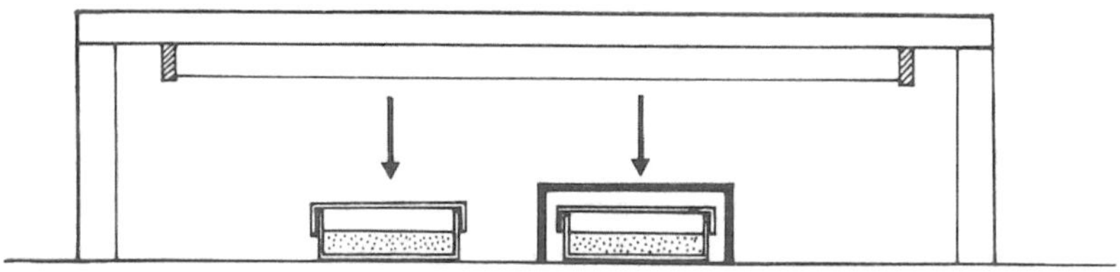

4. **Samples irradiated with controls in darkness under the lamp to reproduce temperature effects.**

5. **After leaving in darkness for set 30 minute period, the samples are centrifuged and the supernatants treated with Drabkins reagent, changing haemoglobin of both oxidation states to single form.**

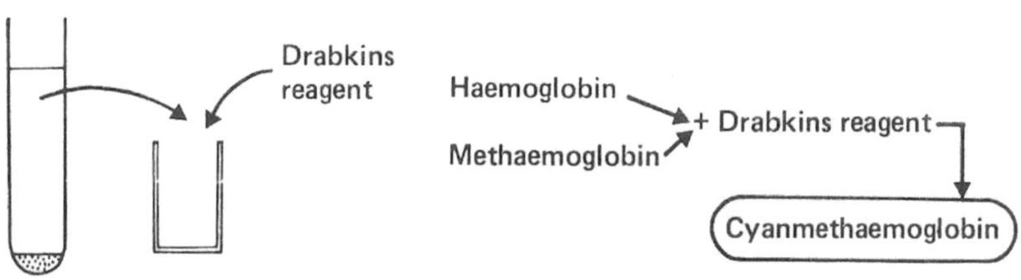

6. **The concentration of cyan-methaemoglobin is measured by spectrophotometry, reading at 420 nm against reagent blank. Degree of photohaemolysis is determined by comparison of results from irradiated and dark control samples with 100% obtained from distilled water dilution.**

Fig. 4. Schema 2 for photohaemolysis method. From: Hetherington and Johnson [13] with permission

mixture for at least 30 min and reading the absorbance at 420 nm against a reagent blank of 2 ml buffered saline plus Drabkin's solution. This chemical determination in which all forms of haemoglobin are converted to cyanmethaemoglobin, rather than direct optical density measurements on the supernatant, is required because the irradiation may lead to the oxidation product, methaemoglobin, with absorption characteristics which are very different from those of haemoglobin itself.

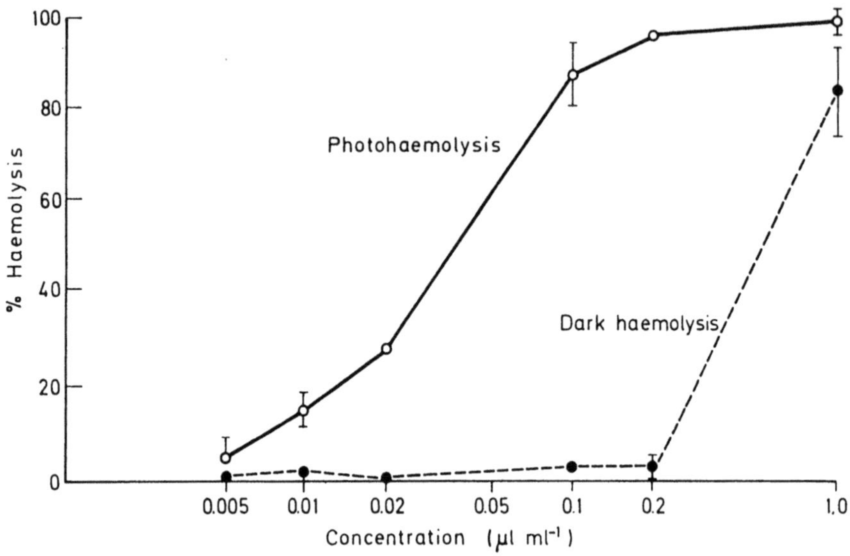

Fig. 5. Photohaemolysis study of a Compositae plant extract with varying amounts of extract and a standard exposure to UV-A

A zero level haemolysis value is obtained using supernatant from an unirradiated RBC suspension without added photosensitizer and a 100% haemolysis value from a final dilution of RBCs of 1 in 1,000 in distilled water.

The degree of photohaemolysis is obtained by calculation:

Sample	Absorbance at 420 nm
0% haemolysis (C)	0.02
100% haemolysis (B)	0.82
Photohaemolysis sample (A)	0.62

$$\% \text{ haemolysis} = \frac{(A-C) \times 100}{B-C} = \frac{(0.62-0.02) \times 100}{0.82-0.02} = 75\%$$

A typical result (Fig. 5) is obtained with an extract from a plant of the Compositae family showing a clear demarcation between phototoxicity and a direct effect of the plant material on the RBC membrane. For other materials, this difference may be more or less significant. Ideally, any quantitative study of photohaemolysis should provide comparative data on dark and photosensitized haemolysis. For screening purposes, we have simply noted whether or not photohaemolysis occurred (Table 4). The outstanding features are the negative results with the psoralens and positive results with extracts from plants of the Compositae, a reversal of the Candida test, a slight increase in the number of drugs which exhibit a phototoxic potential, the finding that a number of photosensitizing drugs still do not demonstrate phototoxicity and the relatively few substances which are phototoxic in both the Candida and photohaemolysis models.

The photohaemolysis model is simple and useful for investigations where the phototoxicity is directed against cell membranes. As shown by the negative results with the

Table 4. Phototoxicity in vitro: photohaemolysis

	Positive	Negative
Plant extracts	Compositae: Chrysanthemum, tansy, yarrow, safebrush, ragweed, dahlia, cornflower, wild feverfew, lettuce, black eyed susan, chicory, golden rod, burdock, cocklebur, dog fennel	Compositae: Sunflower, burweed, sneezeweed Umbelliferae: Giant hogweed, cow parsnip
Chemicals	Alpha-terthienyl, tetrachlorosylicylanilide	8-Methoxypsoralen 5-methoxypsoralen, trimethylpsoralen, alantolactone, isoalantolactone
Fragrance materials	Musk ambrette, 6-methyl coumarin, costus root oil, oak moss, cinnamaldehyde, jasmines, laurel leaf oil, benzyl alcohol, usnic acid, ? amylcinnamaldehyde, ? cinnamyl alcohol, ? hydroxycitronella	Eugenol, isoeugenol, geraniol, methyl salicylate, benzyl benzoate, benzyl salicylate
Dyestuffs	Benzanthrone, anthraquinone, disperse blue 35	
Drugs	Amiodarone, azapropazone, benoxyaprofen, chlorpromazine, protriptyline, ? doxycycline, ? nalidixic acid, ? imipramine, ? tetracycline	Carbamezapine, cimetidine, demethylchlorotetracycline, diflusinal, ethylinoestradiol, frusemide, griseofulvin, hydrochlorthiazide, methyl dopa, oxytetracycline, piroxicam, propranolol, sulphamethoxazole, sulphapyridine, trimethoprin

psoralens, it does not apply when DNA is the primary target for photosensitization. However, it is particularly useful for studies of oxygen dependency [31], temperature effects on phototoxicity [5], and changes in the biochemistry of cell membranes exposed to phototoxic insult [21, 34]. In most instances, photohaemolysis is a typical colloid osmotic haemolysis with early potassium loss following a photosensitized damage to the RBC membrane (Fig. 6). However, for tetrachlorosalicylanilide [17], chlorpromazine [14, 20], and protriptyline [20] the formation of a toxic photoproduct may explain all or part of the haemolysis obtained. Where a photohaemolysis test

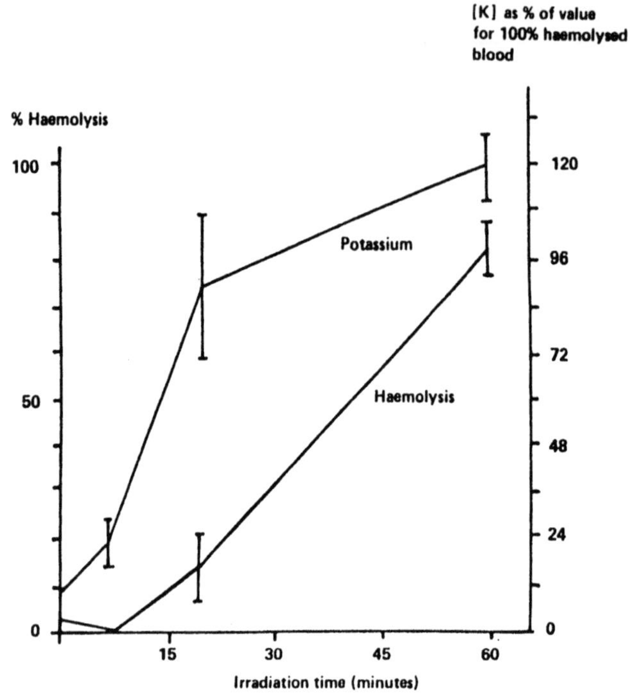

Fig. 6. Photohaemolysis study with the anthraquinone based dyestuff, benzanthrone (1.0 $\mu g\ ml^{-1}$) and varying exposure times with blue light, to illustrate the early potassium loss typical of colloid osmotic haemolysis

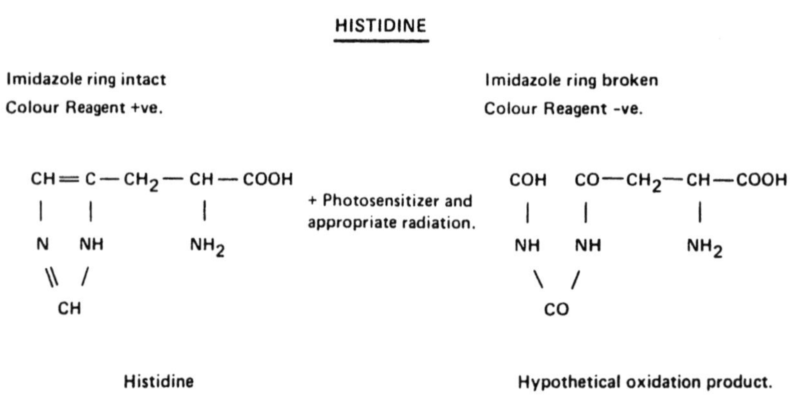

Fig. 7. The principle of the histidine phototoxicity test

proves positive, this possibility should be examined by adding a pre-irradiated solution of photosensitizer to the dilute RBC suspension.

The third model is the photosensitized oxidation of histidine (Fig. 7) in which the imidazole ring of the amino acid is opened [32] and a colour reaction thereby lost. Varying amounts of photosensitizing agent in aqueous or ethanolic solution are added

Table 5. Phototoxicity in vitro: Histidine

	Positive	Negative
Plant extracts	Compositae: Dahlia, yarrow, cocklebur, dog fennel, chicory, ragweed, chrysanthemum, golden rod, lettuce, tansy, sneezeweed, sagebrush	
Chemicals	Kynurenic acid, potassium dichromate, tetrachlorosalicylanilide	8-Methoxypsoralen
Fragrance materials	6-Methyl coumarin, oak moss, amylcinnamaldehyde, musk ambrette, jasmines, colophony	Cinnamaldehyde, cinnamyl alcohol, eugenol, isoeugenol, geraniol, hydroxycitronella, benzyl salicylate, benzyl alcohol, benzyl benzoate, methyl salicylate, costus root oil, laurel oil
Dyestuffs	Methylene blue	
Drugs	Benoxaprofen, diflusinal, nalidixic acid, oxytetracycline propranolol, protriptyline, ?demethylchlorotetracycline, ? frusemide, ? hydrochlorthiazide	Amiodarone, azapropazone, carbamazepine, cimetidine, doxycyline, ethinyloestradiol, griseofulvin, imipramine, methyl dopa, minocycline, piroxicam, sulphamethoxazole, sulphapyridine, tetracycline, trimethoprim

to 0.61 mM L-histidine monochloride monohydrate in 0.01 M phosphate buffer, pH 7.4. The mixtures are irradiated as for photohaemolysis and the histidine remaining is determined by a modified Pauly reaction [29]. For this, 200 microlitres of irradiated solution are made up to 2 ml with phosphate buffer, 200 microlitres of 1% sulphanilic acid in 0.87 N HCl and 200 microlitres of 5% $NaNO_2$ are added and the mixture is left for 10 min. Then, 0.6 ml of 20% $NaCO_3$ is added and after a further 2 min, 2 ml ethyl alcohol. The optical density of the final solution is read at 530 nm against a reagent blank and the histidine determined from a standard curve. Controls of non-irradiated mixture, irradiated histidine alone and photosensitizer alone are required.

The results of this test as a screen for phototoxicity (Table 5) show that the plant extracts which produce photohaemolysis also photosensitize the oxidation of histidine. Significantly, 8-MOP is negative in this test. The fragrance materials which were positive in this test also produced photohaemolysis. However, as shown by the tests with

Table 6. Quantitative studies of histidine phototoxicity

Material	Amount (μl %) to produce 50% destruction of histidine
Compositae extracts (10% w/v)	
Dahlia	27
Yarrow	36
Cocklebur	40
Dog Fennel	42
Chicory	46
Ragweed	49
Chrysanthemum	58
Golden Rod	87
Lettuce	89
Tansy	234
Sneezeweed	269
Sagebrush	380
Kynurenic acid	1.2 (60 μM)

drugs, not all those which produce photohaemolysis are effective in the histidine test, suggesting that some photosensitized damage to cell membranes may be independent of oxygen. The finding that tetrachlorosalicylanilide is effective against histidine in vitro appears to confirm the action of this compound as described by Kochevar [19] in studies of mechanisms for photoallergy and persistent light reactions. It is possible that any phototoxic agent which is effective against histidine might be brought into a general hypothesis to explain the state of persistent light reaction [1] especially as kynurenic acid, a tryptophan metabolite, is very effective in the test (Table 6) and could act as an endogenous photosensitizer [30]. Quantitative studies based on the amount of material required to produce 50% reduction in histidine show that a number of Compositae plant extracts are also very active (Table 6); but it is difficult to devise a satisfactory test of the overall hypothesis and the histidine model should, perhaps, be regarded simply as a test for oxygen dependent photosensitized damage to cellular protein.

The models described so far have been either simple to use or simple in terms of the photosensitized damage or both. The fourth test is more complex using a nucleated mammalian cell in short term culture. However, the cell type chosen, the mouse peritoneal macrophage, is non-dividing and the complications of damage specifically associated with DNA synthesis and cell division are avoided.

The method, derived from those described by Allison et al. [2] and by Freeman et al. [7], consists of irradiating the cells which are cultured on cover slips, 9 x 35 mm, in Leighton tubes (Fig. 8). The medium used is made up as Medium 199 with Hepes buffer and Hank's salts 80 ml, newborn calf serum 20 ml, glutamine 20 mM 2 ml, penicillin/streptomycin 500 I.U./500 microlitres. Five ml of this cooled medium are injected intraperitoneally into a freshly killed mouse after the peritoneum has been exposed (Fig. 8). After one minute, as much fluid is withdrawn as possible, usually 4 ml, and kept cool until ready to be pooled with other specimens for culture as 1 ml per Leighton tube. The cell yield is around 10^6 per ml providing good monolayer cultures. The cells are established after one hour and the medium is then changed and

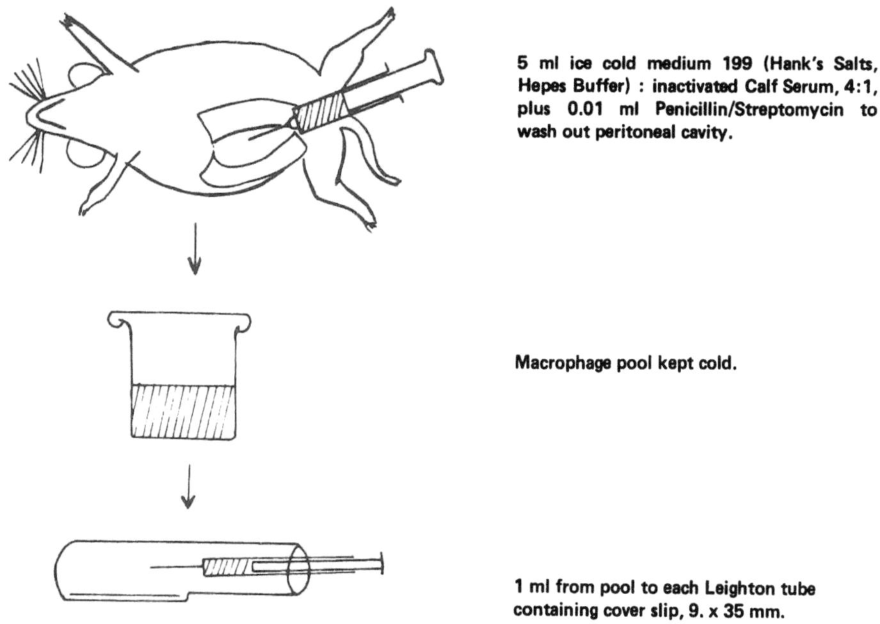

Fig. 8. Mouse peritoneal macrophages: the setting up of cultures

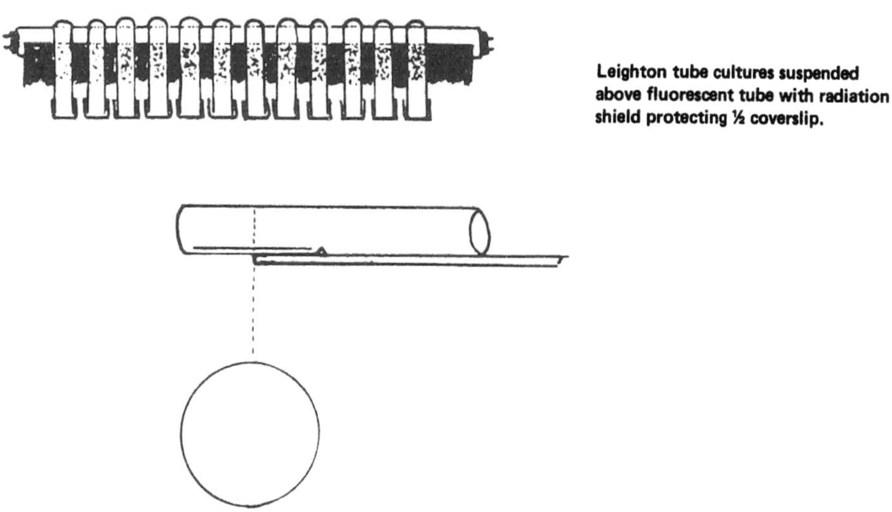

Fig. 9. Mouse peritoneal macrophages: irradiation

replaced with medium containing photosensitizer. Various procedures may now be followed. The cultures may be irradiated, as shown in Fig. 9, with half the culture shielded to provide dark controls, at any time interval after the addition. Alternatively, the medium may be changed again to investigate the location of the phototoxic action. If the preferred site is intracellular, phototoxicity should still be obtained after this

Table 7. Photosensitization of mouse macrophages by benzanthrone. Means of numbers of cells per field, from 3 tubes, 10 fields per culture. Magnification: 400 x. Standard error, and t-test for significance of difference between dark and irradiated ends of coverslips

Treatment	Dark (mean)	S.E.	Irradiated (mean)	S.E.	t	p
Ethanol	6.97	0.62	5.2	0.5	2.22	0.02–0.05
$0.05\ \mu g\ ml^{-1}$ Benzanthrone	5.8	0.58	2.1	0.54	4.67	< 0.001
$1.0\ \mu g\ ml^{-1}$	5.77	0.41	0.67	0.21	11.07	< 0.001
$10\ \mu g\ ml^{-1}$	0.43	0.16	0.8	0.21	1.40	> 0.1

washing procedure. If the action is on the plasma membrane, the photosensitizer is probably lost in the washing process and the phototoxicity is no longer obtained.

For a standard test of phototoxicity, the culture is continued for 24 h after irradiation and then the cells are washed with buffered saline, fixed in formol saline, stained with Giemsa and the cover slip is then mounted on a slide for microscopy. The phototoxic potential of a compound may be quantified by studies of the effects of varying concentrations and varying exposure doses, cell survival being assessed by cell counts in the irradiated and shielded halves of the cover slips (Table 7).

The test may be used as a screening test but we have used it in a limited manner to examine the phototoxicity due to chlorpromazine, benzanthrone and amiodarone. All three were shown to be phototoxic in this model. Cytological and cytochemical studies showed that with chlorpromazine the typical form of phototoxic damage appeared to be a disruption of the plasma membrane with blebbing and eventual complete cellular disintegration. Cytochemical investigations showed that acid phosphatase activity remained discrete in the presence of extreme cytoplasmic blebbing so that the disruption of lysosomes would appear to have no part in this phototoxic reaction. With benzanthrone, the phototoxicity also appears to be directed at the plasma membrane but with amiodarone the major feature is a vacuolation of the cytoplasm which may derive from a primary effect on the lysosome membrane. Different types of cellular damage due to phototoxicity are therefore illustrated with this model. With rose bengal and eosin, which do not enter cells, the surface blebbing reaction is typical and similar effects are obtained with kynurenic acid [33]. Anthracene, uroporphyrin, neutral red, acridine and acridine orange enter the cell and are concentrated primarily within lysosomes, and it is here that they have their phototoxic action [2, 12]. Some compounds may well act at both locations [8]. The major effect of the psoralens appears to be in nuclear DNA.

The final model in this discussion is more complex, both in the type of cell used, the human peripheral blood lymphocyte, and in the methodology required. It is essentially a standard immunocytochemical method for determining the response of lymphocytes to the mitogen, phytohaemagglutinin (PHA) adapted by Scherer et al. [27] to study psoralen phototoxicity and used more recently by Morrison et al. [23] as a phototoxicity screening test. Phototoxicity is assessed by the inhibition of the PHA induced DNA synthesis, measured by 3H-thymidine uptake, in microlitre lymphocyte

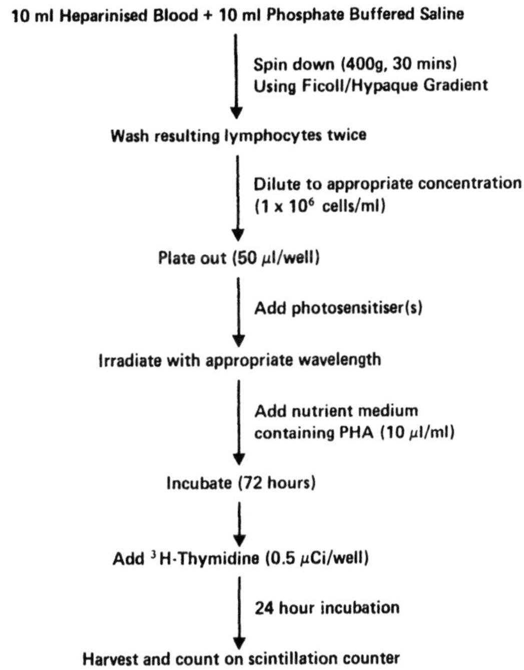

Fig. 10. Schema for the lymphocyte phototoxicity test

cultures set up in microlitre plates. Because a larger number of individual cultures may be set up at any one time, photosensitizer concentration efffects, exposure dose responses and, by using cut-off filters, wavelength dependence, may all be studied with ease.

We have used the method (Fig. 10) to screen the phototoxic potential of drugs, and to compare the phototoxicity of different psoralens. The majority of drugs studied were positive in this test (Table 8). Both Pohl and Christophers [25] and Morrison et al. [23] found this method to be too variable to obtain reliable quantitation. This is not surprising in view of the variability of the lymphocyte response to PHA both inter and intra-individually. However, in a study with three psoralens (Fig. 11) which showed 8-MOP and 5-MOP to be equally phototoxic and 3-carbethoxypsoralen, a monovalent compound, to be so, only at high concentration, it was found that thc use of an inbuilt control of 8-MOP overcame this difficulty. Moreover, when a no drug control was built into the study of drug phototoxicity, to assess the effect of the radiation and culture medium alone, this again worked to check on the variation in lymphocyte response (Table 9). A typical result for benzanthrone (Table 10) shows that three orders of magnitude of concentration variation cover the range of no effect, phototoxicity and toxicity.

Scherer et al. [28] emphasized the usefulness of this model for studying the effects of phototoxicity at different stages of the cell cycle and confirmed previous findings with cells from long term culture that the cells are more sensitive to psoralen phototoxicity in the middle of the DNA synthesis or "S" phase of the cycle. It would be particularly interesting to use this model in such a way to see whether membrane specific phototoxic agents have the same variation. Moreover, if phototoxicity mediated through damage to DNA is amplified through cell division, a comparison of the phototoxic potential of substances against the macrophages with that against lymphocytes

Table 8. Phototoxicity in vitro: Lymphocytes

	Positive	Negative
Chemicals	8-Methoxypsoralen, 5-methoxypsoralen, 3-carbethoxypsoralen, angelicin	
Dyestuffs	Benzanthrone	
Drugs	Amiodarone, azapropazone, demethylchlorotetracycline, doxycycline, frusemide, griseofulvin, hydrochlorthiazide, methyl DOPA, nalidixic acid, oxytetracycline, piroxicam, protriptyline, sulphapyridine, tetracycline, trimethoprim, ? imipramine, ? minocycline	Carbemazepine, chlorpropamide, diflunisal, propranolol, sulphamethoxazole

would give an indication of the major site of phototoxic action for any given photosensitizer.

The five model systems for cutaneous phototoxicity used in our laboratory have proved to give positive results for the majority of photosensitizers tested. Where the results are negative, there is no evidence for cutaneous photosensitivity anyway, e.g., alantolactone or the benzyl fragrance materials, or, in the case of the drugs, a metabolic process may be required for the phototoxicity to be manifest. Some of the materials shown to be positive in vitro, such as the Compositae extracts and certain fragrance materials, have a limited clinical history of phototoxicity although positive photopatch tests have been obtained in human subjects and guinea pigs, but this may be due to the failure of the active compounds to penetrate the intact stratum corneum.

The site of action of any phototoxic agent is still not clear from results with these tests. Nonetheless, where photohaemolysis is very efficient and the yeast test is negative, a cell membrane effect is indicated, possibly confirmed with the cytological studies with macrophages. Where both yeast and lymphocyte tests are effective but the photohaemolysis test is negative, an effect on nuclear DNA should be considered. Where the membrane effect is most marked, the pitch smart, type of cutaneous reaction is most probably the major feature of the cutaneous response. If lysosomal damage is involved, the reaction may be of the exaggerated sunburn type. Where DNA is the major target for phototoxicity, the psoralen type of delayed erythema, blistering and intense hyperpigmentation may be expected to occur.

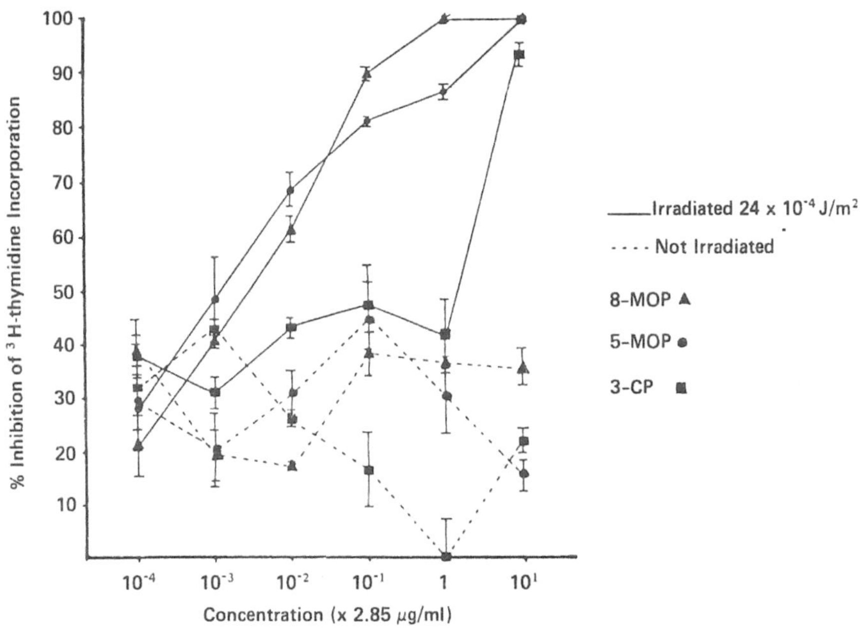

Fig. 11. A comparative study of phototoxicity of different psoralens against lymphocytes. 8-MOP – 8-methoxypsoralen; 5-MOP – 5-methoxypsoralen; 3-CP – 3-carbethoxypsoralen. (From a study by D. M. Hepburn)

Table 9. Phototoxicity in vitro: Lymphocyte results (average counts)

Exposure time (min)	0	15	30	60
Growth medium	10,012	9,662	8,699	5,448
Demethylchlorotetracycline				
5 µg/ml	10,836 (100%)		7,153 (82%)	3,055 (56%)
50 µg/ml	5,622 (56%)	2,055 (21%)	1,730 (20%)	65 (1%)

Table 10. Phototoxicity in vitro: Benzanthrone against lymphocytes

	Average counts	
	Dark control	Blue light 1.39 J/cm²
Ethanol	6,749	6,049
Benzanthrone 0.01 µl/ml	9,175	6,832
Benzanthrone 0.1 µg/ml	7,296	495
Benzanthrone 1.0 µg/ml	278	256

References

1. Addo HA, Ferguson J, Johnson BE, Frain-Bell W (1982) The relationship between exposure to fragrance materials and persistent light reaction in the photosensitivity dermatitis with actinic reticuloid syndrome. Br J Dermatol 107:261–274
2. Allison AC, Magnus IA, Young MR (1966) Role of lysosomes and of cell membranes in photosensitization. Nature 209:874–878
3. Blum HF (1964) Photodynamic action and diseases caused by light. Hafner Publishing Company, New York
4. Daniels F (1965) A simple microbiological method for demonstrating phototoxic compounds. J Invest Dermatol 44:259–263
5. Dubbelman TMAR, Haasnoot C, Van Steveninck J (1980) Temperature dependence of photodynamic red cell membrane damage. Biochim Biophys Acta 601:220–227
6. Epstein SS, Small M, Falk HL, Mantel N (1964) On the association between photodynamic and carcinogenic activities in polycyclic compounds. Cancer Res 24:855–862
7. Freeman RG, Murtishaw W, Knox JM (1970) Tissue culture technics in the study of cell photobiology and phototoxicity. J Invest Dermatol 54:164–169
8. Fritsch P, Gschnait F, Hönigsmann H, Wolff K (1976) Protective action of beta-carotene against lethal photosensition of fibroblasts in vitro. Br J Dermatol 94:263–271
9. Gumar AWS (1976) A quantitative study of phytophotodermatitis. MSc Thesis Dundee Univ Scotland
10. Hatchard CG, Parker CA (1956) A new sensitive chemical actinometer. II. Potassium ferrioxalate as a standard chemical actinometer. Proc Roy Soc (London) A235:518–536
11. Harber LC, Fleischer AS, Baer RL (1964) Erythropoietic protoporphyria and photohaemolysis. J Amer Med Ass 189:191–194
12. Hawkins HK, Ericsson JLE, Biberfield P, Trump BF (1972) Lysosomes and phagosome stability in lethal cell injury. Am J Pathol 68:255–258
13. Hetherington AM, Johnson BE (1984) Photohemolysis. Photodermatoly 1:255–260
14. Johnson BE (1974) Cellular mechanisms of chlorpromazine photosensitivity. Proc Roy Soc (London) 871:873
15. Kagan J, Gabriel R (1980) Candida utilis as a convenient and safe substitute for the pathogenic yeast C. albicans in Daniels' phototoxicity test. Experentia 36:587–588
16. Kahn G, Fleischaker B (1971) Red blood cell haemolysis by photosensitzing compounds. J Invest Dermatol 56:85–90
17. Kahn G, Fleischaker B (1971) Evaluation of phototoxicity of salicylanilides and similar compounds by photohaemolysis. J Invest Dermatol 56:91–97
18. Kavli G, Raa J, Johnson BE, Volden G, Hangsbø S (1983) Furocoumarins of Heracleum Laciniatum: Isolation, phototoxicity, absorption and action spectra studies. Contact Derm 9:257–262
19. Kochevar IE (1979) Photoallergic responses to chemicals. Photochem Photobiol 30:437–442
20. Kochevar IE, Lamola AA (1979) Chlorpromazine and protriptyline phototoxicity: Photosensitized, oxygen independent red cell haemolysis. Photocehm Photobiol 29:791–796
21. Lamola AA (1977) Photodegradation of biomembranes. In: Research in Photobiology. Castellani A (ed) Plenum Press New York 53–63
22. Möller H (1978) Phototoxicity of dictamnus alba. Contact Derm 4:264–269
23. Morrison WL, McAuliffe DJ, Parrish JA, Bloch KJ (1982) In vitro assay for phototoxic chemicals. J Invest Dermatol 78:460–463
24. Mourelatos D, Faed MJW, Gould PW, Johnson BE, Frain-Bell W (1977) Sister chromatid exchanges in lymphocytes of psoriatics after treatment with 8-methoxypsoralen and long wave ultraviolet radiation. Br J Dermatol 97:649–654
25. Pohl J, Christophers E (1979) Dose-effects of 8-methoxypsoralen and long wave UV-light in 3T3 cells: evaluation of a phototoxic index. Experientia 35:247–248
26. Raab O (1900) Über die Wirkung fluoreszierender Stoffe auf Infusorien. Z Biol 39:524–546

27. Scherer R, Kern B, Braun-Falco O (1975) The human peripheral lymphocyte – a model system for studying the combined effect of psoralen plus black light. Klin Wschr 55:137–140
28. Scherer R, Kern B, Braun-Falco O (1977) UVA-induced inhibition of proliferation of PHA-stimulated lymphocytes from humans treated with 8-methoxypsoralen. Br J Dermatol 97: 519–528
29. Sluyterman LA (1960) The effect of oxygen upon the micro-determination of histidine with the aid of the Pauly reaction. Biochim Biophys Acta 38:218–221
30. Swanbeck G, Wennersten G (1973) Evidence for kynurenic acid as a possible photosensitizer in actinic reticuloid. Acta Dermato (Stockholm) 53:109–113
31. Swanbeck G, Wennersten G, Nilsson R (1974) Participation of singlet excited oxygen in photohaemolysis induced by kynurenic acid. Acta Dermato (Stockholm) 54:433–436
32. Weil L (1965) On the mechanism of the photo-oxidation of amino acids sensitized by methylene blue. Arch Biochem Biophys 110:57–68
33. Wenneersten G, Brunk U (1977) Cellular aspects of phototoxic reactions induced by kynurenic acid. I. Establishment of an experimental model utilising in vitro cultivated cells. Acta Dermato (Stockholm) 57:201–209
34. Yamamoto E, Wat CK, MacRae WD, Towers GHN (1979) Photoinactivation of human erythrocyte enzymes by α-terthienyl and phenylheptatriyne, naturally occurring compounds in the Aster aceae. FEBS letters 107:134–136
35. Young AR, Barth J (1982) Comparative studies on the photosensitizing potency of 5-methoxypsoralen and 8-methoxypsoralen as measured by cytolysis in Paramecium caudatum and tetrahymena pyriformis, and growth inhibition and survival in candida albicans. Photochem Photobiol 35:83–88
36. Zaynoun ST, Johnson BE, Frain-Bell W (1977) A study of oil of bergamot and its importance as a phototocix agent. Br J Dermatol 96:475–482
37. Zaynoun ST (1978) The quantitative analysis of bergapten in perfumes. J Soc Cosmet Chem 29:247–263

HET (Hen's Egg Test) in Toxicological Research

N. P. Luepke

Ever since that industrial revolution, man has been subjecting the earth's biosphere to an increasing variety of chemical insults [5]. The increasingly large number of chemicals introduced in the market and also in environment each year has necessitated the monitoring of environmental materials and specimen banking as well as the development of rapid and reliable methods for the evaluation of their toxicity [7]. Furthermore it must be realized that many new agents have been and are now being introduced without adequate or any toxicological evaluation. Toxicologic studies of all these chemicals by the "usual" methods using laboratory animals, especially small rodents, are very difficult and expensive in time and money, and such studies sometimes give inconclusive results; ethical and legal issues (e.g. animal protection laws) must also be taken into account [1].

Chicken embryo models are well known as basic investigational tools for embryotoxicity testing and have many advantages, but they also possess disadvantages. These

Table 1. Frame of HET-embryotoxicity-testing

Embryotoxicity – Testing by HET

1. Lethality
   LD_{50}
2. Retardation
   hatching weight
   bone lengths (humerus, ulna, femur, tibia, metatarsus, skull)
   a : b correlation
   organ weights
3. Teratogenicity
   macroscopic-anatomical
   clearing and staining of skeleton
4. Systemic effects
   blood-chemical parameters (Na, K, Ca, Cl, inorg. PO_4, proteins, creatinine, uric acid, glucose, cholesterol, triglycerides, GOT, GPT, LDH, AP, γ-GT, bilirubine)
   haematological parameters (haemoglobin, haematocrit, blood cells, coagulation)
   organ weights and histopathology
5. Immunpathology
   thymus
   bursa of Fabricius
6. 1–5 investigations after application on day 1 (before incubation) and on day 5 (96 h after beginning)

basic tests were extended and standardized in our laboratories [9, 10] in recent years by the development of the HET (Hen's Egg Test, Hühner-Embryonen-Test). Hen's egg tests give information regarding embryolethality, embryonic and growth development, teratogenicity and systemic toxicity, including immunpathology [6] and metabolic pathways. In HET it is possible to test chemical substances with known structures and concentrations at various stages of embryonic development and organogenesis (Table 1). This HET embryotoxicity test is now extended and standardized in the development of HET chorionallantoic membrane testing as a mucous membrane irritation test. An important point is that testing incubated hen's eggs is in a borderline area between in-vivo and in-vitro systems and does not conflict with ethical and legal considerations or the animal protection laws. The use of chicken embryos could not replace all currently used toxicological tests with mammals in these fields. They should be used to select from among the broad array of new and accumulated agents those most in need of further testing. As a screening test the HET can diminish tests in mammals and assign materials to categories of toxicity.

Materials and Methods

The fertility and hatchability of eggs and the survival of chickens are dependent on a complex interrelationship of ecological factors, among which are the genetic background and the age of the mated birds, the nutritional status and general management of the flock, and in part on seasonal variations. The White Leghorn has been widely used in experimental studies in the past. The data accumulated during the last three decades in our laboratories show that the hatchability of White Leghorn eggs (Shaver Starcross 288A; Lohmann Selected Leghorn LSL) is consistent, reproducible and very high and that this breed of chicken is free of known hereditary defects. Eggs to be treated are first "candled" in order to discard those that are defective; after the elimination of such defective eggs and only use of fresh, fertile eggs, the hatch of control eggs averages 95%. A further restriction is based on the weight of the eggs: all those weighing less than 50 g or more than 60 g are rejected. The eggs used for HET chorionallantoic membrane testing are put in the incubator trays with the large end up; the trays are placed in the incubator, which automatically rotates and is maintained at an optimum temperature of 37.5 °C (±0.5 °C) and a relative humidity of 62.5% (±7.5%). The eggs are "candled" on the fifth day of incubation and every day thereafter; nonviable embryos were removed. On the 10th day of incubation the egg shell is scratched around the air cell by a rotating dentist-sawblade and then removed. After careful removal of the inner egg membranes the vascular chorionallantoic membrane is layed open (Fig. 1a–d); this is a modified technique to that described by Kemper [2]. The test substance, dissolved or suspended, is dropped on to the membrane in a volume of 0.2 ml; in the case of solid testmaterials 0.1 g is applied on the vascular chorionallantois and irrigated after 20 s with 5 ml of warm water. In every case a series of four eggs is used; two eggs, treated with vehicle only, serve as controls. After the application of the test substance, the chorionallantoic membrane, the blood vessels, including capillary system, and the albumen are examined and scored

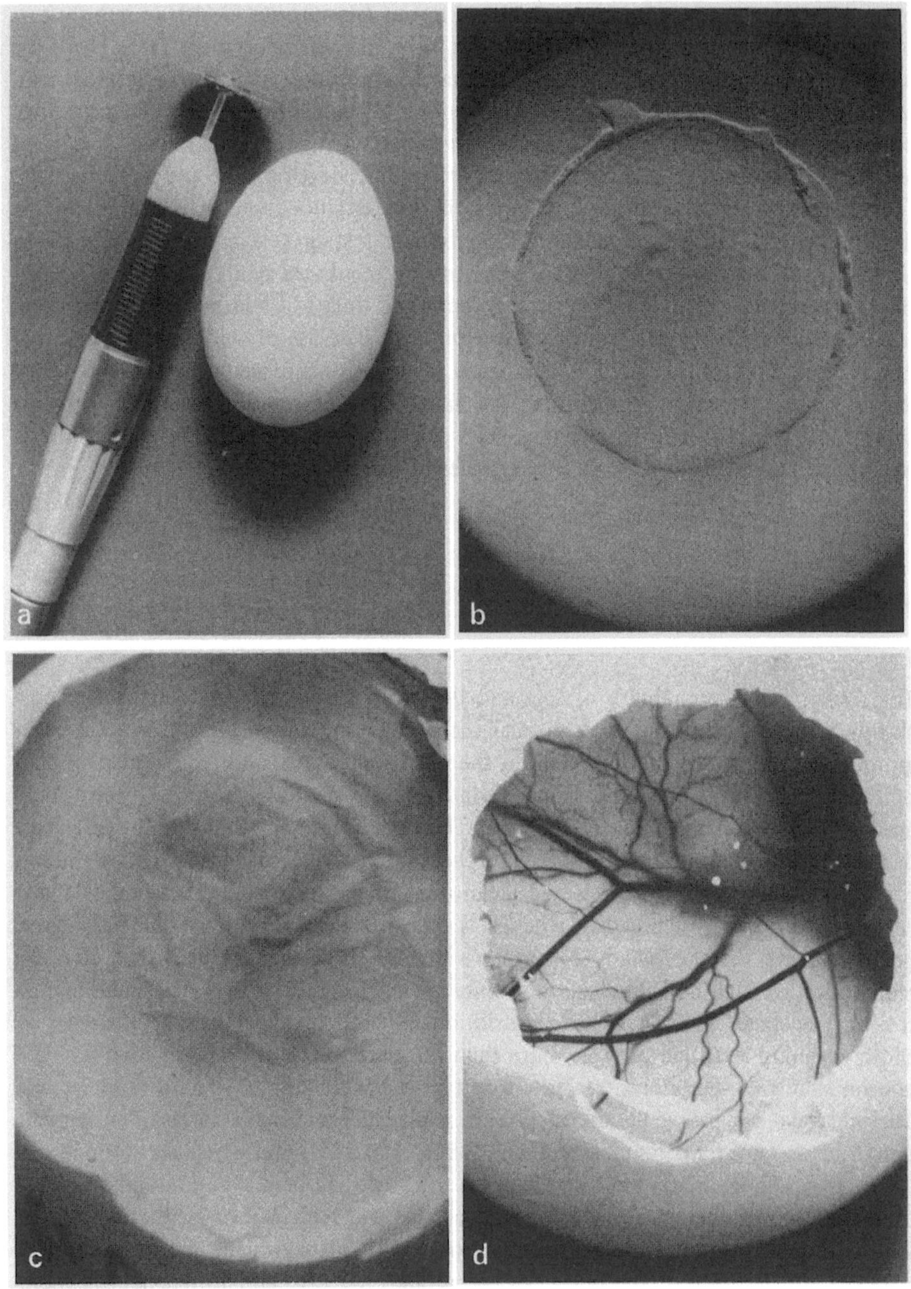

Fig. 1a–d. a egg and dentist-sawblade, b scratched egg, c inner egg membrane, d vascular chorionallantois

for irritant effects (vascular injection, haemorrhages, coagulation) at 0.5, 2 and 5 min after treatment; longer observation times give no additional important information, but need further incubation and a humid chamber. The numerical time dependent scores for vascular injection, haemorrhage and coagulation are summed to give a single numerical value indicating the irritating potential of the test substance on a scale with a maximum value of 21 (Table 2). The mean value of 4 tests makes possible an assessment by a classification scheme analogous to the Draize categories (Table 2). Among others four groups of test materials were selected in order of chemical structure and comparatively tested under the above mentioned conditions of HET chorionallantoic membrane testing (Table 3):

a) vehicles
b) antimicrobial agents
c) oxidation dyes
d) commercial shampoos.

Results

An overview of HET assessments for mucous membrane irritation is given in Table 4. The Figs. 2a–d show an example of a progressive HET vascular chorionallantoic membrane examination. In this case the application of 0.2 ml Na-pyrithione solution (20% in Aqua dest.) on the chorionallantois results in vascular injection and haemor-

Table 2. HET – score and classification scheme

HET – mucous membrane irritation

Score for scaling

Effect / time (min)	0.5	2	5
Injection	5	3	1
Haemorrhage	7	5	3
Coagulation	9	7	5

Classification scheme

Cumulative score (mean of 4 tests)	Assessment
–0.9	practically no irritant
1–4.9	slight irritant
5–8.9	moderate irritant
9–	strong irritant

Table 3. List of test materials

HET – mucous membrane irritation List of test substances	
Vehicles	*Oxidation dyes* (Colipa No)
Aqua dest.	A 5
Propyleneglycole	A 6
NaCl-solution (0.9%)	A 11
	A 14
Antimicrobial agents	A 16
Zn-pyrithione	A 25
Na-pyrithione	A 39
Pyrithionedisulfide	A 42
Pyrithionedisulfide x $MgSO_4$	A 84
Oxypyrion	B 25
Piroctone	B 26
Climbazol	
Isothiazolinones (techn.)	*"Shampoos"*
Isothiazolinones (cosmet.)	I–VIII

Table 4. Overview of HET-assessments

HET – mucous membrane irritation

Overview of results

Substance	Concentration (%)	HET-assessment
Aqua dest.	pure	no irritant
Propyleneglycole	5.0	no irritant
NaCl-solution	0.9	no irritant
Zn-pyrithione	50	strong
Na-pyrithione	20	strong
	1	no irritant
Pyrithionedisulfide	pure	strong
Pyrithionedisulfide x $MgSO_4$	pure	strong
	1	no irritant
Piroctone	0.1	slight
Climbazol	0.5	moderate
Isothiazolinones	15	strong
	1.5	strong
	0.03	no irritant
A 5	1	slight
A 6	1	slight
A 11	0.5	no irritant
A 14	1.5	slight
A 16	0.5	no irritant
A 25	0.5	no irritant
A 39	0.5	no irritant
A 42	2	no irritant
A 84	1	no irritant
B 25	1	no irritant
B 26	1	no irritant

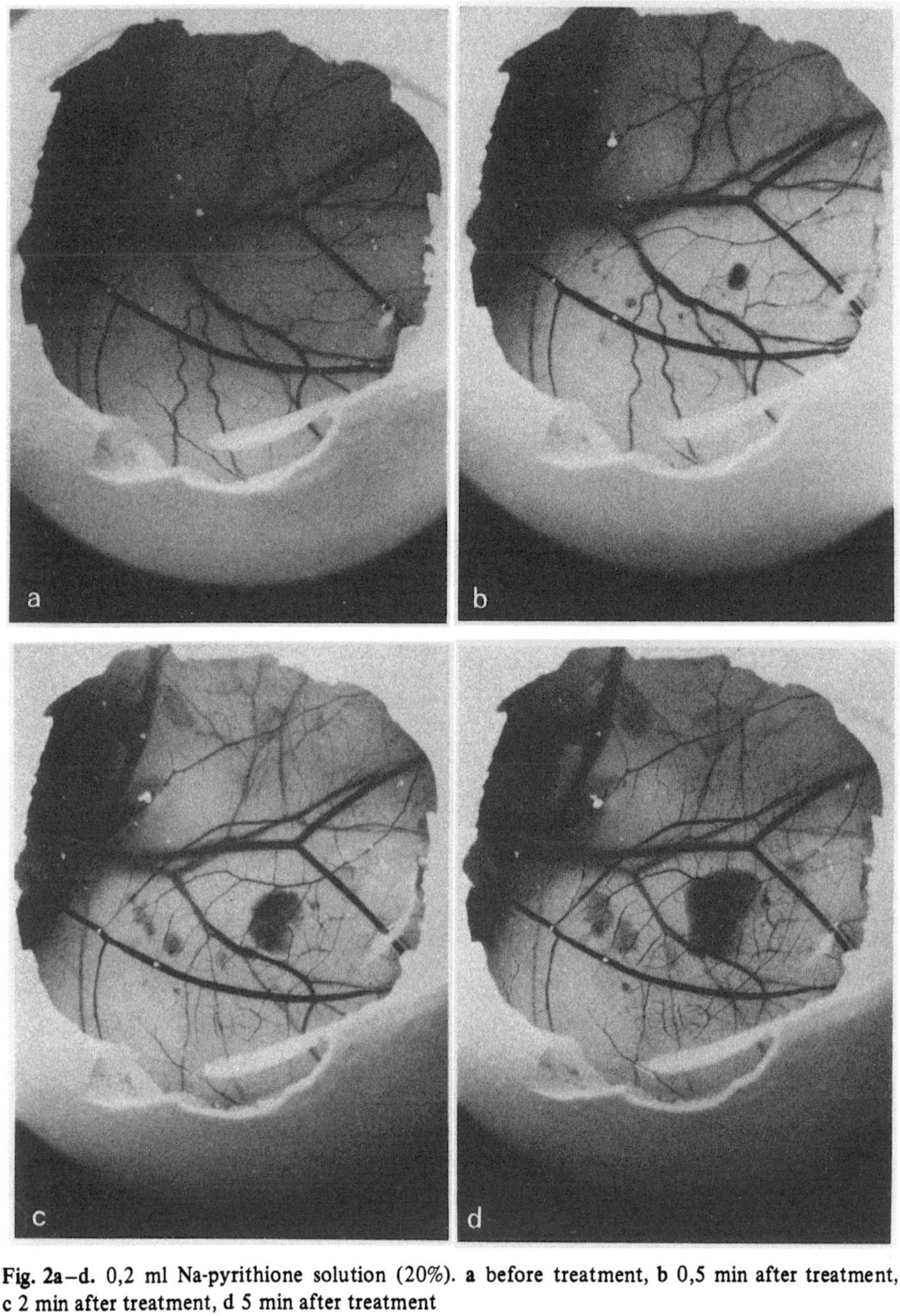

Fig. 2a–d. 0,2 ml Na-pyrithione solution (20%). **a** before treatment, **b** 0,5 min after treatment, **c** 2 min after treatment, **d** 5 min after treatment

rhages without coagulation; these effects can be detected at the first reading point (0.5 min), giving a cumulative score of 12 and, considering the parallel test results, to the HET assessment of "strong irritant". The testing of commercial shampoos under HET-conditions results in slight or less moderate effects, generally in the form of vascular injections (Table 5).

In order to assess the value of both, HET chorionallantois testing and in-vivo eye irritancy testing, to indicate potential hazards it seems useful to compare the assessments of various substances by HET and results reported in literature based on Draize tests.

Table 5. HET-assessments of commercial shampoos

HET – mucous membrane irritation

Shampoos

Shampoo No	HET-score	HET-assessment
I	4.5	slight
II	5	moderate
III*	5	moderate
IV	6.5	moderate
V	5	moderate
VI	5	moderate
VII**	3.5	slight
VIII	5	moderate

* sold as "baby-mild"
** sold as "mild"

Table 6. HET-Draize-comparison: pyrithiones

HET - mucous membrane irritation

Pyrithiones

Substance	Concentration %	HET		Literature assessment
		Score	Assessment	
Zn-pyrithione	50	14	strong	strong
Na-pyrithione	20	12	strong	
	1	0.5	no irritant	no irritant
Pyrithionedisulfide	pure	11	strong	strong
Pyrithionedisulfide x $MgSO_4$	pure	10	strong	strong
	1	0.25	no irritant	no irritant

Table 7. HET-Draize-comparison: phenols

HET – mucous membrane irritation

Phenols

Substance	Concentration %	HET Score	HET Assessment	Literature assessment
Resorcinol	10	14	strong	
	5	8	moderate	
	2.5	4	slight	slight
	0.5	0.5	no irritant	
o-aminophenol	1.5	4	slight	
	1			slight
	0.3	0.75	no irritant	

Table 8. HET-Draize-comparison: isothiazolinones

HET – mucous membrane irritation

Isothiazolinones*

Concentration % a. i.	HET Score	HET Assessment	Literature assessment
(1) 15	20	strong	strong
(2) 1.5	19	strong	strong
(2) 0.3	8	moderate	moderate
(2) 0.075	4	slight	
(2) 0.03	0.5	no irritant	no irritant

* (1) Kathon 886; (2) Kathon CG

Such a corresponding comparison regarding the pyrithiones is given in Table 6 and shows a good correlation between assessments by HET and the reported data [8, 11] based on Draize rabbit eye tests.

This is also valid for the comparisons of the phenols resorcinol and o-aminophenol (Table 7); literature data reported by [3], [12] and isothiazolinones (Table 8). Both chemical classes, especially resorcinol and isothiazolinones, show in higher concentrations under HET-conditions vascular injections, haemorrhages and coagulation. These corrosive effects are comparable to reported results based on in-vivo eye tests.

Discussion

The chicken embryo has been used to assess toxicity and the damaging effects of a wide variety of chemical and physical agents for more than a century by numerous investigators (at present we have found more than 1,000 publications regarding at least 900 chemicals. In these investigations there have been many variations with respect to time and route of application, length of development and method of examination and analysis. We recommend now an extended and standardized chicken embryo model in the form of HET embryotoxicity testing and in the special form of HET vascular chorionallantoic membrane testing. The results described here and other investigations [4, 13, 14] correlate to findings in mammalian systems and indicate that the hen's egg test is capable of demonstrating teratogenic, embryotoxic, systemic and immunopathological potential as well as the mucous membrane irritating potencies of chemical substances and that it is selective and does not respond nonspecifically to any agent tested up to now under these conditions. For this reason the HET is useful for screening large numbers of compounds, e.g., for selecting those that possibly need to be subjected to further specific evaluations.

In the special field of mucous membrane irritation testing a specific score and classification scheme was developed for HET, which allows risk assessments analogous to the Draize scheme. Both the Draize rabbit eye test and HET chorionallantoic membrane test are dependent on the personal interpretation of tissue reactions; numerical scores are "sufficient" for legal and administrative measurements, but generally noninformative for toxicologists. Therefore it is recommended to use in parallel a standard of test substances and, if possible, to give a supplementary description of the classification used.

Conclusions

HET (Hen's Egg Test, Hühner-Embryonen-Test) is a rapid, sensitive and inexpensive toxicity test and gives information too on the embryotoxicity, teratogenicity, systemic and immunpathological effects, metabolic pathways and mucous membrane irritation potencies of chemical substances.

Testing incubated hen's eggs is borderline between in-vivo and in-vitro systems and does not conflict with ethical and legal issues or the animal protection laws.

The results of both, HET embryotoxicity testing and HET chorionallantoic membrane testing, correlate well with findings in mammalian systems.

HET embryotoxicity testing as well as HET mucous membrane irritation testing cannot completely replace currently used toxicological tests with mammals.

References

1. Balls M, Riddell RJ, Worden AN (1983) Animals and alternatives in toxicity testing. Academic Press, London New York
2. Kemper FH (1958) Studien zur Wirkung von Chlorophyll. Fette Seifen Anstrichmittel 60: 830–832
3. Kemper FH, Luepke NP (1983) Toxikologische Charakterisierung von aromatischen Amino-, Hydroxy- und Nitroverbindungen, 3. Aufl. Eigenverlag, Münster, S 98–114
4. Kemper FH, Luepke NP (1983) Pthalsäuredialkylester – pharmakologische und toxikologische Aspekte. VKE-Schriftenreihe, Frankfurt
5. Kemper FH, Luepke NP (1984) General aspects of monitoring and banking of human biological specimens. In: Lewis RA, Stein N, Lewis CW (eds) Environmental specimen banking and monitoring as related to banking. M Nijhoff, Den Haag, pp 67–73
6. Kemper FH, Luepke NP, Renhof M, Weiss U (1982) Fatty acid anilides and the toxic oil syndrome. Lancet I:98–99
7. Luepke NP (1979) Monitoring environmental materials and specimen banking. M Nijhoff, Den Haag, pp 403–409
8. Luepke NP (1979) Wirkstoffe in Antischuppen-Kosmetika. Ärztl Kosmetol 9:174–180
9. Luepke NP (1982) Embryotoxicity-testing of cosmetic ingredients by HET. Preprints 12th Int Congr IFSCC Paris, II:231–248
10. Luepke NP, Kemper FH (1983) Embryotoxicity-testing of aromatic amino-hydroxy- and nitrocompounds by HET. In: Loprieno N (ed) Problemi di tossicologia dei prodotti cosmetici. ETS, Pisa, pp 77–93
11. Luepke NP, Preusser P (1978) Antischuppen-Kosmetika-Wirkung und Toxikologie. Ärztl Kosmetol 8:269–280
12. Luepke NP, Preusser P (1970) Resorcin-Wirkstoff in kosmetischen Mitteln? Ärztl Kosmetol 9:341–346
13. Verret JJ, Scott WF, Reynaldo EF, Altermann EK, Thomas CA (1980) Toxicity and teratogenicity of food additive chemicals in the developing chicken embryo. Tox Appl Pharmacol 56?265–273
14. Wilson JG, Fraser FC (1978) Handbook of teratology. Plenum Press, New York London, IV:147–151

Use of Uninvolved Psoriatic Epidermis as an In Vitro Model for Testing the Anti-Phospholipase Activity of Glucocorticoids

A. Ilchyshyn, E. Ilderton, J. Kingsbury, J. F. B. Norris, R. Summerly and H. J. Yardley

Anti-inflammatory steroids are thought to rely for their pharmacological action on their ability to cause cells to synthesize and/or release anti-phospholipase peptides [1, 2]. We have recently shown [3] that steroid creams applied to the symptomless (uninvolved) epidermis of psoriatic patients cause the high levels of phospholipase A_2 activity present in this tissue [4] to be reduced to normal values. This method of assessing pharmacological activity suffers from the fact that many biopsies are required to establish the time course of the action of the steroid. We have, therefore, developed an in vitro method in which epidermal phospholipase A_2 activity is monitored during culture of a skin slice.

Methods

Skin was removed from lesion-free areas of psoriatic patients using a Davies dermatome (C. R. Thackray, Park Street, Leeds, U.K.). Skin was not taken within 5 cm of any lesion and was macroscopically normal. The skin was rinsed in cold saline and cut into 4 mm x 4 mm squares.

The culture medium was Gibco RPMI-1640 containing 25 mM HEPES and 2 mM L-glutamine. Steroids were dissolved in DMSO and 10 μl portions added to 0.54 ml medium; controls contained DMSO alone. The squares of skin were floated on the medium and incubated at 37 °C. After incubation, each square was immersed in 2M potassium bromide for 30 min at 37 °C, rinsed and the epidermis removed with fine forceps [5]. The epidermis was homogenized in an all-glass tissue grinder in 200 μl 0.25 M sucrose. The homogenate was centrifuged at 150 x g for 3 min, the pellet rehomogenized in a further 200 μl sucrose, centrifuged again and the combined supernatants centrifuged at 15,000 x g for 10 min. This final supernatant was used as an enzyme source.

Phospholipase A_2 was determined using a procedure based on that of Freinkel and Traczyk [6]. Each incubation mixture contained in 120 μl, epidermal supernatant (10–20 μg protein), 1-palmitoyl-2-[1-^{14}C]oleoyl-sn-glycero-3-phosphocholine (0.02 μCi), egg phosphatidylcholine (45 μg), ^{3}H-oleic acid (0.002 μCi, 0.4 μg), sodium taurodeoxycholate (675 μg), Ca^{++} (1 μmole) and glycine/NaOH buffer pH 8.5 (10 μmole). After a 2 h incubation, the radioactivity in the fatty acids was measured by liquid scintillation counting after separation by thin layer chromatography. Inclusion of ^{3}H oleic acid in the incubation medium enabled the recovery of fatty acid to be calculated. Protein was estimated by the Lowry method.

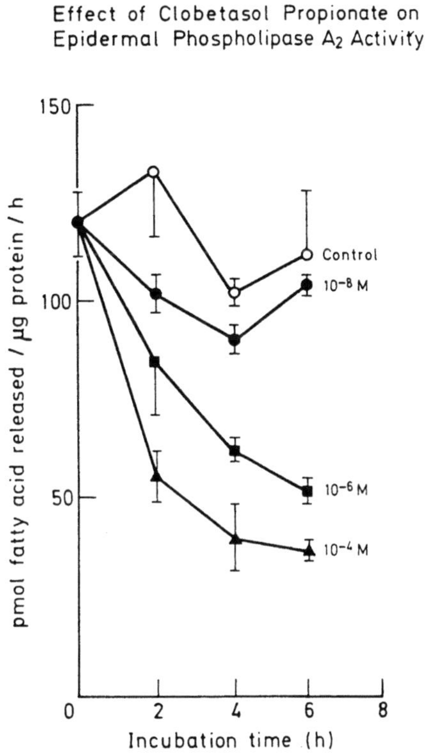

Fig. 1. Skin slices from an uninvolved area of a patient suffering from psoriasis were incubatet in media containing the stated concentrations of clobetasol propionate. At the indicated times, incubation was stopped and the epidermal phospholipase A_2 activity determined. Each point represents the mean of three determinations on one piece of epidermis, and each error bar represents the standard error of the mean

Results

Epidermal phospholipase A_2 activity in normal volunteers ranges from 34–68 (mean ± SD) pmol fatty acid released per μg protein per h (Forster, Ilderton, Norris, Summerly and Yardley, unpublished data). Figure 1 shows that a specimen of symptomless psoriatic epidermis with a phospholipase A_2 activity of 120 units had this activity reduced to the normal range following culture in a medium containing 10^{-5} M clobetasol propionate for two hours. At a concentration of 10^{-6} M, the rate of reduction of activity was slower, and the normal range of activity was not reached until after four hours of culture. At 10^{-8} M, the range of normal values was not entered and, indeed, the phospholipase A_2 activity of the steroid treated epidermis did not differ significantly from the controls cultured in the absence of steroid ($P > 0.1$, Student's t-test).

Figure 2 shows the different responses to culture in 10^{-6} M clobetasol propionate shown by uninvolved psoriatic epidermis having very high (patient A) and high-normal activities (patient B). In the case of patient A, four hours in culture reduced the enzymic activity by a factor of nearly two, thus bringing the activity into the normal range. Epidermis from patient B, however, started with an enzyme activity already within the normal range and treatment with steroid produced no significant effect. Figure 3 shows that clobetasol propionate at a concentration of 10^{-5} M is very much more effecitve in reducing the phospholipase A_2 activity of uninvolved psoriatic epidermis compared with hydrocortisone hemisuccinate at the same concentration.

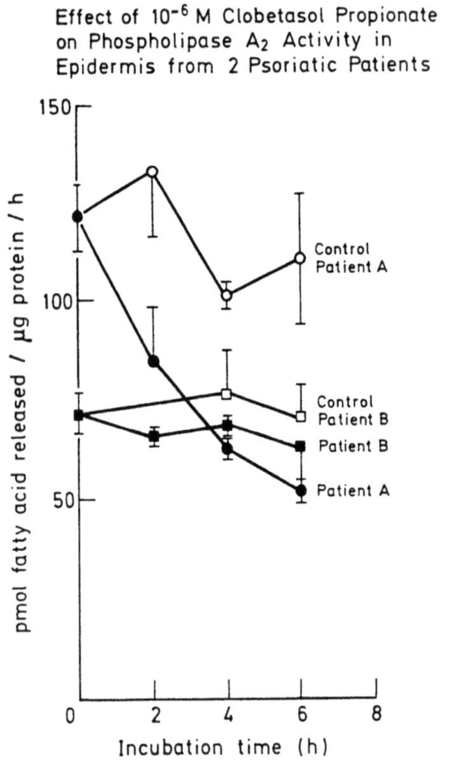

Fig. 2. Skin was taken from the uninvolved areas of two psoriatic patients. Patient A had a very high level of epidermal phospholipase A_2 activity. The phospholipase A_2 activity in the epidermis of pateint B was only 1.2 standard deviations above the mean value found in normal epidermis. All skin samples were incubated in the presence of 10^{-6} M clobetasol propionate, and the epidermal phospholipase A_2 activity determined at the indicated times. Other details are as given in Fig. 1

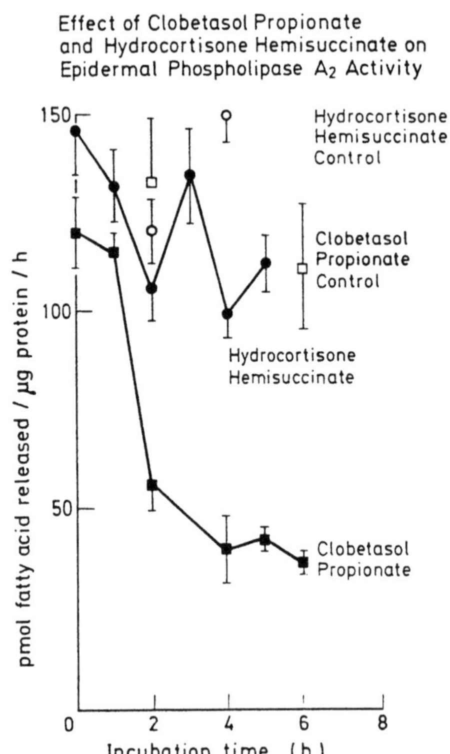

Fig. 3. Skin was taken from the uninvolved areas of two psoriatic patients, both of whom had very high levels of epidermal phospholipase A_2 activity. The skin of one patient was incubated in the presence of hydrocortisone hemisuccinate, while that of the other was incubated in the presence of clobetasol propionate. Epidermal phospholipase A_2 activity was determined at the indicated times. Other details are as given in Fig. 1

Discussion

We have demonstrated inhibition of phospholipase A_2 activity in epidermis cultured in the presence of anti-inflammatory steroids: the culture medium used contained the full complement of amino acids in the expectation that enzyme inhibition would be preceded by steroid induced synthesis of an inhibitory protein.

We have already shown [3] that inhibition of phospholipase A_2 activity caused by the application of steroid creams is more marked in subjects with a high initial enzyme activity; the same phenomenon has now been demonstrated in an in vitro system.

There have been many assay systems described for evaluating topical steroids (see reference 3 for a bibliography). The method described in this communication has the merit – and the drawback – of requiring the use of pathological skin.

The method could easily be extended to investigate the effect of longer incubation times; this would enable any differences in the length of action of the steroids to be investigated.

Conclusion

The method described may find application as an objective in vitro method for assessing the potency of anti-inflammatory steroids. In this method, anti-inflammatory steroids reduce epidermal phospholipase A_2 activity into the normal range. It is therefore necessary to choose epidermis likely to exhibit a high phospholipase A_2 activity, the paradigm being symptomless (i.e. uninvolved) psoriatic epidermis.

Acknowledgement

We gratefully acknowledge the gift of steroids from Glaxo Research Ltd.

Abbreviations

HEPES = N-2-Hydroxyethylpiperazine-N′-2-ethanesulphonic acid; DMSO = Dimethyl sulfoxide

References

1. Blackwell, GJ, Carnuccio R, DiRosa M, Flower RJ, Ivanyi J, Langham CSJ, Parente L, Persico P, Wood J (1983) Suppression of arachidonate oxidation by glucocorticoid-induced antiphospholipase peptides. Samuelsson B, Paoletti R, Ramwell P (eds) Advances in prostaglandin, thromboxane and leukotriene research. Raven Press, New York, 11:65–71
2. Blackwell GJ, Flower RJ (1983) Inhibition of phospholipase. Br Med Bul 39:260–264
3. Norris JFB, Ilderton E, Yardley HJ, Summerly R, Forster S (1984) Ultilisation of epidermal phospholipase A_2 inhibition to monitor topical steroid action. Br J Dermatol 111, Suppl 27: 195–203
4. Forster S, Ilderton E, Summerly R, Yardley HJ (1983) Epidermal phospholipase A_2 activity is raised in the uninvolved skin of psoriasis. Br J Dermatol 109, Suppl 25:30–35
5. Levine N, Hatcher VB, Lazarus G (1976) Proteinases of human epidermis: a possible mechanism for polymorphonuclear leukocyte chemotaxis. Biochim Biophys Acta 452:458–467
6. Freinkel RK, Traczyk TN (1980) The phospholipases A of epidermis. J Invest Dermatol 74: 169–173

Skin Fibroblast Monolayers: A Suitable Case for Treatment?

G. C. Priestley

For over 70 years monolayer cultures of fibroblasts have seemed to offer an ideal system in which the investigator can control all the variables so that different treatments can be compared in identical sister cultures of physiologically normal cells. That ideal is seldom achieved and there are obvious contrasts between a crowded two-dimensional array of rapidly proliferating fibroblasts and the apparently sparsely-populated living dermis. Nonetheless, fibroblast cultures have yielded much useful data on the workings of normal and abnormal cells and on the effects of countless different treatments. When cultures and living skin can be compared directly, the results are encouraging: many characteristics of abnormal cells persist in vitro (e.g. specific enzyme deficiencies in mucopolysaccharidoses, Ehlers-Danlos type VI and dermatosparaxis) and the multistage synthesis of mesenchymal macromolecules proceeds in vitro, in the case of collagen with the same requirements for ascorbic acid and ferrous iron as in the intact tissue. By using fibroblasts we exclude the agency of other systems, like blood vessels and nerves, and the influences of other cell types. The promise and the reality of this technique will be illustrated with the results obtained in our laboratory over the last eight years.

I prefer to avoid the word model, but the fibroblast is a possible target, if not the intended one, for many compounds applied to the skin or taken orally, and we should be aware of its reactions to such treatments. Although it produces and maintains the structural elements of the dermis the fibroblast is only one component of the skin. To use a crude analogy, an architect studying the strength of a building would not base his opinion on the strength of the individual bricks, but he would be unwise to ignore any weakness apparent in them.

The promise of cell culture for dermatology is well illustrated by Berliner's suggestion [3] that inhibition of fibroblast proliferation by synthetic corticosteroids is a useful guide to their relative anti-inflammatory activity. Some impressive results were achieved: log-log plots of cell numbers against corticosteroid concentration gave linear dose-responses and precise estimates of relative potency which agreed with other forms of assessment. But the crucial feature was the use of L-929 fibroblasts, a long-established strain of mouse cells. In several laboratories around the world human skin fibroblasts gave different and less predictable results. The same compounds had little effect, or even stimulated the growth of human cells (Fig. 1), and when inhibition was apparent this required far higher concentrations of steroid than for the L cells [10, 12]. The estimated difference in sensitivity between the mouse and human cells was a factor of 10^3 or 10^4 [11]. The non-linear dose-responses often recorded also showed the greater "noise" inherent in the human system, with more heterogeneity and the effects of in vitro ageing apparent *within* cell lines, and variables such as donor age, body site and

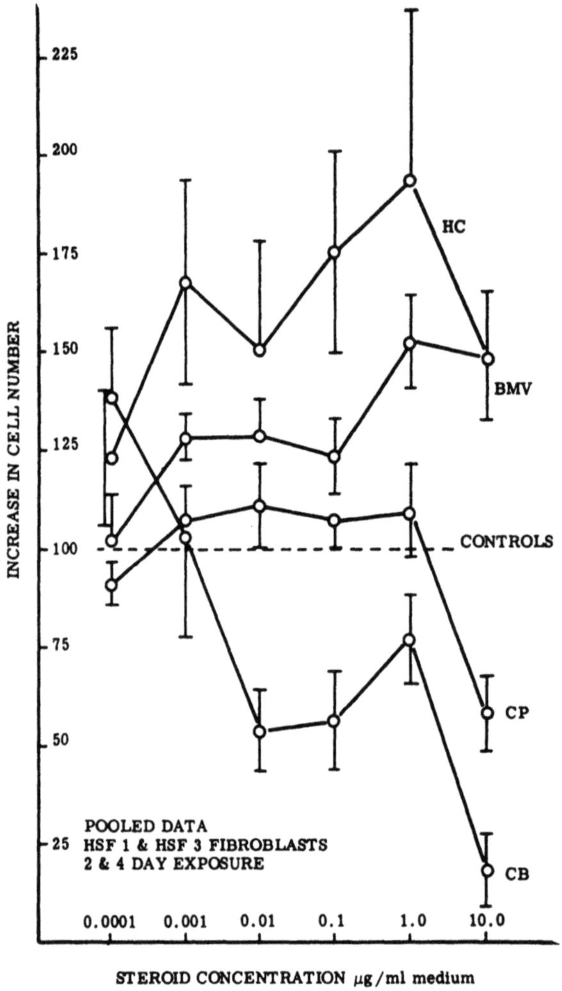

Fig. 1. Effects of 4 corticosteroids on proliferation of two lines of infant foreskin fibroblasts. Values are shown with standard errors (n = 8), for other details see reference 12. HC = hydrocortisone; BMV = betamethasone valerate; CP = clobetasol propionate; CB = clobetasone butyrate

individual genotype all contributing to differences *between* fibroblast lines. Yet it is still possible to say that corticosteroids are generally inhibitory to fibroblast proliferation and metabolism in vitro, as indeed they are in vivo, and to see a broad parallel between clinical performance and potency in vitro for a whole range of compounds from hydrocortisone to clobetasol propionate [8] (Fig. 2).

The experience with corticosteroids demonstrates the need to use human cells, and this is surely the supreme advantage of cell culture in medicine. The ability to use cells from the skin of an individual suffering from a particular disease is also of paramount importance: we cannot assume that cells from normal and abnormal skin have identical properties and reactions to treatment.

We useful fibroblasts from the skin of patients with systemic sclerosis (scleroderma) to study the effect of D-penicillamine on collagen synthesis (as incorporation of ^{3}H-proline into collagenase-sensitive protein) [3]. Starting at 50 μg/ml we gradually increased the concentration of penicillamine in successive experiments, until at 1,600

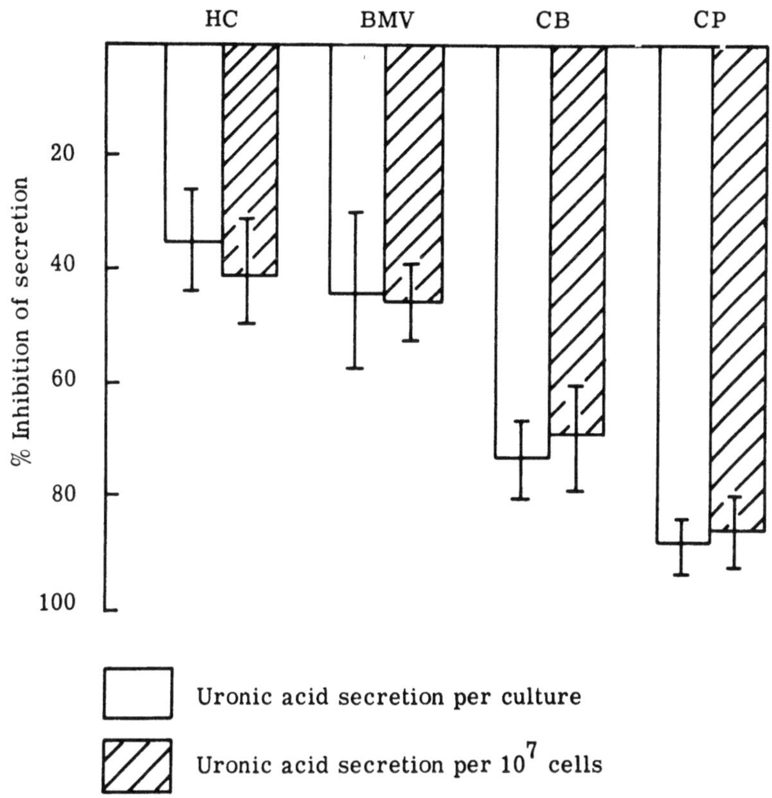

Fig. 2. Inhibition of glycosaminoglycan secretion by scleroderma skin fibroblasts with four corticosteroids at 10 μg/ml. Open columns = uronic acid secretion per culture; hatched columns = secretion per 10^7 cells. Abbreviations as in Fig. 1 [18]

μg/ml, which is about one hundred times the concentration likely to be achieved in the plasma of patients, there was a modest (23%) inhibition. This essentially negative result disagrees with some of the data from animal tissues and with the experience of Herbert and her colleagues [6] who used organ cultures of human skin, but seems more in keeping with clinical experience. In systemic sclerosis only the skin collagen is affected by penicillamine treatment and the target seems to be the aldimine crosslinks, and their consolidation into more permanent bonds, rather than a general effect on polypeptide assembly.

Another proposed treatment for systemic sclerosis is with para-aminobenzoate, again with collagen synthesis as the suggested target [23]. No effect on collagen synthesis by fibroblasts from scleroderma skin [17] was detected even at very high drug concentrations (Fig. 3). However secretion of glycosaminoglycans was diminished at all concentrations, including those attainable in patients. A course of treatment might therefore lead to a lower content of glycosaminoglycans and a decreased water binding in the skin, explaining the greater flexibility and softening of the skin and joints recorded by Zarafonetis [23].

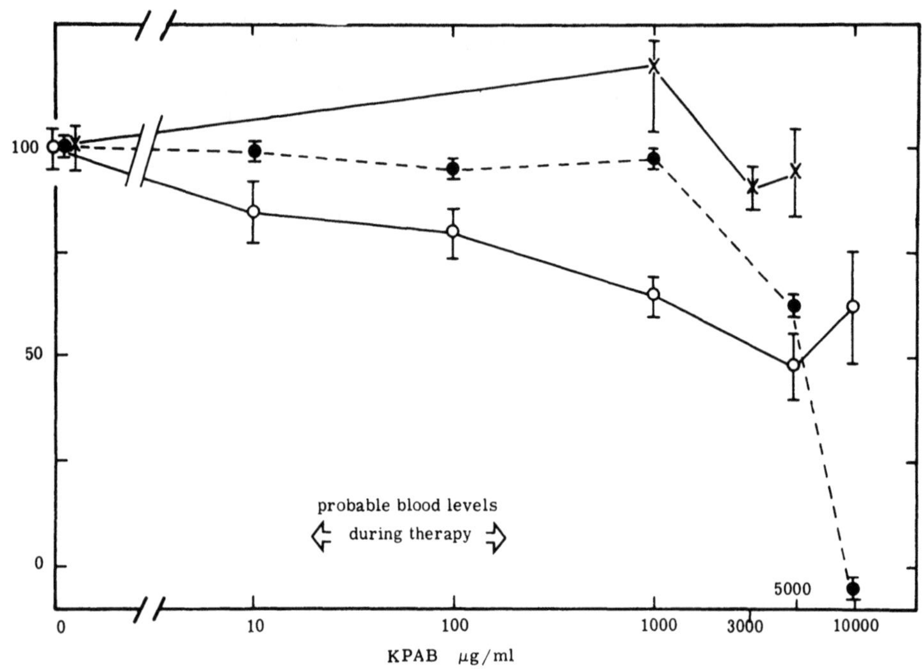

Fig. 3. Effects of potassium p-aminobenzoate on scleroderma skin fibroblasts and rheumatoid synovial cells. Solid circles = cell proliferation; open circles = glycosaminoglycans secretion; crosses = collagen synthesis [17]

An important question in such experiments is the health of the cells. Non-specific toxic effects are of little interest and a prime advantage of cell culture is that the cells can be inspected by phase-contrast microscopy throughout the experiments. In addition, where proliferation is inhibited, removal of the drug may allow it to resume, demonstrating cytostasis rather than cytotoxicity. Such a reversible effect was seen with penicillamine at 1.600 μg/ml [3]. A difference was seen with zinc pyrithione, an anti-microbial compound commonly added to shampoos to combat dandruff. Cells proliferating normally at 0.05 μg/ml were killed within minutes at 0.1 μg/ml [9] (Fig. 4). As thousands of scalps are doused in shampoos containing 1% zinc pyrithione every week without apparent harm, our result might seem to be an abberation of the cell culture system. The explanation must be that very little zinc pyrithione in the shampoos is retained on the scalp, and hence absorbed [22]; injection of the compound, or application in dimethyl sulphoxide has caused paralysis and death in toxicity trials with mice and rabbits [4].

Potent corticosteroids were sometimes cytotoxic, depending on the concentration and length of exposure, and there was a differential effect of clobetasol propionate on fibroblasts grown from psoriatic skin [16]. The fibroblasts from control subjects (NSF in Fig. 5) were only marginally stimulated to proliferate at 0.1 μg/ml compared to fibroblasts from uninvolved psoriatic skin (PSB); while fibroblasts from the patients' involved skin (PSA) were considerably stimulated. At higher steroid levels the cells were progressively inhibited, and at 10 μg/ml the PSA cells were again the most affect-

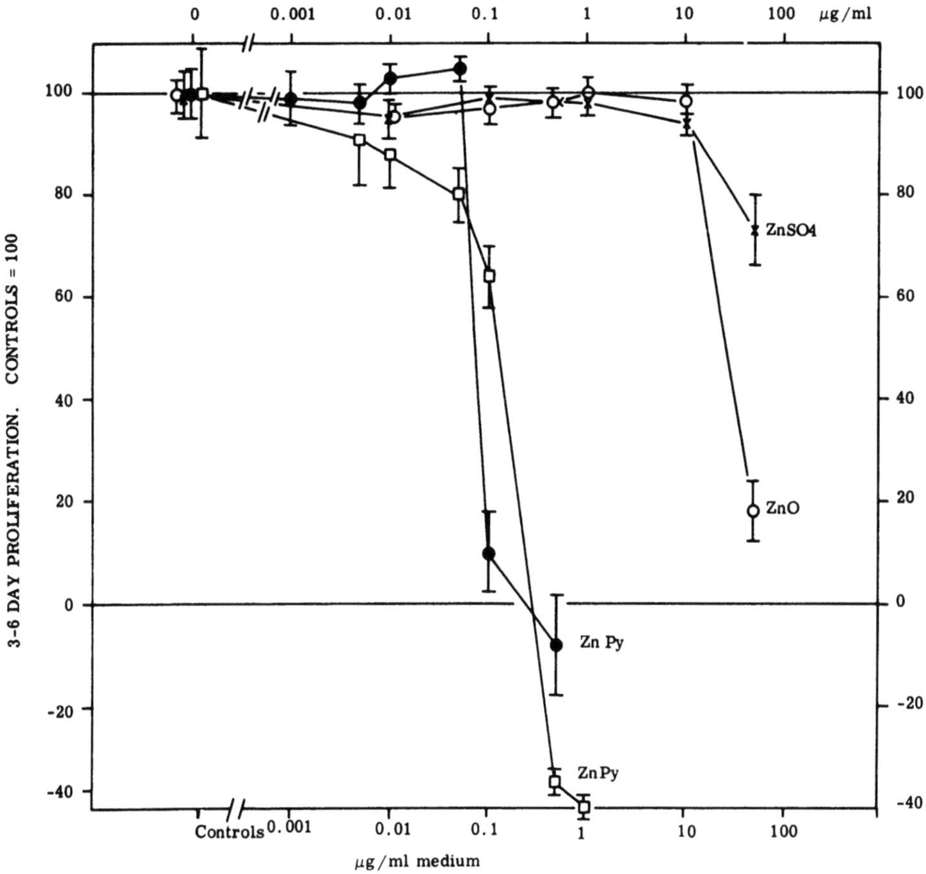

Fig. 4. Effect of zinc compounds on the proliferation of fibroblasts from normal adult human skin (•, ○, x) or NCTC 2544 human epithelial cells (□), ZnPy = zinc pyrithione; $ZnSO_4$ = zinc sulphate; ZnO = zinc oxide [19]

ed and normal fibroblasts the least affected. There was a net loss of the PSA cells and their viability fell to only 50%. Differential effects of hydrocortisone on fibroblasts grown from keloids and from the skin of diabetic patients have also been recorded [20, 21].

Two other features of the psoriatic skin fibroblasts strongly suggest that the hyperproliferation and increased metabolism characteristic of psoriasis are not confined to the epidermis of the lesions. The first evidence of this was the faster proliferation of the psoriatic fibroblasts (recorded as percentage increase in total cells between the third and sixth day of culture). There were large ranges of values in each group [15], but the mean rates were significantly different from the controls ($p < 0.05$); PSA 144 ± 17 (s.e.m.), PSB 134 ± 15, NSF 93 ± 12. These findings complement the data from grafts of psoriatic skin transferred to nude mice [8], where labelling indices in the epidermis were similarly elevated in both the lesional and non-lesional skin of patients compared with grafts from normal subjects. A second finding was that collagen and other protein synthesis was increased equally in both groups of psoriatic fibroblasts [14]. The in-

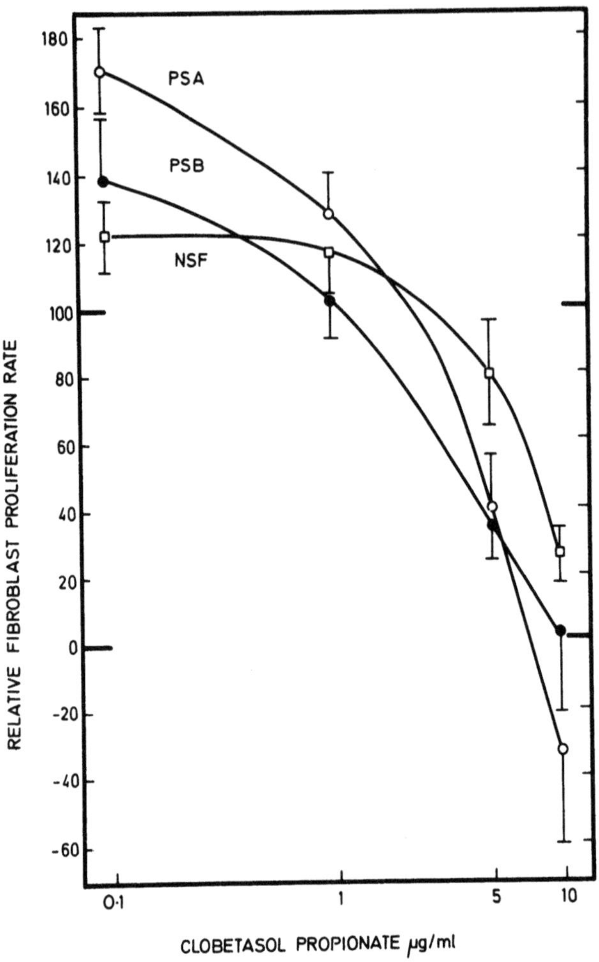

Fig. 5. Differential effect of clobetasol propionate on the proliferation of psoriatic and normal skin fibroblasts. Composite dose-response curves were compiled for 4 cell lines of each type with 4 cultures of each line at each drug concentration. Values shown are means ± s.e.m. The proliferation of untreated culture of each strain is represented as 100; values exceeding 100 show stimulation by the drug, values below 100 indicate inhibition. PSA fibroblasts (○) are from involved psoriatic skin, PSB fibroblasts (●) from uninvolved psoriatic skin and NSD fibroblasts (□) from normal skin [16]

creased levels of procollagen proline hydroxylase in the skin of patients [5] argue against this being an artefact of culture.

Increased synthesis of collagen was also seen with fibroblasts from patients with recessive epidermolysis bullosa (EBDr), and even in cells from epidermolysis bullosa simplex, where no dermal pathology had been suspected [7, 9]. In the recessive disease this may be a response to a greater breakdown of collagen by collagenase. We confirmed the finding of Bauer and Eisen [1] that total collagenase activity was highest in fibroblasts of the EBDr group; but the most significant difference between the dystrophic and simplex fibroblasts was in the amount of free collagenase compared to that

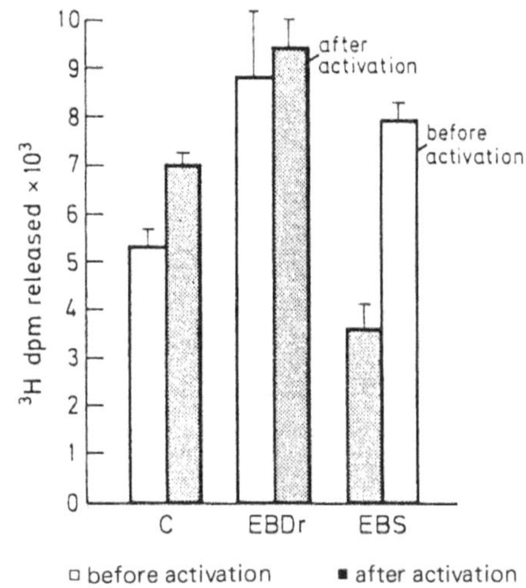

Fig. 6. Collagenase activity released from skin fibroblasts, before or after enzyme activation by Trypsin. C = normal skin fibroblasts; EBDr = dystrophic recessive epidermolysis bullosa; EBS = epidermolysis bullosa simplex. Means from 4 lines are shown (± s.e.m.) [9]

unmasked by prior trypsin treatment (Fig. 6). Most of the enzyme released by simplex fibroblasts required activation, while the trypsin had little effect on the enzyme released by fibroblasts from dystrophic skin [9].

I conclude that despite the attraction of collagen lattices [2], and other refinements to the study of fibroblasts, monolayers will continue to be "a suitable case for treatment". Provided we understand that diploid fibroblast cultures are heterogeneous and that cell lines from several individuals are needed to characterise a particular group or describe the effects of a treatment, then we have a valuable source of information on the cell-specific effects of drugs, and on cellular aspects of human skin disease, which can be pursued with little inconvenience to the patient, or even, in extreme cases, after his death.

References

1. Bauer EA, Eisen AZ (1978) Recessive dystrophic epidermolysis bullosa: evidence for increased collagenase as a genetic characteristic in cell culture. J Exp Med 148:1378–1387
2. Bell E, Scher S, Hull B, Merrill C, Rosen S, Chamson A, Asselineau D, Dubertret L, Coulomb B, Lapiere C, Nusgens B, Neveaux Y (1983) The reconstitution of living skin. J Invest Dermatol 81:2s–10s
3. Berliner DL, Ruhmann AG (1967) Influence of steroids on fibroblasts. I. An *in vitro* fibroblast assay for corticosteroids. J Invest Dermatol 49:117–122
4. Black JG, Howes D (1978) Toxicity of pyrithiones. Clin Toxicol 13:1–26
5. Fleckman PH, Jeffrey JJ, Eisen AZ (1983) A sensitive microassay for prolyl hydroxylase activity in normal and psoriatic skin. J Invest Dermatol 60:46–52
6. Herbert CM, Lindberg KA, Jayson MIV, Vailey AJ (1974) Biosynthesis and maturation of skin collagen in scleroderma and effect of D-penicillamine. Lancet I:187–192
7. Kero M, Palotie A, Peltonen L (1984) Collagen metabolism in two rare forms of epidermolysis bullosa. Br J Dermatol 110:177–184

8. Kreuger GG (1981) Psoriasis: Current concepts of its aetiology and pathogenesis. In: Dobson RL, Thiers BH (eds) 1981 Yearbook of Dermatology. Yearbook Medical Publishers, Chigaco, p 23
9. Oakley CA, Priestley GV (1985) Synthesis and degradation of collagen by epidermolysis bullosa fibroblasts. Acta Dermatovenereol (Stockh) 65:277–281
10. Ponec M (1984) effects of glucocorticoids on cultured skin fibroblasts and keratinocytes. Int J Dermatol 23:11–24
11. Ponec M, Haas de C, Bachra BN, Polano MK (1977) Effects of glucocorticoids on primary human skin fibroblasts. I. Inhibiton of the proliferation of cultured primary human skin and mouse L929 fibroblasts. Arch Dermatol Res 259:117–123
12. Priestley GC (1978) Effects of corticosteroids on the growth and metabolism of fibroblasts cultured from human skin. Br J Deermatol 99:253–261
13. Priestley GC (1980) Changes in the growth and metabolism of cells cultured from normal, sclerotic and rheumatoid connective tissue brought about by D-penicillamine and by sodium salicylate. J Invest Dermatol 74:413–417
14. Priestley GC (1983) Hyperactivity of fibroblasts cultured from psoriatic skin. II. Synthesis of macromolecules. Brit J Dermatol 109:157–164
15. Priestley GC, Adams LW (1983) Hyperactivity of fibroblasts cultured from psoriatic skin. I. Hyperproliferation and effect of serum withdrawal. Brit J Dermatol 109:149–156
16. Priestley GC, Adams LW, Oakley CA, Prescott RJ (1983) Proliferative responses of fibroblasts from psoriatic and normal skin to clobetasol propionate. Acta Dermatovenereol (Stockh) 63:393–396
17. Priestley GC, Brown CJ (1979) Effects of potassium paraaminobenzoate on growth and macromolecule synthesis in fibroblasts cultured from normal and sclerodermatous human skin and rheumatoid synovial cells. J Invest Dermatol 72:161–164
18. Priestley GC, Brown CJ (1980) Effects of corticosteroids on the proliferation of normal and abnormal human connective tissue cells. Br J Dermatol 102:35–41
19. Priestley GC, Brown CJ (1980) Acute toxicity of zinc pyrithione to human skin cells *in vitro.* Acta Dermatovenereol (Stock) 60:145–148
20. Rowe DW, Starman BJ, Fujimoto WY, Williams RH (1977) Abnormalities in proliferation and protein synthesis in skin fibroblast cultures from patients with diabetes mellitus. Diabetes 26: 284–290
21. Russell JD, Russell SB, Trupin TK (1978) Differential effects of hydrocortisone on both growth and collagen metabolism of human fibroblasts from normal and keloid tissue. J Cell Physiol 97:221–230
22. Rutherford T, Black JG (1969) The use of autoradiography to study the location of germicides in skin. Br J Dermatol 81, Suppl 4:75–87
23. Zarafonetis CJS (1959) Treatment of scleroderma. Ann Int Med 50:343–365

Model for Biochemical Studies on the Mechanisms Underlying the Regulation of Cholesterol Synthesis

M. Ponec, J. Kempenaar, L. Havekes and B. J. Vermeer

Cultured cells have been shown to be a good model for investigation of the mechanism underlying the regulation of cholesterol synthesis [3, 9]. Such studies have shown that in most extrahepatic cells, intracellular cholesterol synthesis is regulated by the presence of the low density lipoprotein (LDL), which carries most of the cholesterol in human plasma. This LDL is taken up by the cells by a process called receptor-mediated endocytosis. The LDL binds to the cell surface receptor (LDL receptor) in regions of the plasma membrane called coated pits, and the LDL-receptor complex enters the cells by absorptive endocytosis. During this process the coated pits invaginate, forming endocytic vesicles that carry LDL to lysosomes. There, the LDL is hydrolysed which liberates amino acids from the protein component of LDL and the free cholesterol from the lipoprotein cholesteryl esters.

The free cholesterol then regulates the activity of two microsomal enzymes:

1) It suppresses the activity of 3-hydroxy-3-methylglutaryl coenzyme A (HMGCoA) reductase which is involved in the rate-limiting step of the cholesterol biosynthesis;
2) It activates an acyl CoA : cholesteryl acyltransferase, which reesterifies the excess cholesterol for storage as cholesteryl ester droplets; and furthermore
3) It suppresses the production of LDL receptors and thus prevents intracellular overaccumulation of cholesterol introduced via the receptor pathway.

For the study of receptor-mediated uptake of LDL either biochemical or morphological methods can be utilized. In this paper the biochemical approach will be discussed and the accompaning paper (page 315) will deal with the morphological approach.

In biochemical studies the LDL-induced regulation of cholesterol synthesis can be followed by, for instance, assessing the de novo cholesterol synthesis, using the incorporation of ^{14}C-acetate into cholesterol as a measure of cholesterol synthesis. For such studies, the cells are first conditioned for 24 h in medium containing either complete or lipoprotein-deficient serum (LPDS). Incubation of the cells in medium supplemented with LPDS deprives the cells of exogenous sources of cholesterol, and this leads to stimulation of intracellular cholesterol synthesis. After incubation in a medium containing LPDS, the cells are re-incubated for 6 h in the same medium to which LDL has been added in various concentrations. Next, ^{14}C-acetate is added and the cells are incubated overnight. After the lipid extraction and the separation of extracted lipids by thin layer chromatography [6], the incorporation of ^{14}C-acetate into cholesterol is then measured in both the cell layer and the medium.

The metabolism of LDL by the cells can be studied by assessing the binding, internalization, and degradation of ^{125}I-LDL [5]. For this purpose, the cells are first incubated in medium supplemented with LPDS to maximize the expression of LDL receptors and then incubated for 3 h at 37 °C in the presence of increasing amounts of ^{125}I-LDL. For the assessment of non-specific binding, parallel cultures are incubated in the presence of both ^{125}I-LDL and an excess of non-labeled LDL. Next, the cells are chilled to 4 °C, washed with BSA-containing buffer to remove the non-specifically bound LDL, and incubated for 1 h in the presence of heparin, which leads to the dissociation of LDL bound to the membrane receptor. The fraction of LDL that remains bound to the cells represents the cell-associated LDL fraction. Internalization of the LDL-receptor complex is followed by rapid degradation of protein components of LDL in lysosomes, after which free amino acids are excreted into the medium. The amount of TCA-soluble ^{125}I-labeled material present in the medium represents the fraction of LDL that has been already degraded by the cell.

Comparison of the processes involved in the regulation of cholesterol synthesis in normal human skin keratinocytes and fibroblasts showed marked differences between these two cell types [11], as follows:

1) The rate of cholesterol synthesis was 10 to 20 times higher in keratinocytes than in fibroblasts. In keratinocytes, the excretion of newly synthesized cholesterol was very low, whereas fibroblasts excreted a large proportion of the cholesterol formed (Fig. 1).
2) Changes in the extracellular lipoprotein concentration did not affect the rate of cholesterol synthesis in keratinocytes: when the extracellular environment was

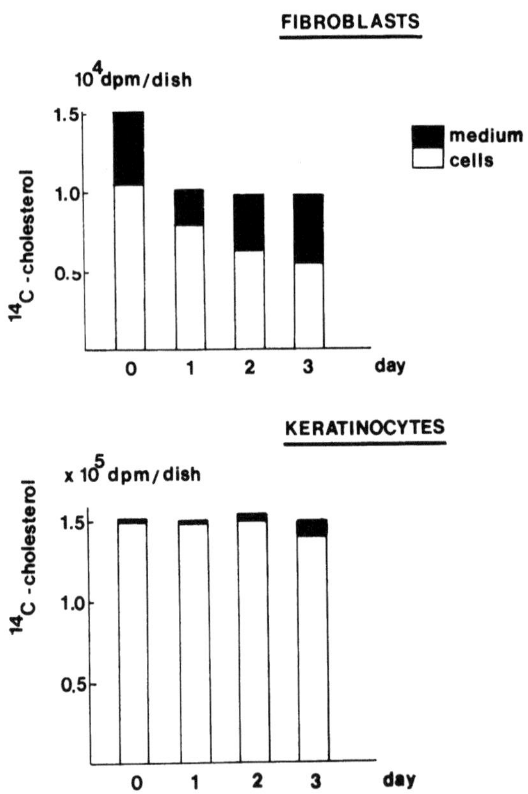

Fig. 1. The "fate" of the de novo synthesized cholesterol. The confluent cultures of fibroblasts and keratinocytes were incubated overnight in medium supplemented with 15% newborn calf serum and with ^{14}C-acetate (5 μC$_i$/dish). Thereafter in two dishes the amount of the de novo synthesized cholesterol present in the cells and in the medium was determined (day 0). In other dishes, the cells were washed 3 times with PBS (to remove the extracellularly present ^{14}C-acetate) and incubated further for 1–3 days in medium supplemented with 15% NBCS, and the amount of ^{14}C-cholesterol present in cells and in medium was determined. All determinations were performed in duplicate

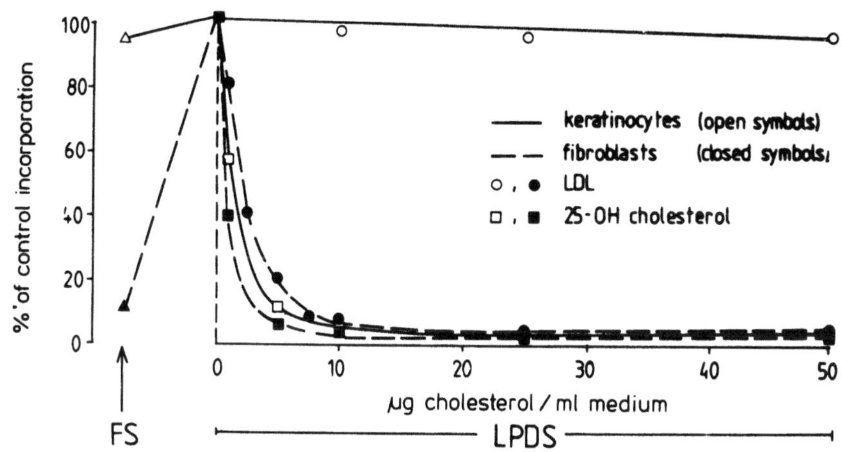

Fig. 2. Effect of culture conditions, LDL and 25-hydroxycholesterol on intracellular cholesterol synthesis in keratinocyte and fibroblast cultures. The confluent cultures of fibroblasts (closed symbols) and keratinocytes (open symbols) were first incubated for 24 h in medium supplemented with 15% full serum (FS) (△, ▲) or 15% lipoprotein-deficient serum (LPDS) and subsequently the latter were incubated for 6 h in medium of the same composition but in the presence of increasing amounts of LDL (○, ●) or 25-dehydroxycholesterol (□, ■). Thereafter, ^{14}C-acetate (5 μC$_i$/dish) was added to all dishes and after the overnight incubation the synthesized ^{14}C-cholesterol was analyzed in the medium and the cell layer. The values shown represent the sum of the ^{14}C-cholesterol present in cells and medium, and are the mean values of duplicate determinations. Cholesterol synthesis measured in cells preconditioned in LPDS (and in the absence of LDL) is expressed as 100%, and results are plotted as percentage of this synthesis

deprived of cholesterol, the intracellular synthesis remained virtually unchanged in keratinocytes but increased markedly in fibroblasts. Furthermore, in keratinocytes the presence of LDL in the growth medium did not affect endogenous cholesterol synthesis, whereas fibroblasts showed a marked decrease in this respect (Fig. 2). However, when the cholesterol was administered in a non-lipoprotein form, i.e., as 25-hydroxycholesterol, a marked suppression of cholesterol synthesis occurred, which showed the presence of the mechanism controlling the activity of HMGCoA reductase (Fig. 2).

3) A defect of LDL metabolism was found to be responsible for the failure of LDL to regulate cholesterol synthesis in keratinocytes: the amount of ^{125}I-LDL bound specifically to the cell membrane receptor, and particularly the amount internalized and degraded by the cells, proved to be much lower in keratinocytes than in fibroblasts (Fig. 3).

The present findings led us to the conclusion that the regulation of cholesterol synthesis in cultured keratinocytes is controlled by an unique mechanism that differs from the one found in most of the other cell types. In stratified squamous epithelia, an equilibrium between cell division and terminal differentiation assures the maintainance of tissue dimensions. In malignant cells this balance is disturbed. Since normal keratinocytes failed to respond to LDL and some malignant cells of epithelial origin have been reported to show a defective response to LDL [1, 4], the possibility

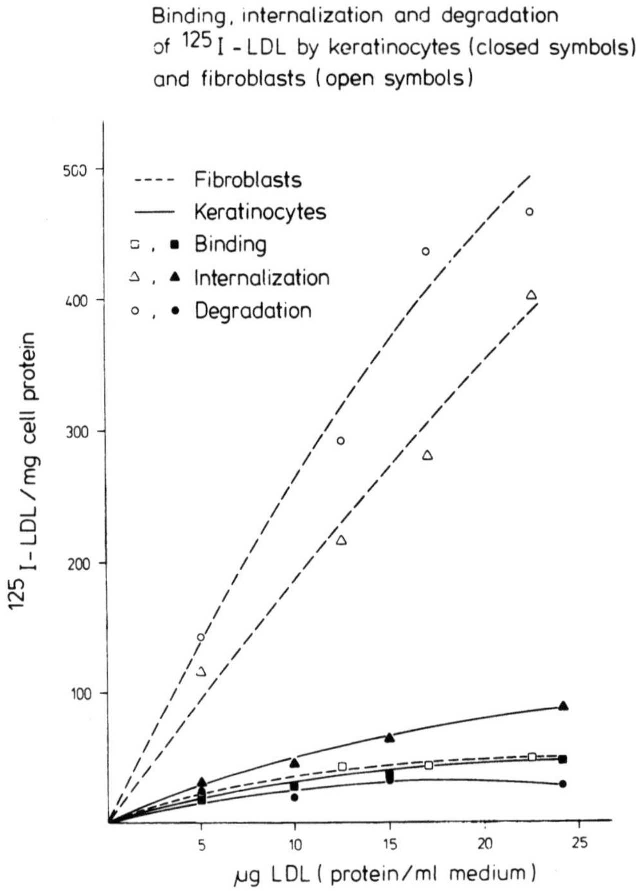

Fig. 3. The binding, internalization, and degradation of ^{125}I-LDL (low density lipoprotein) by cultured human skin keratinocytes and fibroblasts. The cell cultures after 24 h preincubation in medium supplemented with 15% LPDS were further incubated for 3 h at 37 °C in the presence of increasing amounts of ^{125}I-LDL, and thereafter the amount of ^{125}I-LDL bound, internalized and degraded was determined as described in the text. For determination of nonspecifically bound LDL, the cells were incubated with both ^{125}I-LDL and an excess of unlabeled LDL (150 μg/ml). The data presented in the figure represent specific cell binding, internalization or degradation that are calculated by subtraction of nonspecifically bound, internalized and degraded LDL from that totally bound, internalized and degraded

of correlation between the ability of cells to differentiate into squames and the degree of the regulatory effects of LDL on intracellular cholesterol synthesis was investigated. In this study, epithelial cells with and without defective terminal differentiation were compared with respect to the regulation of intracellular cholesterol synthesis by LDL and to the uptake and degradation of LDL. The following types of cell were used: Normal skin fibroblasts, normal keratinocytes and keratinocytes transformed by SV40 (SVK_{14}) [14], as well as a number of squamous carcinoma cell (SCC) lines in which the terminal-differentiation defect was found to occur in the following order: SCC-12F2 $<$ SCC-25 $\leqslant$ SCC-15 $<$ SCC-12B2 $<$ SCC-4 [13].

Table 1. Comparison of the de novo cholesterol synthesis in various confluent cell cultures

Cell	Cholesterol synthesized (nmol/mg protein in the presence of		Ratio* LPDS/NBCS
	15% NBCS	15% LPDS	
Fibroblast	0.9	7.1	8.0
SCC-4	2.5	19.5	7.8
SCC-15	4.3	20.6	4.8
SCC-25	3.4	8.8	2.6
SCC-12B2	5.6	12.3	2.2
SCC-12F2	2.4	5.2	2.2
SVK_{14}	21.4	23.5	1.1
Keratinocyte	19.8	21.8	1.1

* Ratio of the amount of cholesterol synthesized by cells incubated in medium supplemented with lipoprotein-deficient newborn calf serum (LPDS) to that synthesized by cells incubated in medium supplemented with newborn calf serum (NBCS)

The confluent cell cultures were preincubated for 24 h in medium supplemented with either 15% NBCS or 15% LPDS. For the measurement of the de novo cholesterol synthesis ^{14}C-acetate (2.5 μC_i/ml medium; 0.5 mM) was added and the cells were further incubated for 18 h. Subsequently, ^{14}C-cholesterol was measured in both cell layer and medium

Compared with normal skin fibroblasts, most of the cell types under study showed a defective response to changes in extracellular lipoprotein concentrations. Both the inducibility of cholesterol synthesis in cells deprived of extracellular sources of cholesterol (Table 1) and the LDL-induced suppression of the intracellular cholesterol synthesis in cells preincubated in medium supplemented with lipoprotein-deficient serum showed the following sequence: fibroblasts > SCC-4 > SCC-15 ≃ SCC-25 > SCC-12F2 ≃ SCC-12B2 > SVK_{14} ≃ normal keratinocytes (Fig. 4) [12].

A defect in LDL metabolism was responsible for the failure of LDL to regulate the cholesterol metabolism (Fig. 5) because a marked suppression of cholesterol synthesis was observed in all cell types when sterol was delivered in artificial non-lipoprotein form, as 25-hydroxycholesterol (data not shown). Compared with the situation in fibroblasts, the metabolism of ^{125}I-LDL was found to be impaired in SCC-25 cells and virtually absent in SCC-12F2, SVK_{14} and normal keratinocytes [12].

SV40 virus-induced transformation of keratinocytes did not lead to any change in the response of the cells to changes in the extracellular LDL concentration, since both normal and transformed keratinocytes showed the same absence of a response to LDL. For all tested SCC lines except SCC-12B2, good correlation was consistently observed between the degree of LDL-induced suppression of cholesterol synthesis and the decreasing ability of cells to differentiate into squames. When this point in the study was reached interesting features started to appear.

A good parallelism was found between the degree of the defect in the terminal differentiation program and the ability of lipoproteins to regulate the intracellular cholesterol synthesis when normal and malignant keratinocytes were compared.

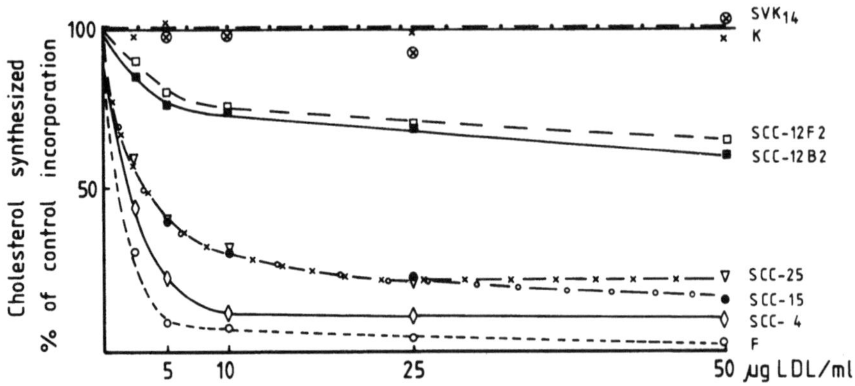

Fig. 4. The effect of LDL on the incorporation of ^{14}C-acetate into ^{14}C-cholesterol by various cultured cells. The confluent cell cultures were preconditioned for 24 h in medium supplemented with 15% LPDS and subsequently incubated for the next 6 h in medium of the same composition but in the presence of the increasing amounts of LDL (0–50 μg protein/ml), after which ^{14}C-acetate (Amersham, 59 mC_i/mMol, 2.5 μC_i/ml medium, 0.5 mM) was added. After the overnight incubation the synthesized ^{14}C-cholesterol was measured in the medium and the cells separately. The values given in the figure ⊗ represent the sum of ^{14}C-cholesterol present in the cells and medium and are the means of duplicate determinations. (K (keratinocyte) (X), SVK_{14} (⊗), SCC-12F2 (□, SCC-12B2 (■), SCC-25 (▽), SCC-15 (●), SCC-4 (◇) and F (fibroblasts (○)) [12]

The behaviour of malignant keratinocytes with a severely defective terminal differentiation with respect to the regulation of cholesterol synthesis resembles that of fibroblasts (and other non-differentiating cells) more than that shown by normal keratinocytes.

The postulated connection between the regulation of cholesterol metabolism by keratinocytes and the ability of these cells to differentiate into squames of course required confirmation by studies on the suppression of normal keratinocytes differentiation. Fortunately, the ability to differentiate into squames can be manipulated by culturing keratinocytes in media containing calcium in various concentrations. The extracellular concentration of calcium ions has been shown to be a regulator of differentiation in cultured human [2, 7] and murine [8, 15] keratinocytes. In medium with a low calcium level proliferation is favoured and differentiation is retarded. This led us to perform a number of experiments with the following results.

Marked differences in the processes involved in the regulation of cholesterol synthesis were observed in keratinocytes cultured in medium containing low (0.07 mM) and high (1.8 mM) levels of calcium.

The rate of cholesterol synthesis in keratinocytes cultured in low-Ca medium was 20–30 times lower than that in high-Ca medium (Table 2). This suggests that differentiating cells need more cholesterol for the assemblage of the cell membrane.

Unlike those cultured in medium with a high calcium level, cells given little calcium show a sharply increased rate of cholesterol synthesis when incubated in cholesterol-depleted medium (Table 2). Furthermore, the addition of increasing concentrations of LDL leads to marked decrease of cholesterol synthesis (Fig. 6). The high degree of binding, internalization, and degradation of LDL was responsible for the good

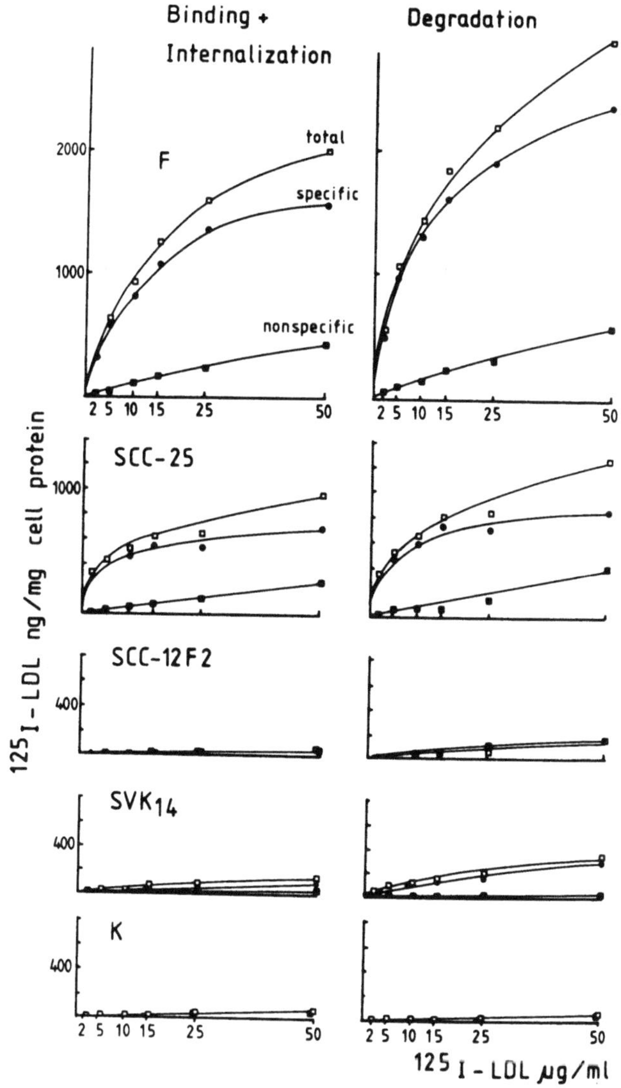

Fig. 5. The cell association (binding plus internalization) and degradation of ^{125}I-LDL by various cultured cells. After 24 h preincubation in medium supplemented with 15% LPDS the confluent cell cultures were further incubated for 3 h at 37 °C in the presence of increasing amounts of ^{125}I-LDL (0–50 µg/ml). The degradation of ^{125}I-LDL was measured as free-iodine trichloroacetic acid-soluble radioactivity in the medium. The binding plus internalization of ^{125}I-LDL was taken as the total amount of cell associated radioactivity. Each value is the mean of duplicate determinations. □—□ total cell association or degradation; ■—■ nonspecific cell association or degradation measured in the presence of 200 µg/ml unlabeled LDL; ●—● specific cell association or degradation as calculated by subtraction of nonspecific cell association or degradation from resp. total cell association or degradation [12]

Table 2. Effect of extracellular calcium concentration on cholesterol synthesis in keratinocytes

Culture condition*	Calcium switch**	Cholesterol synthesized (nmoles/mg protein) in the presence of***	
		5% FS	5% LPDS
High Ca	–	27.8	27.2
Low Ca	–	0.9	9.4
High Ca	low Ca	3.74	10.5
High Ca	traces Ca	0.7	10.8
Low Ca	high Ca	27.0	28.2

* The cells were allowed to grow to confluence in medium with a low (0.07 mM) or high (1.8 mM) calcium level

** The confluent cultures grown either in the presence of high or low calcium concentrations were further incubated for 2 to 3 days in medium with either low or high calcium levels

*** The confluent cell cultures were preincubated for 24 h in medium supplemented with either 5% FS or LPDS. For the measurement of the de novo cholesterol synthesis ^{14}C-acetate (2.5 μC_i/ml medium; 0.5 mM) was added and the cells were further incubated for 18 h. Subsequently, ^{14}C-cholesterol was measured in both cell layer and medium

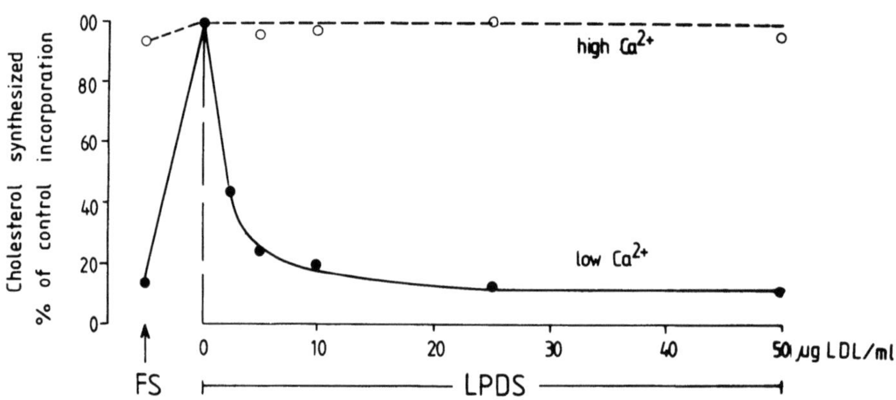

Fig. 6. The effect of culture condition and of LDL on the intracellular cholesterol synthesis in keratinocytes grown in medium containing high (○) or low (●) calcium concentrations. The confluent cultures were first incubated for 24 h in culture medium of the same composition as used for culturing the cells supplemented either with 5% full serum (FS) or 5% lipoprotein-deficient serum (LPDS). The later cell cultures were subsequently incubated for 6 h in medium of the same composition but in the presence of increasing amounts of LDL. Thereafter, ^{14}C-acetate (5 μC_i/dish) was added to all dishes, and after overnight incubation the synthesized ^{14}C-cholesterol was analyzed [11]

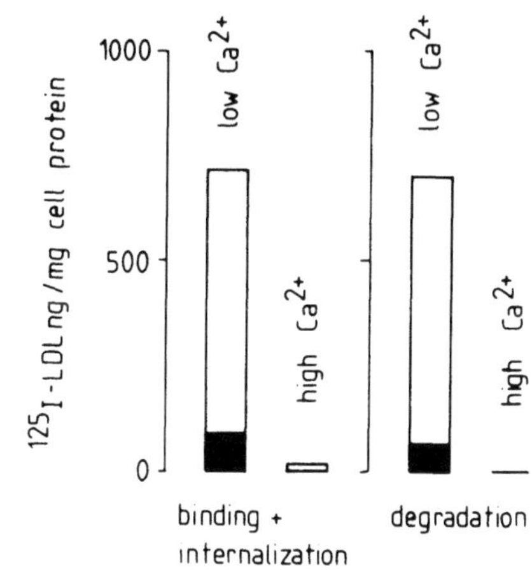

Fig. 7. The cell association (binding plus internalization) and degradation of ^{125}I-LDL by keratinocytes cultured in medium containing high (1.8 mM) and low (0.07 mM) concentration of calcium. After 24 h preincubation in medium supplemented with 5% LPDS the confluent cell cultures were further incubated for 3 h at 37 °C in the presence of 17.1 μg ^{125}I-LDL/ml medium and in the absence or presence of an excess of non-labeled LDL (150 μg/ml). The degradation of ^{125}I-LDL was measured as free-iodine trichloroacetic acid-soluble radioactivity in the medium. The binding plus internalization of ^{125}I-LDL was taken as the total amount of cell-associated radioactivity. The open part of the column represents the specific binding and the black part the non-specific binding

responsiveness of these cells to changes in the concentration of extracellular lipoproteins (Fig. 7).

The changes of both the rate of cholesterol synthesis and the sensitivity of the cells to changes in the extracellular lipoprotein concentration were reversible. When the cells were allowed to grow to confluence in medium with a low calcium level and subsequently incubated in medium with a high calcium level, the rate of cholesterol synthesis increased until it approximated that of cells grown in medium with a high calcium level. Furthermore, after a switch to high-calcium medium, the cells did not respond to changes in the extracellular lipoprotein concentration (Table 2). The opposite effects were seen after a switch from high- to low-calcium medium.

The results presented in this paper show that membrane functions undergo modulation during the terminal differentiation process. These alterations might be accompanied by changes in cholesterol synthesis, since differentiating cells showed a very high rate of cholesterol synthesis. This high rate of cholesterol synthesis might lead to changes in cell membrane composition, since cholesterol is known to be an important component of the cell membrane. Besides the changes observed during the terminal differentiation process, i.e., in the binding, internalization, and degradation of LDL accompanied by an altered responsiveness of cells to LDL-induced effects on cholesterol synthesis, alterations in the binding of other surface ligands, such as EGF, concavalin A and somatomedin C, have been reported in the literature [10]. This suggest that the changes occurring in the cell membrane during the terminal differentiation process are not very specific but quite general alterations.

The following points are the most relevant for the conclusions to be drawn from the present results: With respect to the mechanism for the regulation of cholesterol synthesis, the differentiating normal keratinocytes differ considerably from fibroblasts and all other non-differentiating cells. Malignant keratinocytes with a defective terminal differentation program take a position between these two extremes. Also keratinocytes in which differentiation is strongly inhibited by lowering of the calcium concentration in the culture medium behave almost like fibroblasts or other non-differentiating cells. These findings justify the conclusion that there is a clear relationship between the mechanism underlying the regulation of intracellular cholesterol synthesis and the ability of cells to differentiate into squames.

References

1. Anderson JW, Brown MS, Goldstein JL (1982) Inefficient internalization of receptor-bound low density lipoprotein in human carcinoma A-431 cells. J Cell Biol 88:441–452
2. Boyce ST, Ham RG (1983) Calcium-regulated differentiation of normal human epidermal keratinocytes in chemically defined clonal culture and serum-free serial culture. J Invest Dermatol 81:33s–39s
3. Brown MS, Kovanen PT, Goldstein JL (1981) Regulation of plasma cholesterol by lipoprotein receptors. Science 212:628–635
4. Gal D, Simpson ER, Porter JC, Snyder JM (1982) Effective internalization of low density lipoprotein in epidermal cervical cancer cells. J Cell Biol 92:597–603
5. Goldstein JL, Brown MS (1974) Binding and degradation of low density lipoproteins by cultured human fibroblasts. J Biol Chem 149:5153–5162
6. Goldstein JL, Dana SE, Brown MS (1974) Esterification of low density lipoprotein cholesterol in human fibroblasts and its absence in homogenous familiar hypercholesterolemia. Proc Natl Acad Sci 71:4288–4294
7. Hawley-Nelson P, Sullivan JE, Kung M, Hennings H, Yuspa SH (1980) Optimized conditions for the growth of human epidermal cells in culture. J Invest Dermatol 75:176–182
8. Hennings H, Michael D, Cheng C, Steinert P, Holbrook K, Yuspa SH (1980) Calcium regulation of growth and differentiation of mouse epidermal cells in culture. Cell 19:245–254
9. Mahley RW, Innerarity TL (1983) Lipoprotein receptors and cholesterol homeostasia. Biochim Biophys Acta 737:197–222
10. O'Keefe EJ, Payne RE Jr (1983) Modulation of the epidermal growth factor receptor of human keratinocytes by calcium ion. J Invest Dermatol 81:231–235
11. Ponec M, Havekes L, Kempenaar JA, Vermeer BJ (1983) Cultured human skin fibroblasts and keratinocytes: differences in the regulation of cholesterol synthesis. J Invest Dermatol 81:125–130
12. Ponec M, Havekes L, Kempenaar J, Lavrijsen S, Vermeer BJ (1984) Defective low-density lipoprotein metabolism in cultured normal, transformed and malignant keratinocytes. J Invest Dermatol 83:436–440
13. Rheinwald JG, Beckett MA (1981) Tumorigenic keratinocytes lines requiring anchorage and fibroblast support cultured from human squamous cell carcinoma. Cancer Res 41:1657–1663
14. Taylor-Papadimitriou J, Purkis P, Lane EB, McKay I, Chang SE (1982) Effects of SV40 transformation on the cytoskeleton and behavioural properties of human keratinocytes. Cell Differ 11:169–180
15. Yuspa SH, Koehler BA, Kulesz-Martin M, Hennings H (1981) Clonal growth of mouse epidermal cells in medium with reduced calcium concentration. J Invest Dermatol 76:144–146

A Model for Morphological Studies on Ligand-Receptor Complexes

B. J. Vermeer, A. M. Mommaas-Kienhuis, M. C. Wijsman, J. J. Emeis and M. Ponec

Effector molecules (e.g. growth factors, hormones, peptides) are often bound to a specific plasma membrane site by a high affinity saturable process. This phenomenon is called ligand-receptor binding. The receptor-bound material is then internalized into the cell by a receptor-mediated process called receptor-mediated endocytosis [16]. This process of internalization often takes place via the coated pit-vesicle system [7, 13, 17].

The low density lipoprotein (LDL) receptor pathway represents a good model for the study of plasma membrane receptors and receptor mediated endocytosis. The effects of LDL on the cellular metabolism, e.g., suppresion of endogenous cholesterol synthesis and LDL receptor synthesis are only then installed when internalization and degradation of LDL in the lysosomes have taken place [8]. The cellular mechanisms that play a role in binding and endocytosis of ligand-receptor complexes, e.g., LDL receptors can be investigated by morphological studies [3, 20, 22].

For visualization of the ligand-receptor interaction two different methods can be employed: conjugation of the ligand, e.g., LDL to a marker, and immunological procedures. A prerequisite to the conjugation method is that the biological activity of the ligand remains unaltered. For immunological methods the antigens must be preserved, and therefore only weak fixatives can be used. Moreover this technique has limited value for studying the internalization process [22].

In this study we describe morphological studies of LDL-recepttor binding at 4 °C and LDL-receptor mediated uptake at 37 °C in cultured fibroblasts aand epithelial tumor cells. For fluorescence microscopy we used the indirect immunological method for binding studies and the conjugation technique for internalization studies. For electron microscopy we visualized both binding and internalization with the direct conjugation method. In all experiments the cells were incubated with low concentrations of LDL (25–50 μg protein/ml). At this concentration the vast majority of LDL is handled by the high affinity pathway in fibroblasts and other eukaryotic cells [19].

Materials and Methods

Lipoproteins

Human LDL was isolated from plasma of healthy individuals by density gradient ultracentifugation as described elsewhere [15]. Lipoprotein deficient serum (LPDS) was prepared by ultracentrifugation.

Cell Cultures

Fibroblasts were cultured in Petri dishes containing Ham's F10 medium supplemented with 15% newborn calf serum (NBCS). Human squamous carcinoma cell lines SCC12F2 (originating from facial epidermis), SCC25 and SCC15 (derived from SCC of the tongue) were kindly provided by Dr. J. Rheinwald. Under normal conditions 1×10^5 cells/ 7.5 cm^2 were plated together with 4×10^4 lethally irradiated 3T3 cells/cm^2 in Petri dishes containing Dulbecco-Vogt medium, 5% fetal calf serum (FCS) and 0.4 μg/ml hydrocortisone. After one week of culture the experimental procedures started.

Conjugation Method of LDL

LDL was labeled with the fluorescent probe 1,1′dioctadecyl 3,3,3′3′-tetramethyl-indocarbocyanide (DIL, molecular probes) according to Pitas et al. [12].

For electron microscopy studies LDL was conjugated to colloidal gold as described by Handley et al. [9]. In short, to 5 ml of 20 nm colloidal gold 0.5 ml of LDL (200 μg protein/ml) in 0.05 M EDTA at pH 5.5 was added and mixed by manual shaking. Unlabeled LDL was removed by centrifugation at 9,000 x g for 1 h against a 40% sucrose cushion.

Experimental Procedures

Prior to cell incubations with LDL the cells were preconditioned for 24 h in medium containing 15% LPDS.

Binding studies: Prechilled cells were incubated for 2 h at 4 °C with LDL-DIL or LDL-gold (25–50 μg protein/ml). For the immunological method prechilled cells were incubated for 2 h at 4 °C with LDL, fixed in 1% paraformaldehyde and a classical two-step immunofluorescene technique was performed. In the first step rabbit anti-human apoprotein B 1:20 and in the second step goat anti-rabbit FITC 1:80 (Nordic, Tilburg, The Netherlands) was applied [23].

Receptor-mediated endocytosis: Cells were incubated continuously for 30 min at 37 °C with LDL-DIL or LDL-gold (25–50 μg protein/ml). Pulse-chase experiments were performed with fibroblasts, SCC15 and 12F2 cells: the conjugated LDL was bound to prechilled cells for 2 h at 4 °C and after a washing procedure the cells were warmed and exposed for varying periods to 37 °C for internalization of plasma membrane bound material.

Effect of cell density on the binding of LDL to epithelial tumor cells: A varying number 25×10^3, 50×10^3, 1×10^5, 1.5×10^5, 2×10^5 cells/7.5 cm^2 of epithelial tumor cells (SCC15, 12F2) were plated and cultured for one week. Thereafter an incubation with LDL for 2 h at 4 °C and the immunofluorescence technique as described earlier were performed.

Controls: for the immunofluorescence technique a blocking-test was performed. The conjugated LDL technique was checked by competition experiments with an excess of unlabeled LDL (600 μg protein/ml). Cultured fibroblasts without LDL receptors served as an additional control [3].

Fluorescence Microscopy

After the incubation procedures the cells cultured on glass coverslips were fixed in 1% paraformaldehyde in PBS pH 7.4 for 30 min at room temperature, rinsed in PBS and covered with PPDA-containing mounting fluid and a glass coverslip [23]. The cells were viewed with a Leitz microscope with epifluorescent illumination using a rhodamine filter package for DIL and a fluorescene filter for immunofluorescene techniques. Photographs were taken using a Vario XL Agfa film at 800 ASA.

Electron Microscopy

After the incubation procedures the fibroblasts and the SCC25 cells were fixed with 1.5% glutaraldehyde in cacodylate buffer for 10 min at room temperature and postfixed with 1% OsO_4 in phosphate buffer for 30 min at 4 °C. The SCC15 and 12F2 cells were fixed with 2% glutaraldehyde and 2.5% paraformaldehyde for 3 h at room temperature and postfixed with 1% OsO_4 containing $K_4[Fe(CN)_6]$ for 1 h at room temperature [21]. Following dehydration in a graded ethanol series up to 70% the cells were embedded in situ. Ultrathin sections were made parallel to the plane of the surface of the culture dish, stained with uranyl acetate and lead hydroxide and viewed with a Philips EM 300 electron microscope at 80 kV.

Results

Fluorescence Microscopy

LDL binding and endocytosis in SCC25 cells: At 4 °C the LDL-receptor binding is characterized by clustered LDL particles on the surface of the SCC25 cells (Fig. 1). No linear arrangement of LDL is found on these cells as was described for fibroblasts [3]. When the cells were incubated with LDL-DIL for 30 min at 37 °C an extensive

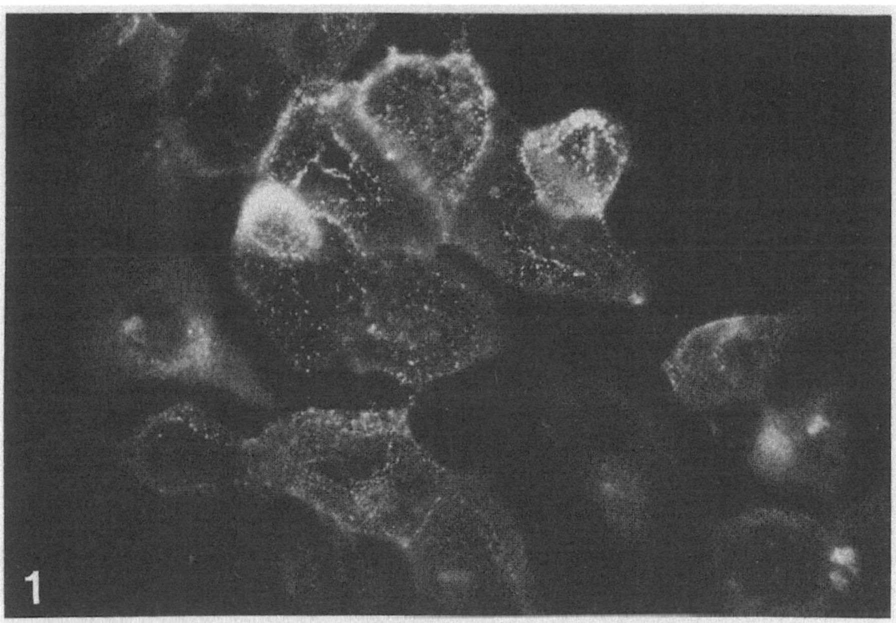

Fig. 1. SCC25 cells incubated with LDL, 2 h at 4 °C. Binding of LDL. x400

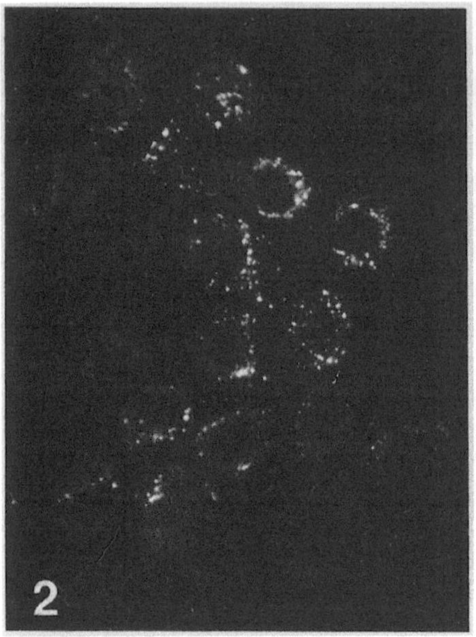

Fig. 2. SCC25 cells incubated with LDL-DIL for 30′ at 37 °C. Continuous internalization. x400

Fig. 3. SCC25 cells pulse-chase experiment; 2 h at 4 °C with LDL-DIL, 30′ 37 °C without LDL. Internalization of membrane-bound LDL. x400

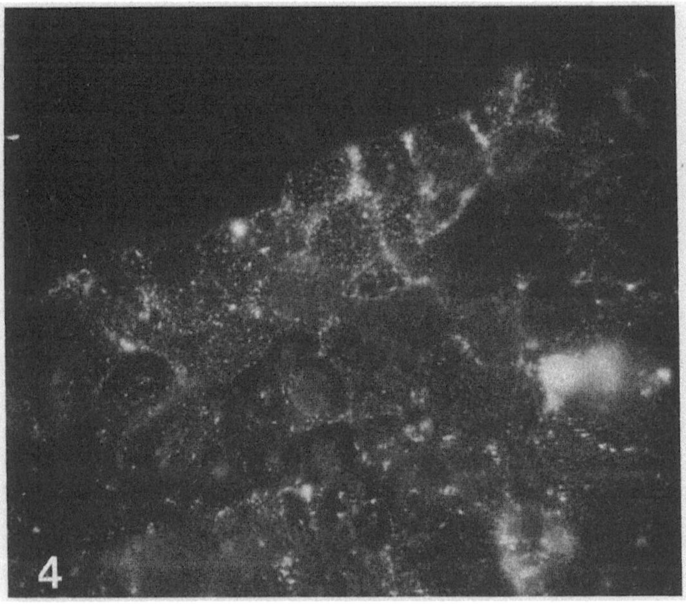

Fig. 4. SCC15 cells, 50 x 10^3 cells plated, 2 h at 4 °C + LDL. All cells have LDL binding sites. x400

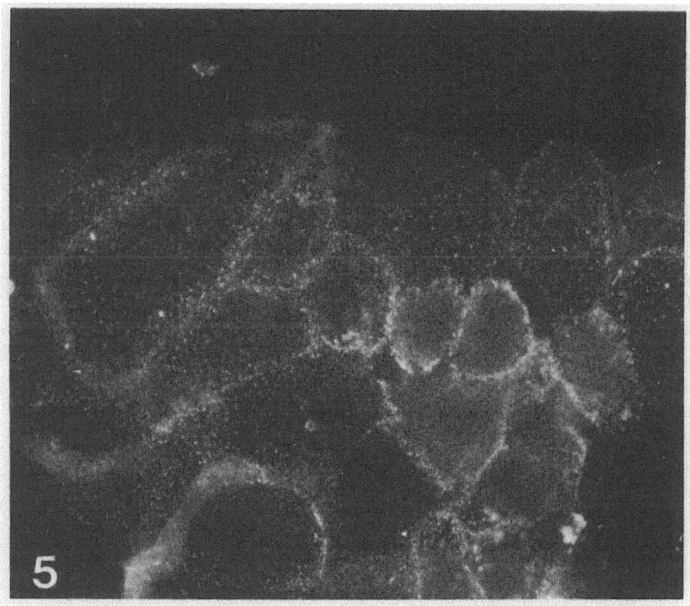

Fig. 5. 12F2 cells, 50 x 10^3 cells plated, 2 h at 4 °C + LDL. All cells have LDL binding sites. x400

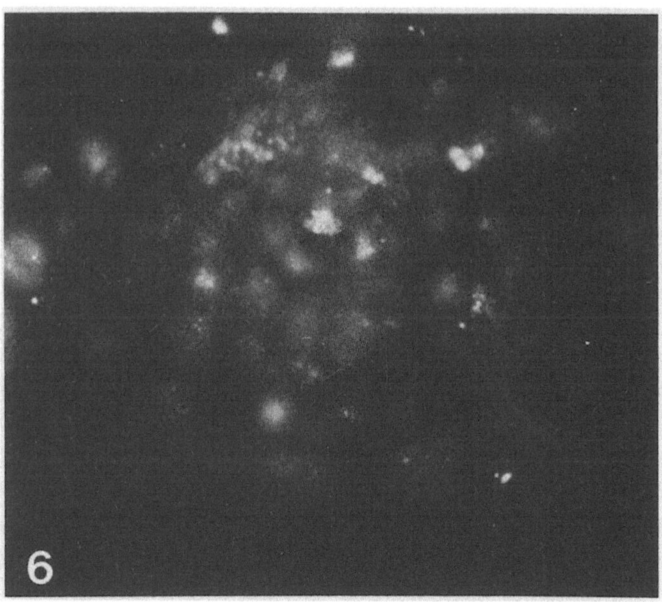

Fig. 6. SCC15 cells, 2 x 10^5 cells plated, 2 h at 4 °C + LDL. In an area of keratinization no LDL binding is found. x400

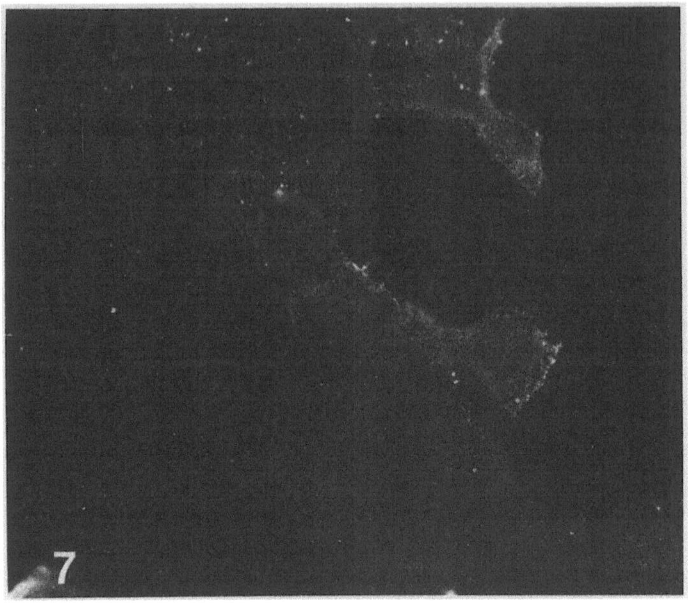

Fig. 7. 12F2 cells, 2 x 10^5 cells plated, inc. 2 h at 4 °C + LDL. LDL is only present in cells on the outermost region of the culture. x400

uptake of LDL-DIL occurred (Fig. 2). However, when pulse-chase experiments were performed with LDL-DIL for 30 min at 37 °C, the amount of inernalized LDL-DIL was not as pronounced as in continuous labeling experiments (Fig. 3).

Effect of cell density on LDL binding: When epithelial tumor cells (SCC15, 12F2) were plated at low density and cultured for one week, nearly all cells had a similar appearance. No obvious differentiation was present and LDL was bound to all cells in culture (Figs. 4, 5). However, in cells plated at higher density and cultured for one week several areas of differentiation, were found. In the regions of keratinization no LDL binding was present (Fig. 6). In these cell cultures the LDL binding was only observed in the outermost regions of the cell colonies (Fig. 7).

Electron Microscopy

Binding and endocytosis of LDL-gold: SCC25 cells incubated for 2 h at 4 °C with LDL-gold showed some LDL in coated pits, but the majority of the excessive LDL binding was located outside coated pit structures (Fig. 8). The LDL-gold was preferentially located on microvillous projections (Figs. 8–10). When SCC25 cells were incubated for 5 min at 37 °C with LDL-gold, the LDL-gold particles were present in small endocytotic vesicles nearby the plasma membrane (Fig. 9). After an incubation period of 30 min at 37 °C, some LDL-gold particles were found in electron lucent vacuoles and dense bodies (Fig. 10). The internalization process took place via the coated pit-vesicle system (Fig. 10, detail).

Pulse-chase experiments with LDL-gold: When LDL-gold was bound to the plasma membrane of fibroblasts at 4 °C and subsequently internalized for 30 min at 37 °C

Fig. 8. SCC25 cells incubated with LDL-gold, 2 h at 4 °C. LDL is present in abundance on the plasma membrane. x25,000

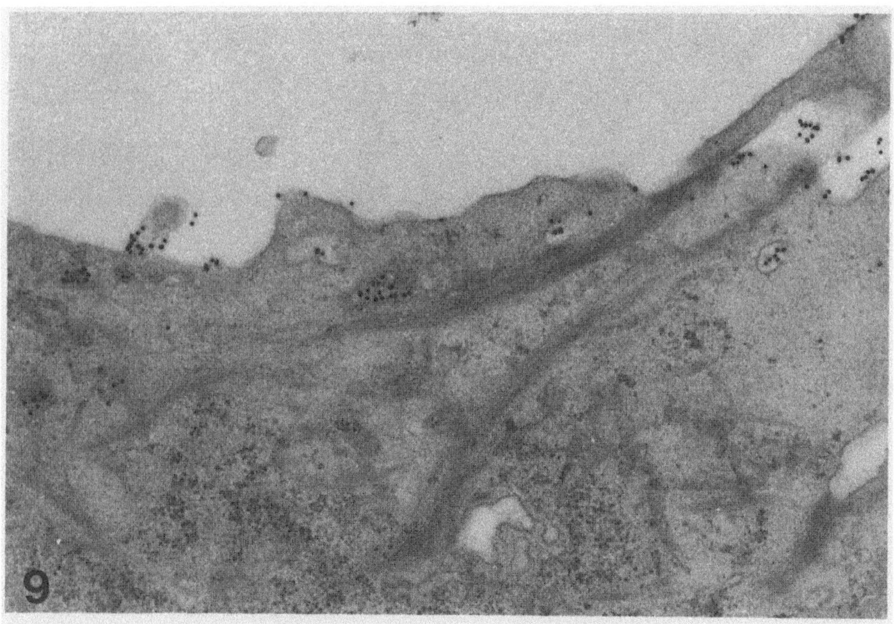

Fig. 9. SCC25 cells incubated with LDL-gold 5′ at 37 °C. The LDL is also found in small vesicles. x25,000

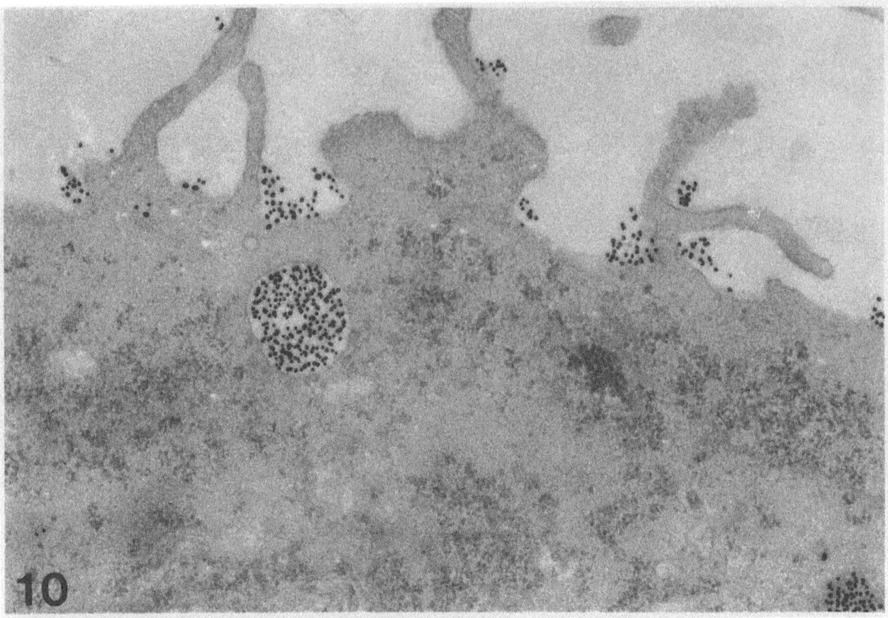

Fig. 10. SCC25 cells incubated with LDL-gold 30′ at 37 °C. The internalized LDL is present in electronlucent vacuoles and dense bodies. x25,000

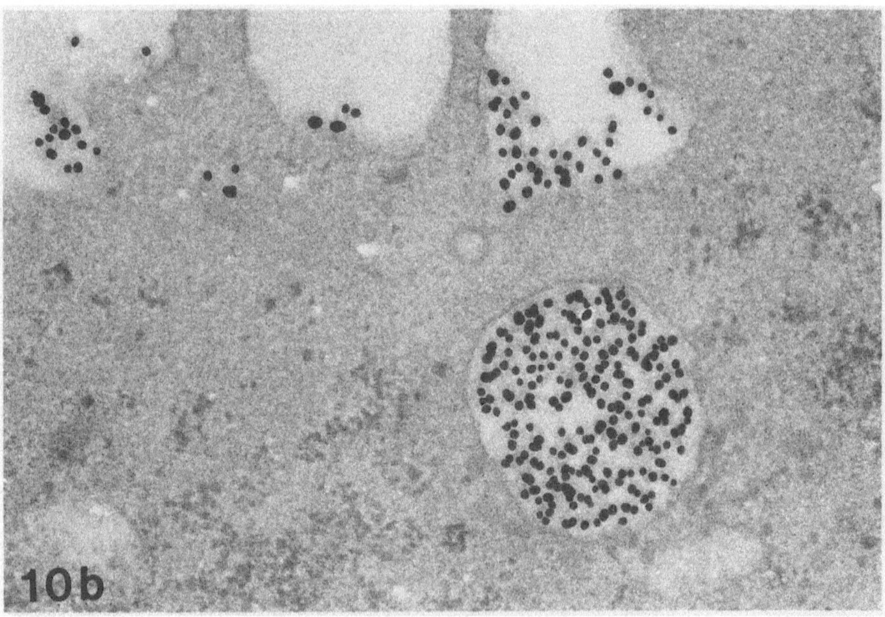

Fig. 10b. A detail of the receptor-mediated endocytosis of LDL, taking place via coated pit-vesicle system. x100,000

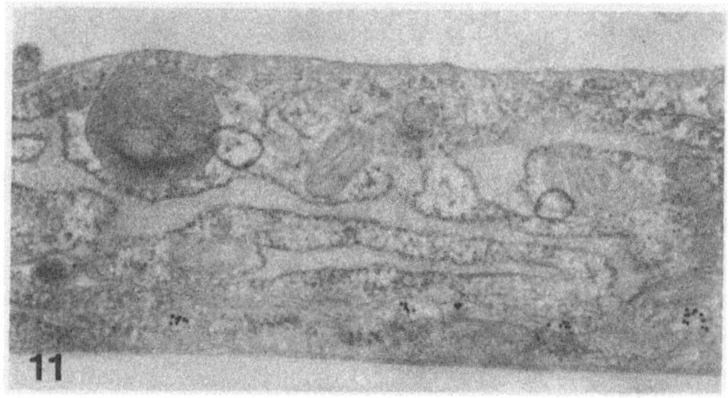

Fig. 11. Fibroblasts, pulse-chase experiment, incubated 2 h at 4 °C + LDL-gold 30′ 37 °C without LDL-gold. The plasma membrane bound LDL is internalized into the cell. x25,000

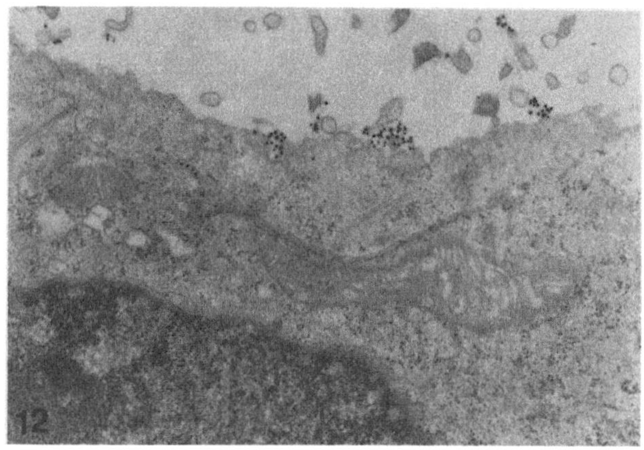

Fig. 12. SCC15 cells, pulse-chase experiment, incubated 2 h at 4 °C + LDL-gold, 2 h at 37 °C without LDL-gold. LDL still remains on the surface, a minority is internalized, x25,000

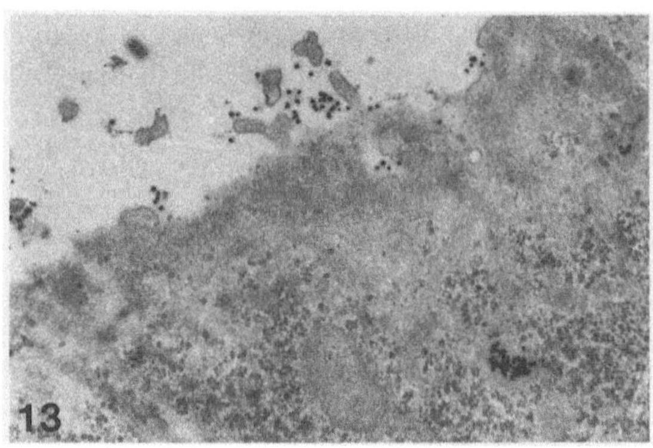

Fig. 13. 12F2 cells, pulse-chase experiment, inc. 2 h at 4 °C + LDL-gold, 2 h at 37 °C without LDL-gold. LDL still remains on the surface. x25,000

all plasma membrane bound LDL-gold was taken up by the cell (Fig. 11). However, only a minority of the plasma membrane bound LDL-gold was internalized in SCC15 and 12F2 cells, using the same pulse-chase methods. Even in case the internalization period was prolonged up till 2–4 h at 37 °C most of the LDL-gold particles remained on the cell surface (Figs. 12, 13).

Discussion

In comparison with fibroblasts and other eukaryotic cells, the epithelial tumor cells show an excessive binding of LDL to the plasma membrane and a defective LDL-receptor mediated endocytosis. Similar findings have been described by other authors for A431 cells and cervical epidermoid cancer cells [2, 5]. The defective endocytosis of LDL can explain the little effect LDL exert on the cellular cholesterol synthesis in epithelial cells [14]. If the excessive binding of LDL to the plasma membrane of these tumor cells only takes place to specific LDL receptors is not yet known. Preliminary data of experiments using monoclonal antibodies against LDL receptors [4] at the light microscope level have shown that these receptors are present on the surface of epithelial tumor cells. These studies have to be extended to immuno-electron microscopy to determine whether the abundant presence of LDL receptors on the cell surface outside coated pits can also be demonstrated with this method. Up till now it is very speculative to suppose a correlation between the two defects in receptor metabolism, e.g., defective endocytosis, excessive number of binding sites on the cell surface. There is a substantial amount of evidence that plasma membranes recycle rapidly and that during receptor mediated endocytosis recirculation of receptors takes place [18]. It is assumed that the LDL receptors are uncoupled in a compartment for uncoupling of receptor and ligand (CURL) as described for asialoglycoproteins [6], and that the uncoupled LDL receptors recirculate to coated pits [1]. The much larger amount of LDL internalized during a continuous incubation at 37 °C than during a pulse-chase experiment by fibroblasts and SCC25 cells (Figs. 2, 3, 10, 11) also indirectly indicates that recirculation of LDL receptors takes place. Because of the complexity of the receptor mediated endocytosis a defect in this uptake may be related to various abnormalities in the cellular mechanisms of this process as shown in Fig. 14. Moreover, the question can be raised to which extent the defects described in epithelial tumor cells are related to the process of differentiation, especially keratinization. With our standard culture conditions no LDL receptors could be demonstrated on cultured keratinocytes [14].

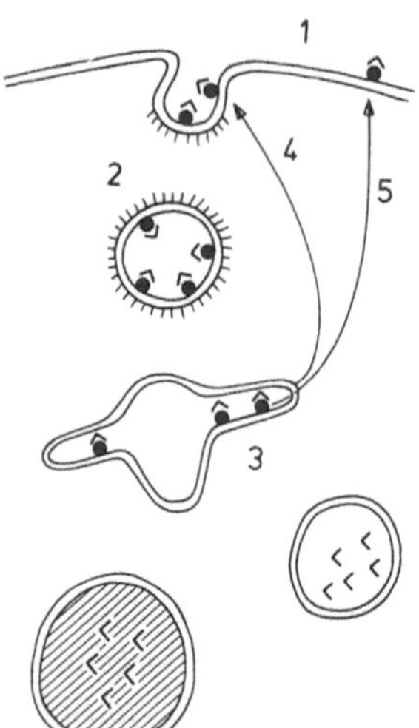

Fig. 14. The various localizations of defects in cellular mechanisms which can influence the receptor mediated endocytosis.
1 = receptor clustering (diffusion)
2 = internalization via coated pit-vesicle system
3 = uncoupling of ligands from receptor
4 = recirculation receptors to coated pits
5 = recirculation receptors to plasma membrane outside coated pits

In addition, epithelial tumor cells, cultured under similar conditions, showed no LDL binding on those areas of the cell culture where differentiation, especially keratinization occurred (Fig. 6). When these epithelial tumor cells were plated at low density and no differentiation had taken place, more LDL binding was observed (Fig. 4, 5), [24]. These data have recently been extended by the observations that keratinocytes cultured in medium containing low calcium, in which differentiation is retarded [10] do have LDL receptors. These latter findings are similar to the studies of O'Keefe et al. [11] on receptor binding of EGF in cultured kerationcytes. Based on these results we postulate that the described defects in LDL receptor metabolism might represent a more general defect in receptor metabolism and that this defect might be related to the process of differentiation. An important question emerges, if the recycling of plasma membrane and fluidity of plasma membrane are also related to this process.

Parts of these studies were supported by grants no. 83/096 from "De Nederlandse Hartstichting" and no. 28.826 from the Praeventie Fonds.

References

1. Anderson RGW, Brown MS, Beisiegel U, Goldstein JL (1982) Surface distribution and recycling of the low density lipoprotein receptor as visualized with antireceptor antibodies. J Cell Biol 93:523–531
2. Anderson RGW, Brown MS, Goldstein JL (1981) Inefficient internalization of receptor-bound low density lipoprotein in human carcinoma A-431 cells. J Cell Biol 88:441–452
3. Anderson RGW, Goldstein JL, Brown MS (1976) Localization of low density lipoprotein receptors on plasma membrane of normal human fibroblats and their absence in cells from a familial hypercholesterolaemia homozygote. Proc Natal Acad Sci 73:2434–2438
4. Beisiegel U, Schneider WJ, Goldstein JL, Anderson RGW, Brown MS (1981) Monoclonal antibodies to low density lipoprotein receptor as probes for study of receptor-mediated endocytosis and the genetics of familial hypercholesterolamia. J Biol Chem 256:11923–11931
5. Gal D, Simpson ER, Porter JC, Snyder JM (1982) Defective internalization of low density lipoprotein in epidermoid cervical cancer cells. J Cell Biol 92:597–603
6. Geuze JH, Slot JW, Strous GJAM, Lodish HF, Schwartz AL (1983) Intracellular site of asialoglycoprotein receptor-ligand uncoupling: double-label immunoelectron microscopy during receptor-mediated endocytosis. Cell 32:277–287
7. Goldstein JL, Anderson RGW, Brown MS (1979) Coated pits, coated vesicles, and receptor-mediated endocytosis. Nature 279:679–685
8. Goldstein JL, Brown MS (1977) Low density lipoprotein pathway and its relation to atherosclerosis. Ann Rev Biochem 46:897–930
9. Handley DAC, Arbeeny CM, Eder HA, Chien S (1981) Hepatic binding and internalization of gold low density lipoprotein conjugates in perfused livers of 172-ethinyl estradiol treated rats. J Cell Biol 90:778–788
10. Hennings H, Michael D, Cheng C, Steinert P, Holbrook K, Yuspa SH (1980) Calcium regulation of growth and differentiation of mouse epidermal cells in culture. Cell 19:245–245
11. O'Keefe EJ, Payne RE (1983) Modulation of the epidermal growth factor receptor of human keratinocytes by calcium ion. J Invest Dermatol 81:231–235
12. Pitas RE, Innerarity ThL, Weinstein JN, Mahley RW (1981) Acetoacetylated lipoproteins used to distinguish fibroblasts from macrophages in vitro by fluorescence microscopy. Arteriosclerosis 1:177–185

13. Pastan J, Willingham MC (1983) Receptor-mediated endocytosis: coated pits, receptosomes and the Golgi. Trends Biochem Sci 8:250–253
14. Ponec M, Havekes L, Kempenaar J, Vermeer BJ (1983) Cultured human skin fibroblasts and keratinocytes: Differences in the regulation of cholesterol synthesis. J Invest Dermatol 81:125–130
15. Redgrave TG, Roberts DCK, West C (1975) Separation of plasma lipid proteins by density-gradient ultracentrifugation. Anal Biochem 65:42–49
16. Roth Th, Woods JW (1982) Fundamental questions in receptor-mediated endocytosis in differentiation and function of haematopoietic cell surfaces. Alan R Liss, New York, pp 163–181
17. Steer CJ, Klausner RD (1983) Clathrin-coated pits and coated vesicles: Functional and structural studies. Hepathology 3:437–454
18. Steinman RM, Mellman JB, Muller WA, Cohn ZA (1983) Endocytosis and the recycling of plasma membrane. J Cell Biol 96:1–27
19. Van Hinsbergh VWM, Havekes L, Emeis JJ, Van Corven E, Scheffer M (1984) Metabolism of low density lipoprotein and acetylated LDL by confluent endothelial cells from human umbilical cord, arteries and veins. Arteriosclerosis 3:547–559
20. Vasile E, Simionescu M, Simionescu N (1983) Visualization of the binding, endocytosis and transcytosis of low density lipoprotein in the arterial endothelium in situ. J Cell Biol 96: 1677–1689
21. Vermeer BJ, de Bruijn WEC, van Gent CM, de Winter CPM (1978) Ultrastructural findings on lipoproteins in vitro and in xanthomatous tissue. Histochem J 10:299–307
22. Vermeer BJ, Havekes L, Wijsman MC, Emeis JJ (1980) Immunoelectronmicroscopical investigations on the absorptive endocytosis of low density lipoproteins by human fibroblats. Exp Cell Res 129:201–210
23. Vermeer BJ, Reman FC, Emeis JJ, de Haas- vd, Poel CAC (1978) Immunoenzyme histochemical demonstration of the binding of low density lipoproteins to cultured human fibroblasts. Histochemistry 56:197–201
24. Vermeer BJ, Wijsman MC, Mommaas-Kienhuis AM, Ponec M, Havekes L (1985) Modulation of low density lipoprotein receptor activity in squamous carcinoma cells by variation in cell density. Eur J Cell Biol 38:353–360

Use of In Vivo and In Vitro Test Systems as Predictors of the Cutaneous Irritancy of Synthetic Detergents (Syndets)

P. J. Dykes, D. L. Williams, L. A. Jenner and R. Marks

There have been several attempts to develop laboratory or animal based test systems [6, 8, 12, 14] which might prove useful either in the prediction of irritancy of synthetic detergents (syndets) in humans or in discriminating between different syndets. The degrees of success of these methods have been variable and the search continues for more suitable techniques.

In this study we have used two laboratory based methods to examine a series of syndets and attempted to correlate the findings with irritancy studies in normal human subjects. The two methods were chosen to model two important aspects of skin-syndet interaction. The first test system employed was an in vivo forced desquamation test which we believe simulates the initial solubilising effect of the syndet on the stratum corneum. The second test used was a cell lysis assay using human skin fibroblasts, which we believe represents the inherent toxicity of the syndets for unprotected mammalian cells.

Syndets

Samples of five liquid syndets were made available by Johnson & Johnson GmbH, Düsseldorf, FRG. They were received coded. All dilutions of syndets were made on a volume to volume basis from the "in use" concentration. That is, no correction was made for the inherent differences in the active content of the different syndets.

Forced Desquamation Test

The forced desquamation test was performed on the forearms of 12 normal volunteers who had given their informed consent. The test was performed using a Desquamator (Cutech Ltd., England) as described by Roberts and Marks [13]. This device is designed to deliver a standardised mechanical stimulus to the skin surface by means of a rotating perspex blade. In this study a stimulus of torque 4 (40 g) was applied for 30 s in the presence of 0.05 M phosphate buffer pH 7.2 containing 0.1% (v/v) of the syndet. Treatments were allocated in a random fashion to different sites on the flexor aspect of the forearm of each volunteer. After the stimulus, samples of buffer were removed and the number of corneocytes released estimated using a haemacytometer. An aliquot of the buffer was also centrifuged (12,000 g x 5 min) and the protein content of the supernatant estimated using a modified Lowry assay [4].

Fibroblast Lysis Assay

Normal human fibroblast cultures were established by routine methods from foreskin specimens removed during routine surgical circumcision. Cells were maintained in Dulbecco's modification of Eagle's medium containing 10% newborn calf serum. Confluent cultures were trypsinised and resuspended and aliquots of 10^4 cells transferred to the wells of a 96-well tissue culture plate (Flow Laboratories, Scotland). After three to four days, when the wells were confluent, the cells were washed with phosphate buffered saline and incubated with either medium alone or medium containing syndet. The syndet concentration ranged from 0.0125 to 0.6% (v/v). After one hour at 37 °C the medium was removed and the cells fixed in phosphate buffered formalin. Cells were stained with crystal violet and and the density of colour due to cellular uptake was assessed using a variation of the densitometric method of Matthews [10]. Briefly, fixed, stained cells were solubilised with glacial acetic acid and the optical density measured at 540 nm using a Titertek Multiskan plate reader (Flow Laboratories, Scotland). Results were expressed as a percentage of the optical density of the control (medium alone) well.

Soap Chamber Test

The soap chamber test was performed as described by Frosch and Kligman [7], with the following modifications: no preselection of volunteers was performed, and the back, rather than the forearm, was used, and the concentration of syndet was 10% (v/v) in water.

In this study 20 normal volunteers, who had given their informed consent, were used. Aliquots of diluted syndet (65 μl) were pipetted onto filter paper discs in 12 mm Duhring chambers. These were then applied to the skin of the back for a 24 h period. This was followed by intermittent application consisting of a 6 h application followed by an 18 h rest period for each of the next 4 days. Assessments of erythema were made on days 1, 2, 3, 4 and 7, using the 0–4 scale described by Frosch and Kligman [7] (Fig. 3).

Results

Forced Desquamation Test

The results of the forced desquamation test are summarised in Fig. 1. Little variation was observed in the mean number of corneocytes released by each syndet. However, the results for protein released did appear to show some differences. This was confirmed by using a non parametric Friedman two-way analysis of variance [3]. Comparison of individual pairs of treatments indicated that at $p < 0.05$, A differed from B, C and E; B differed from C, D and E; C differed from E; and D differed from E. Thus with confidence a rank order, from least to most effective, of

$$B < A < D = C < 3$$

can be assigned.

Fibroblast Lysis Assay

The results of the fibroblast lysis assay are shown in Fig. 2. This shows the optical density of each well expressed as a percentage of the control well for each dilution of

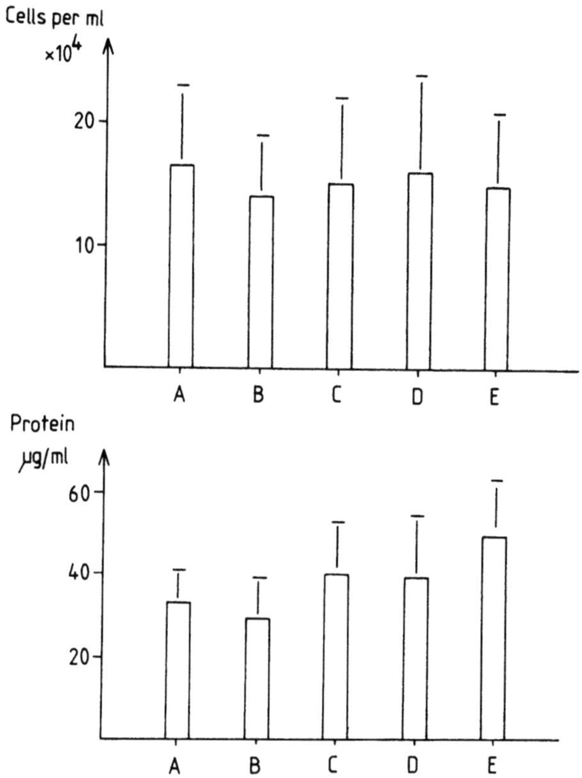

Fig. 1. Effect of different syndets on the number of corneocytes and protein released from the skin surface during the forced desquamation test. Results shown are the mean ± standard deviation for 12 volunteers tested with the syndets A, B, C, D and E

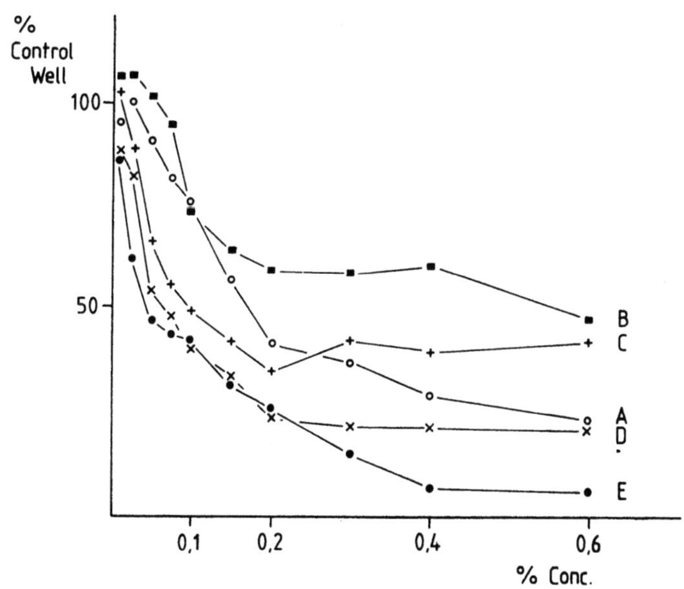

Fig. 2. Effect of increasing concentrations of syndet on fibroblast lysis *in vitro*. Results shown are for the optical density of each well at 540 nm expressed as a percentage of the control (untreated) well

Table 1. Comparison of the 50% lysis value of syndets in two separate experiments

Syndet	50% Lysis value*	
	Experiment 1	Experiment 2
B	0.73	0.52
A	0.31	0.20
C	0.18	0.15
D	0.11	0.08
E	0.07	0.06

* Concentration of syndet at which 50% of control well value is reached

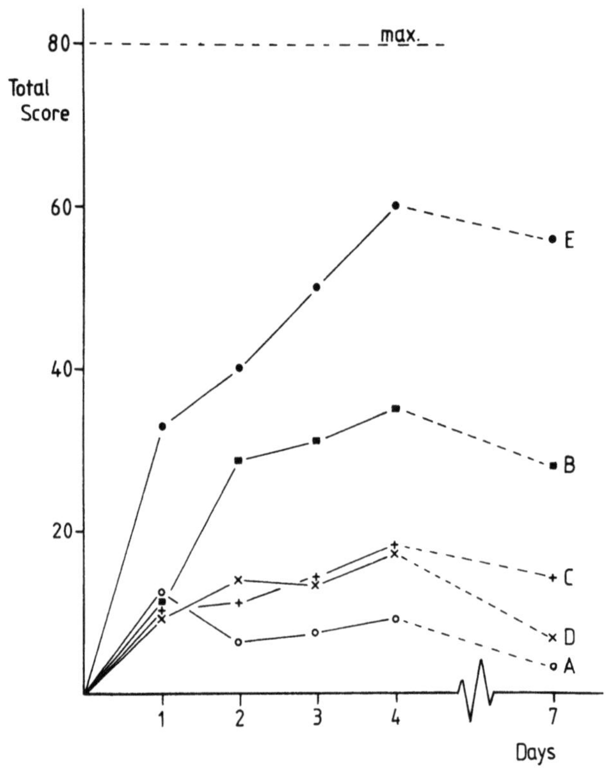

Fig. 3. Effect of different syndets on the erythema score in the soap chamber test. The results are expressed as the total score for erythema over all 20 volunteers at days 1, 2, 3, 4 and 7. The maximum possible score per site = 4, therefore the maximum possible total score per syndet is 4 x 20 = 80. The erythema score was determined according to the following scale; 0 = normal skin; 1 = slight redness, spotty or diffuse; 2 = moderate, uniform redness; 3 = intense redness; 4 = fiery red with edema

Table 2. Comparison of the Rank Orders of Syndets in the three test systems

Test	Least ⟶ Most effective syndet
Cell culture	B < A < C < D < E
Desquamator (protein released)	B < A < D = C < E
Human chamber test	A < D = C < B < E

syndet. The range of syndet concentration was 0.6 to 0.0125%. The data was transformed logarithmically and regression analysis performed to give a line of best fit. From this the concentration of syndet corresponding to 50% of the control well value (50% lysis value) was obtained. The results of two separate experiments are given in Table 1.

As can be seen, this assay distinguishes between the different syndets and although the absolute values differ between the two experiments, the rank order is maintained. The least to most effective rank order of syndets is B < A < C < D < E, which is the same rank order as in the Desquamator test.

Figure 3 shows the result for the soap chamber test. The results are expressed as the total erythema scores for each treatment at days 1, 2, 3, 4 and 7. Differences in reactivity were apparent at day 2 and were maintained over the rest of the study. Syndet E was the most irritant and A the least. Statistical analysis of the results at day 4 using the Friedman test indicated that A, B and E differed from each other and the other treatments and that there was no difference between C and D. With some confidence we can assign a rank order. least to most irritant, of

A < D = C < B < E

A comparison of the rank orders for the three test systems is given in Table 2.

Discussion

The irritancy of a substance after application to the skin will depend, amongst other factors, on the rate of penetration across the stratum corneum, the initial damage to the stratum corneum, the ability to damage epidermal and dermal cell membranes and cellular functions, and the inherent sensitivity of the tissues concerned. In this study tests are described which we believe serve as models for two important aspects of these interactions. Firstly the forced desquamation test, which models the initial solubilisation of the stratum corneum. Secondly, the fibroblast lysis assay, which models for the ability of a substance to damage cell membranes and interfere with cellular processes. Both of these test systems are able to discriminate between the different syndets used. In addition, despite the widely differing nature of the tests, they give the same rank order (Table 2). The reason for this close correlation is not clear but may be related to the similar mechanisms involved in solubilising the plasma membrane of the fibroblast and the lipid envelope of the corneocyte.

The soap chamber test also discriminates between the syndets used. The rank order is different from that observed with the other test systems. Although the rank order shows some similarities in that E is the most effective and A tends to be the least effective, syndet B shows a large difference. In the two test systems B is the least effective, whereas in the soap chamber test it is the second most effective irritant. A good correlation betwen the soap chamber test and antecubital and facial wash tests has been found [5]. There were, however, some discrepancies, and it is not clear whether the differences observed here reflect similar discrepancies or whether the two test systems are not useful as predictors of irritancy.

A variety of test systems have been used to characterize the toxicity of substances that come into contact with the skin [2, 9, 11] but few are suitable for evaluation of soaps, shampoos and detergents. Previously these materials have been assessed on animal eyes in vivo [1] or in vitro tests [12] or in human use type irritancy tests. None to date have been completely satisfactory and we suggest that our approach or some modification of this may prove more adequately predictive.

References

1. Beckley JH, Russell TJ, Rubin LF (1969) Use of the rhesus monkey for predicting human response to eye irritants. Toxicol App Pharmacol 15:1–8
2. Bronaugh RL, Maibach HI (1982) Evaluation of skin irritation: correlation between animals and humans. In: Kligman AM, Leydon JJ (eds) Safety and efficacy of topical drugs and cosmetics. Grune & Stratton, New York, pp 51–62
3. Conover WJ (1980) Practical non parametric statistics, 2nd edn. John Wiley & Sons, New York, pp 299–303
4. Dully JR, Gneve PA (1975) A simple technique for eliminating interference by detergents in the Lowry method of protein determination. Analyt Biochem 64:136–141
5. Frosch PJ (1982) Irritancy of soaps and detergent bars. In: Frost P (ed) Principles of cosmetics for the dermatologist. CV Mosby Co, St Louis, pp 5–12
6. Frosch PJ, Kligman AM (1976) The chamber scarification test of irritancy. Contact Dermatitis 2:314–324
7. Frosch PJ, Kligman AM (1979) The soap chamber test. A new method for assessing the irritancy of soaps. J Amer Acad Dermatol 1:35–41
8. Harrold SP (1959) Denaturation of epidermal keratin by surface active agents. J Invest Dermatol 32:581–588
9. Kligman AM, Wooding WM (1967) A method for the measurement and evaluation of iritants on human skin. J Invest Dermatol 49:78–94
10. Matthews N (1979) Tumour necrosis factor from rabbit. III. Relationship to interferon. Brit J Cancer 40:534–539
11. Philips L. Steinberg M, Maibach HI, Akers WA (1972) A comparison of rabbit and human skin response to certain irritants. Toxicol Appl Pharmacol 21:369–382
12. Prottey C, Ferguson T (1975) Factors which determine the skin irritation potential of soaps and detergents. J Soc Cosmet Chem 26:29–46
13. Roberts D, Marks R (1980) The determination of regional and age variation in the rate of desquamation: A comparison of four techniques. J Invest Dermatol 74:13–16
14. Scaife MC (1982) An investigation of detergent action on cells in vitro and possible correlation with in vivo data. Int J Cosmet Sci 4:179–193

Lectin Induction of Pemphigus and Pemphigoid-Like Bullae in Organ Cultured Human Skin

M. Fitzmaurice and S. D. Deodhar

Previous studies of the pathogenesis of pemphigus vulgaris in an *in vitro* organ culture model [2, 5, 6, 9–12] have shown that deposition of pemphigus antibodies within the intercellular substance (ICS) is responsible for the production of acantholytic suprabasalar bullae in the epidermis of human skin. These studies also suggest that pemphigus antibodies induce acantholysis and the resulting bullae, not by the binding and activation of complement, but rather by the activation and release of epidermal cell proteinases which digest the intercellular cement. Although there is no experimental evidence in the *in vitro* organ culture model to date to support the idea [6, 9], it has also been suggested that the pathogenesis of bullous pemphigoid may be such that deposition of pemphigoid antibodies within the basement membrane (BM) zone is responsible for the induction of subbasalar bullae in the epidermis of human skin. It might, therefore, also be suggested that pemphigoid antibodies induce bullae by the activation and release of basal cell proteinases which digest the lamina lucida of the BM.

It appears, then, that the ability of pemphigus and pemphigoid antibodies to induce intraepidermal bullae may not be as dependent upon their being immunoglobulins (with the attendent functions of complement-binding, etc.) as upon their binding specificity for the glycoprotein antigens of the ICS and BM, which may act as epidermal cell surface receptors. If this is the case, other nonimmunoglobulin substances with similar binding specificities may also be capable of induction of bullae.

One such class of substances is the plant and animal lectins, which have antibody-like binding specificity for monosaccharides as they are found in glycoproteins [3]. A number of lectins have been shown previously to bind to glycoproteins in the ICS and BM of human skin, where pemphigus and pemphigoid antigens are expressed [1, 4, 7, 8, 13]. In fact, we have previously used the patterns of lecitin binding to the epidermis to map the distribution of monosaccharides within the carbohydrate moieties of the glycoproteins of the ICS and BM of human skin (unpublished results).

In this study, we have devised an *in vitro* organ culture model to study bullous formation in human skin, using plant and animal lectins which bind to the epidermal ICS and BM. The relation of lectin binding to bullous formation was also studied by direct and indirect fluorescence using fluorescein isothiocyanate conjugated lectins.

List of abbreviations: BM – Basement membrane: Con A – Concanavalin A: FITC Fluorescein isothiocyanate; F/P – Fluorescein/protein ratio; FTA – FTA hemagglutination buffer; H and E – Hematoxylin and eosin; ICS – Intercellular substance; LPA – Limulus polyphemus agglutinin; OD – Optical density; RCA – Ricinus communis agglutinin; UEA Ulcx europaeus agglutinin; WGA – Wheat germ agglutinin

Materials and Methods

Human Skin

Normal human skin for use in the *in vitro* organ culture system was obtained from hair transplant surgery. During the surgical procedure, up to one hundred 4 mm plugs were removed from alopecic scalp to make room for a like number of transplant plugs from nonalopecic scalp. The alopecic scalp plugs, which are usually discarded, were collected and organ cultured immediately.

Lectin and Fluorescein Isothiocyanate (FITC) Lectin Conjugates

FITC-conjugated and unconjugated forms of the following lectins were used in this study: Concanavalin A (2x crystallized) (Con A), Wheat germ agglutinin (WGA), Ricinus communis agglutinin-60 (RCA-60), Ricinus communis agglutinin-120 (RCA-120) (Miles-Yeda, Kiryat-Weizmann, Rehovot, Israel), Limulus polyphemus agglutinin (LPA), Lotus A, and Ulex europaeus agglutinin-I (UEA-I) (E.Y. Laboratories, San Mateo, CA). The plant or animal sources, monosaccharide binding specificities, protein concentrations, OD495/OD280 ratios, and molar fluorescein/protein (F/P) ratios of the lectins and their FITC-conjugates are shown in Table 1. Each of the lectins and their FITC-conjugates were aliquoted and stored at –20 °C, except unconjugated Con A which was stored at room temperature, until used.

These lectins were selected for use in this study because their monosaccharide-binding specificities include those monosaccharides commonly found within the carbohydrate moieties of human glycoproteins and because most had been shown previously to bind to glycoproteins within the ICS or BM region of human skin as shown in Table 2 (unpublished results).

Organ Culture Media

Sterile Nutrient Mixture F-10 (Ham) (GIBCO, Grand Island, NY) supplemented with 10% fetal bovine serum (heat inactivated, mycoplasma tested, virus screened) (GIBCO) and 1% penicillin-streptomycin (10,000 units – 10,000 μg/ml) (GIBCO) was used as the standard organ culture medium. This organ culture medium was then supplemented with 0.1 mg/ml of one of the 7 lectins or their FITC-conjugates.

Organ Culture Method

The human skin was organ cultured by a modification of the lens paper "raft" method of Michel and Ko [6]. First, sterile 1.5 x 1.5 cm lens paper "rafts" (Fisher Scientific Co., Pittsburg, PA) were edged with high vacuum grease (Dow Corning Corp, Midland, MI) and floated on 5 ml of sterile supplemented or unsupplemented organ culture medium in sterile plastic tissue culture dishes (60 x 15 mm) (Falcon Plastics, Oxnard, CA). The scalp plugs were then washed 3 times with sterile Hank's balanced salt solution (calcium and magnesium-free) (GIBCO), trimmed to a thickness of about 1 mm with a sterile surgical blade, and placed dermis-down on the sterile lens paper "rafts." The organ cultures were then covered, incubated at 37 °C in a moist 5% CO_2 atmosphere for up to 96 h, and harvested.

Duplicate cultures were set up for each supplemented and unsupplemented organ culture medium for each length of incubation (24, 48, 72, and 96 h) with 2–3 plugs

Table 1. Characteristics of fluorescein-isothiocyanate lectin conjugates

FITC-Lectin	Source	Monosaccharide binding specificity	Protein concentration (mg/ml)	OD495 / OD280	Molar F/P ratio
Concanavalin A	Jack bean	Mannose, glucose, glucosamine	17.1–18.3	1.38–1.49	2.14–2.38
Wheat germ agglutinin	Wheat germ	Glucosamine	2.0	0.62	4.10
Ricinus communis agglutinin-60	Castor bean	Galactose, galactosamine	2.55–3.0	0.98–1.19	1.36–1.85
Ricinus communis agglutinin-120	Castor bean	Galactose	2.3	0.9	4.30
Limulus polyphemus agglutinin	Horseshoe crab	Sialic acid	1.0–2.0	0.9–1.3	–
Lotus-A	Lotus seed	Fucose	1.0–2.0	0.96–4.11	–
Ulex europaeus agglutinin-I	Gorse seed	Fucose	1.0	1.16–1.34	–

Table 2. Patterns of binding of lectins to human skin

FITC-Lectin	Epithelial regions						
	ICS						BM
	Stratum basale	Stratum spinosum		Stratum granulosum	Stratum lucidum	Stratum corneum	
		lower	upper				
Concanavalin A	+	+	+	+	+	+	+
Wheat germ agglutinin	–	–	+	+	+	+	–
Ricinus communis agglutinin-60	+	+	+	+	+	+	+
Ricinus communis agglutinin-120	+	+	+	+	+	+	+/–
Limulus polyphemus agglutinin	–	–	–	–	–	–	–
Lotus A	–	–	–	–	+	–	–
Ulex europaeus agglutinin-I	–	–	–	+	+	–	+

per "raft" and 2–3 "rafts" per dish. When harvested, one replicate was snap-frozen in liquid nitrogen, mounted with distilled water on a chuck, cut into 4 micron sections using a Lisshaw cryostat, and then stored at –70 °C until used in the fluorescence studies. The other replicate was fixed in Zenker's solution and submitted for routine processing into hematoxylin and eosin (H and E) stained paraffin sections for histopathologic study. An aliquot of spent organ culture medium was also reserved after each length of incubation and stored at –20 °C until used in the fluorescence studies.

Histopathologic Method

Random H and E stained paraffin sections of each organ culture were examined for histopathologic changes by light microscopy using a Bausch and Lomb Series 354 microscope.

Lectin Fluorescence Methods

Lectin binding to human skin organ cultured with FITC-lectin conjugate supplemented organ culture media was detected using a direct fluorescence technique. In this technique, cryostat sections of each organ culture were simply mounted with glycerol (J. T. Baker Chemical Co., Phillipsburg, NJ): FTA hemagglutination buffer (BBL, Cockeysville. MD) (9:1) and read blindly using a Zeiss standard IFG diagnostic microscope with an IV epi-fluorescence condenser and BG12 blue exciter, FT460 chromatic beam splitter. and LP478 yellow barrier filters.

The spent FITC-lectin conjugate-supplemented organ culture media were also screened for the presence of intact FITC-lectin conjugates using an indirect fluorescence technique with uncultured human skin as the tissue substrate. In this technique, cryostat sections of uncultured skin were incubated with 50 μl of each undiluted medium for 60 min at room temperature in a moist chamber. The sections were then washed with 2 changes of FTA for a total of 15 min at room temperautre, mounted with glycerol: FTA, and read blindly using the fluorescent microscope. An autofluorescence control in which the cryostat sections were incubated with FTA alone was included.

Results

Histopathology

The human skin scalp plugs obtained from each donor at hair transplant surgery consisted grossly of epidermis, dermis, and some of the underlying subcutaneous fatty tissue. Once trimmed for use in the organ culture studies, they consisted of only the epidermis and immediately adjacent dermis. Histopathologic examination of the uncultured skin from each donor showed both epidermis, consisting of slightly thinned but otherwise normal keratinized stratified squamous epithelium, and dermis, consisting of dense collagenous connective tissue essentially devoid of hair follicles and accessory glands with occasional sparse patches of chronic inflammatory cells.

Histopathologic examination of skin cultured in the unsupplemented medium showed that the basic architectural patterns of both the epidermis and dermis were retained throughout the culture period. However, certain histopathologic changes, including

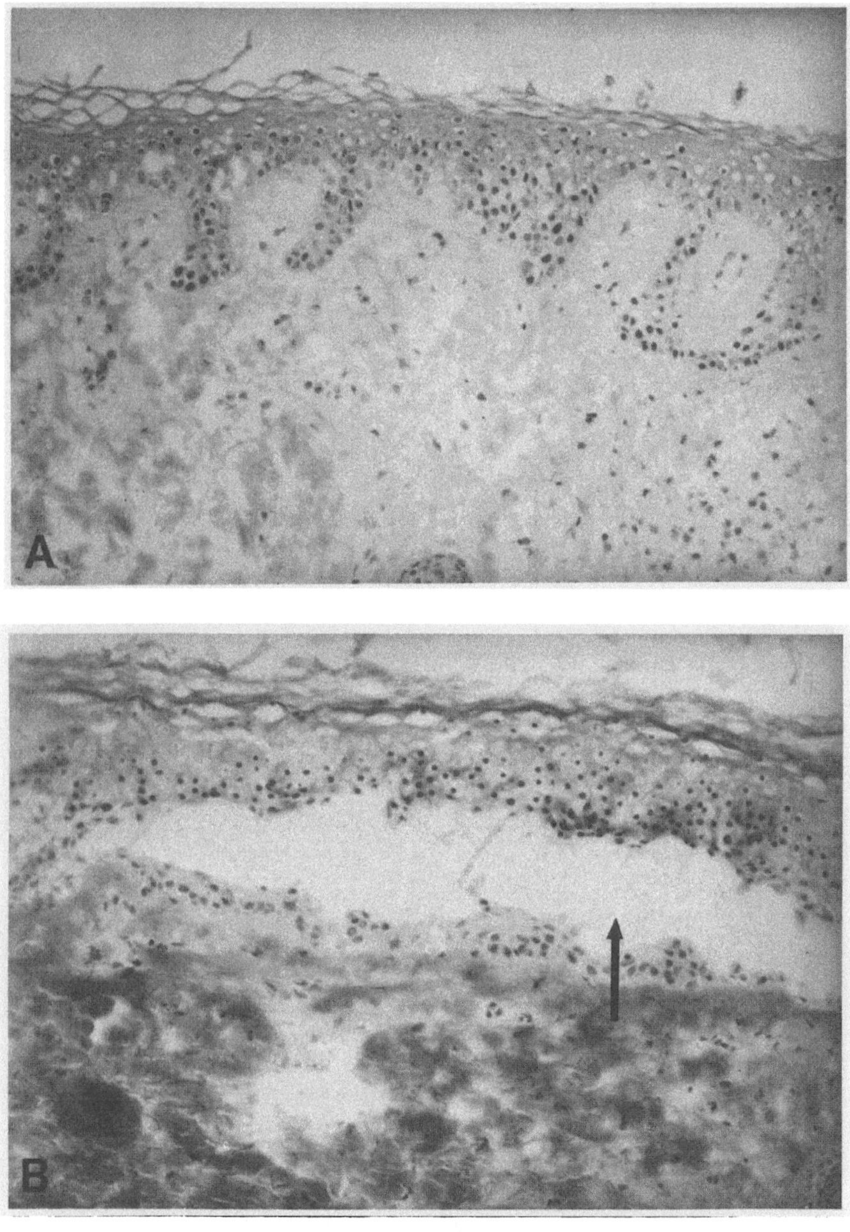

Fig. 1A, B. Histopathology of organ cultured skin: Con A. Histopathology of human skin organ cultured with medium supplemented with 0.1 mg/ml Con A: **A** 24 h culture with no significant histologic change and **B** 72 h culture with non-acantholytic suprabasalar bullous (arrow)

disorganization, intracellular edema, and parakeratosis, were seen in the epidermis. These changes were first seen at 24 to 48 h of culture and became more pronounced with time, so that by 96 h of culture the epidermis consisted of only 3–4 layers of parakeratotic epithelium overlaying the basal cell layer. The histopathology of the dermis was not examined in the same detail as was that of the epidermis. Nevertheless,

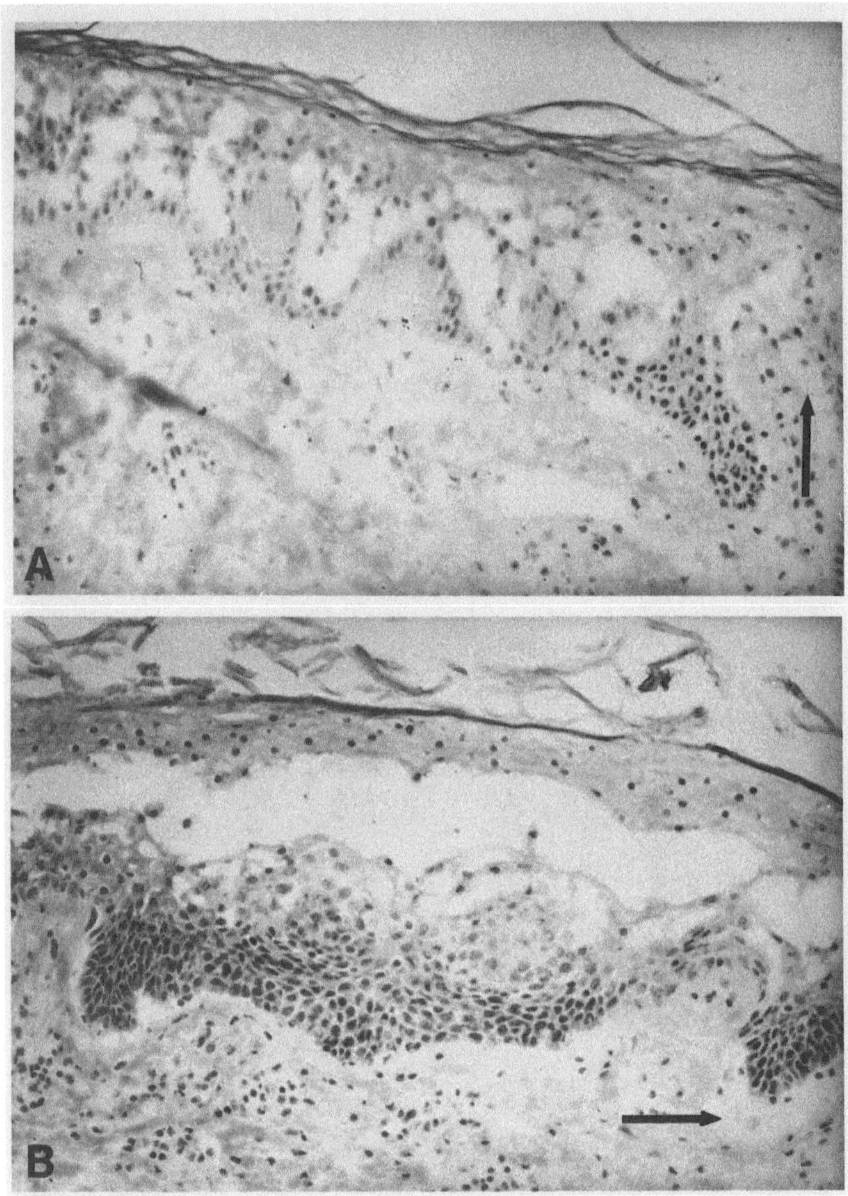

Fig. 2A–C. Histopathology of organ cultures skin: RCA-60, Histopathology of human skin organ cultured with medium supplemented with 0.1 mg/ml RCA-60: **A** 24 h culture with marked suprabasalar acantholysis and early sub-basalar separation (arrow): **B** 48 h culture with both suprabasalar and subbasalar (arrow) bullae: and

the only significant histopathologic change in the dermis was progressive homogenization with loss of detail in the dense collagen bands.

Histopathologic examination of skin cultured in medium supplemented with only 4 of the 7 lectins, Con A, WGA, RCA-60, and RCA-120, showed intraepidermal bullous formation. Specifically, skin cultured in medium supplemented with either FITC-con-

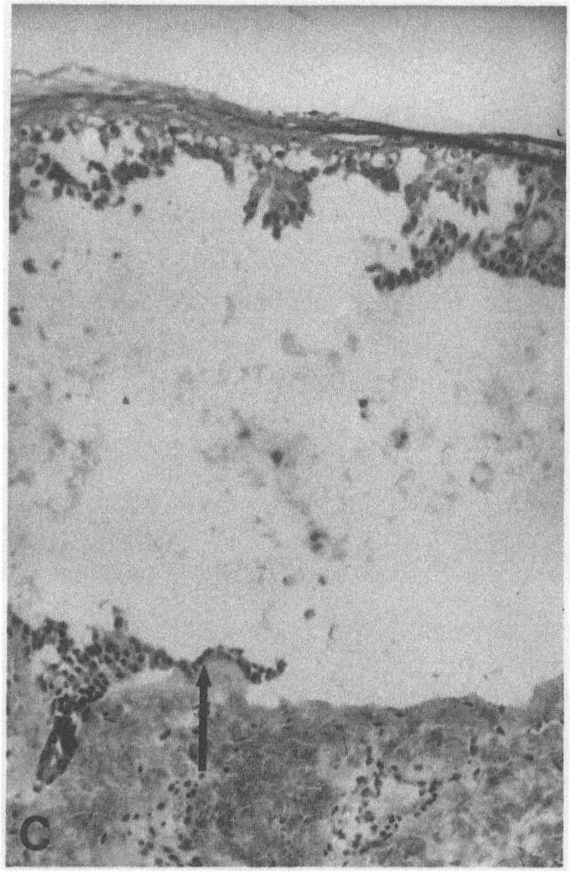

C 72 h culture with coalescent bullous lined in part by tomb-stone like basal cells (arrow)

jugated or unconjugated Con A or WGA showed frank bullous formation at 72–96 h and occasionally complete separation of the upper strata from the lower strata of the epidermis at 96 h of culture. The bullae were suprabasalar, with the stratum basale and sometimes several layers of the stratum spinosum forming the floor of the bullae. The bullae were not generally acantholytic but did rarely contain isolated epidermal cells. The typical histologic appearance of skin cultured in Con A-supplemented medium is shown in Fig. 1. The histologic appearance of skin cultured in WGA-supplemented medium was virtually indistinguishable from that of skin cultured in Con A-supplemented medium.

In contrast, skin cultured in medium supplemented with either FITC-conjugated or unconjugated RCA-60 or RCA-120 showed both progressive intraepidermal acantholysis and subbasalar fissuring beginning at 24–48 h. frank bullous formation at 48–72 h. and frequently complete separation of the epidermis from the dermis at the level of the BM zone at 96 h of culture. There were both suprabasalar bullae and sub-basalar bullae which appeared to coalesce with time in culture, with the floor of the bullae covered at least in part by a tombstone-like row of basal cells. The typical histologic appearance of skin cultured with RCA-60-supplemented medium is shown in Fig. 2. The histologic appearance of skin cultured with RCA-120-supplemented medium was

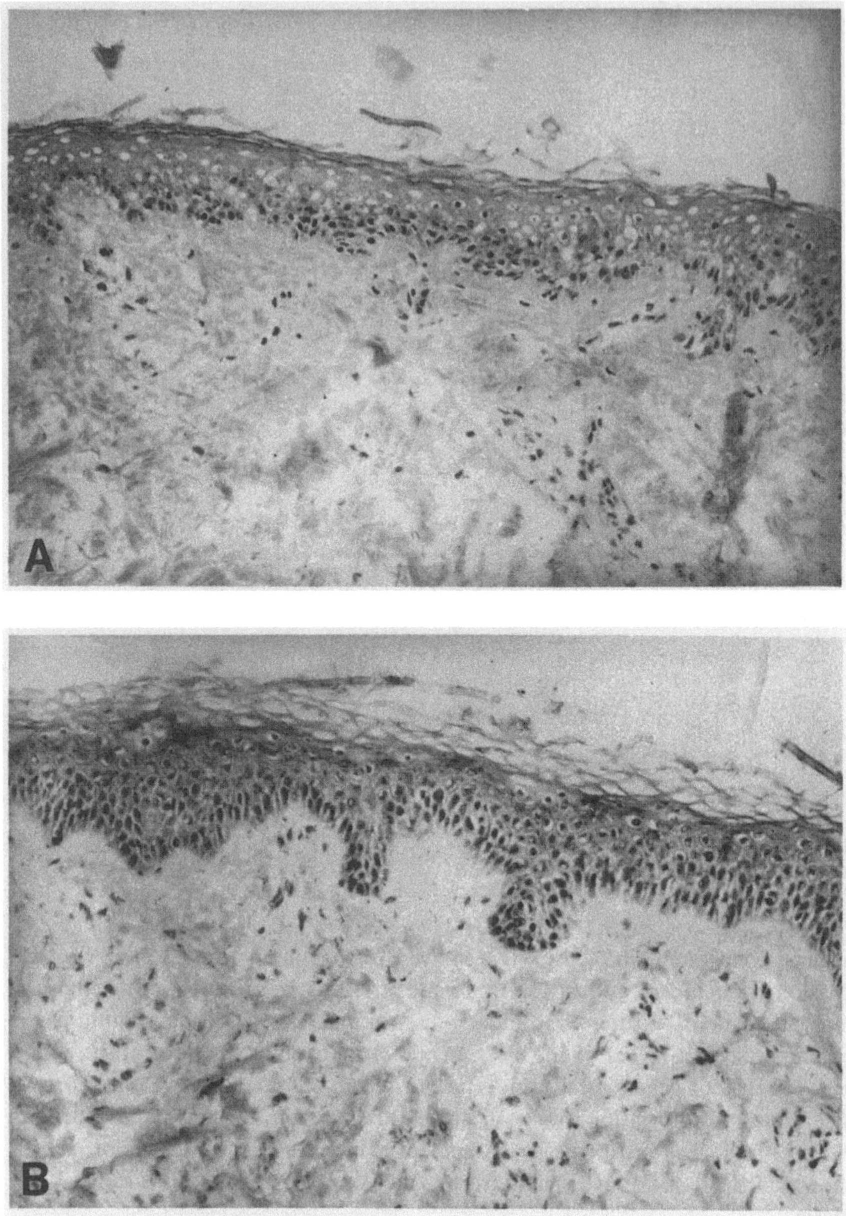

Fig. 3A, B. Histopathology of organ cultured skin: LPA. Histopathology of human skin cultured with medium supplemented with 0.1 mg/ml LPA: **A** 24 h culture and **B** 72 h culture with no significant histological change

virtually indistinguishable from that of skin cultured with RCA-60-supplemented medium.

Finally, skin cultured in medium supplemented with either FITC-conjugated or unconjugated LPA, Lotus A, or UEA-I did not show any bullous-like or other signifi-

Table 3. Histopathology of human skin organ cultures

Culture medium supplement	Culture time (h)	Histopathology			
		Acantholysis	Suprabasalar bullae	subbasalar fissuring	subbasalar bullae
FITC-Con A	0	–	–	–	–
	24	–	–	–	–
	48	–	–	–	–
	72	–	+	–	–
	96	–	+	–	–
FITC-WGA	0	–	–	–	–
	24	–	–	–	–
	48	–	–	–	–
	72	–	+	–	–
	96	–	+	–	–
FITC-RCA-60	0	–	–	–	–
	24	–	–	–	–
	48	+	–	+	–
	72	+	+	+	+
	96	+	+	+	+
FITC-RCA-120	0	–	–	–	–
	24	+	–	–	–
	48	+	+	+	–
	72	+	+	+	+
	96	+	+	+	+
FITC-LPA	0	–	–	–	–
	24	–	–	–	–
	48	–	–	–	–
	72	–	–	–	–
	96	–	–	–	–
FITC-Lotus A	0	–	–	–	–
	24	–	–	–	–
	48	–	–	–	–
	72	–	–	–	–
	96	–	–	–	–
FITC-UEA-I	0	–	–	–	–
	24	–	–	–	–
	48	–	–	–	–
	72	–	–	–	–
	96	–	–	–	–

cant histopathological changes after any length of incubation. The typical histologic appearance of skin cultured with LPA-supplemented medium is shown in Fig. 3. The histologic appearance of skin cultured with Lotus A or UEA-I-supplemented media was virtually indistinguishable from that of skin cultured with LPA-supplemented medium. The histopathology results of a typical experiment are summarized in Table 3.

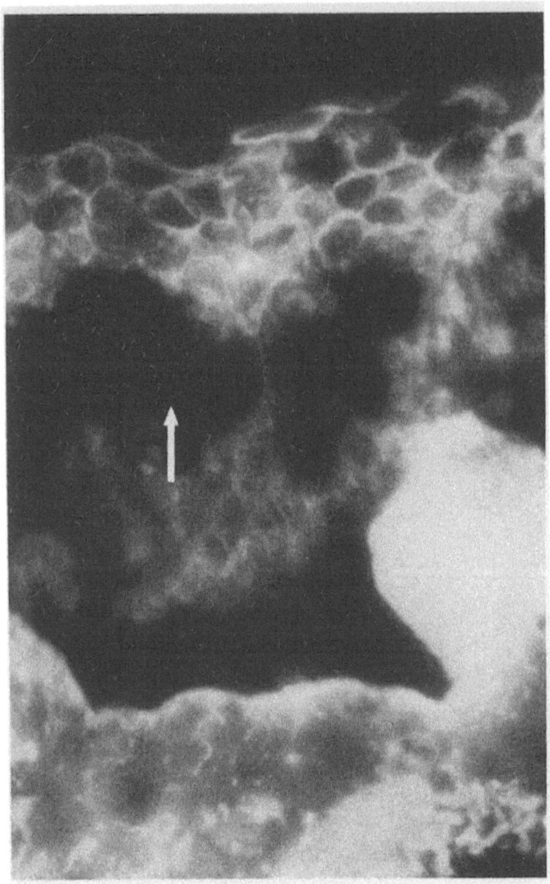

Fig. 4. Direct fluorescence of organ cultured skin: RCA-60. Typical pattern of fluorescence seen in human skin organ cultured with medium supplemented with 0.1 mg/ml FITC-RCA-60 at 48 h. Both ICS and BM fluorescence are seen in association with an intraepidermal bullous (arrow)

Lectin Binding

Lectin binding to the ICS and BM was detected by direct fluorescence in skin cultured with only four of the seven lectins, WGA, RCA-60, Lotus A, and UEA-I. Only two of these four lections, WGA and RCA-60, were among those that produced intraepidermal bullae. Specifically, lectin binding was detected by direct fluorescence in skin cultured in FITC-Con A and FITC-RCA-120 supplemented media, but only at the cut edges of the cultures as a diffuse band of apple-green fluorescence. There was no fluorescence due to FITC-Con A or FITC-RCA-120 binding detectable within the ICS region of the epidermis or along the BM zone between the epidermis and dermis after any length of incubation.

In contrast, lectin binding was detected by direct fluorescence in skin cultured in FITC-WGA and FITC-RCA-60 supplemented media, as either a finely linear, network pattern of apple green fluorescence throughout the ICS region of all strata except the stratum corneum of the epidermis or as a diffuse pattern of apple green fluorescence along the BM zone between the epidermis and dermis. The typical pattern of fluorescence of skin cultured with FITC-RCA-60 is shown in Fig. 4. Fluorescence due to FITC-

Table 4. Fluorescence of human skin organ cultures

Culture medium supplement	Culture time (h)	Direct fluorescence*		Indirect fluorescence**	
		ICS	BM	ICS	BM
FITC-Con A	0	0	0	0	0
	24	0	0	0	1+
	48	0	0	1+	1+
	72	0	0	1+	1+
	96	0	0	0	1+
FITC-WGA	0	0	0	3+	0
	24	0	0	3+	0
	48	2+	0	3+	0
	72	3+	0	3+	0
	96	0	0	3+	0
FITC-RCA-60	0	0	0	3+	3+
	24	0	1+	3+	3+
	48	2+	3+	3+	3+
	72	3+	3+	3+	3+
	96	2+	1+	3+	3+
FITC-RCA-120	0	0	0	3+	0
	24	0	0	3+	0
	48	0	0	3+	0
	72	0	0	3+	0
	96	0	0	3+	0
FITC-LPA	0	0	0	0	0
	24	0	0	0	0
	48	0	0	0	0
	72	0	0	0	0
	96	0	0	0	0
FITC-Lotus A	0	0	0	0	0
	24	0	2+	0	0
	48	0	2+	0	0
	72	0	2+	0	0
	96	0	2+	0	0
FITC-UEA-I	0	0	0	0	0
	24	0	2+	0	0
	48	0	2+	0	0
	72	0	2+	0	0
	96	0	2+	0	0

* Direct fluorescence of skin cultured with FITC-lectin supplemented media
** Indirect fluorescence of FITC-lectin supplemented spent media on uncultured skin

RCA-60 binding was most intense at 48–72 h of culture and decreased thereafter. FITC-WGA binding was frequently no longer detectable by 96 h of culture.

No lectin binding was detected by direct fluorescence in skin cultured in FITC-LPA supplemented media after any length of incubation. Lectin binding was detected by direct fluorescence in skin cultured in FITC-Lotus A and FITC-UEA-I supplemented media as a faint, finely linear pattern of apple green fluorescence along the BM zone between the epidermis and dermis. The direct fluorescence results of a typical experiment are summarized in Table 4.

Intact FITC-lectin conjugates of only four lectins, Con A, WGA, RCA-60, and RCA-120, were detectable by indirect fluorescence in their respective spent organ culture media. Only two of these four lectins, WGA and RCA-60, were among those for which ICS or BM binding could be detected by direct fluorescence. Intact FITC-Con A, FITC-WGA, FITC-RCA-60, and FITC-RCA-120 conjugates were detectable by indirect fluorescence in their respective spent organ culture media after each length of incubation. The intensity of fluorescence due to FITC-Con A, FITC-WGA, FITC-RCA-60, and FITC-RCA-120 binding to the uncultured skin used as tissue substrate did not decrease with the length of incubation of the spent organ culture media. In contrast, intact FITC-LPA, FITC-Lotus A, and FITC-UEA-I conjugates were not detectable by indirect fluorescence in their respective spent organ culture media after any length of incubation. The indirect fluorescence results of a typical experiment are also summarized in Table 4.

Discussion

The histopathologic changes seen here in human skin as a result of the *in vitro* organ culture process itself (epidermal disorganization, interstitial edema, and parakeratosis, etc.) have been described previously [2]. It appears that under the culture conditions used, human skin survives but does not grow, that is, the cells of the upper strata of the epidermis continue to undergo their normal differentiation processes but are no longer regenerated by the cells of the stratum basale. Despite this apparent lack of real growth or genereration, *in vitro* organ cultured human skin has proven a useful model for studying the pathogenesis of both pemphigus vulgaris and bullous pemphigoid.

In our studies, four of the seven lectins were found to be pathogenic to some degree in the *in vitro* organ culture model. Only two of these, RCA-60 and RCA-120, induced bullous histologic changes similar to those seen in pemphigus and pemphigoid. These two lectins induced bullous histologic changes which can best be described as a combination of those seen in both pemphigus vulgaris and bullous pemphigoid. It should also be noted that both of these lectins appeared to induce their suprabasalar bullae via epidermal acantholysis, the histologic hallmark of pemphigus. The other two pathogenic lectins, Con A and WGA, induced bullous histological changes which were quite dissimilar from those seen in pemphigus and pemphigoid. These two lectins induced histologic changes which can best be described as a dissection of the upper epidermis from the lower epidermis along an intercellular plane. Why the suprabasalar bullae induced by RCA-60 and RCA-120 were acantholytic and those induced by Con A and WGA were not is as yet unclear.

All four of the pathogenic lectins that induced suprabasalar bullae (whether acantholytic or not), Con A, WGA, RCA-60, and RCA-120, have been previously shown to bind to the ICS of human skin. Both of the pathogenic lectins that induced subbasalar bullae, RCA-60 and RCA-120, have also been previoulsy shown to bind to the BM of human skin. In contrast, none of the three nonpathogenic lectins, LPA, Lotus A, and UEA-I, have been previously shown to bind to the ICS of human skin, and only one of these lectins, UEA-I, had been previously shown to bind to the BM of human skin.

Lectin binding to the ICS and BM was often difficult to demonstrate by direct fluorescence in human skin cultured with FITC-lectin conjugates. Nevertheless, where seen, lectin binding usually preceeded the induction of bullous histopathologic changes. This suggests, but does not prove, a cause and effect relation between lectin binding and bullous formation. Since binding could not be demonstrated at all for two of the four pathogenic lectins, Con A and WGA, other causes of bullous induction, such as direct toxicity, must still be considered.

It should be noted that direct fluorescence of human skin cultured with FITC-lectin conjugates frequently revealed extensive lectin binding along the cut edges of the culture. Screening of FITC-lectin conjugate-supplemented media by indirect fluorescence also frequently failed to reveal intact FITC-lectin conjugates after even a brief period of culture. This suggests that both trapping of the FITC-lectin conjugates in the dermis as they perfuse up through the culture to the epidermis and dissociation of the FITC-lectin conjugates in the media during the culture process may be responsile, at least in part, for the failure to demonstrate lectin binding in these studies. Further ultrastructural studies with more stable lectin conjugates will be necessary to establish the true relation, if any, between lectin binding and bulla formation.

In conclusion, an *in vitro* organ culture model was devised to study bullous formation in human skin, using plant and animal lectins which bind to the epidermal ICS and BM. This model is a modification of the model first devised by Michel and Ko [6] and applied by others [2, 5, 9–12] to study the pathogenesis of bullous formation in pemphigus vulgaris and bullous pemphigoid, using pemphigus and pemphigoid antibodies. Although several lectins produced intraepidermal bullae, only two produced bullae histologically similar to those seen in pemphigus and pemphigoid. The relation of lectin binding to bullous formation was difficult to assess. The results of these studies suggest that lectin induction of bullae in *in vitro* organ cultured skin is a potential model for studying the pathogenesis of the bullous dermatoses.

References

1. Dabelsteen E, Fejerskov O, Noren O, Mackenzie IC (1978) Concanavalin A and Ricinus communis receptor sites in normal human oral mucosa. J Invest Dermatol 70:11–15
2. Deng J, Beutner EH, Shu S, Chorzelski TP (1977) Pemphigus antibody action on skin explants. Kinetics of acantholytic changes and stability of antigens in tissue cultures of normal monkey skin explants. Arch Dermatol 113:923–926
3. Goldstein IJ (1978) Lectins: Carbohydrate-binding proteins of plants and animals. Adv Carbohy Chem Biochem 35:127–340
4. Hashimoto K, King LE Jr, Yamanishi Y, Beachey EH, Maeyans E (1974) Identification of the substance binding pemphigus antibody and Concanavalin A in the skin. J Invest Dermatol 62:423–435
5. Hu CH, Michel B, Schiltz JR (1978) Epidermal acantholysis induced in vitro by pemphigus autoantibody. Am J Pathol 90:345–351

6. Michel B, Ko CS (1977) An organ culture model for the study of pemphigus acantholysis. Br J Dermatol 96:295–302
7. Nieland ML (1973) Epidermal intercellular staining with fluorescein conjugated phytohemagglutinins. J Invest Dermatol 60:61–66
8. Nishikawa T, Harada T, Hatano H, Ogawa H, Miyazaki H (1975) Epidermal surface saccharides reactive with phytohemagglutinins and pemphigus antigen. Acta Dermatovenereol (Stockh) 55:21–24
9. Pehamberger H, Gschnait F, Konrad K, Holubar K (1980) Bullous pemphigoid, herpes gestationes and linear dermatitis herpetiformis: circulating anti-basement membrane zone antibodies: in vitro studies. J. Invest Dermatol 74:105–108
10. Schiltz JR, Hu CH, Michel B (1979) Corticosteroids, aurothioglucose and soybean trypsin inhibitor do not prevent pemphigus antibody-induced acantholysis in vitro. Br J Dermatol 101:279–283
11. Schiltz JR (1980) Pemphigus acantholysis: A unique immunological injury. I Invest Dermatol 74:359–362
12. Schiltz JR, Michel B, Papay R (1979) Appearance of "pemphigus acantholysis factor" in human skin cultured with pemphigus antibody. J Invest Dermatol 73:575–581
13. Van Lis JMJ, Kalsbeek GL (1975) The interaction of Concanavalin A and the surface coat of stratified squamous epithelium. Br J Dermatol 92:27–35

On the Non-Random Distribution of Dividing Cells

P. M. Gaylarde

The interpretation of results from experiments involving the measurement of the uptake of labelled thymidine in organ cultured skin is difficult since the tissue studied is not homogeneous. Results have been expressed in terms of dry weight, wet weight, surface area, protein content and DNA content. Ideally one would like to know the exact number of cells in the basal layer of the epidermis, but there is no suitable method yet available for such determination. In an attempt to reduce the variability of the results of measurement of changes in the rate of DNA synthesis in response to experimental manipulation, the following strategy was adopted. Explants of skin were cultured in the presence of ^{14}C-thymidine for 5 h, washed, incubated in the presence of the test substances for periods up to 30 h and then incubated in the presence of ^{3}H-thymidine for the final 5 h of culture [6]. The effect of the test substance on the ratio of ^{3}H:^{14}C-thymidine could then be compared to the ratio obtained from comparable control cultures. In this way it was hoped that the uncertainty in the measurements of the dividing cell population could be eliminated.

The anticipated reduction in the variability of the results did not occur, the variance still remaining several times that expected. A second unexpected finding was that the distribution of the results appeared not to be random, the ratios tending to cluster at certain values bearing simple whole number ratios to each other. This suggested that one was dealing with the statistics of a small population. The following experiments were therefore carried out to study this problem.

A sheet of epidermis was obtained by injection of Eagles' minimal essential medium (MEM) into the ear of newly killed guinea pig. The pressure causes the separation of the epidermis and underlying dermis from the cartilage. The skin was then cut to obtain a uniform sheet about 5 cm^2 and was cultured, floating dermis down, on the surface of the medium consisting of 20% foetal calf serum + 80% MEM + 100 μg ml^{-1} penicillin + 50 μg ml^{-1} streptomycin. The sheet was incubated in the presence of ^{14}C-thymidine (20 μC ml^{-1}) for 1 h at the start of the period of culture, washed in MEM, reincubated in label-free medium for 11 h and finally cultured for 1$^1/_2$ h in the presence of ^{3}H-thymidine (20 μC ml^{-1}). The sheet of epidermis was then washed with 5% trichloroacetic acid and methanol and the central portion of the skin sheet was taken and cut into 205 squares each of about $^1/_3$ mm^2. The remainder of the epidermis was discarded since it is reported [12] that the epidermis within up to 3 mm of the cut edge shows increased mitotic acitivty in vitro. It was assumed that the central 0.7 cm^2 portion should be uniform. The small squares of skin were digested in 0.5 ml 1 N NaOH, neutralised with 1 N HCl and counted in a xylene : Triton-X 100 : butyl PBD phosphor mixture.

Table 1. Logarithms of the ratio of ^{3}H-thymidine : ^{14}C-thymidine uptake of skin samples in organ culture

$\bar{1}$.417	$\bar{1}$.667	$\bar{1}$.807	$\bar{1}$.897	0.012
$\bar{1}$.437	$\bar{1}$.685	$\bar{1}$.811	$\bar{1}$.900	0.016
$\bar{1}$.437	$\bar{1}$.693	$\bar{1}$.811	$\bar{1}$.900	0.017
$\bar{1}$.455	$\bar{1}$.695	$\bar{1}$.813	$\bar{1}$.903	0.019
$\bar{1}$.458	$\bar{1}$.698	$\bar{1}$.814	$\bar{1}$.904	0.020
$\bar{1}$.459	$\bar{1}$.703	$\bar{1}$.818	$\bar{1}$.906	0.035
$\bar{1}$.487	$\bar{1}$.710	$\bar{1}$.821	$\bar{1}$.907	0.040
$\bar{1}$.492	$\bar{1}$.714	$\bar{1}$.821	$\bar{1}$.909	0.047
$\bar{1}$.503	$\bar{1}$.722	$\bar{1}$.822	$\bar{1}$.910	0.050
$\bar{1}$.504	$\bar{1}$.723	$\bar{1}$.825	$\bar{1}$.911	0.055
$\bar{1}$.516	$\bar{1}$.726	$\bar{1}$.825	$\bar{1}$.913	0.059
$\bar{1}$.527	$\bar{1}$.726	$\bar{1}$.827	$\bar{1}$.919	0.065
$\bar{1}$.538	$\bar{1}$.728	$\bar{1}$.829	$\bar{1}$.922	0.069
$\bar{1}$.539	$\bar{1}$.729	$\bar{1}$.831	$\bar{1}$.923	0.079
$\bar{1}$.540	$\bar{1}$.732	$\bar{1}$.836	$\bar{1}$.926	0.089
$\bar{1}$.541	$\bar{1}$.742	$\bar{1}$.838	$\bar{1}$.930	0.090
$\bar{1}$.548	$\bar{1}$.743	$\bar{1}$.840	$\bar{1}$.931	0.104
$\bar{1}$.552	$\bar{1}$.748	$\bar{1}$.841	$\bar{1}$.932	0.125
$\bar{1}$.558	$\bar{1}$.751	$\bar{1}$.842	$\bar{1}$.935	0.127
$\bar{1}$.561	$\bar{1}$.754	$\bar{1}$.845	$\bar{1}$.941	0.138
$\bar{1}$.563	$\bar{1}$.754	$\bar{1}$.845	$\bar{1}$.941	0.155
$\bar{1}$.564	$\bar{1}$.754	$\bar{1}$.845	$\bar{1}$.944	0.155
$\bar{1}$.580	$\bar{1}$.760	$\bar{1}$.846	$\bar{1}$.948	0.158
$\bar{1}$.583	$\bar{1}$.761	$\bar{1}$.848	$\bar{1}$.954	0.159
$\bar{1}$.583	$\bar{1}$.763	$\bar{1}$.852	$\bar{1}$.955	0.160
$\bar{1}$.589	$\bar{1}$.764	$\bar{1}$.858	$\bar{1}$.956	0.165
$\bar{1}$.590	$\bar{1}$.769	$\bar{1}$.860	$\bar{1}$.957	0.173
$\bar{1}$.592	$\bar{1}$.771	$\bar{1}$.862	$\bar{1}$.965	0.178
$\bar{1}$.608	$\bar{1}$.774	$\bar{1}$.866	$\bar{1}$.969	0.196
$\bar{1}$.613	$\bar{1}$.778	$\bar{1}$.868	$\bar{1}$.971	0.200
$\bar{1}$.614	$\bar{1}$.779	$\bar{1}$.872	$\bar{1}$.973	0.208
$\bar{1}$.617	$\bar{1}$.784	$\bar{1}$.872	$\bar{1}$.982	0.218
$\bar{1}$.617	$\bar{1}$.789	$\bar{1}$.874	$\bar{1}$.989	0.218
$\bar{1}$.622	$\bar{1}$.789	$\bar{1}$.877	$\bar{1}$.989	0.257
$\bar{1}$.631	$\bar{1}$.794	$\bar{1}$.878	$\bar{1}$.990	0.271
$\bar{1}$.635	$\bar{1}$.798	$\bar{1}$.882	$\bar{1}$.996	0.271
$\bar{1}$.643	$\bar{1}$.800	$\bar{1}$.885	0.002	0.280
$\bar{1}$.645	$\bar{1}$.804	$\bar{1}$.889	0.002	0.312
$\bar{1}$.652	$\bar{1}$.804	$\bar{1}$.891	0.002	0.316
$\bar{1}$.665	$\bar{1}$.806	$\bar{1}$.893	0.005	0.356
$\bar{1}$.666	$\bar{1}$.806	$\bar{1}$.895	0.006	0.373

The ratio of ^{3}H : ^{14}C was $1 : 0.7088 \overset{\times}{\div} 1.604$ (geometric mean $\overset{\times}{\div}$ S.D.) and the median value was 0.700. The results are shown in Table 1 and tests on the distribution of the results show no significant skewness nor does their distribution differ significantly from that of an ideal normal distribution curve using the chi-squared test. The results, however are distinctly leptokurtic. I have provided the reader with the complete data used in the above calculations since the statistics of ratios is not well established in the literature.

Table 2. Distribution of colchicine arrested mitotic figures after treatment of guinea pig skin with coal tar for 3 days

Number of nuclei between adjacent mitoses	Number of times observed	Calculated number expected for random distribution
0	36	7.9
1	17	7.5
2	11	7.1
3	6	6.7
4	5	6.3
5	3	6.0
6	4	5.7
7	6	5.4
8	1	5.1
9	1	4.8
10–19	19	36.0
20–29	10	20.7
30–39	6	11.9
40–49	5	6.8
50+	17	8.9
50–100	12	8.5
100+	5	0.5

Total cell nuclei counted 2,733
Total mitoses counted 147 (5.38%)
Grouping the above data, $x^2 = 174.3$ using Yates correlation; 6 degrees of freedom. The probability that the above distribution arose by chance from a random pattern of distribution $< 10^{-35}$

The skin samples of approximately $^1/_3$ mm^2 contain about 5,000 basal-layer epidermal cells, and autoradiographic studies indicate that the number of basal epidermal cells in DNA synthesis in normal guinea pig ear varies between 3 and 10%. The number of cells in DNA synthesis should therefore be between 150 and 500 cells.

The above figures were used to calculate the expected geometric mean and its standard deviation assuming that DNA-synthesising cells are distributed at random. The model assumes that the rate of DNA-synthesis is constant throughout the S-phase and does not vary from cell to cell. The probable error in measurement of thymidine uptake, resulting from counting and correction errors is less than 0.5% and has been ignored to simplify the calculations.

On a priori grounds it may be argued that the distribution of cells in DNA synthesis at random is binominal and therefore the following formula should describe the distribution of cells in S-phase in each sample of skin:

$$P_n = \frac{x^n (1-x)^{(N-n)} N!}{n! (N-n)!}$$

where P_n = probability of finding n S-phase cells; N = total number of cells in sample; n = number of S-phase cells and x = fraction of cells in S-phase.

Since the result shows that there is no significant difference in the two rates of thymidine uptake (expected ratio 0.6667; measured ratio 0.7088) it has been assumed that the proportion of S-phase cells in each period of measurement is identical.

The standard deviation of the geometric mean of the ratios calculated from the above numbers and the assumed distribution lies between $\overset{\times}{\div}$ 1.06 and $\overset{\times}{\div}$ 1.36. The standard deviation found is about double the worst calculated value and therefore suggests that epidermal cells do not enter S-phase at random but do so as small groups of cells. Measurement of the distribution of mitotic figures in colchicine-treated skin (Table 2) confirms this conclusion.

The distribution of the results (Table 1) cannot be shown to differ significantly from that of a normal distribution curve. The results, however, show an increase at about ± 0.3 from the mean. This increase would be expected in the distribution arising from small numbers, i.e., from grouped cells.

The use of the analysis of variance of geometric means of the ratio of two associated variables is not described in any of the standard texts on statistics. The finding that under certain conditions the geometric mean of the ratios of the variables, their median and their mode all coincide with the ratio of the arithmetic means of the variables suggests that the use of the standard deviation of the geometric mean is appropriate.

Some properties of the geometric mean of the ratios derived from two variables and the distribution of these ratios may be derived by simple algebra.

Consider 2 series of numbers each with y members consisting of the following terms

$$a, b, c, \ldots n \quad \text{and}$$

$$xa, xb, xc, \ldots xn$$

The frequency of each term in both series is similar, so that

$$Aa + Bb + Cc + \ldots + Nn = y\bar{z} \quad \text{and}$$

$$Axa + Bxb + Cxc + \ldots + Nxn = y\overline{xz}$$

where $\bar{z}$ and $\overline{xz}$ are the arithmetic means of each series. The ratio of the means $\frac{\bar{z}}{\overline{xz}} = \frac{1}{x}$.

The permutation of the ratios of the numbers gives rise to a series containing n^2 ratios,

$$\frac{a}{xa}, \frac{b}{xa}, \frac{c}{xa}, \ldots, \frac{n}{xn} \quad \text{and consisting of } y^2 \text{ numbers}$$

The geometric mean of this series of ratios is:

$$\left[\left(\frac{a}{xa}\right)^{AA} \times \left(\frac{b}{xa}\right)^{AB} \times \left(\frac{c}{xa}\right)^{AC} \times \ldots \times \left(\frac{n}{xn}\right)^{NN}\right]^{\frac{1}{y}2} = \frac{1}{x}$$

The geometric mean of the ratios is therefore equal to the ratio of the arithmetic means. The logarithm of the geometric mean of a set of positive numbers is the arithmetic mean of the logarithms of the numbers. Since the terms $\frac{a}{xb}$ and $\frac{b}{xa}$ occur the same number of times and the logarithm of these numbers = $\log \frac{1}{x} \pm \log \frac{a}{b}$, the logarithms of the ratios are therefore distributed symmetrically about the logarithm of the geometric mean.

Since the sum of the frequencies of $\frac{1}{x}$,

$$AA + BB + CC + \ldots + NN$$

is always greater than the sum of the series of frequencies of any other ratio, the mode will correspond to the geometric mean. It therefore follows that when the distribution of two variables is similar, the geometric mean of the ratios, their median and their mode and the ratio of the arithmetic means are all coincident.

The distribution of the logarithms of the ratios is not binominal, Poisson or normal, but depends on the nature of the distribution of the variables. However, empirically it is rare that the distribution can be shown to be significantly different from that of a normal distribution. Using a random number generator, two sets of 10,000 normally distributed numbers were generated. The distribution of the logarithms of the ratios was not significantly different from the anticipated normal distribution. On the other hand, skewed distributions may give rise to ratios, the distribution of the logarithms of which, may be significantly lepto- or platykurtotic.

It is therefore generally satisfactory to use the geometric mean and its variance to test the significance of the difference of two sets of ratios or to test the null hypothesis on normalised data, provided the restrictions concerning distribution are met. It is easier to test whether the median, mode, geometric mean and the ratio of the arithmetic means are coincident.

In applying this method to any set of ratios the reader should be aware of the problems involved and should also use non-parametric methods to establish the validity of any conclusions made using the geometric mean of the ratios. This may seem cumbersome, but since limits of normality may be established, this may prove a valuable method when forecasts are to be made on the basis of the data. The conclusion that S-phase cells are not distributed randomly is supported by the study on the spacing between adjacent mitotic figures. The results of mitotic counts on normal and thickened skin were in no case consistent with random distribution. Since the proportion of mitoses is different for each biopsy, no overall average can be presented. Table 2 shows the actual and calculated distribution of mitotic figures from a typical experiment.

The nature and size of the cell groupings in the epidermis is difficult to estimate but calculations suggest a likely value to be about 5 cells whose DNA synthesis and division are approximately synchronized. The nature of the stimulus organizing synchrony on a local scale is unknown.

Observations on the relationship between Langerhans cells and parakeratosis in mouse tail epidermis indicate that these cells may regulate epidermal cell turnover [12]. The ratio between Langerhans cells and nucleated basal epidermal cells in guinea pig skin is 1:15 [15]. This proportion of Langerhans cells would appear to be consistent with their possible role as initiators of epidermal cell division.

The possibility of cell-to-cell interaction causing synchrony on a local scale as a cause of this effect must be considered. Indirect evidence on this point is provided by the results of variation in cell number on the uptake of thymidine by phytohaemagglutinin-stimulated human lymphocytes in vitro. Beyer and Bowers [1] found a power relationship of 2.5, whereas under different culture conditions we have found the uptake per cell to increase to the power of 1.6 with increasing cell number. They also showed that stimulated lymphocytes could induce the transformation of unstimulated cells, and also showed that the secondarily stimulated responder cells could not in turn stimulate other cells. A similar report of cell-to-cell stimulation was noted by Moorhead (personal communication) who showed that the thymidine uptake of similar numbers of lymphocytes was affected by the geometry of the culture vessel; the smaller the area on which the cells settled, the greater the uptake of thymidine. These results suggest that during the events leading up to DNA synthesis, cells may release factors stimulating their immediate neighbours to react similarly, which might under suitable conditions lead to the formation of an expanding wave of synchronized cells. The release of growth factors into media from cells in culture is well recognised [2] and the nature of the growth factors has been characterised in some instances.

The length of the cell cycle in both prokaryotic and eukaryotic cells has been shown to be related to cell volume in vitro [9, 10]. Models assuming that cell size is critical to processes initiating DNA synthesis and cell division have been proposed and both logarithmic normal and reciprocal normal distributions of cell cycle times are said to be compatible with such models [14]. Fox and Pardee [5] showed no correlation between the duration of G_1 and cell size. We have shown that cell volume and nuclear volume are related to mitotic rates in the skin in vitro, and that the daughter cells are also increased in volume (unpublished observations). The distribution of cell size in the skin does not support the contention that DNA synthesis is initiated by cell volume. The transition probability model of Smith and Martin [13] proposed that a random event in G_1 initiates processes leading to DNA synthesis. The non-random distribution of results in guinea pig skin disproves this hypothesis in its present form. Variability of the duration of the cell cycle would result if the size and membership of the synchronised groups varied from cycle to cycle or if the initiating event travelled through the cell population as an expanding wave.

The results, whilst confirming the relationship between cell size and cell cycle time, do not support either the deterministic or probabilistic hypothesis. These results imply that cell division is initiated by factors outside the cell and that specific agents initiating DNA synthesis should be sought. It appears likely that these same factors control cell volume.

In cell cultures, the phenomenon of contact inhibition is well known. In vivo it has been assumed that similar inhibitory processes are active. The cell density in the epidermis is such that the initiation of cell division will be strongly suppressed by the processes involved in contact inhibition and that therefore an external signal will be required to initiate cell division. The nature of this signal is unknown and may be the result of cooperative phenomena in small cell groups, neurotransmitter substances or autocoids, circulating factors or local factors such as the basement membrane [8]. Cell division and DNA synthesis in normal skin vary diurnally [3, 11], suggesting that central rather than locally controlled factors regulate epidermal cell division. Since post-mitotic basal cells show more tendency to migrate than resting cells [7], the non-uniform distribution of cell division is consistent with the uniform appearance of the skin. It is possible that the non-random mitotic distribution is due to permanently active local clusters of cells surrounded by areas of low mitotic activity. Subsequent migration would then be necessary to maintain the normal epidermal appearance.

The demonstration of non-random cell division in guinea pig skin and the evidence that this is a common phenomenon in cell cultures in vitro are important to our understanding of the regulation of growth and maintenance of tissues. It has long been apparent to me that measurements of the uptake of thymidine by cells in vitro show greater variance than that which would be anticipated from a knowledge of the errors contributed by variation in cell number, pipetting, counting and correction. This excess variation seems common in cell cultures regardless of the cell type or the method of culture used and there are many reports in the literature with coefficients of variation in agreement with the results described here. There are, however, a few reports of experiments whose results indicate the random ungrouped distribution of cell division. The reader may judge for himself whether those quoting such accurate results have been misled by their expectations.

The observation that cells do not divide at random is not new. Fischer [4], studying small colonies of fibroblasts by time lapse photography, noted synchronous cell division but he did not have suitable statistical methods to prove his observations on the non-random distribution of cell division. He also postulated this might be the result of rhythmic behaviour. Subsequent workers disproved the existence of rhythmic cell division in vitro. However, his observations stand, and since the work will not readily be available to the reader, I quote from his conclusion "These experiments seem furthermore to indicate, that the growing tissue cells in vitro, are under the control of the neighbouring cells; we are, in other words, dealing with a partial organism and not of independent cell individuals".

Embryologists have long recognized the importance of cell contact and cell position in the control of cell division and differentiation. This study indicates that cell division and possibly differentiation are also regulated by cell contact in adult tissues and in vitro.

Acknowledgements

I wish to thank the Herbert E. Dunhill Trust for generous financial support.

References

1. Beyer CF, Bowers WE (1975) Lymphocyte stimulation by periodate: evidence for an indirect mechanism. J Cell Biol 67:31A
2. Carrel A (1924) Leucocytic trephones. JAMA 82:255–258
3. Droogleever Fortuyn-van Leijden CE (1917) Some observations on periodic nuclear division in the cat. Proc Sect Sc K Ned Akad Wet 19:38–44
4. Fischer A (1925) Tissue culture. W Heinemann, London
5. Fox TO, Pardee AB (1970) Animal cells: noncorrelation of length of G_1 phase with size after mitosis. Science 167:80–82
6. Gaylarde PM, Brock AP, Sarkany I (1978) Observations on the interactions between non-steroid anti-inflammatory agents and corticosteroids. Austr J Dermatol 19:39–44
7. Gaylarde PM, Sarkany I (1975) Cell migration and DNA synthesis in organ culture of human skin. Br J Dermatol 92:375–380
8. Gerfaux J, Chany-Fournier F, Bardos P, Muh JP, Chany C (1979) Lectin-like activity of components extracted from human glomerular basement membrane. Proc Natl Acad Sci 76: 5129–5133
9. Killander D, Zetterberg A (1965) Quantitative cytochemical studies on interphase growth. Exp Cell Res 38:272–284
10. News and views (1980) Cell cycle control – both deterministic and probabilistic. Nature 286:9–10
11. Pilgrim C, Erb W, Maurer W (1963) Diurnal fluctuations in the numbers of DNA synthesizing nuclei in various mouse tissues. Nature 199:863
12. Schweizer J, Marks F (1977) A developmental study of the distribution and frequency of Langerhans cells in relation to formation of patterning in mouse tail epidermis. J Invest Dermatol 69:198–204
13. Smith JA, Martin L (1973) Do cells cycle? Proc Natl Acad Sci 70:1623–1627
14. Wheals AE (1977) Transition probability and cell cycle initiation in yeast. Nature 267:647–648
15. Wolff K, Winkelmann RK (1967) Quantitative studies on the Langerhans cell population of guinea pig epidermis. J Invest Dermatol 48:504–513

High Doses of Antigen-Nonspecific IgG do not Inhibit Pemphigus Acantholysis in Skin Organ Cultures

Th. Hunziker, U. E. Nydegger, P. J. Späth, H. A. Gerber, M. Hess and U. Wiesmann

A patient suffering from severe pemphigus vulgaris was treated by large volume plasma exchange in combination with an immunosuppressive regimen (corticosteroids and azathioprine [2]. In addition, she was given high doses of polyclonal, polyspecific human *Ig*G (0.5 g Sandoglobulin/kg, Table 1) through the *i.v.* route (IGIV) at the end of each plasma exchange session to restore depleted humoral antibodies and thus reduce the danger of infections. Recent reports show evidence that IGIV protect target plateles in idiopathic thrombocytopenic purpura from attack by anti-plateled auto-antibodies and/or immune complexes [3, 6] and therefore we hoped that this therapeutic measure might yield additional benefits, such as nonspecific displacement of the autoantibody from its epidermal target ("coating" of the antigen) [1], or blocking of anti-epidermal autoantibodies by complex formation with so called public anti-idiotypes present in IGIV [4].

In order to test these hypotheses, studies with a skin organ culture model were carried out using plasma from another pemphigus vulgaris patient who underwent plasma exchange [9] (Table 2). 1 ml of this plasma consistently induced typical acantholysis in native skin explants after 24 to 48 h incubation time and a classical "honeycomb" staining pattern in the basal epidermal zone was detectable by direct immunofluorescence examination (standard techniques) after 24 h incubation time. Acantholysis was never observed with the basic culture medium (Table 2) or Sandoglobulin (30 mg/ml), either alone or in combination. Preincubation of either the skin explants (Table 3) or pemphigus plasma (Table 4) with different concentrations of Sandoglobulin (ranging from 0.15 to 15 mg/ml in the culture medium – i.e., concentrations

Table 1. Selected* immunochemical and biological properties of the immunoglobulin preparation used (Sandoglobulin)

Purified IgG prepared from pooled plasma of more than 1,000 healthy donors	
monomeric	90%
dimeric	3–5%
polymeric	0%
split	5–7%
No complement activation in the ready-for-infusion state	
Activates complement after aggregation by heat or antigen	
Contains 5 parts saccharose to 3 parts IgG for stabilizing purposes	
Active in opsonophagocytic assays	

* from [5], [7], [8]

Table 2. Skin organ culture, methods

Human skin explants 3 x 3 x 0.8 mm on lens paper rafts in plastic Petri dishes	floated on	2 ml of culture medium consisting of – basic culture medium: Eagle's minimal essential medium + 10% fetal calf serum + 200 IU penicillin, 10 μg chlortetracycline, 100 μg streptomycin, 2.5 μg amphotericin B } per ml – 1 ml pemphigus plasma (IIF[1] titer 1 : 128 on human skin) – Sandoglobulin[2] 15, 1.5 or 0.15 mg/ml

Incubated for 24–48 h at 37 °C in a humid atmosphere containing 5% CO_2 in air
Light microscopy after 24 and 48 h incubation time
Direct immunofluorescence after 24 h incubation time

[1] indirect immunofluorescence (standard techniques)
[2] in all experiments Sandoglobulin was used after exhaustive dialysis against phosphate buffered saline to remove saccharose (Table 1)

Table 3. Effect of preincubation of skin explants in Sandoglobulin (37 °C, humid 5% CO_2 atmosphere): I for 6 h, II for 24 h

Sandoglobulin (mg/ml preincubation medium[2])	Acantholysis[1]		Direct immunofluorescence (after 24 h incubation time)	
	I	II	I	II
30	+	+	(+)*	–*
3	+	+	+	(+)*
0,3	+	+	+	+
0 (control)	+	+	+	+

[1] summary of the results after 24 and 48 h incubation time
[2] corresponding to final concentrations of 15, 1.5 and 0.15 mg/ml in the culture medium
– = negative, (+) = focal, + = uniform, * s. footnote

* The finely granular intercellular deposits of pemphigus antibodies became obscured because of homogeneous epidermal staining by high concentrations of Sandoglobulin. This phenomenon was confirmed by indirect immunofluorescence titer studies (standard techniques) including human albumin instead of Sandoglobulin as a control

that may be reached in vivo by i.v. administration) did not prevent or delay acantholysis induced by the pemphigus plasma nor did it inhibit the binding of the specific antibodies visualized by direct immunofluorescence (Table 3, 4).

Thus, the assumption that antigen-nonspecific IgG may coat the pemphigus antigens on epidermal cells making them inaccessible for the pathogenic autoantibodies was

Table 4. Effect of preincubation of pemphigus plasma with Sandoglobulin (1 h, 37 °C)

Sandoglobulin (mg/ml culture medium)	Acantholysis[1]	Direct immunofluorescence (after 24 h incubation time)
15	+	i*
1.5	+	i*
0.15	+	+
0 (control)	+	+

[1] summary of the results after 24 and 48 h incubation time
– = negative, (+) = focal, + = uniform, i = inconclusive, * see footnote in Table 3

not substantiated by our tests in vitro. Likewise, the hypothesis of functionally blocking autoantibody activity by means of idiotype-anti-idiotype interactions with IGIV cannot be supported. Therefore, our earlier considerations on the therapeutic use of IGIV during or after plasma exchange in severe cases of pemphigus have to be revised. The only rationale to retain this procedure at the present time is to restore humoral antibodies in patients who are compromised by immunoglobulin depletion in the course of plasma exchange in combination with an immunosuppressive regimen.

References

1. Gross B, Haessig A, Luescher EF, Nydegger UE, (1983) Monomeric IgG preparations for intravenous use inhibit platelet stimulation by polymeric IgG. Brit J Haematol 53:289–299
2. Hunziker Th, Schwarzenbach HR, Krebs A, Nydegger UE, Camponovo F, Hess M (1981) Plasmaaustausch bei Pemphigus vulgaris. Schweiz Med Wschr 111:1637–1642
3. Imbach P, d'Apuzzo V, Hirt A, Rossi E, Vest M, Barandun S, Baumgartner C, Morell A, Schöni M, Wagner HP (1981) High-dose intravenous gammaglobulin for idiopathic thrombocytopenic purpura in childhood. Lancet I:1228–1231
4. Lambert PH (ed) (1983) Immunopathology of idiotypic interactions. No 1, Vol 6 of Springer Semin Immunopathol
5. Liehl E, Armerding D, Böckmann J (1980) Effektorfunktionen und protektive Wirkung von Standardimmunglobulin und Immunglobulin nach modifizierter milder Säurebehandlung. Infection 8:194–198
6. Nydegger UE, Imbach P, Grau GE. Prospects for therapy of autoimmune disease with immunoglobulins prepared for intravenous use. In: Lambert PH, Izui S, Perrin LH (eds). Recent advances in SLE. Academic Press, London (in press)
7. Römer J, Morgenthaler JJ, Scherz R, Skvaril F (1982) Characterization of various immunoglobulin preparations for intravenous application. I. Protein composition and antibody content. Vox Sang 42:62–73
8. Römer J, Späth PJ, Skvaril F, Nydegger UE (1982) Characterization of various immunoglobulin preparations for intravenous application. II. Complement activation and binding to staphylococcus protein A. Vox Sang 42:74–80
9. Sarkany I, Grice K, Caron GA (1965) Organ culture of adult human skin. Brit J Dermatol 77:65–76

Permeation of Drugs through Human Skin: Method and Design of Diffusion Cells for In Vitro Use

S. A. Akhter and B. W. Barry

Much literature exists on methods for studying percutaneous absorption [6, 8, 9, 15, 20, 25]. Methods of skin investigation can be divided into in vivo and in vitro categories. This discussion is based on the latter type as used in our laboratories.

One of the primary problems facing investigators employing in vitro studies is the design and use of a suitable diffusion cell. Cells have been made from perspex or plastic [22, 23] glass [9] and metal [2]. The major limiting factor controlling the size of a cell is availability of membrane. Synthetic membranes can usually be accommodated in cells with large diffusional areas, e.g., 20 cm^2, but human skin is often in short supply and hence cells with smaller areas are required. However, the diffusional area can only be decreased so far, beyond which the apparatus may produce misleading data since very small skin membranes may be unrepresentative of larger samples. There are also analytical problems with small diffusional areas.

Cell design should also provide for variables such as temperature, compartmental concentrations, sampling, stirring; preferably we should have the flexibility for using a cell under both normal steady state and in vivo mimic situations. It is valuable also if the system is adaptable for use with skin specimens from diverse sources, e.g., human abdominal or scalp skin punch biopsies [2]. Many cell prototypes have been discussed [5, 6, 18, 22, 23, 26]; most can be classified with respect to the similarity or otherwise of the donor and receptor chambers and according to the positioning of the skin membrane – horizontal or vertical.

Zero-Order or Quasi-Steady State Flux

The simplest and most widely reported in vitro method for skin permeation studies employs a membrane (synthetic, e.g., cellulose acetate and polydimethylsiloxane or natural, e.g., excised human or animal skin) which is mounted vertically between two fluid filled stirred chambers. An example of this model is shown in Fig. 1 (quasi-steady state). We used excised human abdominal full thickness skin or stratum corneum clamped across ground glass surfaces, supported by perforated stainless steel discs if necessary to prevent membrane damage. The apparatus is immersed in a thermostated water bath and the two chambers are magnetically stirred from an external source. Drug permeation is monitored from the donor side through to the receptor chamber; the amount of drug permeating is small thus providing an essentially constant donor source. Hence this experimental design is sometimes called the "infinite" dose technique [6,

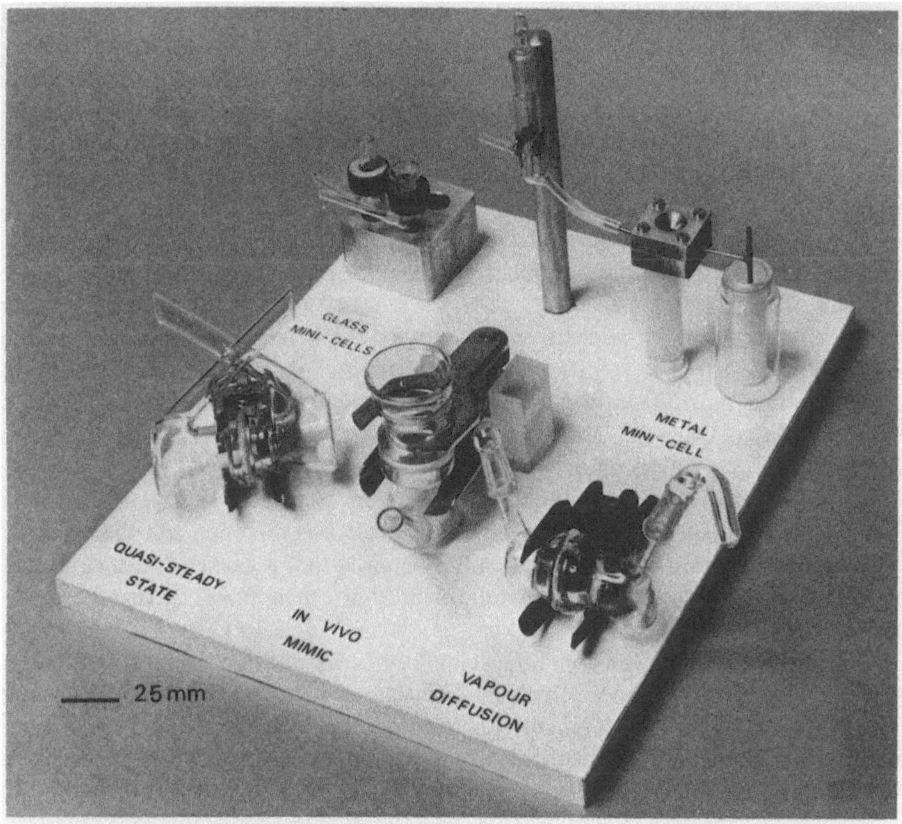

Fig. 1. Experimental model of diffusion cells for use in vitro

18]. The penetration of solute into the receptor chamber starts with an initial lag time, followed by a non-linear phase during which the drug concentration into the receptor increases and finally a linear phase is seen, where the flux is steady or zero-order.

Such quasi-steady state models are useful for deducing physicochemical parameters such as fluxes, partition coefficients and diffusion coefficients. The model represents a closely controlled fundamental physicochemical system, where individual factors which can modify drug penetration can be elucidated.

Design of Steady State Investitagions

In the first type the donor solution may consist of drug dissolved in water (or buffer) and the receptor contains water (or buffer) only. This is the situation of a water conditioned membrane, i.e., the skin is left to equilibrate in contact with water in both chambers.

The donor is next replaced with drug plus water and permeation of the solute is monitored.

The model is also useful in assessing potential penetration enhancer activity. For this the skin is conditioned on both sides with the accelerant or accelerant solution and penetration of solute is then monitored. Under this situation the only concentration gradient across the skin is that of the solute. Southwell and Barry [29] have used the model to study the penetration of octanol, mannitol and caffeine from water conditioned skin and to assess the activity and mode of action of the accelerant 2-pyrrolidone and dimethylformamide, for accelerant conditioned skin. The authors quantified permeation data in terms of permeability, partition and diffusion coefficients. Thus the model allows assessment of the standard parameters which may influence percutaneous penetration. They further suggest that both accelerants used can markedly change the skin membrane under the conditions outlined above. Other workers have used the steady state model similarly [13, 14, 27, 28].

In the second situation the accelerant or accelerant solution is only on the donor side. Thus concentration gradients exist both for the penetration enhancer and the solute. Elfbaum and Laden [13] studied penetration of the picrate ion across guinea pig skin in vitro under the influence of a range of dimethylsulphoxide (DMSO) concentrations. They conclude that although increased DMSO concentration enhances picrate transport, a DMSO concentration gradient is not necessary for enhancement.

In this arrangement the accelerant is only in the receptor compartment. Again a concentration gradient can exist for the accelerant in one direction and the solute in the opposite direction. Sekura and Scala [28] showed that a gradient of dimethylsulphoxide is not a necessary criterion for penetration enhancement of sodium nicotinate.

Finally, the skin may be conditioned on both sides with a penetration enhancer, which is subsequently removed from the diffusion chambers and replaced with a test formulation. Then penetration of the solute is monitored in a "modified" skin membrane, where both drug and accelerant concentrations are allowed to deplete during the experiment.

Advantages of this System

Barry [6] outlined the major advantages of the steady state system as follows: the whole apparatus may be immersed in a thermostated water bath for good control of temperature; air bubbles do not obstruct the passage of solute and thereby reduce the flux; and no hydrostatic head disrupts and possibly ruptures the skin membrane.

Major Disadvantages of this System

The model measures the effects on permeability characteristics of the stratum corneum under the extensive hydration conditions which accompany total immersion of the skin – a situation which is a mis-representation of normal, less hydrated skin. Hydration increases the penetration of most compounds. Scheuplein [25] observed that extended aqueous immersion altered the structure and permeability of the skin. Southwell and Barry [29] have also found that accelerants can markedly change the skin membrane under prolonged contact conditions.

The use of an infinite donor source in this model does not parallel the majority of clinical usage, where the amounts of material applied are a few milligrams of vehicle per square centimetre of skin. However, Anjo et al. [5] suggest that the steady state method duplicates the penetration of substances to which the skin is sometimes exposed, e.g., chronic application of an industrial solvent.

Repetitive sampling and receptor replacement present another disadvantage, especially when an investigation exceeds one week [2].

Relatively large skin samples are required for these glass diffusion cells. This requirement can often present problems where multiple replicates or many diffusion cells are needed, since more than one skin specimen may be necessary; this difficulty becomes an important factor when comparing data from one skin specimen with another [30].

The In Vivo Mimic – Glass Diffusion Cells

To overcome some of the disadvantages outlined previously we employed glass diffusion cells (Fig. 1) enclosing full-thickness cadaverous dermatomed skin, supported on stainless steel perforated discs and mounted horizontally. The skin surface was exposed to controlled room temperature (22 ± 1 °C) and humidity (60 ± 5% RH) and the receptor contained buffer at normal physiological pH, thermostated at body temperature (37 ± 0.5 °C). Before drug or formulation application the skin was equilibrated in the diffusion cells with these conditions for approximately 16 h – this allows the water and temperature gradients to develop across the skin, similar to those found in vivo [18, 26].

Drug application may then take several forms, e.g., as a solid drug film (via solvent evaporation) or a thin liquid, cream or ointment film which is allowed to deplete in concentration. By using this model we can observe the effects of dose size variation, changes in drug concentration and addition of vehicles (e.g. penetration enhancers) on the deposited film. The effect of hydration on drug absorption can also be evaluated by covering the donor chamber [1, 6].

Design of Investigations

Penetration from Solid Films

We have used this system to study the penetration of ibuprofen and flurbiprofen. nonsteroidal anti-inflammatory agents, from deposited films and to observe its usefulness as a model for determining potential penetration enhancers [2, 4].

In these studies ibuprofen was chosen as a model penetrant and applied to full-thickness dermatomed skin from acetone solution. Evaporation of the solvent deposited a solid drug film and the penetration was monitored. The donor chamber was next occluded with Parafilm and finally towards the end of the investigation the effect of adding a small volume of N-methyl-2-pyrrolidone to the deposited film was measured.

Figure 2 shows typical cumulative amount and rate profiles for a 0.7 mg deposited ibuprofen film. The drug was applied in acetone solution, which evaporated to leave a deposited film. During the evaporation period ibuprofen partitioned from the acetone solution, dissolved in the upper layers of the stratum corneum, and eventually reached to produce an initial peak in the rate profile. This peak then fell to a minimum due to dissolution-limited permeation from the drug film. Occlusion had no effect on the absorption from the deposited film; however N-methyl-2-pyrrolidone did dramatically increase the transfer of ibuprofen across the skin – showing its penetration enhancing

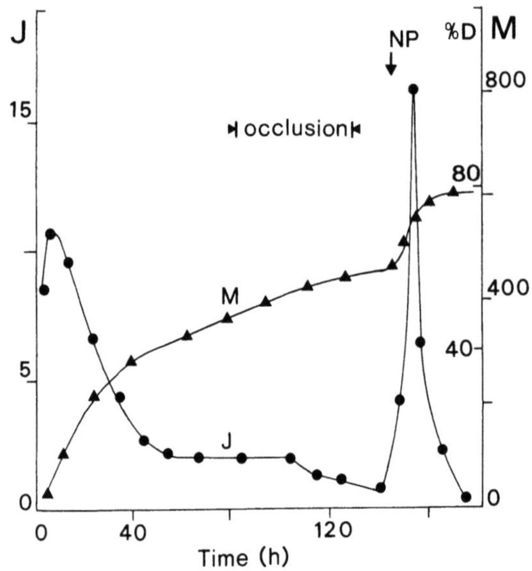

Fig. 2. Penetration of ibuprofen from a 0.7 mg deposited drug film; effect of occlusion and N-methyl-2-pyrrolidine (NP). Data illustrate flux (J, μg (2.27 cm)$^{-2}$ h^{-1}) and cumulative (M, μg (2.27 cm)$^{-2}$; %D, per cent of cumulative) penetration profiles

capacity and also validating the model as a useful tool for determining potential penetration enhancer action.

Thus the experimental design can not only mimic the in vivo situation but can provide useful information of drug penetration kinetics and the effects of different treatments. The model also allows each skin sample to function as its own control. This may be important when considering the biological variability in permeation of human skin within and between specimens [30].

Penetration from Thin Liquid Films

The model maintains simulated in vivo conditions when a thin liquid film is applied to the skin and allowed to deplete in concentration. This procedure is comparable to a single dose application in vivo. The application of liquids can take the form of drug dissolved in a vehicle and either pipetted [23] or painted [16] onto the skin. The drug/vehicle may be mixed with a volatile cosolvent – evaporation of the cosolvent may then produce supersaturated systems having high chemical potential [10]; alternatively evaporation of the cosolvent may be controlled or initially inhibited by covering the donor chamber.

Penetration from a Non-Depleting Source

The in vivo mimic technique may be adapted to determine fundamental physicochemical parameters by providing an essentially constant donor source. This can be done by initially dosing the skin with a drug-vehicle solution and therafter replacing the entire donor at regular intervals, thus maintaining quasi-steady state conditions.

We have investigated the penetration of flurbiprofen from 10% and 100% saturated aqueous solutions [2]. In the design of this study penetration was firstly monitored from an acetone deposited flurbiprofen film. The film was subsequently removed and replaced with a solution of flurbiprofen (1 ml) representing 10% or 100% aqueous saturation, and the penetration profiles were monitored until steady flux was maintained. Under these treatments flurbiprofen flux was approximately 4 fold greater than from the deposited drug film and this indicated that permeation from the drug film was dissolution dependent.

The model again emphasises the possibility of obtaining data with the skin acting as its own control.

Penetration from Mixed Vehicles

When assessing the action of a penetration enhancer, the vehicle system should also be evaluated for potential accelerant acitivity. We have studied the penetration of flurbiprofen from a 10% saturated solution in dimethylisosorbide.

The solution (1 ml) was applied to the skin and permeation monitored for about 70 h. Oleic acid (70 μl) was next introduced into the flurbiprofen: dimethylisosorbide system and further drug absorption was monitored. A similar study using 20 μl azone instead of the oleic acid was also conducted.

The cumulative penetration profile (Fig. 3) indicates that the presence of approximately 7% oleic acid had increased the steady state flux of flurbiprofen 5 fold, when compared to the original flurbiprofen: dimethylisosorbide steady state. Akhter et al. [4] reported that dimethylisosorbide does not act as a penetration enhancer. Thus the oleic acid may have altered the stratum corneum barrier function or modified the availability of flurbiprofen from the vehicle – resulting in a more favourable partitioning of flurbiprofen into the stratum corneum. Oleic acid is reported to assist absorption by the skin of medicaments in ointments [36]. Cooper [11, 12] suggested that

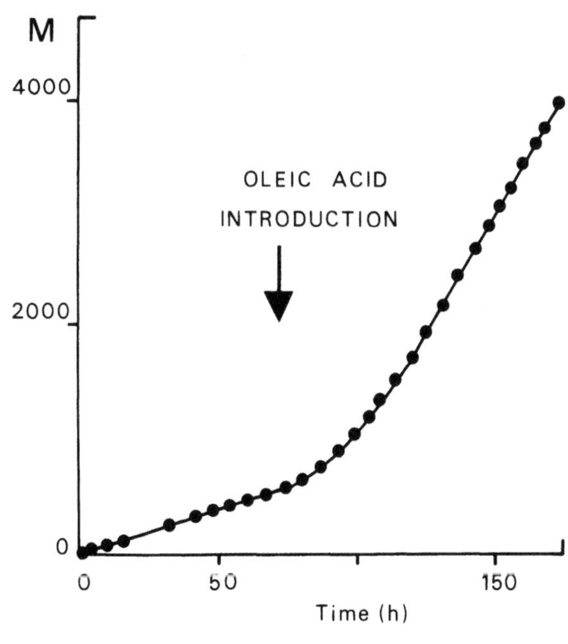

Fig. 3. Penetration of flurbiprofen from a 10% saturated solution in dimethylisosorbide; effect of oleic acid. Data expressed as cumulative penetration (M, μg $(2.27\ \mathrm{cm})^{-2}$) against time

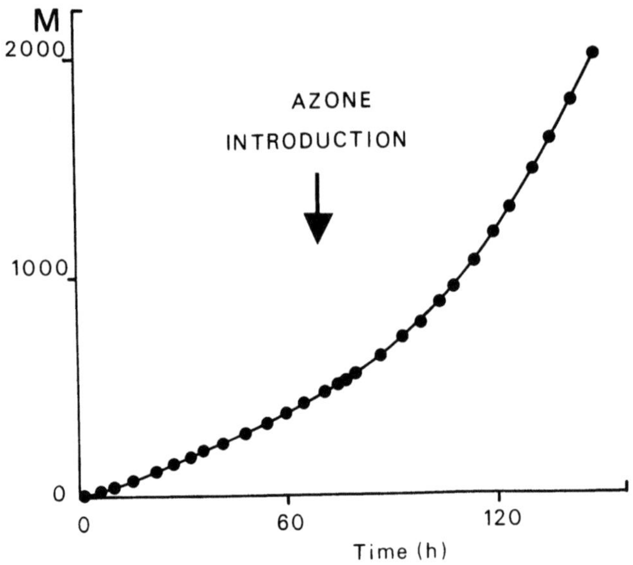

Fig. 4. Penetration of flurbiprofen from a 10% saturated solution in dimethylisosorbide; effect of azone. Data expressed as cumulative penetration (M, μg (2.27 cm)$^{-2}$) against time

small amounts of cis unsaturated fatty acids or alcohols combined with a polar solvent can increase the penetration of non-polar materials such as salicylic acid (at pH 2.6), by about 20 fold, possibly by an increased fluidisation of the stratum corneum lipids.

The presence of approximately 2% azone (1-dodecylazacycloheptan-2-one) increased the flux of flurbiprofen by about 4 fold (Fig. 4) when compared to the original flurbiprofen flux. Azone can enhance the flux of Ara-A by about 10 fold [34] and of Ara-A-2,3′-diacetate by 100–1000 fold [35].

The differences between our flux and reported data may be partially related to the choice of the polar vehicle. Cooper [11, 12] studied absorption from propylene glycol vehicles. Thus our five fold increase in flurbiprofen penetration due to oleic acid may have been greater with a propylene glycol or similar polar vehicle. The author also showed that the flux of salicylic acid fell as the number of ethylene oxide groups increased from the diethylene glycol to the pentaethylene glycol vehicle. Bellantone and Behl [7] observed the necessity of presenting azone in a polar vehicle for maximal penetration enhancement and that the acitvity was reduced as the co-vehicle changed from a polar to a lipophilic system. Stoughton and McClure [33] have also indicated the importance of the vehicle.

To elicit activity the accelerant also has to be present within the skin; thus the partitioning of oleic acid or azone from the dimethylisosorbide vehicle is important. Dimethylisosorbide does not readily penetrate the skin and remains on the surface providing a solution from which the drug/accelerant can partition into the stratum corneum; however, propylene glycol permeates the skin more readily. Hence the penetration of propylene glycol into the skin and its subsequent effect on the stratum corneum may provide for better accelerant (oleic and or azone) passage than from the dimethylisosorbide vehicle, and so produce greater accelerant activity.

Azone is reported to enhance the penetration of 8-bromocyclic AMP by four fold [31, 32], Ara-A species by 170 fold [35], butanol by over 80 fold [7] and 5-flurouracil by 94 fold [33]. We have not observed such dramatic penetration enhancements with this acclerant; however, we have found that N-methyl-2-pyrrolidone increases the absorption of octanol by over 260 fold when the data are reported as total quantity penetrated. Akhter and Barry [1] suggested that expression of data only in terms of total amount penetrated in net time can yield misleading information about penetration enhancer activity.

Advantages and Disadvantages

The advantages of the open cell system are:

- The model mimics in vivo conditions and thus does not suffer from the extensive hydration of the quasi-steady state system.
- Investigations may be designed to assess penetration of drugs in the form and dose most appropriate to their in vivo use.
- Sequential treatments may readily be applied to the same piece of skin – which therefore also acts as its own control. Thus the effect of repeated application (e.g. penetration enhancers) can be evaluated.

The following disadvantages outlined for the quasi-steady state system still remain for the in vivo mimic:

- Repetitive sampling and receptor replacement.
- Relatively large skin samples are required for the glass diffusion cells

The In Vivo Mimic – Miniature Stainless Steel Diffusion Cells

The use of the "open cell" technique can provide cumulative penetration profiles with many data points by regular and frequent sampling [2]. However, experiments are tedious and time consuming and they also require large samples of human skin. To solve these problems we have developed a system of continuous flow through diffusion cells linked to an automated fraction collector [3].

The stainless steel diffusion cells (Fig. 1, metal mini-cell) consists of an upper donor half (diffusional diameter 2 or 4 mm) which opens out in the shape of a cone towards a controlled temperature (22 ± 1 °C) and humidity (60 ± 5% RH); a multichannel peristaltic pump delivers receptor solution through the lower half. Full-thickness skin or stratum corneum is placed between the two halves of the cell, which are screwed together to provide a pressure seal.

The system may be used for in vivo mimic or quasi-steady state procedures.

Table 1. Diffusion of mannitol from saturated dimethylisosorbide: water vehicle (vehicle), deposited drug film (film) and the effect of 2-pyrrolidone (2p) treatment. Data expressed as flux (J, g $cm^{-2}s^{-1}$), the flux ratio (J^*) relative to the vehicle and number of replicates (n)

Treatment	$J \times 10^{11}$	J^*	n
Vehicle	26 ± 4.2	1.0	4
Film	6.9 ± 2.2	0.27	4
2-pyrrolidone	42	1.6	2

Design of Investigations

Penetration from Liquid and Solid Films

We have used the apparatus to study the permeation of the polar model compound mannitol, both under quasi-steady state and in vivo mimic conditions and the effect of 2-pyrrolidone on mannitol penetration from a deposited film [3].

In this study penetration was monitored from a saturated solution of mannitol in an aqueous vehicle with frequent donor replenishment for about 60 h. Evaporation was minimised by covering the donor with glass cover slips; these were then removed and penetration was monitored for a further 60 h but without donor replenishment. During this latter period mannitol was deposited as a film on the skin. Towards the end of the investigation a small volume of 2-pyrrolidone was added to the deposited film.

Results indicated that the initial low penetration of mannitol from the vehicle reached steady state after 25 h. Table 1 shows that this steady state flux then fell to a minimum due to penetration form the mannitol film. Addition of 2-pyrrolidone returned the flux to near its original value, as the crystals redissolved and the mannitol was available for partitioning into the skin.

Penetration from Liquid Films – Assessment of Penetration Enhancer Activity

We have used the automated apparatus to assess the activity of the accelerants N-methyl-2-pyrrolidone and 2-pyrrolidone in two ways [1]. The penetration of octanol was monitored from saturated solution in dimethylisosorbide: water vehicle, then from this vehicle system plus 80% accelerant (on the same skin sample). In the second experiment the skin was initially pretreated with the accelerant for 5 h and then absorption was monitored from a saturated solution of octanol in dimethylisosorbide: water. In both methods one skin specimen was used. Table 2 shows the flux data for octanol absorption, and indicates that both N-methyl-2-pyrrolidone and 2-pyrrolidone enhance the penetration in pretreated skin. Permeation of octanol in normal skin is doubled by N-methyl-2-pyrrolidone; this increase is not as marked as in the pretreated skin.

This indicates the importance of experimental design. The use of both the above procedures provide a better index of accelerant activity.

Table 2. Octanol diffusion in normal and pre-treated skin; effect of 2-pyrrolidone (2p) and n-methyl-2-pyrrolidone (np). Data Shown as steady state rate (J, g $cm^{-2}s^{-1}$) and the flux ratio (J^*) relative to the control treatment

Treatment	Normal		Pretreated	
	$J \times 10^5$	J^*	$J \times 10^5$	J^*
Control	1.4	1.0	0.39	1.0
2-pyrrolidone	1.3	0.93	2.3	5.9
N-methyl-2-pyrrolidone	2.8	2.0	6.6	17

Advantages and Disadvantages

The advantages of the miniature stainless steel cells are:

They can be used for quasi-steady investigations where the donor concentration is constant, with sink receptor conditions, and the experimental design is a typical physicochemical one.

In vivo conditions can also be simulated, where the donor is exposed to a controlled environment and the receptor is maintained at body temperature.

- Many replicates can be run without investigator attendance.
- The diffusion cells require only small quantities of skin.
- The apparatus can be used when tissue is available only in small pieces, e.g., punch biopsies and hair transplant pieces.

Sampling errors are reduced and also the amount of drug required for absorption studies.

The apparatus has the flexicility of using radio isotopes for assay purposes or the diffusion cells may be linked to a spectrophotometer.

Some of the disadvantages of this system are:

- Many scintillation vials and much scintillation fluid are required, especially if the sampling intervals are small, e.g., half hourly.
- Air bubbles lodging under the membrane cannot be observed until the cumulative penetration profile is plotted. Receptor solution should thus be degassed and bubble traps used.

The In Vivo Mimic – Miniature Glass Cells

We have recently developed miniature glass diffusion cells (Fig. 1). These have the advantages discussed above for the stainless steel cells (except that they are not designed for automatic use) and they are useful when only a limited number of cells are required or when the receptor solution is a mixed solvent and so prone to air bubble problems.

Vapour Diffusion Through Human Skin

Figure 1 (vapour diffusion) shows another vertical skin arrangement used to examine the diffusion of vapours through the skin. The donor half of the cell has a well to hold the volatile liquid penetrant away from any contact with the skin except via the vapour state. The donor also has a glass attachment which may be filled with drierite to minimise water molecules building up on the skin durface due to back diffusion from the receptor. The receptor solution is stirred with a teflon coated bar magnet. The complete unit is immersed in a water bath for temperature control.

The apparatus has been used, for example, to study diffusion of benzyl alcohol vapour through dermatomed skin, with a view to relating thermodynamic measurements of a drug in a vehicle to percutaneous absorption. Using benzyl alcohol as a model penetrant and a 50:50 ehtanol:water mixture as the receptor Harrison et al. [19] showed that the steady state vapour flux is proportional to the thermodynamic activity of the vehicle as measured by headspace gas chromatography.

Use of Excised Skin Techniques

The use of excised skin in in vitro studies, as reported here, allows control of the environment and enables the evaluation of factors which determine absorption of compounds and the results may lay the foundations for further in vivo work. A well designed in vitro procedure should produce results which correlate favourably with in vivo studies [17, 18, 21].

Acknowledgements

We thank the SERC for a CASE studentship for SAA, the support of The Boots Co. Ltd.

References

1. Akhter SA, Barry BW (1983) Classification of penetration enhancers for human skin; effect on mannitol and octanol absorption. J Pharm Pharmacol 35:29P
2. Akhter SA, Barry BW (1985) Absorption through human skin of ibuprofen and flurbiprofen; effect of dose variation, deposited drug films, occlusion and the penetration enhancer N-methyl-2-pyrrolidone. J Pharm Pharmacol 37:27–37
3. Akhter SA, Bennett SL, Waller IL, Barry BW (1984) An automated diffusion apparatus for studying skin penetration. Int J Pharm 21:17–26

4. Akhter SA, Meyer MC, Barry BW (1982) Absorption through cadaver skin of ibuprofen, applied as dry films, effect of solvents. J Pharm Pharmacol 34:34P
5. Anjo DM, Feldmann RJ, Maibach HI (1980) Methods of predicting percutaneous absorption in man. In: Mauvis-Jarvis, Vickers CFH, Wepiere J (eds) Percutanous absorption of steroids. Academic Press, London, pp 31–51
6. Barry BW (1983) Dermatological formulations, percutaneous absorption. Dekker, New York Basel
7. Bellantone NH, Behl CR (1983) Azone (dodecylazacycloheptane-2-one) altered skin permeability of model compounds, part II. Am Pharm Assoc, Acad Pharm Sci, 35th meeting, Miami Beach, Florida, USA 13:148
8. Bronaugh RL, Congdon ER, Scheuplein RJ (1981) The effect of cosmetic vehicles on the penetration of N-nitrosodiethanolamine through excised human skin. J Invest Dermatol 76: 94–96
9. Chandrasekaran SK, Shaw JE (1978) Factors influencing the percutaneous absorption of drugs. Curr Probl Dermatol 7:142–155
10. Coldman MF, Poulsen BJ, Higuchi T (1969) Enhancement of percutaneous absorption by the use of volatile : nonvolatile systems as vehicles. J Pharm Sci 58:1098–1102
11. Cooper ER (1982) Increased skin permeability for lipophilic molecules. Am Pharm Assoc, Acad Pharm Sci, 33rd meeting, San Diego, CA, USA 12:129
12. Cooper ER (1984) Increased skin permeability for lipophilic molecules. J Pharm Sci 73: 1153–1156
13. Elfbaum SG, Laden K (1968) The effect of DMSO on percutaneous absorption. A mechanistic study, part I. J Soc Cosmet Chem 19:119–127
14. Elfbaum SG, Laden K (1968) The effect of DMSO on percutaneous absorption. A mechnistic study, part III. J Soc Cosmet Chem 19:163–172
15. Flynn GL, Smith EW (1971) Membrane diffusion. I:Design and testing of a new multi-featured diffusion cell. J Pharm Sci 60:1713–1717
16. Foreman MI, Clanachan I, Kelly IP (1983) Diffusion barriers in skin – a new method of comparison. Br J Dermatol 108:549–553
17. Franz TJ (1975) Percutaneous absorption. On the relevance of in vitro data. J Invest Dermatol 64:190–195
18. Franz TJ (1978) The finite dose technique as a valid in vitro model for the study of percutaneous absorption in man. Curr Probl Dermatol 7:58–68
19. Harrison SM, Barry BW, Dugard PH (1982) Benzyl alcohol vapour diffusion through human skin:dependence on thermodynamic activity in the vehicle. J Pharm Pharmacol 34:36P
20. Idson B (1971) Percutaneous absorption. In: Rabinowitz JL, Myersen RM (eds) Absorption phenomena, Topics in medicinal chemistry, vol 4. Wiley, NY, pp 181–224
21. Marzulli FM, Brown DWC, Maibach HI (1969) Techniques for studying skin penetration. Toxicol Appl Pharmacol Suppl 3:76–83
22. Patel NK, Foss NE (1964) Interaction of some pharmaceuticals with macromolecules I. J Pharm Sci 54:94–97
23. Polano MK, Ponec M (1976) Dependence of corticosteroid penetration on the vehicle. Arch Dermatol 112:675–680
24. Poulsen BJ (1970) The use of models in estimating vehicle effects on the activity of topical corticosteroid formulations. Br J Derm Suppl 6/82:49–52
25. Scheuplein RJ (1965) Mechanims of percutaneous absorption. I. Routes of penetration and influence of solubility. J Invest Dermatol 45:334–346
26. Scheuplein RJ, Ross L (1974) Mechanism of percutaneous absorption. V. percutaneous absorption of solvent deposited solids. J Invest Dermatol 62:353–360
27. Scheuplein RJ, Ross L (1970) Effect of surfactants and solvents on the permeability of the epidermis. J Soc Cosmet Chem 21:853–873
28. Sekura DL, Scala J (1972) The percutaneous absorption of alkyl methyl sulfoxides. In: Montagna W, Van Scott EJ, Stoughton RB (eds) Advances in biology of the skin, vol XII. Appleton-Century-Crofts, NY, pp 257–269

29. Southwell D, Barry BW (1983) Penetration enhancers for human skin: mode of action of 2-pyrrolidone and dimethylformamide on partition and diffusion of model compounds water, n-alcohols and caffeine. J Invest Dermatol 80:507–514
30. Southwell D, Barry BW, Woodford R (1984) Variations in permeability of human skin within and between specimens. Int J Pharm 18:299–309
31. Stoughton RB (1981) Azone TM (1-dodecylazacycloheptan-2-one) enhances percutaneous penetration. III International symposium on psoriasis, Stanford July 13–17, 1981
31. Stoughton RB (1982) Enhanced percutaneous penetration with 1-dodecylazacycloheptan-2-one. Arch Dermatol 118:474–477
33. Stoughton RB, McClure WO (1983) Azone: A new non-toxic enhancer of cutaneous penetration. Drug Devel Ind Ph 9:725–744
34. Sugibayashi K, Higuchi WI, Foy JL, Baker DC, Shannon WM (1983) Influence of 1-dodecylazacycloheptan-2-one (Azone) on the topical delivery of Ara-A-2′,3′-Diacetate (ARA-ADA) in hairless mouse skin. Am Pharm Assoc, Acad Pharm Sci, 35th meeting, Miami Beach, Florida, USA 19:42
35. Vaidyanathan R, Flynn GL, Higuchi WI (1982) Dodecylazacycloheptan-2-one (Azone) enhanced delivery of drugs into the epidermis, I: effect of enhancer concentration on the permeability of selected compounds. Am Pharm Assoc, Acad Pharm Sci, 33rd meeting, San Diego, CA, USA 126:97
36. West A (ed) (1977) Oleic acid. In: Martindale The extra pharmacopoeia, 27th edn. The pharmaceutical press, p 743

Mathematical and Physical Models

Modelling the Scattering and Absorption of Light by the Skin

J. C. Barbenel and F. W. Turnbull

In modelling, a relevant part of a complex process or system is identified and isolated, and replaced by a simpler representation which provides an adequate description but may be analysed or investigated more easily. The representation may be physical or, as in the present chapter, idealised as a set of mathematical equations. There is an extensive literature on mathematical models and modelling [2], but the way in which such descriptions are developed remains largely subjective and empirical. There is, however, a logical progression of ideas which leads to formulation of the usable mathematical model.

Verbal Description

The verbal description of the process to be modelled clarifies what relevant features are to be contained in the model, and the underlying mechanisms which will be transformed into mathematical terms.

Light Scattering and Absorption by the Tissues

Light transmitted or reflected from tissues perfused with bloods is different, both in colour and intensity, from the incident radiation. The difference depends on the optical properties of the tissues and the amount of reduced and oxygenated haemoglobin present. Optical methods can, in principle, be used to measure the blood content of the tissues and the state of oxygenation.

Transcutaneous optical devices which utilise the transmitted light have been described [3] and are commercially available; their use is limited to naturally occurring skin flaps such as the ear lobe, or to raised skin folds. At many sites such folds cannot be raised and reflection methods have a more widespread applicability. Such devices have been described, but these are either qualitative or rely on semi-empirical relationships between reflection and oxygen saturation, requiring continual recalibration for each site and subject [11].

In almost all devices the light illuminating the skin is collimated, that is, it consists of a parallel beam of photons moving in the same direction. The reflected light which re-emerges from the skin surface is, however, diffuse, with the photons moving in a variety of directions. A model of the process of light reflection by the tissues must contain two separate components – the conversion of the collimated light into a source of diffuse photons and a description of how the diffuse photons are scattered to re-emerge from the surface of the skin.

Scattering and Absorption

Light crossing the boundaries between two media of different refractive index will change direction unless the angle of incidence is normal to the boundary. This scattering is the basic mechanism by which collimated light is rendered diffuse by a turbid medium.

The epidermis does not greatly scatter the incident light [8] because of it's thinness. Scattering may be further reduced by making the incident radiation normal to the skin surface. The dermis is a highly effective scattering medium. The complicated arrangement of collagen and elastin fibres, and the presence of blood vessels and the contained cells lead to multiple scattering of the incident radiation.

Light absorption in the skin is due to the presence of pigments within the tissues. Melanin and carotenoids are present in the epidermis, haemoglobin and oxy-haemoglobin in the dermis. The absorption of both forms of haemoglobin is wavelength dependant [1], and by making measurements of absorption at two wavelengths, it is possible to assess the total amount of blood in the tissues and it's saturation.

Mathematical Model

The scattering and absorption process is stochastic in that it depends on random events. Many large scale random processes can, however, be successfully modelled as deterministic and this was the approach used in this study.

There is a close formal analogy between the optical properties of a turbid medium and the electronic properties of semiconductors. McKelvey et al. [9] combined the diffusion and continuity equations where the mean free absorption and scattering path lengths are comparable. The optical formulation of this relationship was used by Cohen and Longini [6] to describe the optical behaviour of the tissue.

The basic equation which defines steady state randomly directed photon density (P) is:

$$D\nabla^2 P = \frac{P}{\tau} - \gamma \tag{1}$$

where γ is the source strength, D the diffusion coefficient of the medium and τ the mean photon lifetime before absorption. Both D and τ are a function of the absorption (ω) and scattering (k) coefficient of the medium for diffuse light.

The flux of photons (F) at any point in the medium given by:

$$F = -D\nabla P \tag{2}$$

The superficial tissues were considered to be an optically isotropic and homogenous infinite half space with a plane boundary. The boundary was treated as a perfect absorber for back scattered photons. Over the area of the source illumination at the boundary, the flux was directed into the medium, outside this source area the flux was emergent.

The light illuminating the skin surface and generation of diffuse photons was modelled in two ways. The first described it as an uniform disc of dipoles, the dipole moment being a function of the optical properties of the medium. The solution for this model involved an area integral which was reduced to a single integral and evaluated numerically.

The second model assumed that the source term, γ, was non-zero inside a cylinder which penetrated into the medium, but the intensity of which decreased exponentially with depth of penetration. The solution for the back scattered flux involved a volume integral which was reducible to a double integral, which may in turn be approximated by a single integral. Full mathematical details will be found in Eason et al. [7].

Predictions were obtained for each model of the intensity of the normal back scattered flux at the surface of the medium away from the source. These predictions depended on the intensity of illumination, the absorption and scattering coefficients of the medium and the distance from a central source. The predicted flux was calculated for concentric detectors at a radius of 6, 8 and 10 mm from the source.

Validation

The mathematical model gives rise to relationships between the back-scattered flux and the optical properties of the medium. Before the model results can be accepted and used, they must be validated. This requires comparison with experimental results in which those parameters which are to be obtained from the model are known and controlled. The source-collector separation and intensity of illumination can be controlled by the experimental system and the variable input parameters for the model are the scattering and absorption coefficients of the tissues. The model predictions were compared to measurements of back scattered flux obtained with fully oxygenated citrated blood samples in vitro. The optical coefficients were vaired by altering the haemoglobin concentration within the range 0.5–13.1 g/100 ml.

The flux measurements were made with the blood contained in a closed cylinder. The fully oxygenated blood was warmed to 36.8 °C and circulated through the apparatus at 200 ml/min by means of a peristaltic roller pump.

The optical system was inset into the flat lid of the cylinder. Light was delivered to the interface between the blood and the cylinder top via a 3.5 mm diameter fibre – optic light guide. Around the input there were three concentric circular fibre optic bundles spaced at 6, 8 and 10 mm radius. The fibre optics of each ring were united to form three separate 3.5 mm bundles, each of which was split using a Y junction and the output filtered at either 633 or 805 nm by filters having a band width of approximately 50 nm. Input illumination was obtained from a quartz iodine bulb, and the intensity of the filtered backscattered light was measured using a p-i-n photodiode. The magnitude of the recovered signal was very low and the input was mechanically chopped at 800 Hz and a phaselock amplifier system used.

Blood samples were withdrawn from the system and an oximeter used to measure haemoglobin concentration and saturation; a centrifugal capillary tube method was used to measure haematocrit. Each complete experiment involved 16 haemoglobin concentrations.

The absorption and scattering coefficients are functions of the haematocrit. For 805 nm

$$\omega = 35\ \omega_0 H \tag{3}$$

where ω_0 is the absorption coefficient when the haematocrit (H) is 0.029 and

$$k = 43.7\ H^3 - 99.7\ H^2 + 56.1\ H - 015 \quad [10]. \tag{4}$$

The back scattered flux was calculated for each haematocrit and source-collector separation. The optical system was designed to measure low intensities of flux and it was not possible to measure the high source intensity used experimentally. All back scattered fluxes were, therefore, measured relative to flux values obtained for a specific haematocrit and the theoretical predictions similarly scaled. Typical results are shown in Fig. 1, with the solid dot being the normalising value.

The results show that agreement between prediction and experimental results is good in the middle range of haematocirts, but falls at both high and low values.

The sudden fall at the low haematocrit occurs when the mean free path of the photons become comparable with the source – collector separation. The agreement between theory and experimental at these low values cannot be improved by modification of the method because photodiffusion theory is not applicable and is no longer an accurate representation of the scattering and absorption process. The discrepancy at high haematocrits is largely due to the low intensity of back scattered flux being

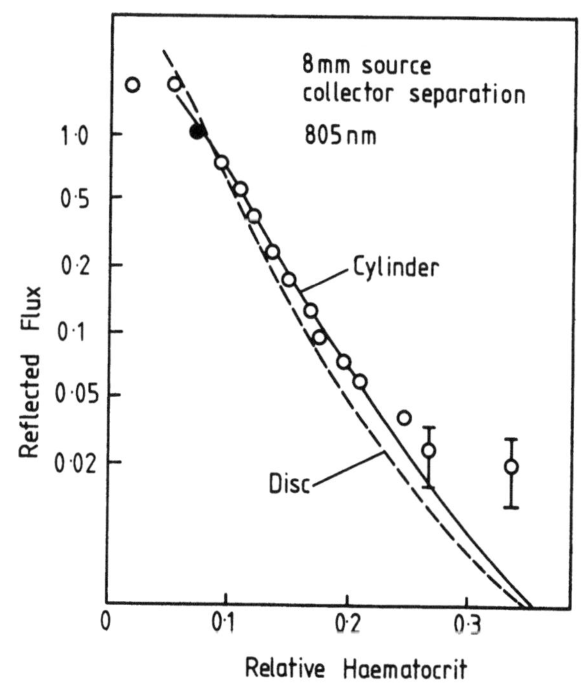

Fig. 1. Comparison of normalised values of predicted and measured back scattered flux – *in vitro* blood

measured. At these low values small zero offsets of either the detector or the amplifying system will produce large errors.

The agreement between the experimental points and the predictions using the cylindrical source model were always better than the disc source (Fig. 1). A more detailed analysis [4] suggested that relative changes in the optical coefficients could be measured more accurately than the coefficients themselves.

Use – In Vivo Measurement of Blood Content and Oxygenation

Measurements were made of the back scattered flux at the surface of the skin of normal volunteers, who were subjected to a test programme designed to change the blood content and oxygenation of the skin and superficial tissues in a predictable although qualitative way. The resulting alterations in blood content and oxygenation were calculated using the model containing the cylindrical source.

Experimental Method

The optical system used in the in vitro validation test was redesigned to be applicable to the skin. The fibre optic bundles were located in a truncated stainless steel cone. The end at which the fibre optics were exposed was flat and had a cross sectional diameter of 12 mm. In order to minimise pressure effects at the probe-skin interface, the area of contact was increased tenfold by surrounding the end of the probe with a Perspex annulus.

Tests were conducted on the skin over the volar aspect of the forearm, care being taken to ensure that the optical probe was not placed over large superficial blood vessels. The limb was held firmly in a stable position within a flexible plastic cast.

The limb was first rendered as bloodless as possible by wrapping the arm in a rubber Esmarch bandage, beginning at the finger tips and firmly winding up the arm towards the axilla. A pneumatic cuff was placed around the upper arm just above the end of the bandage and inflated to above the subject's systolic blood pressure. The bandage was then removed, the cuff maintaining the exsanguinated condition. Base line readings of back scattered flux were made for both wavelengths and all three source-collector separations.

The pressure within the cuff was then varied in a predetermined way at specified time intervals in order to change the blood content and oxygenation of the tissues and optical measurements were made. Table 1 details the test programme and expected changes.

Analysis and Results

The absorption (ω_b) and scattering (k_b) coefficients of whole blood are determined by the oxygen saturation (X) and fractional volume occupied by the red cells (C) and are given by:

$$\omega_b = C(1 - C)(54.1 - 39C) \quad \text{and} \tag{5}$$

Table 1. Test programme to vary skin blood content and oxygenation of the arm

Cuff pressure	Tissue condition		Stage	Measurements after start of stage
	Blood content	Oxygenation		
Greater than systolic	Bloodless		A	15 s after bandage removal
Zero	Normal	Normal	B	3 min
Between diastolic and systolic	Increased	Reduced	C	2 min
Greater than systolic	Increased	Reduced	D_1	3 min
			D_2	6 min
Zero	Hyperaemic	Reduced	E_1	1 min
	Normal	Normal	E_2	3 min

Table 2. Calculated tissue haematocrit and oxygen saturation

Stage	Oxygen saturation				Tissue haematocrit x 10^2			
	6 mm	8 mm	10 mm	Mean	6 mm	8 mm	10 mm	Mean
A								
B	0.84	0.91	0.95	0.90	0.86	0.74	0.80	0.80
C	0.71	0.67	0.76	0.71	2.1	1.9	2.2	2.07
D_1	0.58	0.54	0.65	0.59	2.4	2.1	2.5	2.33
D_2	0.53	0.47	0.58	0.53	2.4	2.1	2.5	2.33
E_1	0.81	0.84	0.91	0.85	1.5	1.2	1.2	1.30
E_2	0.81	0.83	0.90	0.85	1.5	1.3	1.2	1.33

$$k_b = 35[(1 - X)\omega_0 + X\omega_r] \quad (6)$$

where ω_0 and ω_r are the absorption coefficients per unit concentration of oxygenated and reduced blood.

In perfused tissues the optical properties depend on both the blood and tissue coefficients. The absorption coefficient for perfused tissue can be expressed as the sum of the separate absorption coefficients for blood and tissues, as can the scattering coefficients. The analysis was carried out by first calculating the scattering and absorption coefficients of the bloodless tissue. The predicted flux is a complicated nonlinear function of these coefficients, and a single flux value may be provided by several pairs of combinations of the coefficient. The three source-collector separations and the two wavelengths produce six sets of results, which allowed overlapping values to be identified and extraneous values of the coefficients to be rejected.

The second stage of the analysis was the estimation of a fractional volume of tissue occupied by the red cells. With the knowledge of the reference bloodless tissue coefficients the oxygenation insensitivity of the isobestic 805 nm wavelength was used to calculate C, which is the single variable in equations (5) and (6). This stage of the analysis, involving a single variable allowed independent assessments to be made for each of the different source-collector separations.

The oxygen saturation (X) affects only the propagation at the 633 nm wavelength. Using the tissue coefficients and tissue haematocrit the saturation can be determined from equation (6), and once again independent determination can be made for all the source-collector distances.

The results of such an analysis from a typical experiment are shown in Table 2.

Discussion and Conclusions

The model of light transmission in the skin presented here contains major simplifications in order to make any analysis possible.

The treatment of the skin as an optically homogeneous and isotropic material neglects the effects of the known skin structure. The structure exists as discrete layers – the epidermis and the dermis – together with systematic structural variations within each layer. In addition, the capillaries form a relatively regular array. These inhomogeneities can only be included with difficulty, greatly complicating the modelling solution and the analysis of the experimental results. Unfortunately the structure of the skin varies and differs with body site and between subjects. It is unlikely that an idealised multilayer model would be appropriate for any single real test and it seems most unlikely that the advantages produced by this further refinement of the skin structure modelling would be of major value compared with enormous extra complexity introduced.

The tissues were assumed to be semi-infinite in extent, ignoring the real skin thickness. Measurements on excised skin specimens which were systematically reduced in thickness showed that the reflected flux was not significantly altered until the skin thickness was reduced to between 5 and 7 mm. The equivalent skin thickness for perfused tissue will be less than this, and at most sites the semi-infinite approximation will be satisfactory. In those areas where the skin is particularly thin the effect of the optical boundary with the underlying tissue may be significant, particularly if the skin overlies bone.

In the cylindrical source model the photons leaving the fibre optic were assumed to be perfectly collimated. Measurements of the actual illumination produced by the fibre optic source showed the presence of a proportion of diffuse photons. It is possible to construct a hybrid model which incorporates both types of photons in the input. The complexity of the model solution increases with triple integrals appearing, and it is not clear that these can be simplified for numerical evaluation without loss of accuracy.

The measured changes in tissue haematocrit and oxygenation were in broad qualitative agreement with the expected alterations. Quantitative values are more difficult to assess. The measured tissue haematocrit typically varied between 1 and 3.5%. Burton [5] suggested that the capillaries, arterioles and venules contained about 500 ml of blood. If this were uniformly distributed throughout the body tissues the skin haematocrit would be of the order of 0.5%. In fact the blood supply to the skin has an important thermoregulatory function and is greater than might be expected. Thus the experimental values would appear to be not unrealistic.

The range of oxygen saturation values, and the oxygen partial pressures they imply, are not unreasonable. Three tests were made in which capillary blood samples were drawn from the skin of the thumb while the arm was subjected to the standard test programme. The oxygen saturation of the withdrawn samples were measured in a micro-blood gas analyser. The calculated saturation values were similar to the optically obtained results, except that the fall during stages D and E was generally larger when measured optically.

The proposed model of light propagation in the superficial tissues gives good agreement between prediction and experiment when validated *in vitro*. The *in vivo* test results are qualitatively reasonable, although there appears to be no other values for comparison.

Acknowledgement

The initial stages of instrument development was supported by a grant from the Scottish Home and Health Department. F. W. Turnbull was in receipt of an SERC Research Studentship.

The authors are grateful to R. Nisbet, University of Strathclyde, for advice and assistance in computational aspects of the work.

References

1. Anderson RR, Hu J, Parrish JA (1981) Optical radiation transfer in the human skin and application in in vivo remittance spectroscopy. In: Marks R, Payne PA (eds) Bioengineering and the skin. MTP Press Ltd, Lancester, pp 253–265
2. Aris R (1978) Mathematical modelling techniques. Pitman, London
3. Barbenel JC, Gibson F, Turnbull F (1976) Optical assessment of skin blood content and oxygenation. In: Kenedi RM, Cowen JM, Scales JT (eds) Bedsore biomechanics. Macmillan, London, pp 83–93
4. Barbenel JC, Turnbull FW, Nisbet RM (1979) Backscattering of light by red cell suspensions. Med Biol Eng Comput 17:763–768
5. Burton AC (1965) Physiology and biophysics of the circulation. Year Book Medical Publishers, Chicago, p 53
6. Cohen A, Longini RL (1971) Theoretical determination of the blood's relative saturation *in vivo*. Med Biol Eng 10:385–391
7. Eason G, Veitch AR, Nisbet RM, Turnbull FW (1978) The theory of backscattering of light by blood. J Phys D 11:1463–1479
8. Hardy JD, Hammell HT, Murgatroyd D (1956) Spectral transmittance and reflectance of excised human skin. J Appl Physiol 9:257–264
9. McKelvey JP, Longini RL, Brody TP (1961) Alternative approach to the solution of added carrier transport problems in semiconductors. Phys Rev 123:51–57
10. Pisharoty NR (1971) Optical scattering in blood. PhD Thesis Carnegie-Mellon University, Pittsburgh, USA
11. Zijlstra WG, Mook GA (1962) Medical reflection photometry. Van Gorcum, Assen, Netherlands

Mathematical Models for the Ultraviolet Optics of Human Epidermis

B. L. Diffey

The photobiological effects following exposure of human skin to ultraviolet radiation (defined here as electromagnetic radiation in the wavelength interval 200 to 400 nm) include erythema, melanogenesis and photocarcinogenesis. Also, ultraviolet irradiation of patients has recently gained renewed acceptance in dermatology as the basis of various phototherapies for skin diseases. In all of these effects the absorbing molecule(s) or chromophore(s) will be situated in the viable cells of the epidermis or in the dermis. A quantitative understanding of the optical properties of human skin, particularly the epidermis, is essential, therefore, for a proper appreciation of the fraction of ultraviolet radiation (UVR) incident on the surface of the skin which reaches different cell layers within the skin. In principle, the ability to account for the modifying influence of various cell layers on the quantity and spectral quality of the UVR reaching the appropriate chromophore for a given biological response should allow our rapidly increasing knowledge of cellular and molecular photobiology gained from in vitro studies to be related to observed responses in vivo.

There have been several experimental studies on the transmission and reflection of ultraviolet radiation through excised human skin [1, 3, 6, 10–13]. However, there has been relatively little theoretical work on UVR optics in skin. This is perhaps not surprising in view of the complex structure of the organ, which includes such derivatives as hair follicles, sweat glands and sebaceous glands. The skin contains structures whose size are of the order of the wavelength of optical radiation, in which electrical conductivity and dielectric constant have very steep gradients, in which birefringence plays an important role, and which have great shape anisotropy [5]. It appears, therefore, that a rigorous theoretical treatment of skin optics is formidable, if not impossible.

The increased interest in photobiological responses in man requires some understanding of the optical properties of the skin together with the ability to estimate, albeit approximately, the transmission of UVR of different wavelengths through various layers of the skin. This paper reviews two mathematical models which have been developed for describing the optical properties of the epidermis, and compares calculated values of the penetration of ultraviolet radiation of various wavelengths into the epidermis with recent experimental measurements. The models which have been described are both based on radiative transfer theory.

Radiative Transfer Models

There are numerous models of radiative transfer which range from the simple Bouger-Lambert law, which states that there is an exponential decrease in radiation intensity with the depth of penetration of radiation through a homogeneous medium, to sophisticated models of absorption and scattering which have found applications principally in astrophysics [18].

The fundamental problem of the ultraviolet optics of skin is to determine the wavelength and angle photon-number spectrum of the UV photons which flow in each direction at each point in the skin, given the distribution of radiation incident upon the skin. The equation governing the transport of UVR is derived by a continuity principle which is represented in the time-independent transport equation. This is a linear, partial integro-differential equation of the first order, which, in general, depends upon six variables; three position co-ordinates, two direction co-ordinates and one wavelength co-ordinate [7].

In the equilibrium state, the rate at which photons leave a volume element in photon phase space is exactly equal to the rate at which the photons enter the volume element. For monochromatic radiation of wavelength λ nm, the transport equation can be expressed mathematically as follows:

$$\nabla \cdot \underline{\Omega} N(\underline{r}, \underline{\Omega}) + \mu N(\underline{r}, \underline{\Omega}) = \int_{4\pi} N(\underline{r}, \underline{\Omega}') \sigma(\underline{\Omega}', \underline{\Omega}) d\Omega' \quad (1)$$

where $N(\underline{r}, \Omega)d\Omega$ is the number of photons moving in the direction of the unit vector Ω in the element of solid angle $d\Omega$, which cross a unit area located at the point $\underline{r}$, whose normal is in the direction $\underline{\Omega}$.

μ is the total attenuation coefficient of radiation of wavelength λ nm, and is equal to the sum of the absorption coefficient K, and the scattering coefficient, S.

$\sigma(\underline{\Omega}', \underline{\Omega})$ is the differential scattering cross section with respect to angle such that

$$S = \int_{4\pi} \sigma(\underline{\Omega}', \underline{\Omega}) d\Omega' \quad (2)$$

Under conditions of simple geometry, the number of variables may be considerably reduced. Applications of radiative transfer theory to the optical properties of skin have all been restricted to uniform plane sources of infinite sideways extent. In such cases the photon distribution depends upon one position co-ordinate only, the normal distance from the incident face of the skin which is the plane x = 0. If the UVR source is symmetric, the radiation is isotropic in azimuth at all depths, and one direction variable only is involved which is the angle of inclination to the x-axis.

The Kubelka-Munk Model

The Kubelka-Munk (KM) solution to the transport equation (15) has been applied to model the optical properties of both the dermis [1] and the epidermis [19]. The KM solution, which is described in detail elsewhere [14] is based on a two-stream approximation first applied in astrophysics by Schuster [17]. For a plane parallel sample of thickness d irradiated by diffuse, monochromatic radiation of unit flux, the intensity at depth x in the medium for forward moving (I) and backward moving (J) radiation respectively is given by:

$$I = A(1 - b)e^{ax} + B(1 + b)e^{-ax} \tag{3}$$

$$J = A(1 + b)e^{ax} + B(1 - b)e^{-ax} \quad \text{where} \tag{4}$$

$$a = \{k(k + 2s)\}^{1/2} \quad \text{and} \tag{5}$$

$$b = \{k/(k + 2s)\}^{1/2} \tag{6}$$

The parameters k and s are the absorption and backscattering coefficients, respectively, for the diffuse radiation fluxes I and J. It may be shown [14] that for diffuse radiation, the average path length of the radiation within an infinitesimal layer of thickness dx is equal to 2 dx and consequently the absorption and backscattering coefficients per unit length (denoted by K and S respectively) are given by:

$$K = k/2 \quad \text{and} \tag{7}$$

$$S = s/2 \tag{8}$$

The interaction coefficients (K and S) may be expressed in terms of the measured remittance (R) and transmittance (T) as

$$K/S = \{(1 + R^2 - T^2)/2R\} - 1 \tag{9}$$

$$S = \frac{1}{2d}\{K/S(K/S + 2)\}^{-1/2} \coth^{-1} \frac{1 - R(K/S + 1)}{\{K/S(K/S + 2)\}^{1/2} R} \tag{10}$$

Absorption (K) and scattering (S) coefficients for the epidermis have been calculated by Wan et al. [19] from their experimental measurements in vitro of epidermal remittance and transmittance.

Unfortunately, these authors apparently failed to correct for the regular reflectance of the radiation from the surface of the skin which results from the difference in refractive index (n) between air (n = 1.0) and stratum corneum (n = 1.55). In practice, this regular reflectance component is about 5–7% of the incident radiation for perpendicularly-incident radiation. The reason why this component is slightly higher than the theoretical regular reflectance of 4.7% [14] is attributed to the non-planar surface of the skin which produces off-normal angles of incidence [1]. If a correction for regular reflectance is made to the diffuse reflectance of Caucasian epidermis published by Wan et al. [19] and Anderson et al. [1] then it becomes evident from equation (9) that much higher values for the absorption coefficient K will be obtained than were calculated by Wan and his colleagues. Values of the absorption and scattering coefficients calculated after taking into account regular reflectance are shown in Fig. 1 for representative fair-skinned Caucasian epidermis. In calculating these interaction coefficients a net diffuse remittance (after accounting for regular

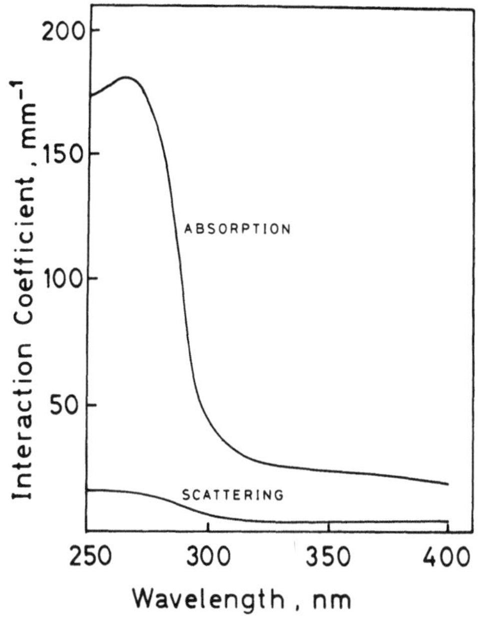

Fig. 1. Scattering and absorption coefficients for Caucasian epidermis calculated by the Kubelka-Munk model

reflectance) of 0.04 was employed for all wavelengths from 250 to 400 nm. Inspection of these coefficients would suggest that absorption is more important than scattering in the ultraviolet region, particularly at wavelengths less than 300 nm.

Recent measurements of the scattering properties of human epidermis [6] indicate a forward-oriented scattering mechanism where the angular profiles of scattering are steep and far from diffuse (Fig. 2). These data show that 80% or more of UVR transmitted through the stratum corneum lie within ± 22.5° from the normal, with only a slight wavelength dependency. After traversing the Malpighian layer about 50% of the transmitted radiation lies within this acceptance angle. Bruls [6] concludes, therefore, that application of the Kubelka-Munk model to the ultraviolet optics of the epidermis is not justified because the scattering is far from diffuse.

A Forward-Oriented Scattering Model

Since the Kubelka-Munk model neglects to account for forward scattering, which has been shown to be a serious omission in considering the ultraviolet optics of the epidermis [6], an alternative model is that described by Diffey [8]. The main feature of this forward-oriented scattering model is that the ratio of forward to backward scattering is an adjustable parameter which can be estimated from molecular light-scattering theory. In common with other theoretical treatments of skin optics, the present model neglects the effects of polarization and assumes that the skin and irradiation field are of infinite sideways extent so that edge effects can be neglected. Also the skin is assumed to be isotropic and the scattering particles, e.g., melanosomes are sufficiently far apart that they scatter independently of each other. In this method the transport equation is converted from a continuous to a discrete form and the solution is obtained by treating the continuous variables representing angle and depth of penetration in a stepwise manner.

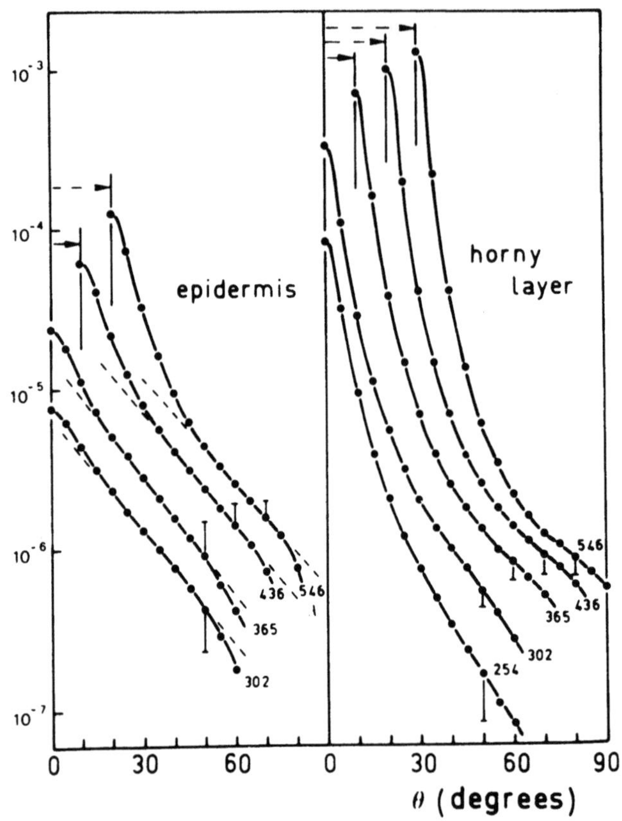

Fig. 2. Fraction of incident radiation at various wavelengths transmitted per unit solid angle ($\Delta\omega$) in the direction θ for epidermis and stratum corneum (horny layer) from the lower back [6]. Profiles denoted by broken arrows have been shifted along the θ-axis to avoid overlapping

The medium is regarded as an assembly of a large number of elementary scattering layers, each of thickness δx. In common with the Kubelka-Munk theory, the scattered radiation inside the medium is regarded as composed of two diffuse fluxes, $\vec{N}(x)$ and $\overleftarrow{N}(x)$, moving in opposite directions. In addition, the model considers forward-directed ("pseudo-collimated") radiation which undergoes small-angle scattering as a result of considerations of the optics of the epidermis [6, 8]. This component may be expressed as

$$\hat{N}(x) = \hat{N}(x - \delta x)\{(1 - \mu\delta x) + \beta S\delta x\} \tag{11}$$

where $\hat{N}(x)$ is the radiation flux at depth x in the medium travelling in a peaked forward direction, and β is the fraction of scattered radiation that is scattered in the forward direction. Consideration of the Mie scattering properties of the epidermis [6] indicated that single particle scattering will be predominantly in the forward direction and led to a value of $\beta = 0.84$ being adopted [8]. The fraction of backward scattering is therefore $(1 - \beta)$.

The diffuse radiation flux moving in a forward direction at depth x is given by

$$\vec{N}(x) = \vec{N}(x - \delta x)\{(1 - \mu\alpha\delta x) + \beta S\alpha\delta x\} + \overleftarrow{N}(x + \delta x)(1 - \beta)S\alpha\delta x \tag{12}$$

Similarly the diffuse radiation flux at depth x travelling in a backward direction is

$$\begin{aligned}\overleftarrow{N}(x) = \overleftarrow{N}(x + \delta x)\{(1 - \mu\alpha\delta x) + \beta S\alpha\delta x\} + \vec{N}(x - \delta x)(1 - \beta)S\alpha\delta x \\ + \hat{N}(x - \delta x)(1 - \beta)S\delta x\end{aligned} \tag{13}$$

The coefficient α is the ratio of the average path length of the diffuse beam in each elementary scattering layer to the layer thickness δx. As mentioned previously, α is equal to two (14).

The transport equation, in the form of equations (11), (12) and (13) is solved in a stepwise manner in which the radiation fluxes $\hat{N}(x)$, $\vec{N}(x)$ and $\overleftarrow{N}(x)$, are calculated at each layer in turn throughout the medium, first in the forward direction starting with the entry layer, then in a reverse direction with the emergent layer. This passage forwards and backwards through the medium is continued until the sets of fluxes at each layer converge. In practice this is achieved after only four passages through the medium, two in the forward direction and two in the reverse direction.

The boundary conditions are that for a plane, parallel beam of UVR of unit flux normally incident upon a medium of thickness d

$$\hat{N}(0) = 1, \vec{N}(0) = 0 \quad \text{and} \quad \overleftarrow{N}(d) = 0$$

Furthermore there is a change in refractive index at the front and back surfaces of the medium which modifies the transmitted (T) and reflected (R) fluxes at the boundaries as follows:

$$T = (1 - r_0)\{\vec{N}(d)(1 - r) + \hat{N}(d)(1 - \hat{r}) + N(0) \exp(-\mu d)(\hat{r} - r_0)\} \tag{14}$$

$$R = \hat{N}(0)r_0 + \overleftarrow{N}(0)(1 - r) \tag{15}$$

The coefficients r_0, $\hat{r}$ and r are the reflection coefficients for the incident radiation, the radiation moving in a peaked forward direction at the exit surface of the medium, and the diffuse radiation at the entry and exit faces of the medium respectively. For stratum corneum the numerical values of r_0, $\hat{r}$ and r are estimated to be 0.05, 0.1 and 0.5, respectively [8].

The ultraviolet radiation transmission and reflection characteristics of various human epidermal specimens were measured by Everett et al. [10] using a recording spectrophotometer with integrating sphere input optics. These experimental data have been used as the basis for calculating the absorption (K) and scattering (S) coefficients for human stratum corneum by an iterative technique using the forward-oriented scattering model [8]. The two interaction coefficients derived from this model are illustrated in Fig. 3 and lead to the following conclusions:

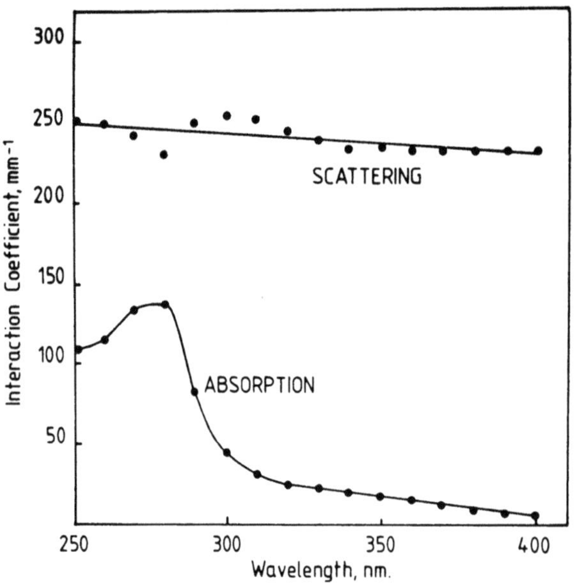

Fig. 3. Scattering and absorption coefficients for Caucasian stratum corneum calculated by the forward-oriented scattering model. A smooth curve (——) has been drawn through the calculated interaction coefficients (•) in each case

- Scattering is more dominant than absorption in describing the optical properties of the stratum corneum.
- The scattering coefficient is approximately independent of wavelength in the range 250 to 400 nm. This finding favours Mie scattering rather than Rayleigh scattering as the main scattering process since Rayleigh scattering varies inversely with the fourth power of the wavelength of the incident radiation.
- The absorption coefficient peaks around 280 nm, corresponding to the absorption maxima of the aromatic amino acids [4] and decreases rapidly in the spectral range 290 to 310 nm. The absorption coefficient falls slowly in the UV-A region reflecting the absorption spectrum of melanin [9].
- Furthermore the absorption coefficient shown in Fig. 3 is in excellent qualitative agreement with the representative photoacoustic spectrum of newborn rat stratum corneum determined by Rosencwaig and Pines [16].

Comparison of Predictions Using Theoretical Models with Experimental Measurements

Experimental measurements on the transmission of optical radiation through human epidermis have been carried out by Bruls [6]. From these data Bruls was able to estimate the approximate in vivo half-value layers of fair-skinned Caucasian stratum corneum and epidermis as a function of wavelength. The half-value layers (HVLs) at various wavelengths predicted by the Kubelka-Munk method were determined by use of the equation (2)

$$\mathrm{HVL} = \ln(4/(1 + b))/a \qquad (16)$$

Table 1. Approximate in vivo half-value layer of fair-skinned Caucasian full thickness epidermis

Wavelength (nm)	Half value layer (μm)	
	Measured (Bruls [6])	Calculated using KM model (Wan et al. [19])
254	6.1	1.8
280	5.8	2.2
297	12.7	6.6
313	20.4	11.1
365	29	13
405	37	15

Table 2. Approximate in vivo half-value layer of fair-skinned Caucasian stratum corneum

Wavelength (nm)	Half value layer (μm)	
	Measured (Bruls [6])	Calculated using forward-oriented scattering model (Diffey [8])
254	8.0	5.0
280	5.5	4.2
297	9.9	8.5
313	17.1	14.1
365	31	22
405	38	31

where a and b are as defined in equations (5) and (6). The coefficients k and s are calculated from the interaction coefficients per unit length (K and S) from equation (7) and (8), and values of K and S at the appropriate wavelengths are determined from Fig. 1. A comparison of the measured and calculated in vivo half-value layers for full-thickness epidermis at selected wavelengths in the ultraviolet spectrum are shown in Table 1. It may be seen that the KM model underestimates the transmission of UVR through the epidermis, particularly in the UV-C region. This probably arises from the assumption inherent in the KM model that the scattering in the epidermis is diffuse. Consequently the application of this model to the optics of the epidermis necessitates the introduction of a large absorption to account for the small observed remittance which leads to an underestimate for values of the HVLs.

The HVLs at selected wavelengths in the ultraviolet spectrum were also calculated for fair-skinned Caucasian stratum corneum using the forward-oriented scattering model. In this instance a value of 100 μm was arbitrarily chosen for the thickness of the stratum corneum, and that value of depth (x μm) noted when the sum of the radiation fluxes, $\hat{N}(x) + \vec{N}(x) + \overleftarrow{N}(x)$, was equal to one half of the incident radiation flux ($\hat{N}(0)$). The HVLs calculated in this way are compared with measured values of the approximate in vivo HVLs in stratum corneum [6] in Table 2. It may be seen there is good agreement between the two sets of HVLs, although the calculated values consistently underestimate the measured values by an average of 23%. Nevertheless this is considerably closer than the HVLs for the epidermis calculated by the KM model where, on average, the calculated values are only 43% of the measured values.

Discussion

It is suggested that the observed reflectance or remittance of UVR from human skin in vivo results from scattering of radiation in a mainly forward direction through the epidermis, followed by more symmetrical scattering in the dermis. This follows from consideration of Mie scattering since the dimensions of cell organelles situated in the epidermis are of the order of the wavelength of optical radiation whereas the scattering centres in the dermis are probably collagen fibres with dimensions of the order of 0.1 μm. This proposal is in opposition to that of Anderson et al. [1] who suggested that the observed reflectance is attributed to minimal scattering within the epidermis. These authors arrived at this conclusion by applying the Kubelka-Munk model to measured values of reflectance and total transmittance through various layers of the skin. However this model does not take account of forward scattering and indicates that the epidermal absorption coefficient should be much greater in magnitude than the scattering coefficient. The numerical consequence of this finding shows the fraction of radiation directly transmitted through the epidermis to be very similar to that totally transmitted. This assertion does not support the experimental findings of Everett et al. [10] that the total UVR transmitted through either the stratum corneum or whole epidermis is several times that directly transmitted in the spectral range 250 to 400 nm.

On the other hand, the forward-oriented scattering model correctly accounts for the large differences between the directly transmitted (i.e. unattenuated) and totally transmitted radiation through the epidermis, and incorporates a forward-scattering component which has been shown to be important in epidermal optics [6]. It is suggested that this model is preferable to the KM model for describing the optical properties of the epidermis. A more precise and realistic model of epidermal optics is still needed.

References

1. Anderson RR, Hu J, Parrish JA (1981) Optical radiation transfer in the human skin and application in in vivo remittance spectroscopy. In: Marks R, Payne PA (eds) Bioengineering and the skin. MPT Press Ltd, Lancester, pp 253–265
2. Anderson RR, Parrish JA (1982) Optical properties of human skin. In: Regan JD, Parrish JA (eds) The science of photomedicine. Plenum Press, New York, pp 147–194

3. Bachem A, Reed CI (1930) The penetration of ultraviolet light through the human skin. Arch Phys Ther 11:49–56
4. Beaven GH, Holiday ER (1952) Ultraviolet absorption spectra of proteins and amino acids. Adv Protein Chem 7:319–386
5. Blois MS (1969) Discussion on skin optics. In: Urbach F (ed) The biologic effects of ultraviolet radiation with emphasis on the skin. Pergamon Press, Oxford, p 170
6. Bruls WG (1984) Optical properties of human epidermis. PhD Thesis, University of Utrecht, The Netherlands
7. Chandrasekhar S (1960) Radiative transer. Dover, New York
8. Diffey BL (1983) A mathematical model for ultraviolet optics in skin. Phys Med Biol 28: 647–657
9. Edwards EA, Finkelstein NA, Duntley SQ (1951) Spectrophotometry of living human skin in the ultraviolet range. J Invest Dermatol 16:311–321
10. Everett MA, Yeargers E, Sayre RM, Olson RL (1966) Penetration of epidermis by ultraviolet rays. Photochem Photobiol 5:533–542
11. Hardy JD, Hammell HT, Murgatroyd D (1956) Spectral transmittance and reflectance of excised human skin. J Appl Physiol 9:257–264
12. Kaidbey KH, Agin PP, Sayre RM, Kligman AM (1979) Photoprotection by melanin: a comparison of black and caucasian skin. J Am Dermatol 1:249–260
13. Kirby-Smith JS, Blum HF, Grady HG (1942) Penetration of ultraviolet radiation into skin as a factor in carcinogenesis. J Natl Canc Inst 2:403–412
14. Kortüm G (1969) Reflectance spectroscopy. Springer, Berlin
15. Kubelka P, Munk F (1931) Ein Beitrag zur Optik der Farbanstriche. Z Technische Physik 12: 593–601
16. Rosencwaig A, Pines E (1977) Stratum corneum studies with photoacoustic spectroscopy. J Invest Dermatol 69:296–298
17. Schuster A (1905) Radiation through a foggy atmosphere. Astrophys J 21:1–22
18. van de Hulst HC (1980) Multiple light scattering. Academic Press, New York
19. Wan S, Anderson RR, Parrish JA (1981) Analytical modelling for the optical properties of the skin with in vitro and in vivo applications. Photochem Photobiol 34:493–499

Simulation of Diffusion in Skin

M. I. Foreman

There remains much current interest in trans-dermal drug delivery systems, as one possible means, for example, of minimising toxicity by local delivery of the drug to the site of trauma. Whilst there have been major advances in understanding the nature of skin as a barrier, experimental approaches to the problem are limited by the nature of the mathematics involved. The difficulty arises because the basic diffusion equations are second order differential equations, for which solutions only exist under a restricted number of boundary conditions [1]. The actual process of molecular diffusion is, however, extremely simple; it can be closely approximated by random particle motion with a mean path length in unit time in one or more directions. Such a process is easily simulated on a digital computer, requiring only that the boundary conditions of interest are superimposed onto the basic process. This greatly simplifies the study of diffusion phenomena by freeing the experimenter from the tyrrany of the boundary conditions imposed by analytical mathematics.

An elegant example concerns the situation where diffusing material is applied as a thin surface film, which then diffuses through the membrane into an effectively infinite sink [2]. This simulates closely the case where the drug is applied to a skin as a surface film, which eventually penetrates through to the vasculature, and is one example for which analytical solutions are not available. Simulation of the process by digital computer is extremely easy. The predicted behaviour is shown in Fig. 1 for the cumulative amount of material penetrating a plane membrane as a function of time. The validity of such simulations has been checked against cases where the boundary conditions do permit comparison against analytical solutions. Further, experimentally observed diffusion curves for this situation fit the predicted behaviour very well, and it is possible, by fitting the experimental and predicted curves, to obtain an estimate of the diffusion constant for the experimental case. This is simply done by obtaining an estimate of the factor S_t which relates real time for an experimentally observed membrane to "computer time" (Fig. 1).

The simulation can, of course, be used to study other aspects of the diffusion process. For example, the flux of material through the lower face of the membrane can be predicted. The resulting curve shows striking similarities to the response-time profile of the human vasoconstrictor response to topically applied corticosteroids. This raises the possibility that the human vasoconstrictor response may be determined by the rate at which the drug is presented to the appropriate receptors, rather than by the local drug concentration.

One major advantage of the use of this set of boundary conditions in experimental design is that it is no longer necessary to have the skin sample mounted between two solutions, as is the case for the more traditional methodology. This avoids undue

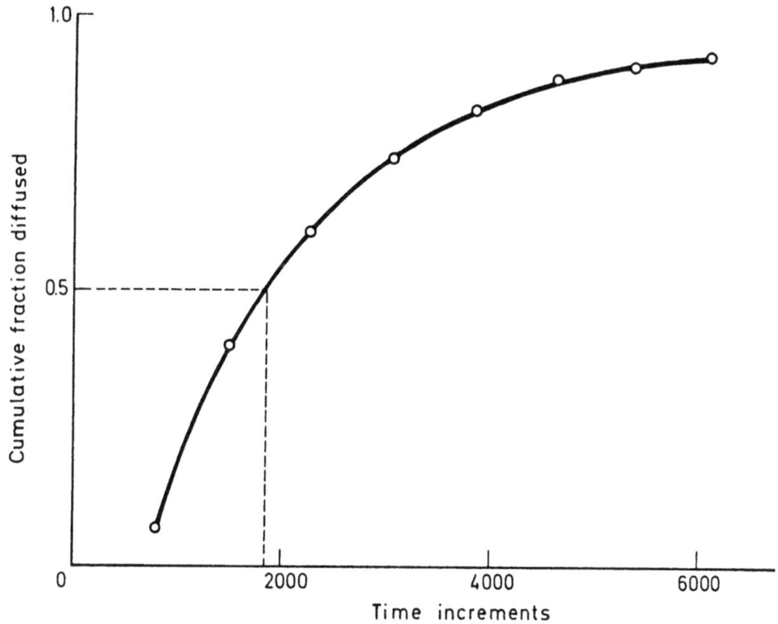

Fig. 1. Simulated diffusion curve for penetration of material applied as a thin surface film through a plane membrane into an infinite sink.

Time increments = real time x S_t; $S_t = \frac{(L/50)^2}{2D}$

$T^D_{\frac{1}{2}}$ = time for one half of the diffusant to penetrate the membrane = 1,855 x S_t (from the simulation) = $0.371\ L^2/D$

L = membrane thickness, D = diffusion constant

hydration of the tissue, which has long been known to have a marked effect on the barrier property of the skin. By studying steroid diffusion in skin under both occluded and non-occluded conditions, it has been found that skin, most probably the stratum corneum in fact, has the ability to "irreversibly" bind a certain proportion of the applied diffusant. This binding ability is markedly reduced by tissue hydration which results from occlusion, and the whole process is clearly akin to the steroid reservoir phenomenon reported previously [3]. It now seems possible to investigate this process quantitatively for the first time. For example, Fig. 2 shows the effect of occlusion on the binding ability of skin for a selection of steroids. A reservoir effect of this kind has not been found with two non-steroids studied so far; urea and dimethylsulphoxide.

An attempt has also been made to investigate the thermodynamics of the processes involved in skin diffusion. Using nandrolone as a model steroid, measurements of the diffusion constant at three temperatures have provided an estimate of the activation energy for diffusion in non-occluded skin, which appears to be 7 to 10 kilocalories per mole. By making some fairly reasonable assumptions concerning the binding process for steroids within the stratum corneum, it can be argued that there exist binding sites which normally are hydrated. These sites lose the bound water on exposure to steroids, which themselves become bound. The free energy change for this process can

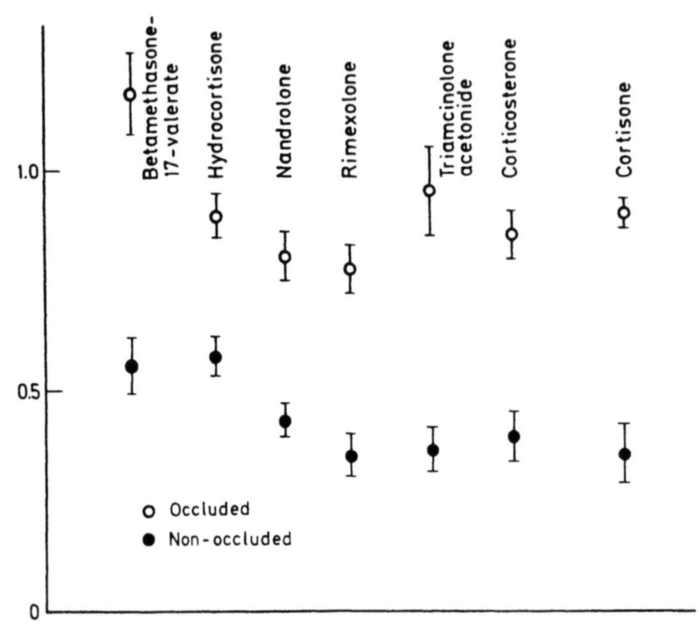

Fig. 2. Effect of occlusion on the fraction of material (F) free to diffuse. The fraction bound in the stratum corneum is 1 – F

be shown, from the appropriate Arrhenius plot, to be 17 to 20 kilocalories per mole. This value is entirely consistent with chemisorption processes.

Most work to date treats skin as a homogenous membrane. The diffusion constants quoted are therefore "apparent" rather than real, and represent a sort of aggregate behaviour of the various contributing layers. It would be a major advantage to be able to identify the actual contribution of the separate layers. By noting that the diffusion curve observed for a laminated system is experimentally identical to that predicted for a homogenous membrane, it is possible to argue that the "apparent" diffusion constant D_{app} is approximately related to the diffusion constants D_1 and D_2 by the equation;

$$L/\sqrt{D_{app}} = L_1/\sqrt{D_1} + L_2/\sqrt{D_2}$$

Where L, L_1 and L_2 are the lengths of the whole membrane and the component layers respectively. From data available to date for diffusion of various steroids in skin, this equation appears to fit the experimental situation within the limits of error. One difficulty with such an equation is that non-unit partition coefficients within the membrane are ignored. Nor is it clear how such a situation would affect the diffusion curves observed. This is, in fact, a further area of skin diffusion which remains virtually unexplored, and which is particularly amenable to study by computer simulation.

References

1. Crank J (1970) The mathematics of diffusion. Oxford University Press
2. Foreman MI, Kelly I, Lukowiecki GA (1977) A method for the measurement of diffusion constants suitable for occluded and non-occluded skin. J Pharm Pharmacol 29:108–109
3. Vickers CFH (1963) Existence of reservoir in the stratum corneum-experimental proof. Arch Dermatol 88:20–23

An Interactive Computer Model as a Laboratory Tool for Research on Epidermal Cellular Interactions

E. Mitrani

We have been involved for several years in computer modeling of epidermal cellular interactions in skin. Since microcomputers have become readily available, we have now adapted our three-dimensional computer simulation of the skin so that it can be run on a standard Apple II or IBM PC microcomputer with the aim that this type of model might become a standard tool in skin research laboratories. In this article an attempt is made to present a general idea of what a computer simulation of skin might entail. The understanding of what the model actually is may help one to realize the possible significance of the results that are obtained. No attempt is made to give a detailed description of the model. Several aspects describing details of the model as well as results have been given elsewhere [1, 2, 6–9].

Description of the Model

A game of chess is a good example of a simulation which essentially entails three aspects: first one builds the acutal component units (pawns etc.) within a structure (the checkerboard), second, one specifies the rules (i.e. a bishop moves diagonally) and finally one has to play the game, i.e., the structure develops in time whenever a move is made. Similarly, in the simplest form the computer simulation of skin can be described by three analogous aspects. 1) The cells constitute the component units and are defined within a structure which is determined by the three dimensional space in which cells are located. 2) Rules are given at the cellular and supracellular level and every attempt is made to provide rules which are derived from biological observations. Yet, from the model point of view rules could be altered in the same way as one could conceive a chess game in which different rules governing the behaviour of the figures could be given, i.e., what would happen if the black pawns are allowed to move backwards as well as forward. 3) The model develops in time by increasing the age of each component unit in fixed constant increments; it then determines how this affects the system behaviour and makes the necessary changes.

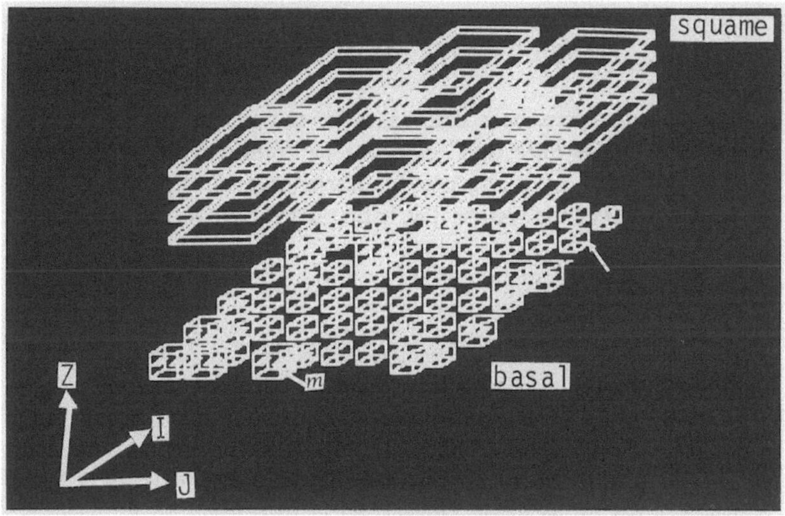

Fig. 1. Three dimensional diagram of skin as simulated by the computer. In the basal layer, some large cells (M) are about to divide. Other basal cells which are being displaced from the basal layer (arrow) can be seen. The squames in higher layers overlay a larger number of basal cells

Structure and Component Units

The structure is an idealized epidermal tissue. A number of layers of cells lie over a substrate of connective tissue almost in a lattice array. Mitosis takes place in the basal layer and cells are extruded from the basal layer, progressing upwards towards the surface of the epidermis, from which they are eventually lost sometime after death.

Figure 1 illustrates a three dimensional diagram of skin as simulated by the computer. It is relatively easy to create this type of structure on the computer. Essentially one has to define a set of three dimensional matrices of the form TIM(I, J, Z). I and J are variables which represent the X and Y coordinates of the horizontal plane, and the variable Z represents the vertical component. Each cell, represented by a cube, occupies a position I, J, Z in space and the matrix TIM represents the age of each particular cell. For example if in position whose coordinates are 3, 2, 5, there is a 50-h old cell, then this will be represented in the matrix TIM as TIM(3, 2, 5) = 50. Positions may or may not be occupied by cells. Positions not occupied by cells are given an arbitrary negative number, i.e., if position 3, 3, 5 is empty then TIM(3, 3, 5) = − 1.

In summary, the first aspect of the model creates the cubes which are physical entities representing cells and it takes into account the positions that the cubes occupy in the three dimensional lattice.

Cell and System Rules

The program rules have been defined, as far as possible, on the basis of experimental observations. Those rules that could not be defined with certainty were dealt with, by systematically introducing possible alternatives and examining their effect on the overall behaviour of the system. The basic rules currently incorporated into the model, are as follows:

Cell Rules

The fundamental process on which the system depends is the cell cycle. A newly formed basal cell will progress, unless stopped from doing so, through the various cycle phases G1, S, G2 and M to produce two daugther cells. G1 has been subdivided into G1a and G1b.

1) The phases G1a, S, G2 and M are assumed to have short constant durations of the order of ten hours.
2) The length of the phase G1b is assumed to be variable (2 to 200 h) since it is postulated that entry into S phase is determined by a probability function which is time independent [11].
3) A cell cannot leave the basal layer whilst in G1a, S, G2 and M phases [1].
4) Detachment during G1b is mediated by a viscous grip and is qualified according to the viscosity constant ck. Detachment of cells is therefore a process which depends on pressure and on time [5].
5) The nearer a cell is to the basal layer the higher the chances of entering S phase.
6) A cell has a chance of becoming mitotic up to a given age. After that, it flattens and the further away it is from the basal layer, the shorter its life span.

System Rules

Within the structure cells interact with each other. It is postulated that a dividing cell gives over its location to one of its two daugther cells. The second daughter cell will divide in the vertical plane if no neighbouring cell is in G1b. Otherwise, an attempt will be made to extrude the neighbouring cell from the basal layer leaving the place to be occupied by the daughter cell. If the situation arises that no position is available and no neighbouring cell can be extruded, then the computer proceeds to create extra spaces between the cells in the basal layer and thus contributes to the crowding of the basal layer.

It is also postulated that basal cells are continuously being submitted to two types of forces: one local, resulting from the nearby mitosis (Fm), and another global which may result from an excess number of cells in the basal layer (Fn). Consequently the model incorporates folding of the basal layer which may result from overcrowding, and thus comprises a basal layer which can expand and contract dynamically. The time needed to release a new daughter cell should be equal to TG1a + ck/Fn. If n9 is the number of basal cells per unit length in normal conditions and n8 is the number of basal cells when the basal layer is folded, then the force Fn will be proportional to (n8 – n9)/n9. The detachment fraction experienced by G1b cells due to the force generated by mitosis Fm is proportional to 1/ck.

Time

The third element of the simulation is time and this element is responsible for the evolution of the system. At the start only a few basal cells are created by assigning to them ages corresponding to stages randomly distributed round the cell cycle. Time is implemented by creating a counter. What the computer program does is as follows. For the first iteraction it sets a counter equal to 1, i.e., COUNTER = 1 or TIME = 1. This means that the first unit of time has elapsed and therefore the program will scan every cell in the system in order to update it, i.e., cells are made one unit of time older. Once the program has updated all the cells in time it then checks for each cell its CELL RULES and according to the rules prescribed, the age and the position the cell occupies, it makes the necessary changes. For instance if any one cell was found to be in the previous iteration, one hour before completing mitosis it should now divide to create two daughter cells. The computer program according to the rules specified under CELL RULES will proceed to produce two new cells, i.e., it will create a new cube with age = 0 in the position occupied by the mother cell and the second one according to the rules specified under CELL RULES for allocation of cells. If a neighbouring cell has to leave the layer, then the nearest empty position in the next higher layer is allocated to the removed neighbouring cell and if no position is available shifting continues in the higher layers until an empty position is found or until a new layer is created to accomodate the last shifted cell.

Once every single cell has been examined and updated the computer will increase the counter by 1 and it will proceed to repeat the updating cycle for each cell once again. By performing this exercise over and over again, a situation that evolves with time is obtained. One single basal cell at the begining of the simulation can generate the whole three dimensional structure after a sufficient number of iterations have evolved.

Experimentation

It is important to realize that within the simulation each cell constitutes a physical entity which occupies a defined space and which evolves in time. Furthermore, the computer cells are transparent, which means that all the information which is being generated throughout the simulation is available and can be recorded. It is possible to continuously monitor for analysis every single cell in the system from the moment it is created until the moment it is shed from the epidermis. Quantitatively it is possible to record any parameter. Those that are continuously recorded in our program are:

Mitotic rate, cell cycle time, epidermal thickness, average age of cells at time of death, cell transit times, total cell number, percentage of vertical mitosis, degree of overcrowding in the basal layer (stress), average age at time of extrusion from the basal layer. Labelling experiments: Pulse labelling; continuous labelling; percentage labelled mitoses curves, mitosis frequency distribution histogram. One can determine how any of these output parameters is altered when one changes any of the input parameters or any of the basic postulates.

The microcomputer version of the computer model incorporates a graphics routine in addition, which displays continuously an evolving three-dimensional picture (Fig. 1) of the simulated epidermis. This provides visual display for the system as it evolves with time. All basal cells are displayed of the same size but as cells enter the M phase the cells are displayed larger in size so that the mitotic spindle can be visualized. M cells double their size by the time they divide to become two daughter cells. The graphics routine also displays G1b cells as they are displaced upwards with respect to the basal layer by amounts corresponding to their detachment fractions. It is also possible to see that postmitotic cells, after a given age, become flattened and overlay a certain number of basal cells. Nuclei of basal cells not in G1b are displayed so that those cells can be distinguished from other cells.

Clearly the rules that have been postulated above can be tested in the model versus the biological system, modified and new rules added. The computer program has been written such that with a minimum amount of programming experience one can ask "WHAT IF" questions not only about quantitative but also about qualitative aspects and see how these aspects or alternative rules may affect the behaviour of the model vis a vis the biological system. Some of the problems which have been analyzed with the model, are briefly described here.

Entry into S Phase

It was originally assumed that the time of the cell cycle is fixed. Experimentation with the model on this assumption led to the observation that cell divisions would become synchronized, which meant that all the basal cells tended to come into the same phase or stage of the cell cycle. A group of cells which happened to have a slight advantage in number would have that advantage continously reinforced until they would eventually overtake the whole structure [6]. To avoid such synchrony it was found necessary to introduce the assumption that the cell cycle time, or at least the intermitotic time, is not deterministically dictated but includes one element of probability. In agreement with the results of Smith and Martin [11] it was therefore postulated that the entry from the latter part of G1 (which we have called Gib) to S is determined by a probability function which is time independent.

Migration of Cells from the Basal Layer

An analysis has been made of the possible alternatives for cells to leave the basal layer distally in steady state conditions [9]. Results from that analysis challenge the concept that cells actively migrate from the basal layer. Although there is evidence for active migration of cells from the basal layer in circumstances such as wound healing [12], only the work of Etoh et al. [3] constitutes a serious attempt to demonstrate that active migration from the basal layer may take place in normal epidermis. Etoh et al experiments consisted in inhibiting mitosis with β-irradiation and recording the number of basal cells that left the basal layer after irraditation. Our computer simulations of Etoh et al.'s experiments showed that if migration is an active process, the rate of cell migration consistent with the kinetics of the process being examined, would yield a depopulation value considerably higher than the one observed experimentally. We found, by contrast, that rates of migration comparable to those obtained experimentally after inhibition of mitosis could be obtained if it was assumed instead that cells are passively removed from the basal layer as a result of forces generated, directly or indirectly, by dividing cells.

The Plane of Division of Basal Cells

Vertical mitoses are known to occur in mammalian epidermis. When vertical mitoses occur, one daughter cell is automatically cut off from the basal layer and therefore loses its grip to the baseline. Bullough and Mitrani [1] have reviewed evidence to show that, under the lateral tension generated locally by a mitotic division, not all the basal cells are equally liable to loose their baseline grip and so to move distally. No cell loses its grip throughout the whole mitotic sequence, mainly during S, G2 and M phases, or during the first part of G1 which we have named G1a and which lasts about 10 h. Thus only cells in the second part of G1 (Gib) can be forced to relinquish their basal grip. Normally Gib occupies a significant fraction of the cell cycle. However, in hyperplasia the considerable decrease in cycle duration is almost totally due to shortening of the Gib phase.

We have used our computer model to determine the probability of a dividing cell finding within a given neighbourhood, a Gib cell that can be extruded from the basal layer [7]. Those experiments suggest that as the mitotic rate increases, stress in the basal layer may build up as less cells are found suitable to be extruded. In mouse and guinea pig ear epidermis the stress that could have been caused by a high birth rate do not materialize since instead of cells crowding up in the basal layer the potential stress is released by cells dividing in the vertical plane. We have established that in those conditions the increase in vertical mitoses is sufficient to release the basal stress that otherwise would have been created [2]. Similarly the model has predicted that in a system with a very high proliferation rate in which the duration of Gib approaches zero, such as the hair follicle, in order for stress not to build up in the basal layer, the proportion of vertical mitoses should be 100%. We have examined the angle of mitosis in the active hair bulb and found that indeed the number of vertical mitoses in this system approaches 100% [2].

Conditions for Stability

There are many conditions of mammalian skin in which during hyperplasia the percentage of vertical mitoses remains fairly low and consequently stress in the basal layer does build up [7]. This stress will be transmitted to the dermis in the form of tension. Experiments with the computer model on this aspect of dermal-epidermal interaction has led us to propose the following. A hyperproliferating system may reach a situation in which basal stress will build up if there are not a sufficient number of vertical mitoses. This stress will be transmitted to the dermis as tension but nonetheless the system will remain stable. The change in mitotic rate is absorbed by folding of the baseline, basal stress remain high, but the system does achieve a balance between cell production and cell loss. If the dermis responds to this tension by synthesis of new dermal components, the balance of cells produced to cells extruded will be upset and no equilibrium point will be reached. New dermal tissue formation will only relieve the stress temporarily. Soon the demand for places in the basal layer (created by the decrease in number of extruded cells) will increase so that a new stress situation will be created, new dermal growth will arise as a response to the tension and therefore a non-stable, self expanding situation will arise [7]. The conclusions from those experiments are that stability cannot be achieved in chronic epidermal hyperplasia if the dermis responds by new connective tissue synthesis.

Conclusions

The main advantage of the computer simulation as described here is that it works as an "interactor" between the various elements that may constitute the system. Biological processes are by essence complex and it is therefore difficult to examine the repercussions that a particular hypothesis may have on the whole process. A computer simulation is most suitable for interacting the constituent hypothesis of a system and for determining the possible outcomes of such interactions. The overall effects of one particular hypothesis can be further tested by performing what is called a sensitivity analysis. In this type of analysis, variations in each hypothesis, either quantitative or qualitative, are systematically introduced into the system and the effect on the output of the system is recorded. This allows one to determine how sensitive the system is to fluctuations of its components and which type of results or observations are most dependent on a particular hypothesis.

We have so far based our computer analysis on the central assumption that epidermis is constituted by a homogeneous cell population. There have been reports that at least two distinct populations of basal keratinocytes coexist in mammalian epidermis [4,

10]. It requires little effort to make the necessary modifications in order to consider in our computer model an heterogenous basal cell population.

References

1. Bullough WS, Mitrani E (1976) An analysis of the epidermal control mechanism. In: Houck JC (ed) Chalones. North-Holland Publishing Co, Oxford, p 70
2. Bullough WS, Mitrani E (1978) The significance of vertical mitosis in epidermis. Br J Dermatol 99:603–610
3. Etoh H, Taguchi YH, Tabachnick J (1975) Movement of beta-irradiated epidermal basal cells to the spinous-granular layers in the absence of cell division. J Invest Dermatol 64:431–435
4. Lavker RM, Sun TT (1982) Heterogeneity in epidermal basal kerationocytes: morphological and functional correlations. Science 215:1239–1241
5. Leun GC van der, Lowe LB, Beerens EGJ (1974) The influence of skin termperature on dermal epidermal adherence. J Invest Dermatol 62:42–46
6. Mitrani E (1976) A theoretical analysis of cellular interactions leading to epidermal homeostasis in the adult mammal. Doctorate Thesis, London University, London
7. Mitrani E (1978) Possible role of connective tissue in epidermal neoplasia. Br J Dermatol 99: 233–244
8. Mitrani E (1981) A computer model for epidermal cellular interactions. Biosystems 14:179–191
9. Mitrani E (1983) Is upward basal cell movement independent of mitosis in the normal epidermis? Br J Dermatol 109:635–642
10. Potten CS (1981) Cell replacement in epidermis (keratopoiesis) via discrete units of proliferation. Int Rev Cytol 69:271
11. Smith JA, Martin L (1973) Do cells cycle? Proc Nat Acad Sci 70:1263–1267
12. Winter GD (1972) Epidermal regeneration in the domestic pig. In: Maibach HI, Rovee DT (eds) Epidermal wound healing. Chicago Year Book Medical Publishers, Chicago, III, p 71

Sound Skin Models – Acoustic Properties of Epidermis and Dermis

P. A. Payne, C. Edwards and C. J. Hacking

Ultrasound has been used extensively for obtaining dimension data from human and non-human skin in vivo. Early work [1, 6] used the A-scan methods at relatively low frequencies of about 15 MHz. More recently [5] higher frequencies have been employed in order to yield better axial resolution and it is now possible to obtain both dermal and epidermal thickness data routinely.

A-scan or pulse echo mode techniques can be difficult to interpret especially where abnormal structures such as tumours are under investigation. Following the developments in ultrasound use established in other fields, such as obstetrics, attention has recently focused on obtaining B-scan or cross-sectional image data [4].

In both these modalities, A and B-scan imaging, the fundamental properties of the various tissues involved determine the form or character of the data obtained. For example, if ultrasound energy is directed towards a junction between two types of tissue, say epidermis and dermis, then assuming the interfaces to be flat, it is the acoustic impedance of the two different tissues which determines the amplitude of the echo obtained from the boundary. There is no reported data for the acoustic impedance of the various components of the skin. In contrast, values have been established for most other body tissues and organs [8]. Another fundamental parameter of vital importance in determining dimensional data for skin is the velocity of sound. Some work on establishing this parameter has been reported [2] but this is only for whole skin. If a measurement of acoustic impedance can be made for each of the components of skin, then an assessment of acoustic velocity can also be made given a measurement also of the density of the tissue. This follows from the definition of acoustic impedance Z.

$$Z = \rho c \ (\mathrm{kg\ m^{-2}\ s^{-1}}) \qquad (1)$$

where ρ is density (kg m^{-3}) and c is velocity (ms^{-1}). 1 rayl = 1 (kg m^{-2} s^{-1}).

A direct measurement of Z is possible by measurement of ρ and c, but this is difficult except on samples of excised tissue. In this case, c may be established by making a thickness measurement by micrometer and then using an A-scan technique to establish the transit time for ultrasound energy. One problem is that tissue is much changed by removal from its normal surroundings and the data so obtained are not reliable. In this paper an alternative approach has been employed and is presented below. The results reported represent early work in this important area and much further work needs to be done. They show, however, that this data can be established and indicate that there may be important diagnostic information available from such measurements.

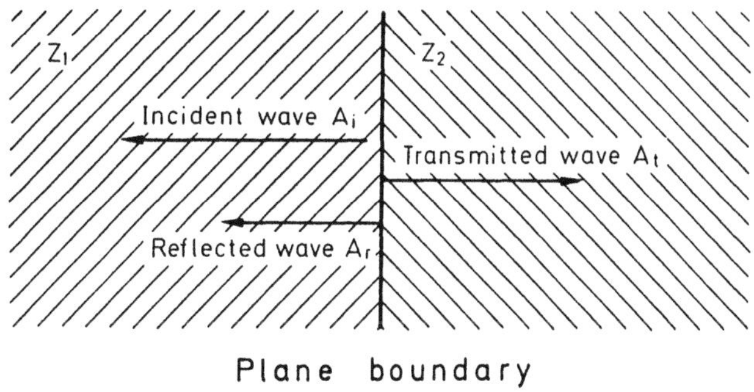

Fig. 1. Transmission and reflection at a plane boundary

Theory

At a plane boundary between two acoustically different media of acoustic impedance Z_1 and Z_2 the amplitude reflectivity, R is given by:

$$R = \frac{A_r}{A_i} = \frac{Z_2 - Z_1}{Z_2 + Z_1} \qquad (2)$$

where A_i is the amplitude of the incident energy and A_r the amplitude of the reflected energy. Figure 1 illustrates this concept.

The amplitude of an incident wave is difficult to establish experimentally in a direct manner. We may measure A_i indirectly by examination of A_r from a range of materials whose acoustic impedance is known and this experiment has been performed in distilled, degassed water at 20 °C, the acoustic impedance of which is also well documented [3, 7]. The experiment may then be repeated on skin and given a measurement of A_r we may infer a value of Z_{skin}.

Method

A 10 mm diameter, 25 μm thick polyvinylidene fluoride transducer was employed having a focal length of 38 mm. By measurement this transducer produces ultrasound energy centred at 18 MHz with a bandwidth (−6 dB points) from 10 MHz to 28 MHz. Since the ultrasound energy field produced by any transducer is complex and variable along the axis, great care was taken to place all the samples examined at exactly the same point. This ensures a constant value of A_i. A range of materials of various acoustic impedances was then examined and following careful alignment to ensure normal incidence and reflection the value of A_r was recorded using a Tektronix 7854 digitising oscilloscope. The experimental arrangement is shown in Fig. 2. A total

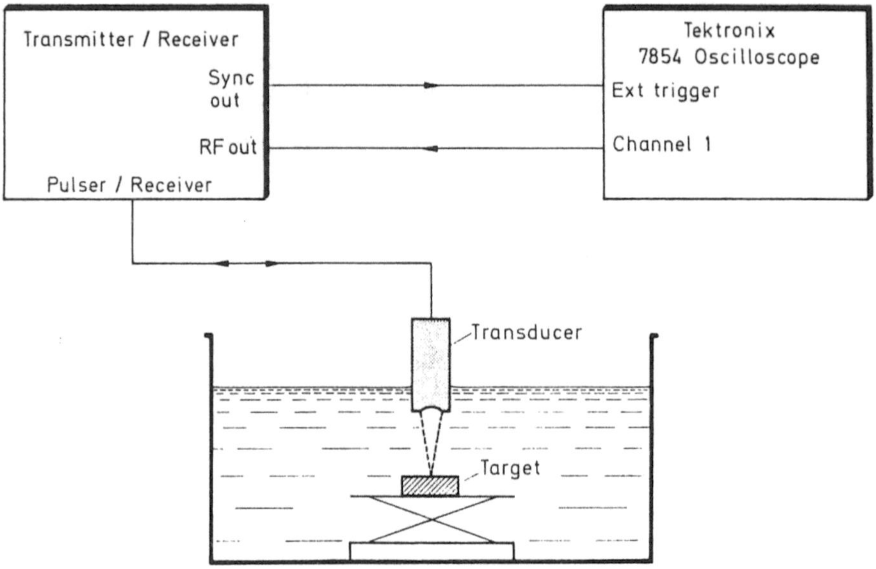

Fig. 2. Experimental schematic

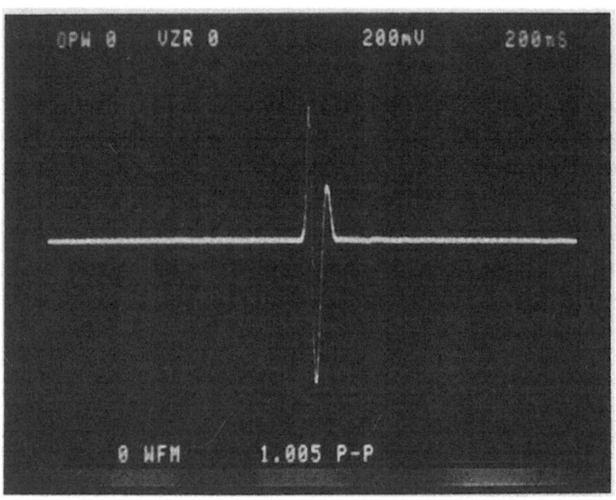

Fig. 3. Typical measured result for a steel target. Peak reflection = 1,005 mV (16 dB attenuation)

of twelve different materials were employed to calibrate the measurement system. Each material consisted of flat parallel sided samples having "polished" faces to minimise the effects of scattering at the reflecting surface. Typical results are shown in Figs. 3, 4.

Since the acoustic impedance of any given material is a sensitive measure of its constituent materials and also the manner of preparation, each sample was carefully

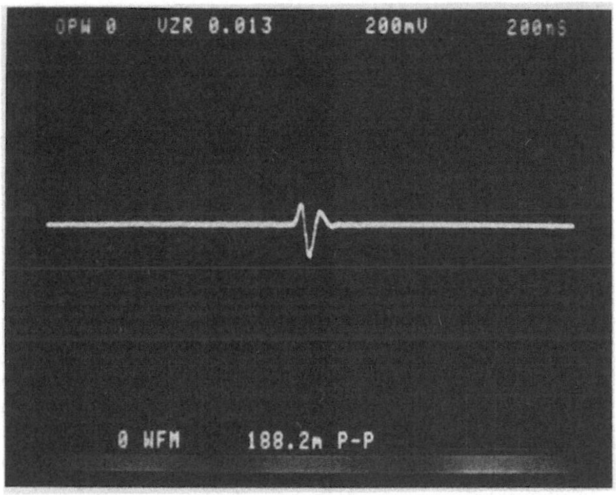

Fig. 4. Typical measured result for a rubber target. Peak reflection = 188.2 mV (16 dB attenuation)

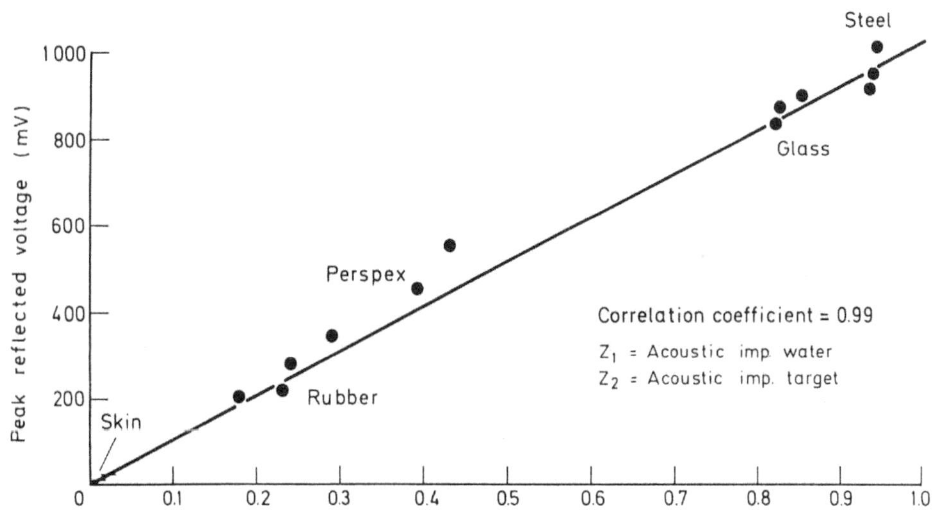

Fig. 5. Peak reflected voltage versus $\frac{Z_2 - Z_1}{Z_2 + Z_1}$ for various materials

subjected to independent measurements of ρ and c. The acoustic impedance was then calculated using equation (1).

From equation (2) it can be seen that if A_r (which is measured) is plotted against $\frac{Z_2 - Z_1}{Z_2 + Z_1}$ the relationship should be a straight line passing through the origin.

Table 1. Comparison of measured data and reference source data for various materials

Target	Density (gm cm^{-3})		Acoustic velocity (ms^{-1})		Acoustic impedance (Rayls)	
	Measured	Data book	Measured	Data book	Measured	Data book
Copper	8.89	8.93	4,941	5,010	43.93	44.5
Brass	9.97	8.6	4,576	4,700	45.62	40.2
Steel	7.82	7.85	6,366	5,960	49.55	47.0
Al. (cyl)	2.71	2.7	6,866	6,420	18.61	17.33
Al. (sheet 1)	2.64	2.7	7,092	6,420	18.72	17.33
Al. (sheet 2)	2.62	2.7	7,193	6,420	18.84	17.33
Perspex	1.21	1.18	2,842	2,680	3.43	3.16
Polystyrene	1.193	1.06	2,288	2,350	2.72	2.49
Neoprene rubber	1.46	1.33	1,654	1,600	2.414	2.13
Butyl rubber	1.252	1.07	1,945	–	2.435	–
Thermoplastic	1.18	–	1,819	–	2.146	–
PVDF	1.77	1.78*	2,112	1,500 to 2,200*	3.74	2.67 to 3.92*
Glass BK7	2.514	2.51*	6,225	5,660	15.65	14.21
Glass soda	2.438	2.5	6,265	–	15.27	–
Water @ 20 °C	–	0.998	–	1,497		1.494

Data from Weast [7] and Kaye and Laby [3] except * which is manufacturer's data

Figure 5 shows the results of this calibration exercise and the data can be seen to lie close to the theoretical line (r = 0.99). Table 1 compares the measured acoustic impedance with data from Kaye and Laby [3] and Weast [7]. This provides further confidence in the experimental methods employed.

The slope of the graph given in Fig. 3 is proportional to A_i, the incident acoustic pressure. We may therefore use the graph to measure Z_{skin} directly, given a measure of the amplitude of A_r. Alternatively, we can compute Z_{skin} from the equation to the graph of Fig. 5, which is:

$$A_r = 1{,}020 \frac{Z_2 - Z_1}{Z_2 + Z_1} - 0.02 \quad (3)$$

where in this case Z_2 is the value for skin and Z_1 that for water (1.494×10^6 kg m^{-2} s^{-1}). A_r is of course the measured reflected signal.

Measurements of skin acoustic impedance were made on a small number of nine volunteers. Three sites were chosen, the palm, the inner forearm midway between wrist and elbow, and a site near the elbow. In each case the skin was first washed with a weak detergent to assist the wetting action of the water coupling medium. It was noted that without this precaution, considerable variability in echo amplitude occurred and this is probably due to very small air bubbles trapped in the surface of the stratum corneum which were eliminated by the detergent wash. A further effect due

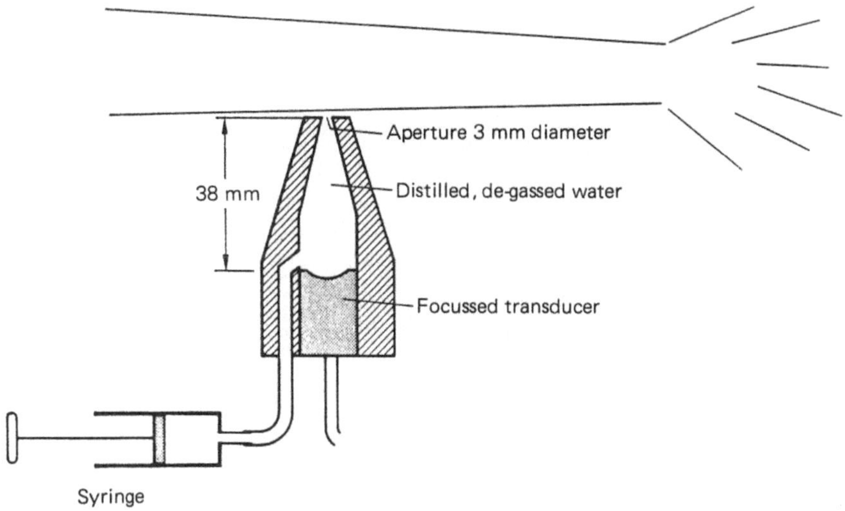

Fig. 6. Measurement arrangement for in vivo work

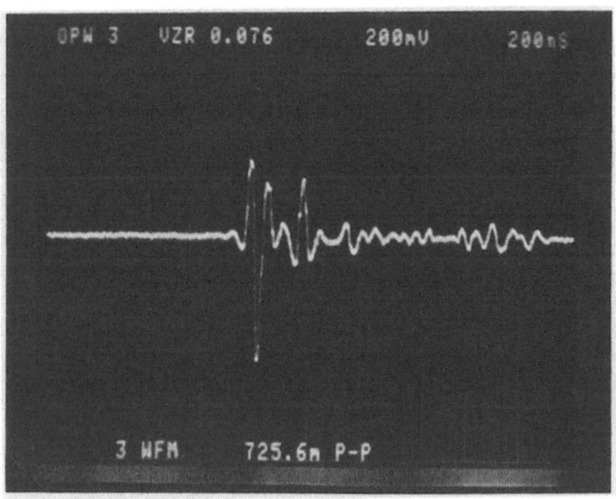

Fig. 7. Typical palm reflection. Peak reflection = 725.6 mV (0 dB attenuation) ≡ 18.2 mV (16 dB attenuation)

to the lipids at the surface may also be present. At each site a water bath and transducer set-up (Fig. 6) was placed on the skin and carefully aligned to produce maximum echo. This was then recorded and the measurement repeated several times. The Tektronix 7854 digitising oscilloscope was then used to store each reading and compute the peak-to-peak value of the signal (Figs. 7, 8).

Following the measurements of stratum corneum properties, the site midway along the forearm was stripped of stratum corneum using repeated applications of adhesive tape. This was done until all the stratum corneum was seen to be removed. A further measurement was then made using the same procedure. The data are shown in Table 2, with a typical result in Fig. 8 and 9.

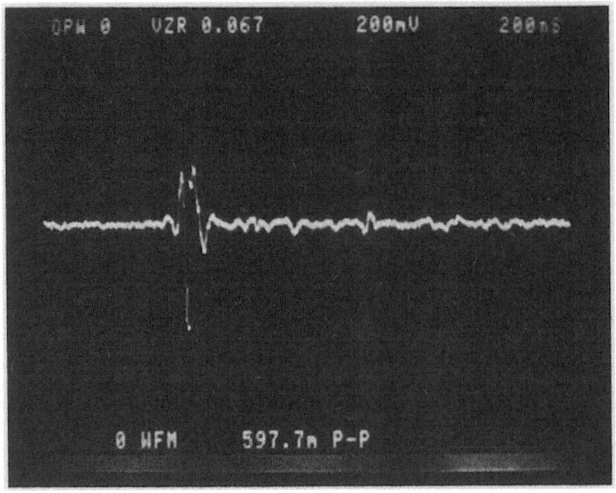

Fig. 8. Subject 2, stratum corneum – middle forearm. Peak reflection = 597.7 mV (0 dB attenuation) ≡ 15 mV (16 dB attenuation)

Table 2. Measured values of stratum corneum and epidermal acoustic impedance for five subjects

Subject	Peak reflection (mV) Epidermis	Peak reflection (mV) Stratum corneum	Acoustic impedance (Rayls) Epidermis	Acoustic impedance (Rayls) Stratum corneum
1	7.4	15.9	1.516	1.542
2	7.7	15.0	1.517	1.540
3	7.1	24.1	1.515	1.568
4	11.0	20.6	1.527	1.556
5	9.3	12.1	1.522	1.530
Mean	8.5	17.54	1.519	1.547
Standard deviation	1.46	4.27	0.004	0.013

Results

The first and most obvious comment on this work is that it is a preliminary study on nine normal male subjects of age range 20 to 30 years. Clearly more data would allow more definite conclusions, but some general comments can be made. The impedance of stratum corneum is just higher than water, and the stratum granulosum impedance is lower in all cases than stratum corneum, as expected. The value of about 1.55 rayls

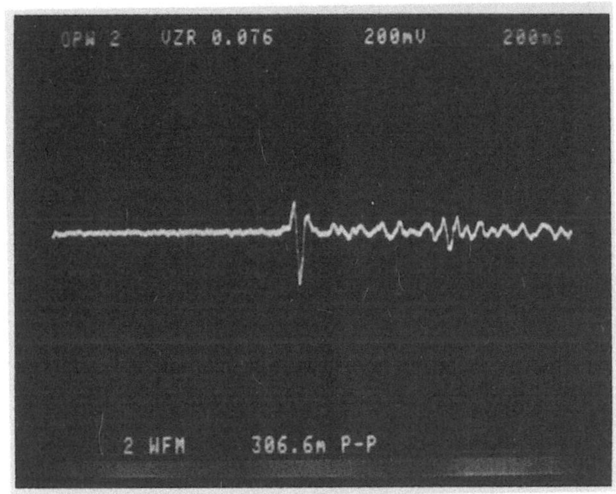

Fig. 9. Subject 2, epidermis – middle forearm. Peak reflection = 306.6 mV (0 dB attenuation) ≡ 7.4 mV (16 dB attenuation)

for stratum corneum and about 1.52 for stratum granulosum seem reasonable considering the water content of the skin, and even with a larger sample we would not expect a significant change in these values, although the standard deviations may well reduce.

A value of density for each of the skin layers may be obtained from the literature, and this may be used for direct non-invasive measurements of the speed of sound of the exposed skin layer. Thus, absolute thickness measurements are possible.

Comments

Scattering Due to Surface Roughness. This experiment assumes specular 180° reflection from plane surfaces. From the measurements on metal sample blocks, the results correspond well enough to be sure that all, or at least a constant proportion of the reflected energy was being detected. Reflection from skin will not be as specular as from polished metal surfaces. The effect of this is to remove energy from the detected reflection beam indicating a lower impedance. If some measure of the amount of energy loss, i.e., the reflected beam spread, can be made, then a correction may be applied. The determining factor for the reflection regime is the relationship between incident wavelength and surface roughness dimensions. In our case, the incident wavelengths range from 50 μm to 150 μm. The skin surface roughness has several distinct orders of roughness, ranging from squame (corneocyte) dimensions (about 1.0 μm thick and about 50 μm diameter) to furrow dimensions. In this case, the reflection will be diffuse, so that most of the reflected energy will occur at 180° to incident, but some will be at different angles and not be collected by the receiver. Therefore,

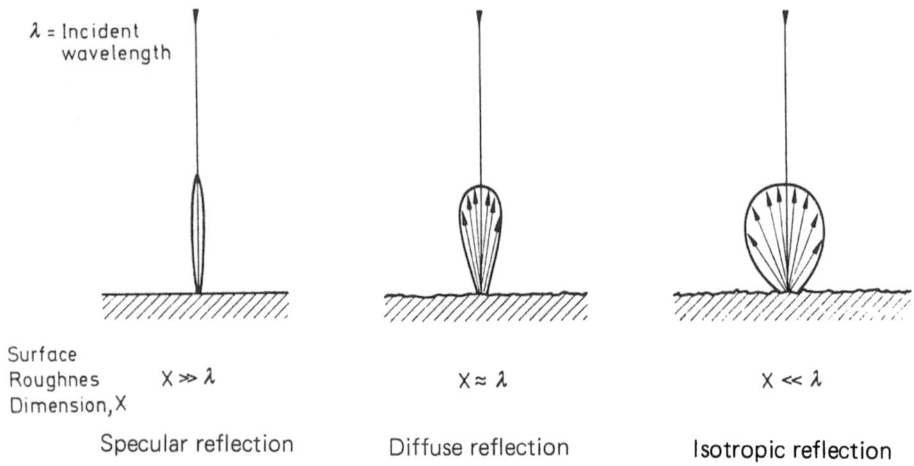

Fig. 10. Surface reflection characteristics

in general, we can say that the characteristic impedance measured by this method will be somewhat lower than the actual value. The amount of correction needed can be estimated by assessing the shape of the reflection intensity versus "angle from normal" diagram ("polar" or "directivity" diagram) (Fig. 10). From the dimensions of wavelength and skin roughness one would expect the correction to be of the order of 10%.

Importance of Characteristic Impedance Measurement. The characteristic acoustic impedance $Z = \rho c$ is a useful measure of the "mechanical" properties of the stratum corneum and/or epidermis in that the speed of sound is dependent on elastic moduli, and the density is an obvious variable of state. The density can be measured independently in vitro, allowing an independent measured estimate of the speed of sound.

Further Work

Clearly a larger sample is required to give more statistical validity to the quoted values of impedance. A study of normals versus diseased, normal variations with age, sex, skin type, etc., and measured impedance versus treatment (topical or systemic) would be of value to build up a body of reliable data for each category.

Variations in the method should be tested. For example, the use of an unfocused beam, or different frequencies should allow alterations in the reflected energy beam-shape, and for each incident beam and skin surface, a calibration factor should be found to allow for non-specular reflection. Careful measurements should also allow estimation of dermal impedance through the epidermis, although this idea has not yet been tested.

References

1. Alexander H, Miller DL (1979) Determining skin thickness with pulsed ultrasound. J Invest Dermatol 72:17–19
2. Goss SA, Johnson RL, Dunn F (1978) Comprehensive compilation of empirical ultrasonic properties of mammalian tissue. J Acoust Soc Am 64:623–657
3. Kaye GWL, Laby TH (1978) Tables of physical and chemical constants, 14th edn. Longmann, London, pp 68–69
4. Payne PA, Grove GL, Alexander H, Quilliam RM, Miller DL (1982) Cross-sectional ultrasonic scanning of skin using plastic film transducer. Bioengineering and the skin newsletter 3:241–246
5. Payne PA, Quilliam RM (1983) The measurement of skin thickness using pulsed ultrasound and PVDF transducers. Bioengineering and the skin newsletter 4:97–104
6. Tan CY, Marks R, Payne P (1981) Comparison of xeroradiographic and ultrasound detection of corticosteroid induced dermal thinning. J Invest Dermatol 76:126–128
7. Weast RC (ed) (1980) CRC handbook of chemistry and physics, 60th edn. CRC Press, Florida, p E47
8. Wells PNT (1977) Biomedical ultrasonics, Academic Press, London, p 136

In Vitro Test Systems for Evaluation of the Physical Properties of the Skin

H. G. Vogel

Despite several attempts [1–12, 15] to describe the mechanical behaviour of skin by a mathematical model none of them has been proven to be sufficient for evaluation of all properties including the history before measurement and the time-dependence during measurement. Therefore several methods have to be applied to investigate different aspects of biorheology.

Table 1 gives a survey for biomechanical test models for skin that can be used in vitro. The basic experiment is to record stress-strain curves. From the stress-strain curves one can read the ultimate values which are dependent on strain rates. Furthermore, stress values at low extension degrees or strain values at low stress degrees can be evaluated.

The anisotropy of skin can be seen very clearly in in vitro experiments from the fact that stress-strain curves cross when taken either perpendicular or longitudinal to the body axis at least in some animal species. The irregularity of the curve perpendicular to the body axis can be explained by the step phenomenon which again has its explanation in the anatomy of rat skin. Further experimental approaches are the hyster-

Table 1. Biomechanical test methods for skin in vitro

Stress-strain curves
Ultimate values
Dependence on strain rate
Stress values at low extension degrees
Strain values at low stress values
Anisotropy of skin
Crossing of stress-strain curves
Step phenomenon
Anatomical background
Hysteresis experiment
Repeated strain
Cyclic strain up to constant extension degree
Cyclic strain with increasing baseline
Constant load
Creeping behaviour
Constant strain
Relaxation
Residual tension after unstretching
Mechanical recovery
Isorheological point

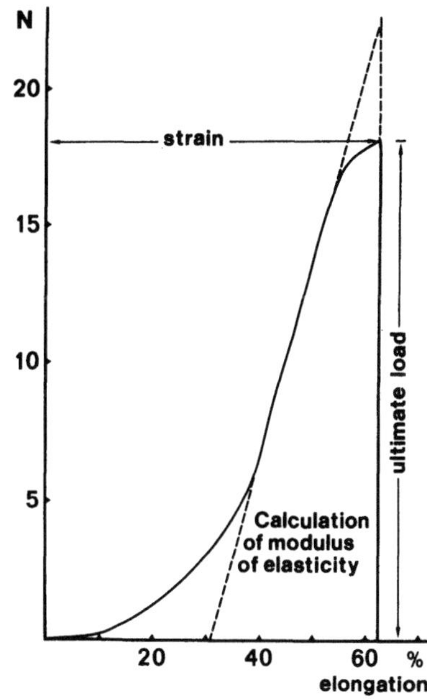

Fig. 1. Original registration of stress strain experiment

esis experiment and the behaviour of skin after repeated strain which can be measured either by cyclic strain up to a constant extension degree or by cyclic strain with increasing baseline. The behaviour after constant load can be measured by the creep experiment. The behaviour under a given strain can be analyzed by the relaxation curve and by the mechanical recovery. Furthermore, the isorheological point gives an indication of the elastic and viscous behaviour of skin.

To analyze the in vitro properties of human cadaver skin or skin from animals a flap of skin is removed and skin thickness is measured by calipers. A dumb-bell shaped specimen is cut from this skin sample and fastened between the clamps of an Instron^{R}-instrument. The samples are processed as fast as possible after death of the animals. In the case of human cadaver skin the samples were kept in a humid atmosphere until testing. Stress-strain curves of skin have a characteristic shape (Fig. 1). During low strain values there is a gradual increase of load, the curves having a concave shape. Afterwards an almost straight part is reached after which some yielding of curves occur which ends in a sudden break of the specimen. From this point one can measure ultimate strain and ultimate load. Tensile strength is calculated by dividing ultimate load by the cross sectional area. From the straight part of the curve the modulus of elasticity is calculated.

What can be investigated with ultimate values? For example, the ultimate values from a study on age dependence of rat skin show that thickness of skin increases with age up to a maximum at 12 months and decreases thereafter. The ultimate extension values follow another shape. Here we find a small increase during puberty, a maximum at 4 months and a decrease thereafter. Ultimate load, tensile strength and modulus of elasticity show a very sharp increase during puberty and a maximum at 12 months [16, 20]. These results are dependent on experimental conditions, such as strain rate [13]. In one experimental series in rats, the tensile strength was determined at various

Table 2. Dependence of stress-strain curves on the direction of the body axis

Anistotropy of skin
Crossing of stress-strain curves
Step phenomenon
Total stress loss due to step phenomenon
Strain again due to step phenomenon
Total work loss due to step phenomenon
Anatomical background

strain rates starting from creep velocities up to velocities approaching free fall. In this case tensile strength was dependent almost linearily from the logarithm of the strain rate.

Furthermore, the values of the stress-strain curve at low extension degrees or at low stress values can be evaluated. One may record the lower part of the stress-strain curves with a higher magnification. When the moduli of elasticity are calculated for the lower parts of the stress-strain curve one finds that those behave quite different from the ultimate modulus of elasticity depending on maturation and age [22, 25].

Especially in rats as well as in other animal species a pronounced dependence of the stress-strain curves on the direction with regard to body axis is found, indicating anisotropy of skin (Table 2). These experiments were also performed in samples of human cadaver skin. In spite of the Langer's lines we did not find such pronounced dependence on the body axis in human cadaver skin, as we found in animals. If stress-strain curves from samples of rat skin were obtained either perpendicular or longitudinal to the body axis, the ultimate stress values are almost the same. Ultimate extension was much higher in samples perpendicular to the body axis. This is only true for the higher part of the stress-strain curves. At low loads the strain values perpendicular to the body axis were even lower than longitudinal to the body axis. This phenomenon was studied in various experimental series and found to be true for almost all age classes. The crossing of the curves was more pronounced at middle ages.

This observation can at least partially be explained by the so-called step-phenomenon. If samples obtained perpendicular to the body axis are extended a gradual increase of load is observed at low degrees of extension which suddenly is interrupted by a decrease of the registered curve. Then the curve increases again, being interrupted by a second or third step. One can measure the decrease of load due to the steps and the increase of strain due to this step-phenomenon [27, 28].

This phenomenon is influenced by age as well as by other factors and can, therefore, be taken as a parameter for test systems in vitro. It can at least partially be explained by the fact, that rat skin has a muscular layer below the dermis, which cannot be separated without altering the mechanical properties of rat skin to a great extent. The muscular fibers are in a direction longitudinal to the body axis. If the samples are obtained perpendicular to the body axis the muscle bundles are cut transversally.

In the hysteresis experiment not only the elastic but also viscous properties are measured (Table 3) [21, 23]. In a sample of skin the usual stress-strain curve up to a given extension degree is obtained (Fig. 2). When the sample is unstretched the unloading curve shows a different pattern. It reaches the baseline much earlier than the curve

Table 3. Measurements of viscous properties

Hysteresis experiment
Values at the end of the loading cycle at various extension degrees
Stress at the end of the loading cycle
Modulus of elasticity at the end of the loading cycle
Energy input during the loading cacle
Energy dissipation indicated by unloading phase
Ratio between energy dissipation and energy input at each hysteresis cycle
Residual extension after the hysteresis cycle

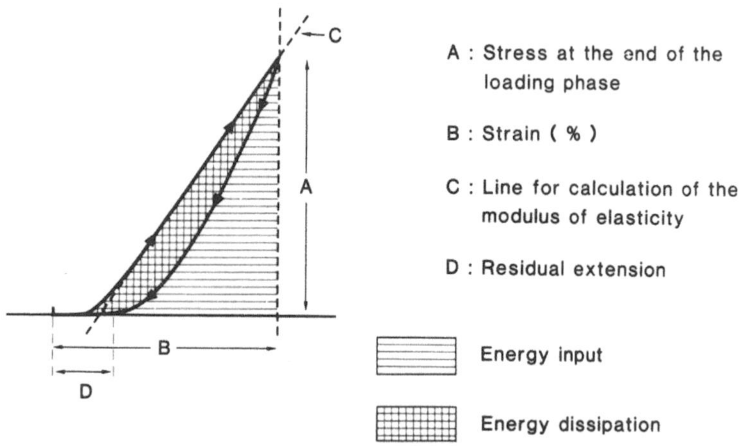

Fig. 2. Original curve of hysteresis experiment

left it during the upward phase. From this curve the stress and the modulus of elasticity at the end of the loading phase at a given strain can be measured. By planimetry from the area below the upward curve the energy input and from the area between the hysteresis loop the energy dissipation can be calculated as well as the ratio between energy dissipation and energy input at each hysteresis cycle and the residual extension after the hysteresis cycle. The ratio between energy dissipation and energy input is much less influenced by maturation and aging compared to the parameters mentioned before. This also holds true for residual extension. Therefore, both parameters deriving from the hysteresis experiments which are indicators of viscosity, behave differently than the parameters of elasticity.

The behaviour after repeated strain can be measured either by extending the specimen with a cyclic strain up to a constant value and measuring stress values [17, 27] or by cyclic strain with application of an increasing baseline [26] (Table 4). In this case the maximal load values at each extension step or the maximal stress values at each extension step are recorded. Furthermore, in this experiment one can calculate the number of cycles per extension step which are necessary until the load valucs are decreased to half of the starting values.

Table 4. Behaviour after repeated strain

Repeated strain
Cyclic strain up to constant extension degree
Cyclic strain with increasing baseline
Maximal load values at each extension step
Maximal stress values at each extension step
Number of cycles per extension step

Table 5. Behaviour under constant load (so-called creep experiment)

Constant load
Creep experiment
Time until break
Ultimate extension rate

Table 6. Constant strain experiment

Constant strain
Relaxation
Initial tension at given strain degrees
Tension after 5 min relaxation period
Tension after 5 min relaxation period as percentage of initial tension
Loss of tension after 5 min relaxation period
Loss of tension after 5 min relaxation period as percentage of initial tension
Ratio between loss of tension after 5 min relaxation period and initial tension
Residual tension after unstretching to 90% of initial strain
Residual tension as percentage of initial tension
Mechanical recovery
Mechanical recovery as percentage of initial tension
Mechanical recovery as percentage of tension after unstretching
Coefficient A_1 of formula $\sigma = A_1 + A_2 \cdot \log t$
Coefficient A_2 of formula $\sigma = A_1 + A_2 \cdot \log t$

The behaviour under constant load needs another experimental set up. This is the so-called creep experiment, in which time to break and ultimate extension rate are measured (Table 5) [19]. In this case the samples were fastened into special clamps. To the lower clamps a rod of an inductive displacement pick up was attached bearing a metal plate at the lower end. The lower clamp, the rod and the plate had together a weight of 100 g. Additional load could be added to the metal plate.

After release of the lower clamp the elongation of the sample due to the constant load was measured over the period until breaking or at least over 1 h. From this curve the ultimate extension rate would be calculated. The ultimate extension rate dropped during maturation but also during aging. The time to break increased during maturation and aging. Such a parameter as measured in the retardation or creep experiment indicates viscosity or plasticity and behaves quite differently from parameters like tensile strength or elasticity as indicated before.

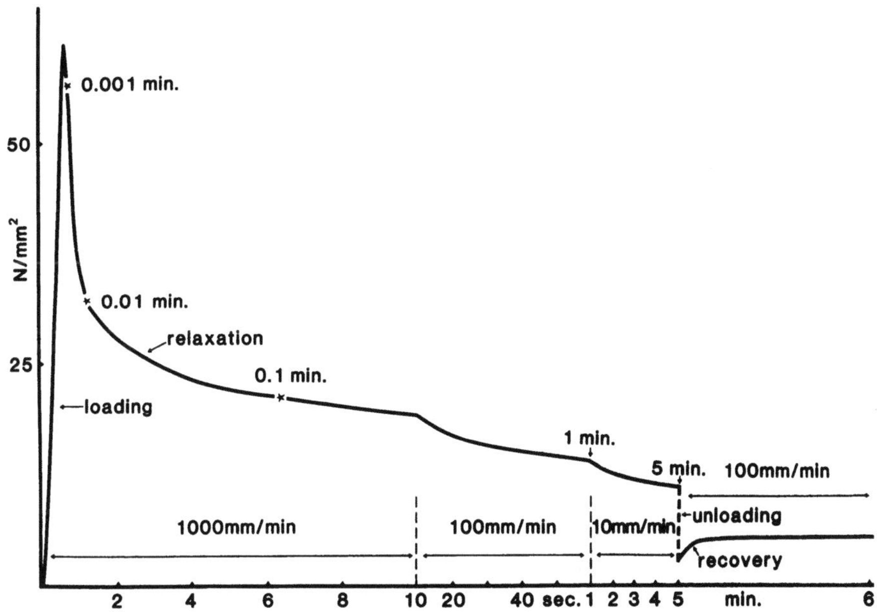

Fig. 3. Original recording of relaxation experiment

One very useful experimental set up is the relaxation experiment (Table 6, Fig. 3) [14, 18]. In this case a sample is extended to a given degree of extension. Then this extension is kept constant and the load is measured over 5 min. Due to the relaxation process the stress or load decreases first rapidly, then at a slower rate. The decrease of load is for the first approximation correlated linearily with the logarithm of time. After 5 min the sample is unstretched to 90% of the original strain. Then one can observe the phenomenon of mechanical recovery, that means that load increases. This relaxation experiment can be evaluated in several ways. One can measure the initial stress at a given strain and the following decrease. Then the tension after a 5 min relaxation period can also be expressed as percentage of initial tension. Furthermore, one can calculate the ratio between loss of tension during the 5 min relaxation period and the initial tension. Another parameter is the residual tension after unstretching to 90% of the initial strain. Again one can calculate this residual tension as percentage of initial tension. Mechanical recovery is another parameter which also can be related either to initial tension or tension after unstretching. Furthermore, one can calculate according to the formula $\sigma = A_1 + A_2 \times \log t$, the coefficients A_1 and A_2. This experiment can be used for analyzing various factors of the mechanical properties of skin in vitro.

A new experimental approach is the determination of the isorheological point [24] (Table 7, Fig. 4). In this experiment initially two relaxation processes are registered which involve the so-called conditioning of the samples. The relaxation processes are different between the first and the second relaxations. Even the third relaxation would show some changes. In order to avoid the influence of long duration of the experiment, the isorheological point is analyzed during the third relaxation process. In this case the load is decreased to one half of the load measured after the second relaxation period.

Table 7. Determination of the isorheological point

Isorheological point
Strain at consecutive loading cycles
Strain at the isorheological point
Load at the isorheological point
Product of percent strain times load at the isorheological point
Steepness of the saw-tooth curve at the isorheological point

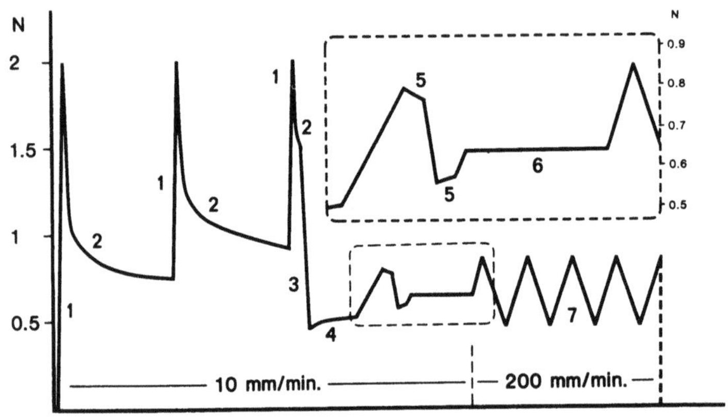

Fig. 4. Repeated relaxation and determination of the isorheological point

Then the increase of load due to mechanical recovery is measured. Going up and down with the cross head of the instrument the point where neither immediate relaxation nor mechanical recovery occurred is achieved. At this point by driving the cross head up and down we can measure a saw-tooth curve from which the steepness is obtained.

The isorheological point which is characterized by the per cent strain and the measured load at the isorheological point and the product of per cent strain times load at the isorheological point were determined. The differences between the values perpendicular and longitudinal to the body axis are rather small, at least for low loads. Furthermore, the age-dependence is almost absent. This parameter, the isorheological point, seems to be an indicator for the mechanical values of connective tissue, which is more or less independent of several factors. Also the steepness of the saw-tooth curve at the isorheological point shows almost no influence of the direction of the body axis and of aging at high initial load whereas at low load some differences were seen.

Conclusions

Detailed analysis of several parameters of mechanical behaviour of skin indicating various dimensions such as tensile strength, elasticity and viscosity indicates that there is no comprehensive model available to describe the biorheology of skin. Special methods give a clear insight into clearly defined physical properties.

References

1. Anna O, Arnold G, Hartung C, Zech M (1974) Biomechanik and Simulation des Zugverhaltens der Chordae tendinae des menschlichen Herzens. J Mat Technol 5:209–217
2. Danielson DA (1973) Human skin as an elastic membrane. J Biomechanics 6:539–546
3. Frisen M, Mägi M, Sonnerup L, Viidik A (1969) Rheological analysis of soft collagenous tissue, Part I: Theoretical considerations. J Biomechanics 2:13–20
4. Frisen M, Mägi M, Sonnerup L, Viidik A (1969) Theological analysis of soft collagenous tissues, Part II: Experimental evaluations and verifications. J Biomechanics 2:21–28
5. Fung YC (1981) Biomechanics. Mechanical properties of living tissue. Springer, New York
6. Hirsch C, Sonnerup L (1968) Macroscopic rheology in collagen material. J Biomechanics 1: 13–18
7. Jamison CE, Marangoni RC, Glaser AA (1968) Viscoelastic properties of soft tissue by discrete model characterization. J Biomechanics 1:33–46
8. Jenkins RB, Little RW (1974) A constitutive equation for parallel-fibered elastic tissue. J Biomechanics 7:397–402
9. Lanir Y, Fung YC (1974) Two-dimensional mechanical properties of rabbit skin. II. Experimental results. J Biomechanics 7:171–182
10. Sanjeevi R (1982) A viscoelastic model for the mechanical properties of bilogical materials. J Biomechanics 2:107–118
11. Soong TT, Huang WN (1973) A stochastic model for biological tissue elasticity in simple elongation. J Biomechanics 6:451–485
12. Veronda DR, Westmann RA (1970) Mechanical characterization of skin-finite deformations. J Biomechanics 3:111–124
13. Vogel HG (1972) Influence of age, treatment with corticosteroids and strain rate on mechanical properties of rat skin. Biochim Biophys Acta 286:79–83
14. Vogel HG (1973) Stress relacation in rat skin after treatment with hormones. J Med 4:19–27
15. Vogel HG (1976) Measurement of some viscoelastic properties of rat skin following repeated load. Conn Tiss Res 4:163–168
16. Vogel HG (1976) Altersabhängige Veränderungen der mechanischen und biochemischen Eigenschafte der Cutis bei Ratten. Akt Gerontol 6:477–487
17. Vogel HG (1976) Mechanical parameters in rat skin following repeated load. Influence of age and corticosteroid treatment. G-I-T 20:113–121
18. Vogel HG (1976) Tensile strength, relaxation and mechanical recovery in rat skin as influenced by maturation and age. J Med 2:177–188
19. Vogel HG (1977) Strain of rat skin at constant load (creep experiments). Influence of age and desmotropic agents. Gerontology 23:77–86
20. Vogel HG (1978) Influence of maturation and age on mechanical and biochemical parameters of connective tissue of various organs in the rat. Conn Tiss Res 6:161–166
21. Vogel HG (1978) Age dependence of mechanical parameters in rat skin following repeated strain. Akt gerontol 11:601–618
22. Vogel HG (1981) Directional variations of mechanical parameters in rat skin depending on maturation and age. J Invest Dermatol 76:493–497
23. Vogel HG (1983) Age dependence of visoelastic properties in rat skin. Directional variations in stress-strain and hysteresis experiments. Bioeng Skin 2:136–155
24. Vogel HG (1985) Repeated relaxation and determination of the isorheological point in skin strips of rats as being influenced by maturation and aging. Bioeng Skin 1:321–335
25. Vogel HG, Hilgner W (1977) Analysis of the low part of stress-strain curves in rat skin. Influence of age and desmotropic drugs. Arch Dermatol Res 258:141–150
26. Vogel HG, Hilgner W (1978) Viscoelastic behaviour or rat skin after repeated and stepwise increased strain. Bioengineering and the skin 2:22–23
27. Vogel HG, Hilgner W (1979) The "step phenomenon" as observed in animal skin. J Biomechanics 12:75–81
28. Vogel HG, Hilgner W (1979) Influence of age and of desmotropic drugs on the step phenomenon observed in rat skin. Arch Dermatol Res 264:225–241

The Use of Statistical Models Employing Analysis of Variance for Efficient Analysis of Data

M. F. Corbett

The use of pairwise statistical tests as part of the analysis of data is, by now, commonplace in dermatology. However in many studies, both laboratory experiments and clinical trials, more than two treatments are compared concurrently. It is then appropriate to use a more general technique, such as analysis of variance, for analysis of the data. The generalisation from t test to analysis of variance and the choice of a suitable model for analysis of variance for a particular example will be discussed.

The t Test

A study of the effect of benzoyl peroxide on the excretion of sebum [5] in 15 patients compared the post-treatment sebum excretion rate with pre-treatment control values. They found that the post treatment sebum excretion rate measured in mg per cm^2 per min was more than 20% higher than the control rate. The two sets of observations were properly compared using a t test (method of paired comparisons) which shows the post-treatment values to be significantly higher.

The t test is appropriate for the comparison of two treatments, or treatment and controls, and tests the hypothesis that their effects are much the same. In outline:

$$\frac{\text{difference between 2 means}}{\text{standard error term}}$$

More Than Two Treatments

A clinical trial may be designed to compare the new with one or more established standard treatments and perhaps with placebo as well. In a paper by Chalker et al. [2] the following treatments were compared: the combination of topical erythromycin and benzoyl peroxide gel versus 3% erythromycin gel versus benzoyl peroxide gel versus the gel base in the treatment of 165 patients with acne. In this example there are four treatments rather than two. If we compared them two by two, using t tests or any pairwise test, we might be misled as to the meaning of the outcome because the true probability of obtaining a relatively extreme t value, even though there is not a substantial difference between any of the pairs of means, increases quite rapidly with the number of tests done. We consider significant a test result which would be rare (i.e. happens by chance with a probability of less than 5%) if there was in truth little difference between the treatments. The risk of getting a test result which we would count as significant when comparing four treatments two by two at the supposed 5% level, when none of the treatments really had an effect, would in fact be 0.18 rather than 0.05.

In a laboratory experiment there may well be more than 3 treatments in all; as there is much better control over the experiment in the laboratory than in a clinical trial there is less need to keep the study really simple. Given a good chance of exactly completing the protocol, it is both more economical and more informative to do a larger and more complicated experiment. For example, Coutts and Greaves [4] reported a study in which six antihistamines were ranked according to their in vitro H1 blocking potency.

Analysis of Variance

In cases like these, when there are more than two treatments to compare, we need a generalisation from the t test to an analysis of variance. The t test is equivalent [3] to the square root of the ratio:

$$\frac{\text{Variance estimate from 2 means}}{\text{Variance estimate within the samples}}$$

which is an example of the type of ratio used in the analysis of variance.

Table 1 shows some of the results of measuring weal diameter after a standard stimulus in patients with symptomatic dermographism from a study by Breathnach et al. [1]. In this study five treatments were compared to see whether taking an antihistamine for 5 days altered the size of weals appearing in response to the standard stimulus.

The variance (i.e. square of standard deviation) of all these values can be calculated in the usual way to give a "total variance" of 2.7. It is not included in the usual analysis of variance table, but it is interesting to consider this value [6]. The data is classified, at first, simply by treatment into a "one-way" classification.

Table 1. Weal diameter caused by stylus pressure 9.4 x 10 Pa in patients with symptomatic dermographism

	Treatments				
	1	2	3	4	5
	2.5	2.5	2.5	2.5	0
	4	2.5	3	3.5	3.5
	8	6	7	10	4.5
	5	5	4	4.5	3.5
	3	4	2.5	0	3
	5.5	4.4	4	4	3.5
	3.5	2	3	2	2.2
	6	4.6	4.4	4.3	5
	6	4.5	4.5	3.4	1.5
	4.5	4	2.5	3.5	3
	4.5	4	4.5	3.5	3.5
	4	4	4	3	1
Mean	4.7	4.0	3.8	3.7	2.9
Standard Deviation	1.5	1.1	1.3	2.3	1.4

Table 2. Analysis of variance

Source	df	ss	ms	F	
Among groups (treatments)	4	21.2	5.3	2.1	$p < 0.25$ (ns)
Within groups (error)	55	139.9	2.5		
Total	59	161.1	(2.7)		

We can also estimate the variance from the means of the treatment groups, which gives us the variance among groups (or "between treatments"), (22.9 + 16 + 14.44 + 13.69 + 9.41 – ((19.1)/5))/4. The variance may be estimated within the groups as well by calculating one inside each group and averaging across the groups (2.29 + 1.29 + 1.64 + 5.4 + 2.08)/5 = 12.7/5 = 2.54. These estimates are shown in the analysis of variance table (Table 2).

In this table we can see that the sums of squares and the degrees of freedom (which show the number of values on which each part of the calculation is based) are additive – the estimates add to from the total. The "within groups" estimate is a weighted average of variance estimates, each calculated inside a treatment group. Any treatment effect acting on every member of a group is a constant within that group and so does not alter the variance calculated inside it. Thus there is no added treatment effect in the "within groups" estimate of the total variance. 2.5 is quite a good estimate of the quantity already calculated as 2.7.

At 5.3 the "among groups" estimate is larger. If it is significantly larger, the increase is due to the inclusion of an added variance due to treatment effect. We test for a treatment effect by checking whether the ratio between the "among groups" and "within groups" estimates is close to 1 (i.e. not significant), or so much larger that we can be fairly sure that our data values reflect an effect of the treatments. With a one-way classification, this data does not show a significantly large treatment effect. Would a further partition of the total sum of squares be possible so as to show a larger ratio between the error term and the term which includes the treatment effects?

Table 3 shows the original data classified by patient as well as by treatment. Now a further partition of variance can be used which takes the variation among patients out of the "within groups" estimate so that a much larger variance ratio will be obtained between the "among groups" and the "within groups" estimates. Table 4 shows the resulting more detailed analysis of variance.

The two-way classification was used in the report published by Breathnach et al, in which some of the treatments were shown to modify the wealing response significantly.

Only part of the data has been considered here so as to demonstrate clearly that, in some circumstances, data which does not show a significant treatment effect when analysed in a one-way table, may show a useful and significant difference if the two-way classification and analysis is used. The sum of squares due to biological differences between subjects is partitioned from the rest of the "within groups" sum of squares; consequently the error variance is smaller and the ratio between treatment variance

Table 3. Weal diameter caused by stylus pressure 9.4 x 10 Pa in patients with symptomatic dermographism

Patient no	Treatments				
	1	2	3	4	5
1	2.5	2.5	2.5	2.5	0
2	4	2.5	3	3.5	3.5
3	8	6	7	10	4.5
4	5	5	4	4.5	3.5
5	3	4	2.5	0	3
6	5.5	4.4	4	4	3.5
7	3.5	2	3	2	2.2
8	6	4.6	4.4	4.3	5
9	6	4.5	4.5	3.4	1.5
10	4.5	4	2.5	3.5	3
11	4.5	4	4.5	3.5	3.5
12	4	4	4	3	1
mean	4.7	4.0	3.8	3.7	2.9
s.d.	1.5	1.1	1.3	2.3	1.4

Table 4. Analysis of variance

Source	df	ss	ms	F	
Among rows (patients)	11	99.5	9.0	9.8	
Among groups (treatments)	4	21.2	5.3	5.8	$p < 0.001$
Within groups (error)	44	40.4	0.9		
Total	59	161.1		(2.7)	

("among groups") and the error variance is larger – possibly large enough for significance.

Conclusion

Although a two way classification will not always be necessary, it is best whenever possible to plan in advance to use a detailed model, and also to make sure that an analysis which models the experiment closely will be practical. Prior consideration is required at the planning stage to ensure that the protocol of the experiment describes a method for which a closely fitting analysis of variance model can be chosen.

It is easy to see that the t test is a special case of the one-way analysis of variance. Analysis of variance comprises a very flexible and powerful family of techniques, one of which may appropriately be chosen for a member of a large range of experimental designs. The best choice for a particular study is often a matter in which expertise can pay large dividends, in terms of savings in experimental materials and resources, and more precise conclusions; thus it is at the planning stage that expert statistical help in choosing a model is the most important.

References

1. Breathnach SM, Allen R, Milford Ward A, Greaves MW (1983) Symptomatic dermographism: Natural history, clinical features, laboratory investigations and response to therapy. Clin Exp Dermatol 8:463–476
2. Chalker DK, Shalita A, Graham Smith J, Swann RW (1983) A double-blind study of the effectiveness of a 3% erythromycin and 5% benzoyl peroxide combination in the treatment of acne vulgaris. J Am Acad Dermatol 9:933–936
3. Colquhoun D (1971) Lectures on biostatistics. Claredon Press, Oxford, pp 179–180
4. Coutts A, Greaves MW (1982) Evaluation of six antihistamines in vitro and in patients with urticaria. Clin Exp Dermatol 7:529–535
5. Cunliffe WJ, Stainton C, Forster RA (1983) Topical benzoyl peroxide increases the sebum excretion rate in patients with acne. Brit J Dermatol 109:577–579
6. Sokal RR, Rohlf FJ (1981) Biometry, 2nd edn. WH Freeman, San Francisco, pp 198–202

Subject Index

MIX
Papier aus verantwortungsvollen Quellen
Paper from responsible sources
FSC® C105338

If you have any concerns about our products,
you can contact us on
ProductSafety@springernature.com

In case Publisher is established outside the EU,
the EU authorized representative is:
Springer Nature Customer Service Center GmbH
Europaplatz 3, 69115 Heidelberg, Germany

Printed by Libri Plureos GmbH
in Hamburg, Germany